Municipal Solid Waste Management (2nd Edition)—Innovative Solutions and Sustainable Strategies

Municipal Solid Waste Management (2nd Edition)—Innovative Solutions and Sustainable Strategies

Guest Editor

Carlos Afonso Teixeira

MDPI

Basel • Beijing • Wuhan • Barcelona • Belgrade • Novi Sad • Cluj • Manchester

Guest Editor
Carlos Afonso Teixeira
Centre for the Research and
Technology of
Agro-Environmental and
Biological Sciences (CITAB)
University of Trás-os-Montes
& Alto Douro
Vila Real
Portugal

Editorial Office
MDPI AG
Grosspeteranlage 5
4052 Basel, Switzerland

This is a reprint of the Special Issue, published open access by the journal *Sustainability* (ISSN 2071-1050), freely accessible at: https://www.mdpi.com/journal/sustainability/special_issues/7PJE2F96CF.

For citation purposes, cite each article independently as indicated on the article page online and as indicated below:

Lastname, A.A.; Lastname, B.B. Article Title. *Journal Name* **Year**, *Volume Number*, Page Range.

ISBN 978-3-7258-7549-8 (Hbk)
ISBN 978-3-7258-7550-4 (PDF)
https://doi.org/10.3390/books978-3-7258-7550-4

Contents

About the Editor

Carlos Afonso Teixeira

Carlos Afonso de Moura Teixeira is an Assistant Professor at the University of Trás-os-Montes e Alto Douro (UTAD), Portugal, where he serves as Director of the PhD Programme in Environmental Sciences and Technologies and of the Master's Programme in Environmental Engineering. He is an integrated researcher at the Centre for the Research and Technology of Agro-Environmental and Biological Sciences (CITAB) and has extensive experience in teaching and research in waste management, environmental impact assessment, life cycle assessment, resource efficiency, and circular economy. He has authored more than 30 peer-reviewed publications indexed in Scopus and Web of Science, as well as several book chapters and books, and holds one registered patent. He has acted as principal investigator in 15 research projects and has supervised or co-supervised numerous Master's and PhD theses in Environmental Engineering and related fields. He currently serves as Scientific Editor of the Elsevier journals *Results in Engineering* and *Next Research* and is Guest Editor of several Special Issues for MDPI journals in the areas of sustainability, environmental engineering, and waste management. His research focuses on sustainable waste systems, decarbonisation strategies, life cycle modelling, and the transition towards circular and low-carbon economies.

Preface

Municipal solid waste management is a key component of sustainable urban development, intersecting environmental protection, public health, governance, and climate mitigation. Despite advances in waste treatment technologies, persistent performance gaps reveal the importance of institutional capacity, governance arrangements, and contextual factors.

This Reprint brings together the contributions of the second edition of the Special Issue 'Municipal Solid Waste Management: Governance, Digitalization, and Technological Innovation for Sustainable Transitions'. Building on the first edition, the volume places greater emphasis on institutional readiness, digitalization as a decision-support instrument, and the conditional effectiveness of technological solutions within circular economy strategies.

The included studies address governance and behavioural aspects, digital and intelligent systems for planning and operations, and waste treatment technologies evaluated through efficiency, risk, and circularity lenses. Together, they support a system-oriented understanding of MSWM and provide a coherent basis for research, practice, and policy development in diverse institutional contexts.

Carlos Afonso Teixeira
Guest Editor

sustainability

MDPI

Editorial

Municipal Solid Waste Management: Governance, Digitalization, and Technological Innovation for Sustainable Transitions

Carlos Afonso Teixeira

Centre for the Research and Technology of Agro-Environmental and Biological Sciences (CITAB), Institute for Innovation, Capacity Building and Sustainability of Agri-Food Production (Inov4Agro), Universidade de Trás-os-Montes e Alto Douro (UTAD), Quinta de Prados, 5000-801 Vila Real, Portugal; cafonso@utad.pt

1. Introduction and Context

Municipal solid waste management (MSWM) is increasingly recognized as a critical socio-technical subsystem within broader sustainability transitions, linking urban metabolism, institutional governance, environmental risk, and climate mitigation. Although technological innovation in waste treatment has expanded rapidly, empirical evidence continues to reveal uneven system performance across contexts, indicating that technical capability alone is insufficient to explain sustainability outcomes.

Building on the first edition of this Special Issue, which framed MSWM through integrated management and environmental performance perspectives, the present second edition advances the discussion by foregrounding institutional capacity, digitalization as a governance instrument, and the conditional nature of technological effectiveness within circular economy strategies. Rather than revisiting established hierarchies or best-practice narratives, this edition examines how MSWM systems operate under conditions of institutional asymmetry, infrastructural constraint, and environmental and health risk.

2. Conceptual Orientation and Objectives

This Special Issue brings together thirteen contributions structured around three interrelated analytical strands: (i) governance, behavior, and institutional capacity; (ii) digitalization and intelligent decision-support systems; and (iii) waste treatment technologies, risk, and circular valorization.

Across these strands, the objective is not to catalogue technical solutions, but to interrogate the conditions under which MSWM interventions deliver effective, scalable, and socially robust outcomes, as well as the circumstances under which they stagnate or generate unintended effects, particularly across heterogeneous institutional and geographical settings.

3. Governance, Behavior, and Institutional Capacity

The governance- and behavior-focused contributions reframe MSWM sustainability as a problem of institutional sequencing rather than technological choice. Empirical analyses demonstrate that internal municipal capacity—encompassing staffing adequacy, financial coordination, service regularity, and planning coherence—accounts for a substantial share of performance variation, exceeding 40% in low-capacity contexts [contribution 1]. This finding challenges technology-first narratives that implicitly assume institutional adaptation will follow technical deployment.

Behavioral analyses further undermine individualistic interpretations of household non-compliance. Evidence showing that more than half of households in underserved municipalities engage in informal disposal practices despite pro-environmental attitudes [contribution 2] indicates that such behaviors are rational responses to infrastructural deficits and service unreliability rather than attitudinal failure. When considered alongside disaster waste management case studies [contribution 3] and multi-level indicator frameworks for zero-waste strategies [contribution 4], a consistent pattern becomes evident: governance fragmentation and misaligned service provision systematically transfer responsibility to households while simultaneously constraining feasible compliance pathways.

MSWM failures thus emerge less from behavioral resistance or technological inadequacy than from misordered institutional development, in which policy ambition persistently outpaces administrative, financial, and operational capacity. This insight has direct implications for policy design, sequencing, and capacity-building strategies.

4. Digitalization, Data, and Intelligent Systems

The digitalization-focused contributions assign a conditional and context-dependent role to intelligent systems in MSWM. Digital tools function primarily as capability amplifiers, with their effectiveness contingent on the presence of robust governance arrangements, data infrastructures, and decision-making routines.

Advanced predictive models applied to incineration processes reduce temperature forecasting errors by approximately 40% under variable waste composition [contribution 5], while automated waste identification systems achieve classification accuracies exceeding 90% in controlled environments [contribution 6]. Data-driven optimization models further demonstrate efficiency gains in construction and demolition waste systems when analytical tools are aligned with operational constraints [contribution 7]. These gains, however, depend critically on data quality, process formalization, and institutional absorptive capacity.

Digitalization does not compensate for weak governance; instead, it magnifies existing institutional strengths and weaknesses. In low-capacity contexts, intelligent systems risk producing technically sophisticated yet operationally marginal or symbolic interventions, offering limited contribution to substantive system transformation.

5. Treatment Technologies, Risk, and Circular Valorization

The technological contributions challenge efficiency-centered interpretations of sustainability in MSWM by exposing the disconnect between treatment performance and broader environmental and health outcomes. Integrated industrial park models demonstrate that coordinated systems can achieve material recovery rates approaching 90% [contribution 8], yet such efficiency gains do not inherently correspond to reduced environmental burden or risk.

Evidence from organic waste treatment illustrates this tension clearly. Despite process optimization, anaerobic digestion reduces antibiotic resistance gene concentrations by less than one logarithmic unit, leaving residual loads between 10^4 and 10^9 gene copies per kilogram of digestate [contribution 11]. These findings raise substantive concerns regarding land application practices and question assumptions that biological treatment aligns automatically with circular economy objectives.

Efficiency gains in treatment and material recovery do not translate into sustainability outcomes when environmental and health risks remain weakly governed, particularly within circular economy strategies that prioritize valorization over risk containment. Analyses of thermochemical treatment [contribution 9], landfill leachate management [contribution 10], and hospital bio-waste valorization [contribution 12] further reinforce the need for assessment frameworks that integrate risk governance alongside efficiency and recovery metrics.

https://doi.org/10.3390/su1805227

This systemic perspective is reinforced by a comprehensive review included in this Special Issue, synthesizing evidence from more than 170 studies and estimating that circular economy integration in the waste sector could reduce projected greenhouse gas emissions growth by 30–39% by 2050 [contribution 13]. Crucially, this mitigation potential depends on coordinated governance, controlled material flows, and explicit management of environmental and health risks, rather than on isolated technological adoption.

6. Geographical Diversity and Transferability

Across all three analytical strands, geographical diversity functions as an explanatory lens rather than a descriptive variable. The contributions demonstrate that MSWM interventions are deeply contingent on institutional maturity, regulatory enforcement, and infrastructural baselines. Strategies that yield efficiency gains in high-capacity settings frequently generate limited or counterproductive outcomes when transferred without adaptation to contexts characterized by governance fragmentation or service deficits.

These findings underscore the importance of context-sensitive policy transfer and caution against the uncritical replication of MSWM models across institutional environments.

7. Concluding Remarks

This second edition of the Special Issue reframes sustainability transitions in municipal solid waste management as fundamentally institutional and conditional processes. Across governance, digitalization, and treatment technologies, the contributions converge on a central insight: technological potential translates into sustainability outcomes only when aligned with institutional sequencing, governance coherence, and explicit risk management.

Progress in MSWM therefore depends less on the continuous introduction of new technologies than on the capacity to govern existing ones under conditions of institutional asymmetry, uncertainty, and environmental and health risk. MSWM should thus be understood not as a technical problem awaiting optimization, but as a socio-technical system requiring deliberate, context-sensitive coordination across policy, infrastructure, and knowledge domains.

Funding: This work is supported by National Funds from FCT—Portuguese Foundation for Science and Technology, through the projects UID/04033/2025: Centre for the Research and Technology of Agro-Environmental and Biological Sciences, and LA/P/0126/2020 (https://doi.org/10.54499/LA/P/0126/2020).

Acknowledgments: The authors would like to thank the CITAB-Inov4Agro and FCT—Portuguese Foundation for Science and Technology for their financial support.

Conflicts of Interest: The authors declare no conflict of interest.

List of Contributions:

1. Tahulela, A.C.; Hashemi, S.; Lourens, M.E. Assessing Strategic GIS Perceptions in Waste Management Planning: A Readiness Model from South Africa's Vhembe District. *Sustainability* **2025**, *17*, 10626. https://doi.org/10.3390/su172310626.
2. Tahulela, A.C.; Hashemi, S.; Lourens, M.E. Household Waste Disposal Under Structural and Behavioral Constraints: A Multivariate Analysis from Vhembe District, South Africa. *Sustainability* **2025**, *17*, 7429. https://doi.org/10.3390/su17167429.
3. Caldera, S.; Jayarathna, C.; Desha, C. Evaluating the Characteristics of Disaster Waste Management in Practice: Case Studies from Queensland and New South Wales, Australia. *Sustainability* **2025**, *17*, 6300. https://doi.org/10.3390/su17146300.
4. Kaur, A.; Thyberg, K.L.; Tonjes, D.J. Zero-Waste Program Success: A Systems Approach to Indicators at the Micro, Meso, and Macro Levels. *Sustainability* **2025**, *17*, 3644. https://doi.org/10.3390/su17083644.

5. Pian, J.; Si, M.; Sun, A.; Tang, J. Sustainability-Oriented Furnace Temperature Prediction for Municipal Solid Waste Incineration Using IWOA-SAGRU. *Sustainability* **2025**, *17*, 8987. https://doi.org/10.3390/su17208987.
6. Giel, R.; Fiedeń, M.; Dąbrowska, A. Real-Time Automatic Identification of Plastic Waste Streams for Advanced Waste Sorting Systems. *Sustainability* **2025**, *17*, 2157. https://doi.org/10.3390/su17052157.
7. Lopes, A.; Teixeira, C.A. Data-Driven Optimization of Construction and Demolition Waste Management: Pattern Recognition and Anomaly Detection. *Sustainability* **2025**, *17*, 4211. https://doi.org/10.3390/su17094211.
8. Yu, L.; Chen, S.; Tan, Z. Analysis of Solid Waste Treatment and Management in Typical Chinese Industrial Parks with the Goal of Sustainable Development and Future Suggestions. *Sustainability* **2024**, *16*, 6731. https://doi.org/10.3390/su16166731.
9. Zulqarnain, R.M.; Wang, H.; Siddique, I.; Ali, R.; Naveed, H.; Virk, S.A.; Ahamad, M.I. Thermochemical Techniques for Disposal of Municipal Solid Waste Based on the Intuitionistic Fuzzy Hypersoft Evaluation Based on the Distance from the Average Solution Technique. *Sustainability* **2025**, *17*, 970. https://doi.org/10.3390/su17030970.
10. Costa, C.; Pinedo, M.L.; Riascos, B.D. Presence of Humic Acids in Landfill Leachate and Treatment by Flocculation at Low pH to Reduce High Pollution of This Liquid. *Sustainability* **2025**, *17*, 481. https://doi.org/10.3390/su17020481.
11. Franchitti, E.; Mafrica, A.; Traversi, D. Tracking of Antibiotic Resistance Genes in Sludge from Treatment Plants for Organic Fraction of Municipal Solid Waste. *Sustainability* **2024**, *16*, 10922. https://doi.org/10.3390/su162410922.
12. Sepetis, A.; Georgantas, K.; Nikolaou, I. A Proposed Circular Economy Model for Hospital Bio-Waste Management in Municipal Settings. *Sustainability* **2025**, *17*, 5. https://doi.org/10.3390/su17010005.
13. Ijimdiya, S.J.; Kumarasamy, M.V.; Adu, J.T.; Pandi, D. Role of the Circular Economy Framework for Sustainable Waste Management and Climate Change Mitigation. *Sustainability* **2026**, *18*, 1946. https://doi.org/10.3390/su18041946.

https://doi.org/10.3390/su1805227

sustainability

MDPI

ticle

ssessing Strategic GIS Perceptions in Waste Management lanning: A Readiness Model from South Africa's 'hembe District

fani Confidence Tahulela [1,*], Shervin Hashemi [2,3] and Melanie Elizabeth Lourens [3]

[1] Department of Public Management, Durban University of Technology, Steve Biko Campus, Durban 4000, South Africa
[2] Institute for Environmental Research, Yonsei University College of Medicine, 50-1 Yonsei, Seodaemun-gu, Seoul 03722, Republic of Korea; shervin@yuhs.ac
[3] Faculty of Management Sciences, Durban University of Technology, Durban 4000, South Africa; shervinh@dut.ac.za (S.H.); melaniel@dut.ac.za (M.E.L.)
* Correspondence: aifanit@dut.ac.za

Abstract

Municipal solid waste management (MSWM) in low-capacity urban contexts is frequently constrained by fragmented governance, limited institutional readiness, and premature implementation of digital technologies. This study investigates how internal operational capacity, external factors, and Geographic Information System (GIS) integration interact sequentially to influence waste governance outcomes in South Africa's Vhembe District. Using survey data from 399 municipal actors and Structural Equation Modeling (SEM), the findings indicate that internal capacity encompassing staffing sufficiency, financial coordination, and service regularity is the most significant determinant of operational performance ($\beta = 0.432$, $p < 0.001$). This improvement in operations subsequently enhances strategic service effectiveness ($\beta = 0.267$, $p < 0.001$). GIS does not directly improve daily waste operations but contributes significantly to long-term planning outcomes, such as route optimization and infrastructure siting ($\beta = 0.130$, $p = 0.017$). External factors, particularly community participation, exhibit limited standalone influence, becoming effective only when foundational systems are stabilized. These insights inform the Municipal Readiness Model for Digital Waste Governance (MRM-DWG), a three-stage framework that aligns interventions with institutional maturity: (1) internal operational readiness, (2) strategic digital integration, and (3) participatory and external engagement. The MRM-DWG advances a sequencing logic rooted in absorptive capacity theory, offering a context-sensitive governance tool applicable to similarly constrained municipalities across the Global South.

Keywords: Geographic Information Systems; municipal solid waste; structural equation modelling; sustainable waste management; Vhembe District

1. Introduction

Effective municipal solid waste management (MSWM) is essential for environmental sustainability, public health protection, and the achievement of Sustainable Development Goals (SDGs), particularly SDG 11 (sustainable cities and communities), SDG 12 (responsible consumption and production), and SDG 13 (climate action) [1,2]. However, in many low- and middle-income countries (LMICs), growing urbanization and rising waste volumes continue to outpace institutional capacity, resulting in service irregularities, infrastructure deficits, and environmental degradation [3–5].

In South Africa, the National Environmental Management: Waste Act (Act No. 59 of 2008) requires municipalities to develop Integrated Waste Management Plans (IWMPs) [6–8]. Yet, implementation is uneven, particularly in rural and peri-urban regions due to financial constraints, limited technical capacity, and sociocultural challenges such as illegal dumping and low environmental awareness. These factors manifest differently across contexts, for instance, Kubanza [9] links them to low-income urban settlements; Raghu and Rodrigues [10] emphasize the behavioral dimensions shaping waste practices; Viljoen et al. [11] document household-level practices and challenges in rural towns; Tahulela et al. [12] highlight the structural and behavioral constraints shaping disposal practices in Vhembe District and Mukwevho et al. [13]. Highlight the institutional barriers that hinder effective implementation within municipalities.

Internal inefficiencies, including underfunded operations, untrained staff, and aging vehicle fleets, contribute to unreliable waste collection and unsanitary conditions.

For instance, Khatoon [14] shows how chronic underfunding weakens municipal waste services, while Zhang et al. [2] highlight logistical inefficiencies such as poor route planning and limited staff capacity. Patel et al. [15] add that outdated vehicle fleets often break down, worsening service reliability and sanitation.

External factors such as limited stakeholder engagement, fragmented intergovernmental coordination, and inconsistent policy enforcement further inhibit effective delivery [15,16].

In response to these structural and operational challenges, governance reforms increasingly advocate for the integration of digital technologies and participatory mechanisms to improve transparency, responsiveness, and spatial coordination in waste management [2,17]. In particular, Geographic Information Systems (GISs) have been widely adopted in global MSWM systems, supporting spatial mapping, route optimization, landfill siting, and real-time monitoring [18,19]. International case studies, such as in Pedro Ruiz Gallo, Peru, demonstrate that participatory GISs can enhance both service delivery and environmental compliance [20]. However, empirical evidence from under-resourced municipalities suggests that GIS effectiveness is conditional on institutional readiness. In environments marked by capacity constraints, digital systems may be adopted but remain underutilized due to limited technical expertise, weak data infrastructure, or lack of integration into routine municipal functions [21,22].

Existing literature has predominantly assessed the effects of internal governance, stakeholder engagement, and technological innovation in isolation [23]. Few studies have examined how these elements interact, particularly under conditions of asymmetric institutional capacity [24,25]. Furthermore, limited attention has been given to the distinction between operational outcomes, such as collection frequency, logistical responsiveness, and strategic outcomes such as infrastructure planning and coverage expansion. This analytical gap may lead to unrealistic expectations that technologies or participatory reforms yield uniform benefits across diverse municipal settings [9,26].

While prior MSWM studies have emphasized the relevance of internal institutional capacity, technological tools, and community participation as important governance drivers, these contributions often conceptualize these mechanisms as parallel and coequal, without interrogating the temporal logic or institutional sequencing that may condition their effectiveness. Zhang et al. [2] examine urban challenges with technology and governance as independent mechanisms. Yukalang et al. [23] emphasize participation and regulatory weaknesses without linking them to institutional sequencing. Viljoen et al. [11] identify household-level practices as standalone drivers. Mukwevho et al. [13] highlight barriers to the implementation of integrated waste management plans in municipalities, treated as independent from governance reforms. Mudzusi et al. [26] point to bureaucratic inefficiencies as isolated barriers. Zwane et al. [27] emphasize climate–resource nexus impacts without considering sequencing. Kalina et al. [28] describe municipal failure as a discrete governance problem.

This study advances the literature by addressing this critical gap. It examines not only whether these governance components matter, but how and when they become effective, arguing that their impact is contingent upon a staged process of institutional maturation. By incorporating absorptive capacity theory [29,30] the study reframes reform effectiveness as a sequential process, where internal operational stability is a prerequisite for both strategic digital integration and external participatory engagement.

An integrated theoretical lens combining Institutional Theory [31] and the Technology Acceptance Model (TAM) [32] is applied to examine how governance structures shape reform outcomes. Institutional Theory highlights the role of organizational routines, legitimacy, and resource alignment, while TAM emphasizes perceived usefulness and systemic fit as determinants of technology adoption. Together, these frameworks suggest that the performance impact of digital and participatory reforms depends not only on adoption but also on the municipal system's absorptive capacity and institutional maturity. Despite extensive documentation of reform challenges in LMIC municipalities, few studies offer a model that sequences interventions according to institutional absorption capacity. This study seeks to address this omission through the Municipal Readiness Model for Digital Waste Governance (MRM-DWG).

Unlike prevailing frameworks such as those proposed by Zhang et al. [2] and Yukalang et al. [23] that treat institutional, digital, and participatory drivers as parallel levers of reform, this study introduces a sequenced and conditional framework rooted in absorptive capacity. The MRM-DWG advances a diagnostic model that links municipal reform success to specific stages of institutional readiness, thereby offering a conceptually distinct and empirically tested alternative for adaptive waste governance. In doing so, it makes an original contribution to debates in urban governance and Information and Communications Technologies for Development (ICT4D) by aligning technological and participatory interventions with context-specific operational thresholds, especially in resource-constrained LMIC settings. Empirical analysis focuses on South Africa's Vhembe District region, where GIS platforms and stakeholder engagement structures have been formally introduced, yet persistent challenges remain, including service irregularities, uncollected waste, and spatial service gaps [26,33]. A SEM approach is employed to investigate how internal municipal conditions, external engagement, and GIS usage impact both operational and strategic MSWM outcomes. Drawing on these relationships, the study develops the Municipal Readiness Model for Digital Waste Governance (MRM-DWG). This sequenced framework aligns reform components with institutional maturity and contributes to ongoing debates on adaptive governance and policy sequencing in resource-constrained municipal systems. This study also employs the household dataset of 399 respondents reported in Tahulela et al. [12], whereas the present analysis focuses on GIS-related variables to develop the MRM-DWG.

The MRM-DWG is proposed to address the limits of existing frameworks that treat governance drivers as parallel and universally effective. By embedding sequencing based on absorptive-capacity thresholds, the model provides a readiness-based approach that links reform outcomes to the maturity of municipal institutions. The objective of the study is to investigate how internal capacity, stakeholder engagement, and GIS integration interact in shaping both operational and strategic MSWM outcomes, and to assess the conditions under which these elements become effective. The framework is empirically tested in the context of South Africa's Vhembe District to demonstrate not only whether governance components matter, but also when and how they become effective in resource-constrained municipal systems.

2. Theoretical Framing: Capacity, Technology, and Reform Readiness

The effectiveness of municipal waste reform is shaped not solely by the availability of policy frameworks or digital innovations but by the institutional ability to absorb, internalize, and operationalize them. This notion of governance readiness is rooted in Institutional Theory, as articulated by DiMaggio and Powell et al. [31] and Godfrey et al. [34], which posits that organizational stability, leadership structures, technical expertise, and bureaucratic routines are preconditions for the successful implementation of externally introduced reforms. In fragile municipal systems characterized by fragmented departments or resource scarcity, such internal attributes often determine whether new interventions are routinized or abandoned.

In parallel, the TAM [32,35] offers insight into how municipal actors perceive and utilize information systems. While TAM emphasizes perceived usefulness and ease of use as antecedents of adoption, it implicitly assumes that decision-making occurs in functionally coherent environments. In many LMIC municipal contexts, however, these assumptions rarely hold. Even where GIS platforms or smart tools are introduced, their actual impact is mediated by variables such as staffing adequacy, data literacy, and financial control, rendering TAM's predictions contingent on foundational institutional maturity.

To bridge these limitations, the present study incorporates absorptive capacity theory [29,30], which foregrounds an institution's ability to recognize, assimilate, and apply new knowledge. This theory asserts that reforms succeed only when the recipient institution possesses the internal architecture, cognitive, technical, and procedural capacity to integrate innovations into routine governance. Accordingly, technologies such as GIS may be formally adopted yet remain symbolically rather than substantively integrated unless underpinned by reliable service routines and skilled personnel.

This conceptual synthesis underpins the MRM-DWG developed in this study. Rather than treating governance tools as parallel drivers of reform, the model hypothesizes a sequenced logic, where mechanisms such as GIS integration or external stakeholder engagement become effective only once internal operational systems are functional. Specifically:

- Stage 1: Internal Operational Readiness characterized by human resource sufficiency, logistical coordination, financial stability, and regularized service delivery.
- Stage 2: Strategic Digital Integration, where digital tools such as GIS support forecasting, spatial planning, and infrastructure mapping.
- Stage 3: Participatory and External Engagement involving community feedback mechanisms, donor involvement, and intergovernmental support, which are only impactful once internal and digital layers are consolidated.

This sequencing follows absorptive capacity theory by Cohen & Levinthal [29] as later extended by Zahra & George [30] and applied in recent governance studies [15,16], which shows that operational stability is a prerequisite for digital adoption, and that both must be in place before participatory reforms become effective [2,23].

These theoretical foundations, including Institutional Theory, Technology Acceptance, and Absorptive Capacity, collectively underpin the logic of the MRM-DWG. Rather than treating governance levers as independent or additive, the model posits that reform effectiveness depends on the functional sequencing of institutional readiness: internal operational capacity precedes the efficacy of digital innovations, which in turn enable meaningful participatory engagement. This logic is consistent with recent empirical evidence from Vhembe District, where household waste disposal practices were shown to be constrained by structural inequities and behavioral intentions in ways that illustrate sequencing dynamics [12]. This approach differs from prevailing models in the MSWM literature, which tend to treat governance mechanisms as coequal components [23].

To operationalize this framework empirically, the next section presents the conceptual structure and hypotheses of the model as tested through SEM.

Conceptual Model Structure and Hypotheses Development

To empirically examine the MRM-DWG framework, we constructed a Structural Equation Model (SEM). This model connects three latent exogenous constructs: Internal Factors (IF), External Factors (EF), and GIS Integration (GIS). It also links these factors to two endogenous outcome constructs: Operational Waste Management (WM) and Strategic Waste Management Effectiveness (SWM). The structural logic is informed directly by the conceptual framework below, see Figure 1, and the model hypothesizes the following paths:

- Hypothesis 1—Internal Factors: Operational Waste Management Municipalities with strong internal capacity demonstrate better routine service delivery [26], consistent with Institutional Theory, which emphasizes that organizational routines, resources, and legitimacy are prerequisites for effective service provision [31,36].
- Hypothesis 2—External Factors: Operational Waste Management's external support mechanisms, such as participation and donor aid, have a limited direct effect unless internal systems are stable [37,38], reflecting absorptive capacity theory, which argues that external inputs are ineffective when organizations cannot adapt and apply them [29].
- Hypothesis 3—GIS Integration: Operational Waste Management GIS alone does not improve operational delivery unless embedded within competent institutional systems [25], aligning with TAM, where perceived usefulness leads to adoption only when organizational readiness supports its integration [32,35].
- Hypothesis 4—GIS Integration: Strategic Waste Management GIS contributes positively to long-term spatial planning and service optimization [18], as absorptive capacity theory suggests that digital tools can drive higher-level reforms once they are consolidated within institutional routines [29,30].
- Hypothesis 5—External Factors: Strategic Waste Management's external inputs enhance strategic performance only when internal and digital capacities are in place [39], consistent with institutional perspectives that highlight how stakeholder engagement and partnerships are most effective when organizational legitimacy and technological infrastructure are established [36].

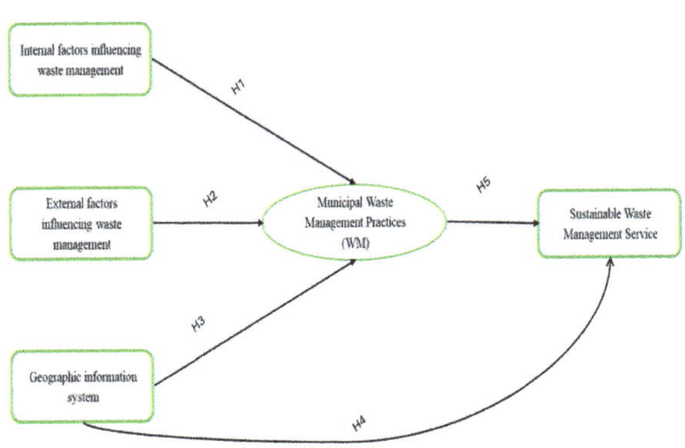

Figure 1. Conceptual framework for the Model of the Municipal Readiness Model for Digital Waste Governance (MRM-DWG).

In addition to these direct hypotheses, the framework anticipates potential indirect effects, including the mediating role of operational waste management in shaping strategic outcomes and the interaction of GIS with external support factors. These extended relationships are examined in the structural analysis and discussed in Sections 4.3 and 4.4.

3. Materials and Methods

3.1. Research Design and Study Area

This study adopted a quantitative, cross-sectional survey design to examine the institutional, technological, and participatory drivers of municipal solid waste management (MSWM) performance. Fieldwork was conducted in four local municipalities, Makhado, Musina, Thulamela, and Collins Chabane, within Vhembe District, Limpopo Province, South Africa. As illustrated in Figure 2.

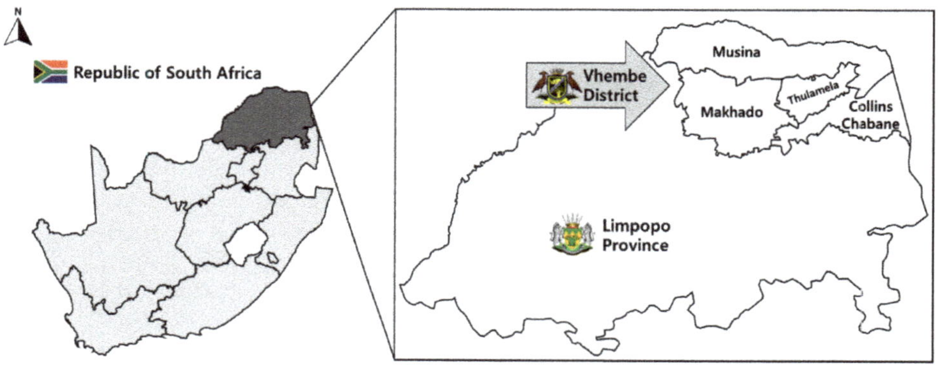

Figure 2. Map of Vhembe District Municipality showing Musina, Makhado, Thulamela, and Collins Chabane Local Municipalities.

These sites were selected for their diversity in settlement typologies (urban, township, rural, and farm-based) and their dual characteristics of digital reform mandates, such as GIS adoption and persistent service delivery gaps. This variation created a fitting empirical context for evaluating the MRM-DWG, which posits that internal operational coherence is a prerequisite for the success of digital and participatory reforms.

3.2. Sampling Strategy and Data Collection

A stratified random sampling approach was used to ensure representativeness across municipalities and settlement types. The four typologies (urban, township, rural, and farm-based) were chosen as they reflect the district's main settlement patterns and service delivery contexts, with households proportionally allocated using municipal population statistics. This approach ensured that the sample reflected the district population structure and was appropriate for testing the proposed framework. The target respondents were adult household members involved in, or knowledgeable about, domestic waste practices.

This study utilized the same household dataset of 399 respondents as in the research conducted by Tahulela et al. [12]. While both studies employed the same survey instrument, Tahulela et al. [12] focused on analyzing behavioral factors. In contrast, this study specifically examines GIS-related variables to develop a readiness model. The questionnaire was administered in English and verbally translated into Tshivenda where necessary. A pilot test with 20 respondents in a neighboring municipality led to refinements in wording and scale alignment.

3.3. Sociodemographic Profile of Respondents

The study analyzed data from 399 respondents, evenly distributed across four municipalities: Makhado, Musina, Collins Chabane, and Thulamela. As shown in Table 1, a substantial proportion resided in rural areas (35.8%) or farm settlements (27.1%). The majority were female (61.4%) and aged between 18 and 39 years (56.1%). Most households consisted of 5–6 members (47.4%), and 34.8% of participants had lived in their communities for more than 20 years. Secondary education was the most commonly reported level (35.6%). Employment status was characterized by self-employment (37.1%) and unemployment (33.1%), while 43.1% of respondents reported monthly household incomes below R10,000 [12]

Table 1. Demographic Characteristics of Respondents [12].

Variables/Items		Frequency (%)
Municipality	Makhado Musina Collins Chabane Thulamela Municipality	99 (24.8) 100 (25.1) 100 (25.1) 100 (25.1)
Residence Area	Rural Area Farm Area City Centre Township	143 (35.8) 108 (27.1) 61 (15.3) 87 (21.8)
Gender	Female Male	245 (61.4) 154 (38.6)
Age	18–29 30–39 40–49 50–59 60–69 ≥70	111 (27.8) 113 (28.3) 93 (23.3) 59 (14.8) 16 (4) 7 (1.8)
Number of Household Members	1–2 3–4 5–6 7–8 ≥9	20 (5) 152 (38.1) 189 (47.4) 25 (6.3) 13 (3.3)
Number of Years Residing in the Area	<1 Year 1–5 Years 5–10 Years 10–15 Years 15–20 Years ≥20 Years	2 (0.5) 43 (10.8) 81 (20.3) 44 (11) 90 (22.6) 139 (34.8)
Education	Never Schooled Primary Secondary Undergraduate University Graduate Other	15 (3.8) 81 (20.3) 142 (35.6) 74 (18.5) 86 (21.6) 1 (0.3)
Occupation	Unemployed Self-Employed Government Employee Farmworker Other	132 (33.1) 148 (37.1) 68 (17) 37 (9.3) 14 (3.5)
Average Monthly Household Income	<R 10,000 R 10,000–R 30,000 R 30,000–R 50,000	172 (43.1) 174 (43.6) 53 (13.3)

3.4. Instrumentation and Construct Development

The questionnaire measured five latent constructs derived from institutional theory and the technology acceptance framework, operationalized through 5-point Likert-type items (1 = Strongly Disagree, 5 = Strongly Agree). To illustrate the operationalization of each latent construct, Table 2 presents example items included in the survey instrument. Items were drawn from validated sources and adapted for relevance to the South African local governance context [40,41]. Each constructs maps directly to components of the MRM-DWG model: internal capacity (Stage 1), GIS integration (Stage 2), and external participation (Stage 3). Reliability and validity of the constructs were confirmed through pilot testing, Cronbach's alpha, and confirmatory factor analysis, ensuring the robustness of the measurement model.

Table 2. Example Items per Construct.

Construct	Sample Items
Internal Factors (IF)	"Municipal collectors arrive when we are not home"; "We do not receive bin bags regularly."
External Factors (EF)	"There are insufficient collection points"; "Lack of technical skills in our area."
GIS Integration (GIS)	"GIS may help improve waste collection planning"; "GIS could lead to corruption."
Operational Waste Practices (WM)	"Distance to disposal points leads to open dumping"; "Waste contaminates water sources."
Strategic Effectiveness (SWM)	"We are unsure when waste will be collected"; "Waste is left to rot before collection."

3.5. Data Analysis Procedures

Data were analyzed using IBM SPSS 29 and AMOS 29, following a structured multi-stage process involving both measurement validation and structural modeling.

3.5.1. Exploratory and Confirmatory Factor Analysis

An Exploratory Factor Analysis (EFA) was first conducted to assess the dimensional structure of the constructs. Sampling adequacy was confirmed via a Kaiser-Meyer-Olkin (KMO) value of 0.798 and Bartlett's test of sphericity (χ^2 = 5728.070; df = 253; $p < 0.001$).

Each construct demonstrated acceptable internal consistency:

- Cronbach's alpha ranged from 0.774 to 0.924.
- Composite Reliability (CR) values exceeded 0.77 for all constructs.

A Confirmatory Factor Analysis (CFA) was then used to validate the measurement model. Convergent validity was confirmed through standardized factor loadings ($\geq$0.66), Average Variance Extracted (AVE) (0.427–0.625), and CR (0.775–0.918). Discriminant validity was established where AVE > MSV for all construct pairs. Model fit indices indicated a strong fit:

$$\text{CFI} = 0.913, \text{IFI} = 0.914$$

$$\text{RMSEA} = 0.076, \text{RMR} = 0.030$$

$$\chi^2/\text{df} = 3.304$$

3.5.2. Structural Equation Modeling (SEM)

The full model, based on the MRM-DWG sequencing logic, was tested using SEM with Maximum Likelihood estimation. Five structural paths were examined, linking IF, EF, and GIS to both WM (operational) and SWM (strategic) outcomes. An additional path tested the mediating effect of WM on SWM. The constructs were measured through respondents' direct involvement in household waste practices, which provided sufficient familiarity with the issues under study, and this informed the evaluation of path significance. Path significance was evaluated via standardized regression weights (β), critical ratios, and p-values.

Model fit remained strong in the structural model:

$$\chi^2 = 720.233, \text{df} = 213, \chi^2/\text{df} = 3.287,$$

$$\text{CFI} = 0.913, \text{IFI} = 0.914, \text{RMSEA} = 0.076, \text{RMR} = 0.031$$

Multicollinearity was assessed via inter-construct correlation matrices, with all correlations < 0.65 and no variance inflation factors (VIFs) exceeding acceptable thresholds.

3.6. Ethical Considerations

Ethical clearance was obtained from the Institutional Research Ethics Committee (IREC) at the Durban University of Technology (Protocol #294/22). Furthermore, the Vhembe District Municipality granted administrative permission. The approval and informed consent process covered both the dataset reported in Tahulela et al. [12] and its reuse in the present analysis. All participants were briefed on the research purpose, confidentiality safeguards, and their right to withdraw. Written informed consent was obtained, and all data were anonymized in compliance with South African data protection legislation.

4. Results

4.1. Analysis of Factors Influencing Waste Management Efficiency

Table 3 presents the factor analysis outcomes, mean scores, and Cronbach's alpha values. The results revealed that 68.7% of the variance in waste management efficiency was attributable to the four identified constructs. Sampling adequacy was confirmed by a KMO value of 0.798, and Bartlett's test of sphericity was significant ($p < 0.001$), indicating the suitability of the factor analysis for the dataset. The factor loadings evaluating the perceived benefits of GIS in waste management ranged from 0.786 to 0.874, demonstrating strong associations among the measured variables. The highest mean score (1.17 ± 0.63) pertained to concerns that using GIS may lead to corruption due to potential financial mismanagement. Nonetheless, there is a strong consensus that GIS can enhance waste collection efficiency, mitigate health hazards, and improve environmental protection. The overall reliability of this construct was high (Cronbach's alpha = 0.924), affirming the consistency of the responses.

The factor loadings measuring the internal factors affecting waste management efficiency ranged from 0.793 to 0.850, indicating strong associations. The highest mean score (2.04 ± 0.715) was related to the issue of waste being scattered by animals before municipal collectors arrive, underscoring a significant inefficiency in waste collection timing. The lowest mean score (1.96 ± 0.650) was associated with financial constraints preventing residents from purchasing bin bags, resulting in open dumping. The construct exhibited high reliability (Cronbach's alpha = 0.867), supporting the consistency of the responses.

The factor loadings assessing external factors affecting waste management efficiency ranged from 0.699 to 0.744, with the highest loading (0.744) linked to a lack of technical waste management skills. The highest mean score (2.04 ± 0.766) was associated with inadequate funding for waste collection, indicating that financial constraints are a major external challenge. The lowest mean score (1.86 ± 0.630) pertains to the lack of technical skills, suggesting the need for capacity building in waste management. The construct's internal reliability was moderate (Cronbach's alpha = 0.785).

Table 3. Factor Loading Coefficient, Mean, and Standard Deviation of the Constructs and Cronbach's Alpha.

Construct	Variable	Description	Factor Loadings	Mean ± SD	Cronbach's Alpha
GIS in waste management	GIS1	Using GIS technology to improve waste collection by our municipality would be highly welcome	0.786	1.07 ± 0.309	0.924
	GIS2	GIS use would alleviate the health hazards caused by waste disposal and collection in our municipality, as it would allow the managers to plan waste collection more appropriately	0.868	1.10 ± 0.428	
	GIS3	Our environment would be more protected and safer if GIS were used to plan our waste collection and disposal	0.839	1.13 ± 0.546	
	GIS4	The money for the GIS could be best used to develop our schools and hospitals	0.874	1.14 ± 0.529	
	GIS5	Using GIS may lead to corruption, as the finances for it may get diverted	0.823	1.17 ± 0.630	
	GIS6	GIS use may lead to job losses, and employment is very high in this area.	0.838	1.14 ± 0.534	
	GIS7	Our municipal rates may increase if we use the GIS	0.826	1.15 ± 0.562	
Internal factors in waste management efficiency	IF4	We do not receive the municipal bin bags	0.793	2.02 ± 0.680	0.867
	IF5	We do not have the finances to buy bin bags, so we are forced to dump open waste at collection points	0.850	1.96 ± 0.650	
	IF6	Municipal collectors come when we are not at home, and if we keep our bins out early, they get scattered by animals and will not be collected.	0.795	2.04 ± 0.715	
	IF7	No recycling programme has been set up in the community by the municipality	0.803	1.96 ± 0.654	
External factors in waste management efficiency	EF1	Lack of technical waste management skills	0.744	1.86 ± 0.630	0.785
	EF2	Lack of waste collection equipment	0.734	1.92 ± 0.702	
	EF3	Lack of adequate collection points	0.699	1.96 ± 0.695	
	EF4	Insufficient waste collectors	0.700	1.89 ± 0.614	
	EF5	Poor funding for waste collection	0.704	2.04 ± 0.766	
Effectiveness of municipal waste management services	SWM3	We are never sure when our waste will be collected	0.882	2.12 ± 0.860	0.854
	SWM4	Our waste is left out to rot and littered before it comes to be collected	0.840	2.17 ± 0.878	
	SWM5	Only big wastes are collected, and our area is left messy as they do not pick up the littered waste	0.798	2.16 ± 0.854	
	SWM6	Our municipality does not have proper waste collection equipment and enough personnel to do a good job	0.777	2.11 ± 0.853	
Waste management practices	HWM1	Waste is not properly collected in my area	0.662	1.9 ± 0.737	0.774
	HWM4	The distance to the waste disposal point is very far for my household, hence we are forced to dump it in the open	0.717	1.99 ± 0.728	
	HWM5	Due to open dumping by people, our natural water sources are contaminated	0.770	1.94 ± 0.67	

Variance = 68.7%. Kaiser-Meyer-Olkin Measure of Sampling Adequacy = 0.798. Bartlett's Test of Sphericity = ($p < 0.001$; $\chi^2 = 5728.1$; df = 253).

The factor loading evaluating the effectiveness of municipal waste management services ranged from 0.777 to 0.882, with the highest loading (0.882) associated with uncertainty about waste collection schedules. The highest mean score (2.17 ± 0.878) corresponded to waste being left to decompose and litter before collection, highlighting serious inefficiencies in service delivery. The construct had a high reliability score (Cronbach's alpha = 0.854), confirming response consistency.

The factor loadings assessing waste management practices ranged from 0.662 to 0.770, with the highest loading (0.770) related to concerns about the contamination of natural water sources due to open dumping. The highest mean score (1.99 ± 0.728) was associated with excessive distance to waste disposal points, compelling households to engage in open dumping. The construct had a moderate reliability score (Cronbach's alpha = 0.774), indicating acceptable internal consistency.

4.2. Scale Reliability and Construct Validity

The CFA was subsequently employed to validate the Measurement Model (MM) by assessing its convergent and discriminant validity. Table 4 delineates the composite reliability, convergent validity, and discriminant validity of the MM. Convergent validity pertains to the extent to which a measure correlates with other measures of the same construct Hair et al. [42] The criteria used to assess convergent validity included standardized loadings, construct (composite) reliability, and AVE. The CR for all constructs ranged from 0.775 to 0.918, indicating good to excellent internal consistency. The AVE measures, which reflect the degree to which items represent their respective constructs, ranged from 0.427 to 0.625. Notably, IF and SWM exhibited the highest AVE measures (0.625 and 0.623, respectively), suggesting that their indicators capture a substantial proportion of the variance in their constructs.

Table 4. Composite Reliability, Average Variance Extracted, and Maximum Shared Square Values.

	CR	AVE	MSV	MaxR(H)	EF	IF	WM	GIS	SWM
EF	0.787	0.427	0.240	0.796	0.654				
IF	0.869	0.625	0.240	0.878	0.490	0.790			
WM	0.775	0.535	0.224	0.783	0.287	0.473	0.732		
GIS	0.918	0.619	0.046	0.930	0.032	−0.215	−0.084	0.786	
SWM	0.868	0.623	0.069	0.873	0.116	0.038	0.263	0.106	0.789

Conversely, discriminant validity refers to the degree to which a construct is empirically distinct from other constructs [43]. Maximum Shared Variance (MSV) is employed to assess the discriminant validity of a scale or construct. The MSV represents the highest correlation between a construct and any other construct within the model. For discriminant validity, the AVE must exceed the MSV. The MSV values ranged from 0.046 to 0.240, with IF and EF sharing the highest MSV (0.240). This indicates a moderate correlation between the two constructs, implying that internal and external factors collectively influence waste management efficiency. The maximum reliability, MaxR(H), ranged from 0.783 to 0.930, with GIS (0.930) demonstrating the highest reliability, thereby reinforcing the robustness of its measurements.

The correlation matrix reveals varying degrees of relationship among the constructs. The strongest correlation was observed between IF and EF ($r = 0.490$), suggesting that internal inefficiencies are closely linked to broader external constraints. The WM also exhibited a moderate correlation with IF ($r = 0.473$), indicating that internal limitations significantly impact waste collection practices. However, GIS shows a weak correlation with all other constructs, with its highest correlation being negative with internal factors ($r = -0.215$), implying that GIS adoption is perceived as independent or even negatively associated with existing internal waste management structures.

4.3. Fitness of the Model

The overall fit of the model was evaluated using multiple criteria. The goodness-of-fit indices are as follows: $\chi^2 = 697.2$, df = 211, $p < 0.001$, $\chi^2/\text{df} = 3.304$, CFI = 0.913, IFI = 0.914, RMR = 0.030, RMSEA = 0.076. These indices indicate that the model appropriately fits the data. The χ^2/df value was below the threshold of 5. The Incremental Fit Index (IFI) = 0.914 and the Comparative Fit Index (CFI) = 0.913 exceeded the recommended threshold of 0.90, demonstrating a strong comparative fit relative to the null model. The RMR value of 0.030 was well below the recommended threshold of <0.06, indicating minimal difference between the observed and predicted covariance matrices. The RMSEA of 0.076 was within the recommended ≤0.08 range, suggesting a reasonable fit with minor room for improvement. Overall, the fit indices confirmed that the model provides a reliable and valid representation of the underlying constructs. Figure 3 illustrates the CFA and model fitness.

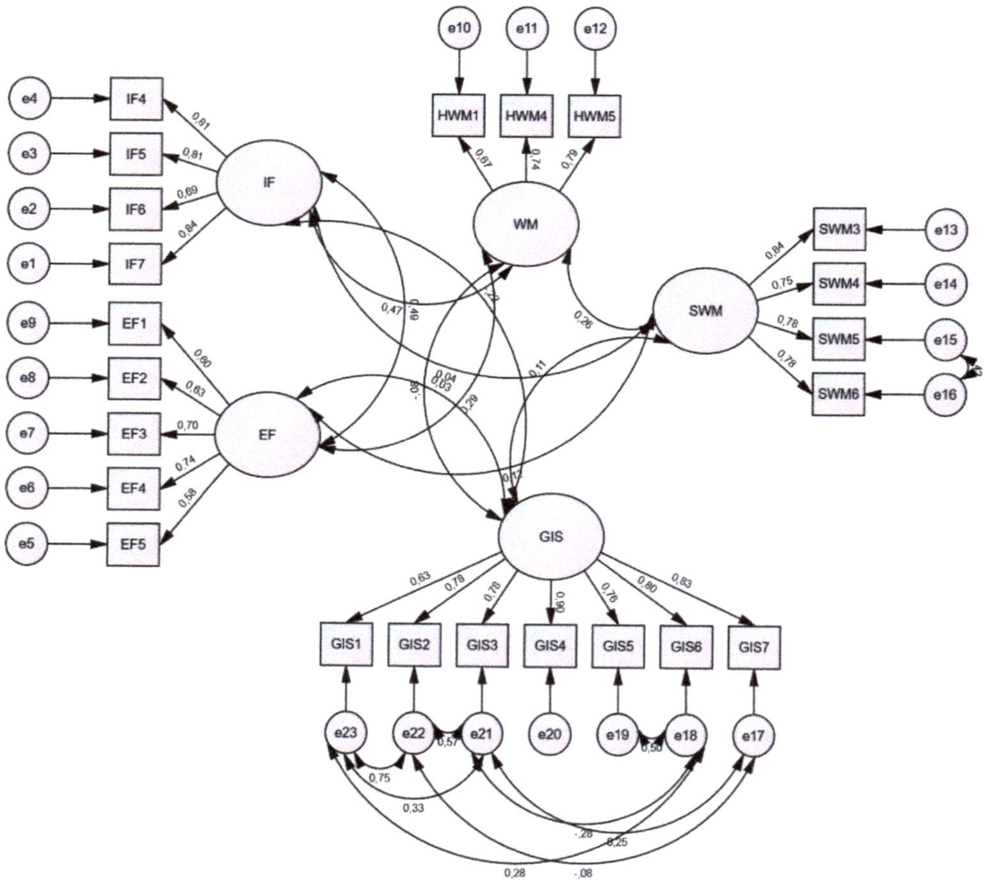

Figure 3. CFA Model of Key Factors Influencing Waste Management.

4.4. Structural Equation Model of Waste Management Variables (Hypothesis Testing)

The MM was subsequently transformed into a path model to elucidate the relationships among latent variables. SEM was employed to evaluate all hypothesized relationships among these latent variables. As depicted in Figure 4, rectangular shapes represent independent and dependent manifest or observed variables, while circular shapes denote the corresponding latent variables for each observed variable. The directional arrows illustrate the linkages between the latent variables in the SEM and the analogous variables, thereby establishing operational interplays among them. The numerical values adjacent to the latent variables in the SEM are factor loadings, referred to as standardized regression weights (SRW). The criteria used to assess the goodness of fit for the MM were similarly applied to the structural model. The goodness-of-fit indices were as follows: $\chi^2 = 720.2$, df = 213, $p < 0.001$, $\chi^2/\text{df} = 3.381$, CFI = 0.913, RMR = 0.031, IFI = 0.914, RMSEA = 0.076, indicating that the measurement model was acceptable. Consequently, the model was found to align with the empirical data.

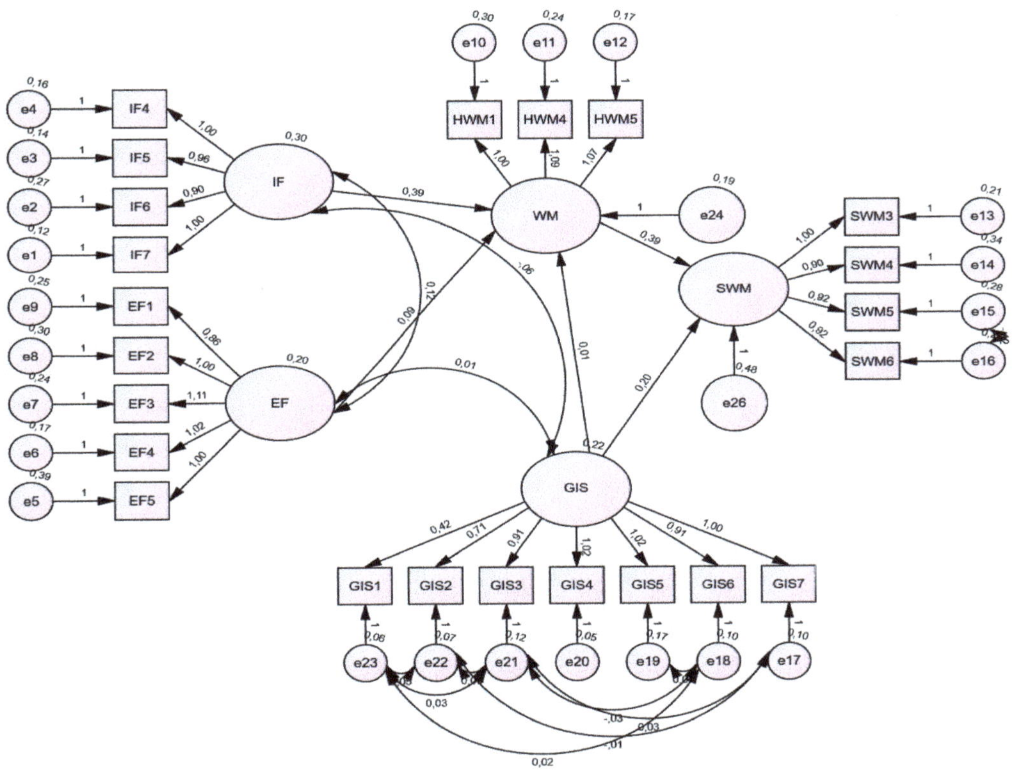

Figure 4. SEM of Factors Influencing Waste Management in Vhembe District.

The relationships between the independent variables (IVs) and their respective dependent variables (DVs) are shown in Table 5. As hypothesized in the conceptual framework, the path from IF to municipal waste management practices was statistically significant ($\beta = 0.432$; CR = 5.653; $p < 0.001$), suggesting that improvements in internal factors lead to significant enhancements in WM. Therefore, the decision rule is to reject the null hypothesis and accept the alternative hypothesis (H1), which posits that IF has a positive impact on WM.

Table 5. Hypothesis conclusions of the Measurement Model (MM).

Hypotheses	Dependent Variable	Independent Variable	Standardized Coefficient (β)	Standard Error	Critical Ratio	p-Value	Decision
H1	WM	IF	0.432	0.068	5.653	<0.001	Supported
H2	WM	EF	0.078	0.080	1.084	0.278	Not Supported
H3	WM	GIS	0.007	0.060	0.119	0.905	Not Supported
H4	SWM	GIS	0.130	0.084	2.395	0.017	Supported
H5	SWM	WM	0.267	0.090	4.385	<0.001	Supported

Conversely, the path from EF to WM was not statistically significant ($\beta = 0.078$; CR = 1.084; $p = 0.278$). This finding indicates that variations in external factors do not significantly influence municipal waste management practices. Thus, the decision rule is to fail to reject the null hypothesis, indicating insufficient evidence to support the alternative hypothesis (H2), which asserts that EF has a positive impact on WM.

Similarly, the path from GIS to municipal waste management practices was not statistically significant ($\beta = 0.007$; CR = 0.119; $p = 0.905$). This suggests that GIS does not directly influence municipal waste management practices. The decision rule is to fail to reject the null hypothesis, as there is no significant relationship supporting the alternative hypothesis (H3) that GIS influences WM.

However, the path from GIS to the SWM was significant ($\beta = 0.130$; CR = 2.395; $p = 0.017$), indicating that GIS positively influences SWM. The decision rule is to reject the null hypothesis and accept the alternative hypothesis (H4), which states that the GIS has a positive effect on SWM.

Similarly, the relationship between municipal waste management practices (WM) and the Effectiveness of Municipal Waste Management Services (SWM) was found to be significant ($\beta = 0.267$; CR = 4.385; $p < 0.001$), indicating that WM positively influences SWM. The decision criterion was to reject the null hypothesis and accept the alternative hypothesis (H5), which posited that WM exerts a positive effect on SWM.

5. Discussion

5.1. Internal Capacity: Foundational Determinant of Waste Governance

The study confirms that internal municipal capacity exerts the most significant influence on routine waste management practices in the study context ($\beta = 0.432$, $p < 0.001$). This construct encompasses staffing adequacy, service regularity, financial coordination, and logistical responsiveness. These internal functions are foundational to governance reform. This interpretation builds on earlier research that documents institutional weaknesses as service delivery bottlenecks [41] by clarifying their temporal and structural role. Internal systems shape the success of other reform mechanisms. In municipalities where internal performance, defined by reliable scheduling, staff presence, and logistical flow, was robust, outcomes in waste collection and routing improved consistently [21]. Consequently, the study proposes that internal readiness constitutes Stage 1 of the MRM-DWG. It must precede digital and participatory interventions. The model embeds institutional thresholds into reform sequencing, contributing to frameworks emphasizing absorptive capacity and reform pacing [44].

Cross-contextual evidence reinforces this logic. In Bangladesh, digital tracking systems failed to improve service delivery, where operational routines remained unstable [4]. In Uganda, participatory platforms delivered limited gains in municipalities lacking baseline budgets and staff continuity [39,45]. These cases echo the pattern observed in Vhembe, reinforcing the model's temporal logic. This evidence contributes to a governance literature that increasingly prioritizes "when" reforms are implemented, not just "what" reforms are chosen. In low-capacity municipalities, digital and participatory innovations should follow, not precede, the consolidation of basic internal functionality.

5.2. External Inputs as Conditional, Not Independent, Enablers

External factors, including stakeholder engagement, intergovernmental collaboration, and funding, did not significantly influence routine waste practices in the SEM model ($\beta = 0.078$, $p = 0.278$). Although normatively important, these mechanisms depend on internal system maturity. This challenges governance models that treat external support as autonomous levers of change. Instead, external inputs act as latent enablers: effective only when logistical and human resource systems are stable. This finding aligns with institutional embeddedness theory, which posits that external reforms only succeed when adapted to local institutional architecture Pritchett et al. [44]. The observed correlation between internal and external factors ($r = 0.490$) suggests that external effectiveness may be mediated by internal improvements. Within the MRM-DWG, external interventions are assigned to Stage 3, serving as scaling mechanisms. They scale but cannot replace foundational delivery systems.

International evidence supports this: a study by Duru et al. [46] in Nigeria, donor-funded reforms failed without internal tracking systems, while Nepal's community initiatives stagnated due to municipal fragmentation [28,47]. An alternative explanation for the weak statistical association may involve perceptions: in low-capacity municipalities, external platforms may appear symbolic or donor-driven rather than integrated into daily operations. This disconnect may explain their limited observed impact. Thus, while external participation and support remain important, their introduction should follow system stabilization to avoid mismatched interventions.

5.3. GIS as a Strategic Enabler, Not an Operational Instrument

The role of GIS varied across waste governance layers. While GIS had no significant impact on daily operations ($\beta = 0.007$, $p = 0.905$), it showed a moderate positive effect on strategic service effectiveness ($\beta = 0.130$, $p = 0.017$). This highlights that GIS is effective for strategic planning but not routine operations. GIS was conceptualized through perceived benefits such as route optimization and infrastructure mapping. However, respondents often lacked direct experience with GIS. This limited deployment likely explains GIS's minimal operational role. The findings challenge linear adoption models like the TAM, which assumes perceived usefulness leads directly to use. Instead, they support a phased logic: digital tools function effectively only after operational baselines are stabilized.

Thus, GIS aligns with Stage 2 of the MRM-DWG as a strategic instrument for data visualization, planning, and forecasting, not for routine collection or fault reporting, which depend on staff and logistics. Experiences from Pakistan and Indonesia confirm this pattern, where GIS adoption failed under weak operational conditions [16,48]. By contrast, GIS applications in Shone Town and Dessie City (Ethiopia) [18,22], as well as in Azuay Province (Ecuador) [49] and Pedro Ruiz Gallo (Peru) [20], only proved successful after municipal authorities had standardized internal service routines and aligned institutional processes. This confirms that GIS effectiveness is conditional, not universal, and depends on timing, training, and system readiness.

5.4. Framing the Municipal Readiness Model for Digital Waste Governance (MRM-DWG)

The MRM-DWG emerges from the study's empirical results as a sequenced framework designed to align reform interventions with a municipality's institutional capacity. Drawing upon layered SEM performance outcomes, the model provides a structured pathway for understanding when and how internal, digital, and participatory mechanisms influence waste governance outcomes.

To visualize this sequencing, Figure 5 presents the MRM-DWG. The model translates SEM results into three-phase logic: internal capacity (Stage 1), digital integration (Stage 2), and participatory engagement (Stage 3). Solid arrows indicate statistically supported paths (IF-WM; GIS-SWM; WM-SWM), while dashed lines represent non-significant or conditional relationships (EF-WM; GIS-WM), emphasizing the importance of sequencing reform interventions. A comparative synthesis of the MRM-DWG against widely cited governance reform models is presented in Table 6. At its core, the MRM-DWG proposes:

- Stage 1—Operational Readiness: Foundational capacity is characterized by adequate staffing, regular collection, logistical coordination, and financial control. This stage showed the strongest predictive relationship with waste performance (β = 0.432, $p < 0.001$).
- Stage 2—Strategic Digital Integration: Once functionality stabilizes, GIS and related tools support planning, forecasting, and spatial optimization. GIS significantly influenced strategic outcomes (β = 0.130, p = 0.017), but not daily operations.
- Stage 3—Participatory and External Engagement: External collaboration becomes meaningful only after foundational and strategic systems are in place. The non-significant effect of external factors (β = 0.078, p = 0.278) underlines their dependence on prior internal capacity.

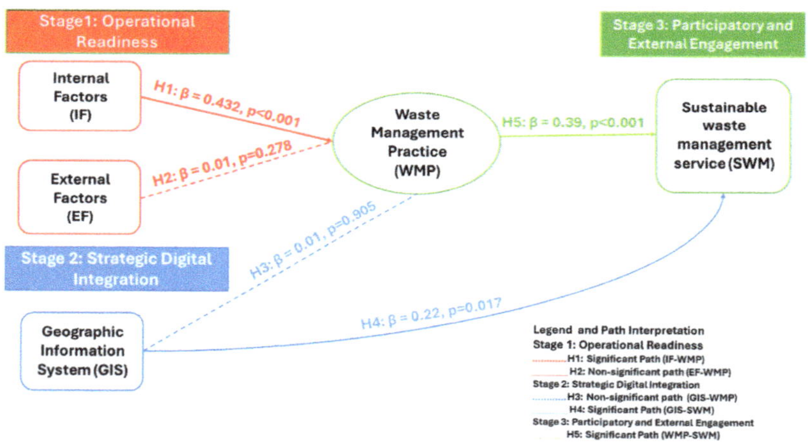

Figure 5. Municipal Readiness Model for Digital Waste Governance (MRM-DWG).

Table 6. Comparative view of the MRM-DWG against prominent governance frameworks.

Framework	Core Assumptions	Limitations	MRM-DWG Distinction
Additive Reform Models	Interventions are most effective when layered together	Ignore sequencing, overburden low-capacity systems	Introduces staged logic based on capacity thresholds
ICT4D (Information and Communication for Development)	Technology drives performance	Often assumes linear adoption; ignores institutional maturity	Aligns digital integration with system maturity, not universal adoption
Participatory Governance	Community involvement increases accountability and efficiency	Presumes readiness for participation	Delays participatory scaling until institutional groundwork is secured
Smart City or Policy Transfer Models	Global solutions are replicable across urban settings	Downplays local variation and embedded constraints	Anchors reform logic in the municipal context, emphasizing readiness diagnostics

This logic replaces additive reform models with a capacity-sensitive sequencing tool, grounded in institutional theory and reform pacing frameworks [44].

5.5. Implications and Policy Recommendations

The findings of this study offer a refined, readiness-based framework for guiding waste governance reforms in resource-constrained municipal systems. Rather than promoting simultaneous interventions, the evidence supports a sequenced strategy, one that acknowledges the varying absorptive capacity of institutions and matches intervention types to distinct phases of governance maturity.

5.5.1. Implication 1: Operational Capacity Must Precede Reform Initiatives

The consistent statistical significance of internal factors ($\beta = 0.432$, $p < 0.001$) emphasizes that reforms should begin with strengthening routine service delivery. Municipalities must first address baseline constraints such as staffing shortfalls, irregular collection, coordination inefficiencies, and logistical breakdowns. These dimensions are often overlooked in digital-first strategies, forming the core of the operational ecosystem upon which all subsequent reforms depend.

- Policy Recommendation: National and provincial governments should invest in diagnostic audits that assess municipal readiness through indicators like equipment uptime, staff-to-service ratios, and collection regularity. Targeted funding for basic delivery systems rather than premature ICT upgrades should be prioritized where internal gaps persist.

5.5.2. Implication 2: Align Digital Tools with Strategic, Not Tactical, Objectives

The empirical finding that GIS had no significant impact on daily operations ($\beta = 0.007$, $p = 0.905$) but positively affected strategic effectiveness ($\beta = 0.130$, $p = 0.017$) suggests the need to reconceptualize the role of technology in waste governance. In municipalities where operational foundations are fragile, digital tools are unlikely to resolve routine inefficiencies but can support planning, forecasting, and long-term infrastructure development.

- Policy Recommendation: GIS and other smart city technologies should be positioned as medium- to long-term investments. Their deployment should follow service stabilization and should be oriented toward spatial optimization tasks such as route design, landfill siting, and equitable infrastructure allocation. This aligns with findings from cities like Kigali and Nairobi, where GIS success was contingent on baseline system reliability [50], and contrasts with failure cases in Pakistan and Indonesia, where digital tools were introduced prematurely [16,48].

5.5.3. Implication 3: Stage Participatory Mechanisms According to Institutional Readiness

The non-significance of external factors in predicting operational performance ($\beta = 0.078$, $p = 0.278$) suggests that stakeholder participation and external funding, while normatively important, are not inherently effective unless introduced within a functioning system. When layered onto unstable systems, participatory mechanisms may be symbolic or lead to stakeholder fatigue.

- Policy Recommendation: Donor agencies and development programs should calibrate their participatory expectations to match municipal capacity levels. Participatory forums, community engagement platforms, and co-production schemes should be introduced only after foundational delivery mechanisms and strategic planning systems are functional. Evidence from Nepal and Ghana supports this sequence, showing that external partnerships failed to yield measurable improvements where internal coherence was lacking [51,52]

5.6. Framework Application: MRM-DWG as a Governance Planning Tool

The MRM-DWG offers practical value beyond the immediate study context. It can serve as a diagnostic tool for multiple actors:

- Local Governments: to assess readiness levels and prioritize capacity-building before reform sequencing.
- Development Agencies: to target interventions (e.g., funding, technical assistance) in alignment with institutional maturity.
- Researchers and Evaluators: to apply the model in comparative contexts across sectors such as water, sanitation, and energy governance.

5.7. Implication: Future GIS Integration Should Be Spatially Grounded

Although this study did not involve primary GIS-based spatial analysis, its findings highlight the importance of GIS integration at the strategic level (Stage 2 of the MRM-DWG). To strengthen this phase, future research and municipal practice in regions such as Vhembe should incorporate spatial diagnostics that go beyond perception-based modeling. For example, the United Nations Environment Program [17] shows how GIS tools can be used to guide landfill siting and broader resource recovery planning. In Nepal, Bharadwaj et al. [47] demonstrate how spatial analysis can optimize waste collection routes and strengthen financial planning for municipalities. Similarly, Nikou et al. [48] illustrate how spatial data can be used to map underserved communities and highlight disparities in access to waste services.

Such approaches have been applied in similar contexts. Cobos-Mora et al. [53] used multi-criteria GIS analysis to select equitable transfer station sites in Ecuador, while Araiza-Aguilar et al. [49] employed spatial modeling to design optimized waste collection networks in peri-urban Latin America. These spatial planning strategies foreground equity, and environmental risk reduction outcomes remain vital for sustainable service delivery.

Embedding these spatial functions into the digital planning phase of the MRM-DWG would enhance both its analytical utility and its relevance to municipalities seeking to align with SDG 11 (inclusive cities) and SDG 13 (climate action). Future studies applying the model should therefore incorporate GIS-based mapping as a decision-support tool to complement the perception-based insights reported here.

5.8. Contextual Generalizability and Scaling of the MRM-DWG

Although this study focuses on South Africa's Vhembe District broadly, the scope of application for the MRM-DWG extends to municipalities in other low- and middle-income contexts where governance fragmentation and capacity constraints hinder reform, including but not limited to regions across Sub-Saharan Africa and Asia. Its sequenced approach, prioritizing operational readiness before digital and participatory interventions, offers a flexible framework adaptable to diverse local contexts.

However, effective transferability depends on contextual diagnostics. Differences in institutional maturity, fiscal autonomy, and civic engagement may shape the model's applicability. Cities such as Kampala, Dhaka, or Kumasi, where digital tools have been introduced in low-capacity environments, could benefit from using the MRM-DWG as a diagnostic tool to recalibrate the sequencing of reforms.

Future applications of this model should follow principles of adaptive policy design [44], emphasizing institutional fit over uniform replication. This enhances its relevance for development partners, local governments, and reform evaluators seeking scalable, phased strategies in resource-constrained urban systems.

5.9. Limitations and Future Research

This study surveyed 399 households across four municipalities using a stratified random sampling design to ensure proportional representation across settlement types. While households were chosen deliberately as the unit of analysis because they are both direct service recipients and active participants in waste practices, the perspectives of municipal officials, private contractors, NGOs, and informal waste workers were not included. Future research should extend the MRM-DWG by incorporating these stakeholders to capture a fuller, multi-actor validation of governance readiness. In addition, comparative assessments with alternative governance frameworks would further clarify the distinct contributions and boundaries of the MRM-DWG. This study utilized the dataset from Tahulela et al. [12]. While different variables were investigated in this analysis, relying on a single survey restricts the ability to validate the findings externally. Future research should replicate the model using independent datasets for more robust results.

6. Conclusions

This study investigated the layered interdependencies between internal capacity, digital integration, and external engagement in shaping municipal waste management outcomes in South Africa's Vhembe District. Employing SEM, the analysis found that internal operational capacity encompassing workforce adequacy, financial coordination, and service regularity was the most significant predictor of effective waste governance ($\beta = 0.432$, $p < 0.001$). In contrast, digital tools such as GIS and external inputs like stakeholder participation exhibited conditional effects, effective only after internal systems had stabilized. The study has achieved its stated objective by demonstrating that operational capacity forms the foundation for reform effectiveness, while GIS integration and participatory engagement yield sustainable impact only when sequenced upon this capacity. In doing so, the study developed and validated the MRM-DWG, providing a structured framework for aligning reform components with institutional maturity.

These findings contribute to a growing body of literature emphasizing context-sensitive, staged reform strategies over parallel or additive models. By empirically distinguishing between operational and strategic performance domains, the study introduces MRM-DWG, a sequential, staged framework aligning interventions with institutional maturity:

- Stage 1: Internal stability forms the precondition for reform success.
- Stage 2: Strategic digital technologies follow operational consolidation.
- Stage 3: Participatory scaling becomes effective only when embedded within stable, functioning systems.

The primary contribution, therefore, is the MRM-DWG's sequential logic, which aligns governance interventions explicitly with absorptive capacity and institutional readiness. Although derived from a single case study, the logic underpinning the MRM-DWG demonstrates clear potential for adaptation across other low- and middle-income municipal contexts characterized by fragmented governance and constrained capacities.

Future research should empirically test and refine the MRM-DWG in comparative and longitudinal settings across other governance domains, such as water, sanitation, and energy, thereby further validating its generalizability. This would enhance its utility as a practical tool for structuring realistic, phased approaches to institutional reform in resource-constrained environments globally.

This study's cross-sectional design precludes causal inference and may not capture temporal dynamics in institutional reform or digital adoption. Further, reliance on self-reported data introduces the potential for perceptual bias. Although measurement validity was rigorously tested, future research could strengthen generalizability through Longitudinal panel studies to track institutional maturity over time

Author Contributions: Conceptualization, A.C.T. and S.H.; methodology, A.C.T.; software, A.C.T.; validation, A.C.T., S.H. and M.E.L.; formal analysis, A.C.T.; investigation, A.C.T.; resources, S.H.; writing—original draft preparation, A.C.T.; writing—review and editing, S.H. and M.E.L.; visualization, A.C.T. and S.H.; funding acquisition, A.C.T. All authors have read and agreed to the published version of the manuscript.

Funding: This work was supported by the Space Science grant from the Department of Science and Technology at the Durban University of Technology.

Institutional Review Board Statement: The study was conducted in accordance with the approval by the Institutional Research Ethics Committee of Durban University of Technology (Ethics Clearance Number: IREC 294/22 and 4 July 2023).

Informed Consent Statement: Informed consent was obtained from all subjects involved in the study.

Data Availability Statement: The data that support the findings of this study, also analyzed in Tahulela et al. [12], are available from the corresponding author, Aifani Confidence Tahulela, upon reasonable request.

Conflicts of Interest: The authors declare no conflicts of interest. The funders had no role in the design of the study, in the collection, analysis, or interpretation of data, in the writing of the manuscript, or in the decision to publish the results.

Abbreviations

The following abbreviations are used in this manuscript:

AVE	Average Variance Extracted
CFA	Confirmatory Factor Analysis
CFI	Comparative Fit Index
CR	Construct (Composite) Reliability
DVs	Dependent Variables
EF	External Factors
EFA	Exploratory Factor Analysis
GIS	Geographic Information Systems
IF	Internal Factors
IVs	Independent Variables
IWMPs	Integrated Waste Management Plans
KMO	Kaiser-Meyer-Olkin
LMICs	Low- and Middle-Income Countries
MaxR(H)	Maximum Reliability
MM	Measurement Model
MRM-DWG	Municipal Readiness Model for Digital Waste Governance
MSWM	Municipal Solid Waste Management
SEM	Structural Equation Modeling
SWM	Strategic (Sustainable) Waste Management
TAM	Technology Acceptance Model
WM	Waste Management
WMP	Waste Management Practice

eferences

Lal, R.; Bouma, J.; Brevik, E.; Dawson, L.; Field, D.J.; Glaser, B.; Hatano, R.; Hartemink, A.E.; Kosaki, T.; Lascelles, B.; et al. Soils and Sustainable Development Goals of the United Nations: An International Union of Soil Sciences Perspective. *Geoderma Reg.* **2021**, *25*, e00398. [CrossRef]

Zhang, Z.; Chen, Z.; Zhang, J.; Liu, Y.; Chen, L.; Yang, M.; Osman, A.I.; Farghali, M.; Liu, E.; Hassan, D.; et al. Municipal Solid Waste Management Challenges in Developing Regions: A Comprehensive Review and Future Perspectives for Asia and Africa. *Sci. Total Environ.* **2024**, *930*, 172794. [CrossRef]

Kaza, S.; Yao, L.; Bhada-Tata, P.; Van Woerden, F. *What Waste 2.0: A Global Snapshot of Solid Waste Management to 2050*; World Bank: Washington, DC, USA, 2018. [CrossRef]

Abir, T.M.; Datta, M.; Saha, S.R. Assessing the Factors Influencing Effective Municipal Solid Waste Management System in Barishal Metropolitan Areas. *J. Geosci. Environ. Prot.* **2023**, *11*, 49–66. [CrossRef]

Ukala, D.C.; Ifeanyi, A.; Owamah, H.I. A Review of Solid Waste Management Practice in Nigeria. *NIPES—J. Sci. Technol. Res.* **2020**, *2*, 169–174. [CrossRef]

Department of Environmental Affairs and Tourism. *National Waste Management Strategy Implementation: Recycling and Extended Producer Responsibility*; Department of Environmental Affairs and Tourism: Pretoria, South Africa, 2005; pp. 1–29. Available online: https://sawic.environment.gov.za/documents/234.pdf (accessed on 1 October 2025).

Makhado Local Municipality. *Integrated Waste Management Plan (IWMP)*; Makhado Local Municipality: Makhado, South Africa, 2017.

National Environmental Management. *National Environmental Management: Waste Act 59 of 2008*; National Environmental Management: Nairobi, Kenya, 2009.

Kubanza, N.S. Analysing the Challenges of Solid Waste Management in Low-Income Communities in South Africa: A Case Study of Alexandra, Johannesburg. *S. Afr. Geogr. J.* **2025**, *107*, 169–189. [CrossRef]

Raghu, S.J.; Rodrigues, L.L.R. Behavioral Aspects of Solid Waste Management: A Systematic Review. *J. Air Waste Manag. Assoc.* **2020**, *70*, 1268–1302. [CrossRef]

Viljoen, J.M.M.; Schenck, C.J.; Volschenk, L.; Blaauw, P.F.; Grobler, L. Household Waste Management Practices and Challenges in a Rural Remote Town in the Hantam Municipality in the Northern Cape, South Africa. *Sustainability* **2021**, *13*, 5903. [CrossRef]

Tahulela, A.C.; Hashemi, S.; Lourens, M.E. Household Waste Disposal Under Structural and Behavioral Constraints: A Multivariate Analysis from Vhembe District, South Africa. *Sustainability* **2025**, *17*, 7429. [CrossRef]

Mukwevho, P.; Radzuma, P.; Roos, C. Exploring Barriers to the Effective Implementation of Integrated Waste Management Plans in Developing Economies: Lessons Learned from South African Municipalities. *Sustainability* **2024**, *16*, 9643. [CrossRef]

Khatoon, A. Waste Management—A Case Study in Nepal. In *Solid Waste Policies and Strategies: Issues, Challenges and Case Studies*; Springer: Singapore, 2020; pp. 185–196.

Patel, J.; Mujumdar, S.; Srivastava, V.K. Municipal Solid Waste Management in India-Current Status, Management Practices, Models, Impacts, Limitations, and Challenges in Future. *Adv. Environ. Res.* **2023**, *12*, 95–111. [CrossRef]

Filimonova, N.; Birchall, S.J. Sustainable Municipal Solid Waste Management: A Comparative Analysis of Enablers and Barriers to Advance Governance in the Arctic. *J. Environ. Manag.* **2024**, *371*, 123111. [CrossRef]

United Nations Environment Programme. *Beyond an Age of Waste Turning Rubbish into a Resource*; United Nations Environment Programme: Nairobi, Kenya, 2024; ISBN 978-92-807-4129-2.

Chandel, A.S.; Weto, A.E.; Bekele, D. Geospatial Technology for Selecting Suitable Sites for Solid Waste Disposal: A Case Study of Shone Town, Central Ethiopia. *Urban Plan Transp. Res.* **2024**, *12*, 2302531. [CrossRef]

Nguyen, A.T.; Nguyen, N.; Phung, P.; Yến-Khanh, N. Residents' Waste Management Practices in a Developing Country: A Social Practice Theory Analysis. *Environ. Chall.* **2023**, *13*, 100770. [CrossRef]

Medina-Medina, A.J.; Salas López, R.; Barboza, E.; Tuesta-Trauco, K.M.; Zabaleta-Santiesteban, J.A.; Guzman, B.K.; Oliva-Cruz, M.; Tariq, A.; Rojas-Briceño, N.B. Participation GIS for the Monitoring of Areas Contaminated by Municipal Solid Waste: A Case Study in the City of Pedro Ruiz Gallo (Peru). *Case Stud. Chem. Environ. Eng.* **2024**, *10*, 100941. [CrossRef]

Hussain, I.; Elomri, A.; Kerbache, L.; El Omri, A. El Smart City Solutions: Comparative Analysis of Waste Management Models in IoT-Enabled Environments Using Multiagent Simulation. *Sustain. Cities Soc.* **2024**, *103*, 105247. [CrossRef]

Kebede, H.A.; Ayenew, W.A. Solid Waste Disposal Site Selection Analysis Using Geospatial Technology in Dessie City Ethiopia. *Heliyon* **2023**, *9*, e18865. [CrossRef]

Yukalang, N.; Clarke, B.; Ross, K. Barriers to Effective Municipal Solid Waste Management in a Rapidly Urbanizing Area in Thailand. *Int. J. Envrion. Res. Public Health* **2017**, *14*, 1013. [CrossRef] [PubMed]

Strydom, W.F. Barriers to Household Waste Recycling: Empirical Evidence from South Africa. *Recycling* **2018**, *3*, 41. [CrossRef]

Shabani, T.; Defe, R.; Mavugara, R.; Mupepi, O.; Shabani, T. Application of Geographic Information Systems (GIS) and Remote Sensing (RS) in Solid Waste Management in Southern Africa: A Review. *SN Soc. Sci.* **2024**, *4*, 39. [CrossRef]

Mudzusi, T.N.; Munzhedzi, P.H.; Mahole, E. Governance Challenges in the Provision of Municipal Services: In the Vhembe District Municipality. *Afr. Public Serv. Deliv. Perform. Rev.* **2024**, *12*, 749. [CrossRef]

27. Zwane, N.; Botai, J.O.; Nozwane, S.H.; Jabe, A.; Botai, C.M.; Dlamini, L.; Nhamo, L.; Mpandeli, S.; Petja, B.; Isaac, M.; et Analysis of the Projected Climate Impacts on the Interlinkages of Water, Energy, and Food Nexus Resources in Narok Coun Kenya, and Vhembe District Municipality, South Africa. *Water* **2025**, *17*, 1449. [CrossRef]
28. Kalina, M.; Makwetu, N.; Tilley, E. "The Rich Will Always Be Able to Dispose of Their Waste": A View from the Frontlines Municipal Failure in Makhanda, South Africa. *Envrion. Dev. Sustain.* **2023**, *26*, 17759–17782. [CrossRef] [PubMed]
29. Cohen, W.M.; Levinthal, D.A. Absorptive Capacity: A New Perspective on Learning and Innovation. *Adm. Sci. Q.* **1990**, *35*, 1: [CrossRef]
30. Zahra, S.A.; George, G. Absorptive Capacity: A Review, Reconceptualization, and Extension. *Acad. Manag. Rev.* **2002**, *27*, 1 [CrossRef]
31. DiMaggio, P.J.; Powell, W.W. The Iron Cage Revisited: Institutional Isomorphism and Collective Rationality in Organization Fields. *Am. Sociol. Rev.* **1983**, *48*, 147. [CrossRef]
32. Davis, F.D. Perceived Usefulness, Perceived Ease of Use, and User Acceptance of Information Technology. *MIS Q.* **1989**, *13*, 3 [CrossRef]
33. Olaniyi, F.C.; Ogola, J.S.; Tshitangano, T.G. Challenges of Effective Management of Medical Waste in Low-Resource Setting Perception of Healthcare Workers in Vhembe District Healthcare Facilities, South Africa. *Trans. R. Soc. S. Afr.* **2021**, *76*, 81– [CrossRef]
34. Godfrey, L.; Scott, D.; Trois, C. Caught between the Global Economy and Local Bureaucracy: The Barriers to Good Was Management Practice in South Africa. *Waste Manag. Res. J. A Sustain. Circ. Econ.* **2013**, *31*, 295–305. [CrossRef]
35. Venkatesh, V.; Davis, F.D. A Theoretical Extension of the Technology Acceptance Model: Four Longitudinal Field Studies. *Man Sci.* **2000**, *46*, 186–204. [CrossRef]
36. Scott, W.R. *Institutions and Organizations: Ideas and Interests*; Sage: Thousand Oaks, CA, USA, 2008.
37. Paul, M.; Bussemaker, M.J. A Web-Based Geographic Interface System to Support Decision Making for Municipal Solid Was Management in England. *J. Clean. Prod.* **2020**, *263*, 121461. [CrossRef]
38. Soukopová, J.; Ochrana, F.; Klimovský, D.; Mikušová Meričková, B. Factors Influencing the Efficiency and Effectiveness Municipal Waste Management Expenditure. *Lex Localis J. Local Self-Gov.* **2016**, *14*, 359–378. [CrossRef]
39. Katusiimeh, M.W.; Nabimanya, B.; Komwangi, D. The Neglected Governance Challenges of Solid Waste Management in Ugand Insights from a Newly Created City of Mbarara. *Afr. J. Gov. Public Leadersh. (AJoGPL)* **2023**, *1*, 28–41.
40. Adedara, M.L.; Taiwo, R.; Bork, H.-R. Municipal Solid Waste Collection and Coverage Rates in Sub-Saharan African Countries: Comprehensive Systematic Review and Meta-Analysis. *Waste* **2023**, *1*, 389–413. [CrossRef]
41. Finocchiaro Castro, M.; Guccio, C.; Romeo, D.; Vidoli, F. How Does Institutional Quality Affect the Efficiency of Local Governmen An Assessment of Italian Municipalities. *Econ Polit* **2025**, *42*, 569–597. [CrossRef]
42. Hair, J.F.; Hult, G.T.M.; Ringle, C.M.; Sarstedt, M.; Danks, N.P.; Ray, S. *Partial Least Squares Structural Equation Modeling (PLS-SEN Using R*; Springer International Publishing: Cham, Switzerland, 2021; ISBN 978-3-030-80518-0.
43. Hair, J.F. *A Primer on Partial Least Squares Structural Equation Modeling (PLS-SEM)*; Sage Publications: Thousand Oaks, CA, USA, 201
44. Andrews, M.; Pritchett, L.; Woolcock, M. *Building State Capability: Evidence, Analysis, Action*; Oxford University Pres Oxford, UK, 2017.
45. Okot-Okumu, J. Solid Waste Management in Uganda. In *Future Directions of Municipal Solid Waste Management in Africa*; Afri Institute of South Africa: Pretoria, South Africa, 2015; pp. 107–135.
46. Duru, R.U.; Ikpeama, E.E.; Ibekwe, J.A. Challenges and Prospects of Plastic Waste Management in Nigeria. *Waste Dispos. Sustai Energy* **2019**, *1*, 117–126. [CrossRef]
47. Bharadwaj, B.; Rai, R.K.; Nepal, M. Sustainable Financing for Municipal Solid Waste Management in Nepal. *PLoS ONE* **202** *15*, e0231933. [CrossRef]
48. Nikou, V.; Sardianou, E. Bridging the Socioeconomic Gap in E-Waste: Evidence from Aggregate Data across 27 European Unio Countries. *Clean. Prod. Lett.* **2023**, *5*, 100052. [CrossRef]
49. Araiza-Aguilar, J.A.; Rojas-Valencia, M.N.; Nájera-Aguilar, H.A.; Gutiérrez-Hernández, R.F.; García-Lara, C.M. Using Spati Analysis to Design a Solid Waste Collection System. *Urban Sci.* **2024**, *8*, 95. [CrossRef]
50. Kabera, T.; Wilson, D.C.; Nishimwe, H. Benchmarking Performance of Solid Waste Management and Recycling Systems in Ea Africa: Comparing Kigali Rwanda with Other Major Cities. *Waste Manag. Res.* **2019**, *37*, 58–72. [CrossRef]
51. Melles, G.; Gautam, S.; Shrestha, R. Circular Economy for Nepal's Sustainable Development Ambitions. *Challenges* **2025**, *16*, [CrossRef]

. Nyanyofio, J.G.T.; Domfeh, K.A.; Buabeng, T.; Maloreh-Nyamekye, T.; Appiah-Agyekum, N.N. Governance and Effectiveness of Public–Private Partnership in Ghana's Rural-Water Sector. *Int. J. Public Sect. Manag.* **2022**, *35*, 709–732. [CrossRef]
. Cobos-Mora, S.L.; Guamán-Aucapiña, J.; Zúñiga-Ruiz, J. Suitable Site Selection for Transfer Stations in a Solid Waste Management System Using Analytical Hierarchy Process as a Multi-Criteria Decision Analysis: A Case Study in Azuay-Ecuador. *Envrion. Dev. Sustain.* **2023**, *25*, 1944–1977. [CrossRef]

sustainability

MDPI

Article

Household Waste Disposal Under Structural and Behavioral Constraints: A Multivariate Analysis from Vhembe District, South Africa

Aifani Confidence Tahulela [1,*], Shervin Hashemi [2,3] and Melanie Elizabeth Lourens [3]

1 Department of Public Management, Durban University of Technology, Steve Biko Campus, Durban 4000, South Africa
2 Institute for Environmental Research, Yonsei University College of Medicine, 50-1 Yonsei, Seodaemun-gu, Seoul 03722, Republic of Korea; shervin@yuhs.ac
3 Faculty of Management Sciences, Durban University of Technology, Steve Biko Campus, Durban 4000, South Africa; shervinh@dut.ac.za (S.H.); melaniel@dut.ac.za (M.E.L.)
* Correspondence: aifanit@dut.ac.za

Abstract

Both behavioral intentions and structural constraints shape household waste disposal in low-resource settings. This study integrates the Theory of Planned Behavior (TPB) with Environmental Justice (EJ) to examine informal waste disposal in Vhembe District, South Africa, a region marked by infrastructural deficits and uneven municipal services. A cross-sectional survey of 399 households across four municipalities assessed five disposal behaviors, including river dumping and domestic burial. Only 8% of households used formal bins, while over 50% engaged in open or roadside dumping. Although education and income were inversely associated with harmful practices, inadequate service access was the most significant constraint on formal disposal. Logistic regression revealed that rural residents and households in underserved municipalities were significantly more likely to engage in hazardous methods, regardless of socioeconomic status. These findings extend TPB by showing that perceived behavioral control reflects not only psychological agency but also material and institutional limitations. By reframing informal disposal as a structurally conditioned response rather than a behavioral deficit, the study advances EJ theory and provides a transferable TPB–EJ framework for decentralized, justice-oriented waste governance. The results underscore the need for Sustainable Development Goal (SDG)-aligned interventions that integrate equitable infrastructure with context-sensitive behavioral strategies.

Keywords: environmental justice; household waste disposal; rural South Africa; sustainable development goals; theory of planned behavior

1. Introduction

The rapid increase in municipal solid waste (MSW) presents a significant environmental and governance challenge globally, particularly in low- and middle-income countries (LMICs), where formal waste management systems are underdeveloped. Projections indicate that global waste generation will rise from 2.01 billion tons in 2016 to 3.40 billion tons by 2050, with the most pronounced growth expected in LMICs [1]. In sub-Saharan Africa, over 60% of the population lacks access to reliable waste collection services, leading to dependence on informal and often hazardous disposal methods such as burning, open dumping, and burial [2–4]. These practices contribute to the transmission of communicable diseases, groundwater contamination, and air pollution, thereby posing threats to human and ecosystem health and hindering progress towards Sustainable Development Goals (SDGs) 3, 6, 11, and 12 [5–7].

In South Africa, these issues are exacerbated by historical and spatial inequalities in waste service provision [8,9]. Only 64% of households receive regular waste removal, with significantly lower coverage in rural provinces such as Limpopo [10,11]. In the Vhembe District, diverse settlement types, ranging from rural villages and farms to peri-urban townships, are characterized by fragmented service delivery, infrastructural undercapacity, and procedural exclusion, particularly in informal and underserved communities [12,13]. These structural constraints render informal waste disposal not merely a matter of choice but also an adaptive response to systemic neglect. This neglect reflects not only local governance gaps but also broader failures in policy implementation. South Africa's waste management is guided by the Waste Act (No. 59 of 2008) and the National Waste Management Strategy [14,15], which promotes sustainable and equitable service delivery. However, gaps persist between policy intent and delivery, especially in marginalized regions, reinforcing the need for frameworks that address both behavioral intent and structural constraints.

Theoretical approaches to household waste behavior have traditionally drawn on psychological frameworks, notably the Theory of Planned Behavior (TPB), which explains individual decision-making through the constructs of attitude, subjective norms, and perceived behavioral control [16–18]. The TPB has demonstrated predictive validity in urban and high-resource contexts, including recent extensions to waste separation in children [19] and willingness to pay for waste services [4,20]. However, in LMIC rural contexts, where infrastructural support is absent or erratic, the model's assumptions regarding volitional control are often untenable [4,21]. Constrained material conditions can suppress behavioral execution, regardless of positive intention, limiting the TPB's utility in these environments.

To address these shortcomings, scholars have increasingly incorporated structural frameworks such as Environmental Justice (EJ), which examines how institutional, spatial, and procedural inequalities mediate environmental exposure and access to services [22–26]. Recent studies have emphasized the value of integrating TPB with EJ, reconceptualizing perceived behavioral control to reflect both individual agency and structural constraints [18]. Despite its conceptual appeal, the empirical application of TPB–EJ integration remains limited, particularly in rural African contexts, where environmental marginality intersects with socio-demographic exclusion.

This study addresses a significant empirical and conceptual gap by employing an integrated TPB–EJ framework to analyze household waste disposal practices in the Vhembe District of Limpopo Province. The district's diverse settlements, infrastructural disparities, and governance asymmetries present a compelling case for examining the intersection of structural and behavioral factors in shaping waste management outcomes. Utilizing survey data from 399 households across four local municipalities, the analysis disaggregated informal disposal behaviors into river dumping, roadside disposal, and burial within compounds. This analysis advances current theoretical debates by operationalizing perceived behavioral control as both a psychological disposition and a function of external material and institutional realities. Furthermore, the findings provide empirical specificity to national policy debates by highlighting underreported disposal behaviors and supporting the call for more inclusive, decentralized waste governance systems aligned with SDGs 11.6 and 12.5.

2. A Review of the Theoretical Framework

Effective household waste management in resource-limited environments is influenced by the complex interplay of individual perceptions, socio-cultural norms, and structural inequalities. Analyzing rural households' decision-making, particularly in the absence of reliable waste collection services, necessitates an analytical framework that considers not only behavioral intent but also the broader spatial and institutional contexts in which these behaviors occur. This study employs two complementary theoretical models: the TPB and the EJ framework.

2.1. Theory of Planned Behavior (TPB)

The TPB, developed by Ajzen [16], has been widely used to understand pro-environmental behaviors, including recycling, litter prevention, energy conservation, and waste sorting. This model suggests that behavior is influenced by three constructs: attitudes towards the behavior, encompassing beliefs about its consequences; subjective norms, referring to perceived social pressures; and perceived behavioral control, which relates to the perceived ease of performing the behavior based on past experiences and obstacles.

In the context of household waste disposal, the TPB has been used to evaluate how individuals form intentions for proper waste management, such as utilizing municipal collection services and avoiding illegal dumping [18,21,27]. However, in low-service environments like Vhembe, perceived behavioral control is constrained by external conditions, such as the absence of waste bins, infrequent collection, or distant disposal sites. These constraints render high behavioral intent ineffective, highlighting a limitation of the TPB in material deprivation contexts.

Recent empirical studies have reinforced TPB's utility in LMICs' waste behavior research. Moeini et al. [17] systematically reviewed TPB applications and found that perceived behavioral control is often the strongest predictor of waste separation when infrastructure exists. In Nigeria, Eremionkhale et al. [28] found that subjective norms and control beliefs significantly influenced urban households' participation in recycling programs. Similarly, Babazadeh et al. [29] and Alam et al. [30] applied the TPB to solid waste management in Iran and showed that interventions focused on enhancing perceived control improved behavioral outcomes.

In rural India, Haque et al. [4] integrated TPB with environmental concern constructs and found that social pressure from community leaders shaped attitudes towards formal disposal methods. These findings reinforce the context-sensitive adaptation of the TPB to settings with fragmented services. Furthermore, Amir et al. [18] applied the TPB in Vietnam and showed that knowledge alone is insufficient; logistical and policy support are needed to activate intention.

2.2. Environmental Justice (EJ)

The EJ framework shifts the focus from individual decision-making to systemic inequalities in environmental risk distribution. Initially developed in the United States to address racialized exposure to environmental hazards, the EJ framework has expanded to examine inequalities related to class, space, ethnicity, and governance [22,23,31,32]. The EJ paradigm rests on three dimensions: distributional justice, concerning the equitable allocation of environmental goods and burdens; procedural justice, addressing fairness in environmental decision-making; and recognitional justice, acknowledging diverse identities and experiences in environmental governance. In rural waste management, EJ reveals how historical, political, and economic processes create spatial inequalities in service access and delivery.

Recognitional justice highlights how waste policies often ignore local knowledge, cultural practices, and traditional leadership. In Vhembe, backyard waste burial may be a practical response to service gaps, not ignorance. Yet such practices are often dismissed, showing that local ways of knowing are undervalued.

In the Vhembe District, communities in remote or informal settlements lack regular municipal waste services (MWSs). This exclusion stems from structural disinvestment, inadequate planning, and institutional fragmentation. The EJ framework reinterprets informal disposal practices as adaptive responses to service gaps rather than deviant behavior, understanding these behaviors in the context of broader social conditions.

Recent studies support the application of EJ in waste governance. The Organization for Economic Co-operation and Development (OECD) [26], the United States Environmental Protection Agency (US-EPA) [33], and the Association of American Medical Colleges (AAMC) Center for Health Justice [34] linked the EJ to health disparities from waste exposure in marginalized communities. In Africa, Akenji et al. [35] argued that the lack of procedural and recognitional justice drives centralized waste policy failures. In Brazil, Schneider et al. [36] showed how informal recyclers face systemic barriers despite their environmental contributions. Studies from South and Southeast Asia, by Urme et al. [37] in Bangladesh and Zulkipli et al. [38] in Malaysia, use EJ principles to reveal the socio-political roots of spatial dumping. These studies confirm that informal waste practices are structurally determined rather than resulting from ignorance alone. While the TPB explains behavioral motivation, EJ contextualizes structural constraints, and their integration enables a multi-level analysis of household disposal behavior. The following section outlines the conceptual structure and operationalization of this integration.

2.3. An Integrated Framework: Behaviour Within Structural Constraint

Although the TPB and EJ have been extensively applied in environmental research, their combined application remains limited in the rural African context. This study proposes the integration of TPB and EJ, acknowledging that behavior is both intentional and structurally conditioned. In this framework, the TPB captures the influence of individual attitudes, perceived norms, and behavioral control on waste-related intentions at the household level. EJ examines the spatial and institutional forces that facilitate or hinder such behaviors.

This integration is particularly relevant to perceived behavioral control. Within TPB, this concept refers to an individual's perception of their ability to perform a specific behavior. From the EJ perspective, perceived control reflects both psychological state and material/institutional realities. For example, a household may be environmentally conscious and willing to use formal disposal services but unable to do so because of a lack of infrastructure or regulatory neglect.

As illustrated in Figure 1, behavioral outcomes are constrained by psychological and structural factors. The operationalization of theoretical constructs is summarized in Table 1, which maps each variable to its position within the integrated TPB–EJ framework, showing the measurement type, justice dimension, and alignment with the SDGs. The framework analyzes the behavioral intentions and systemic constraints of household waste disposal. These constraints directly influence the feasibility of pro-environmental action. By reinterpreting behavioral control through psychological and structural lenses, the model enables a realistic understanding of waste management practices in low-resource settings, including the Philippines, where similar infrastructural constraints and informal disposal behaviors have been empirically observed [39]. This approach aligns with SDGs 11 and 12, which advocate inclusive waste governance systems. This provides a basis for empirical analysis, linking socio-demographic, spatial, and infrastructural variables to waste disposal practices in the Vhembe District.

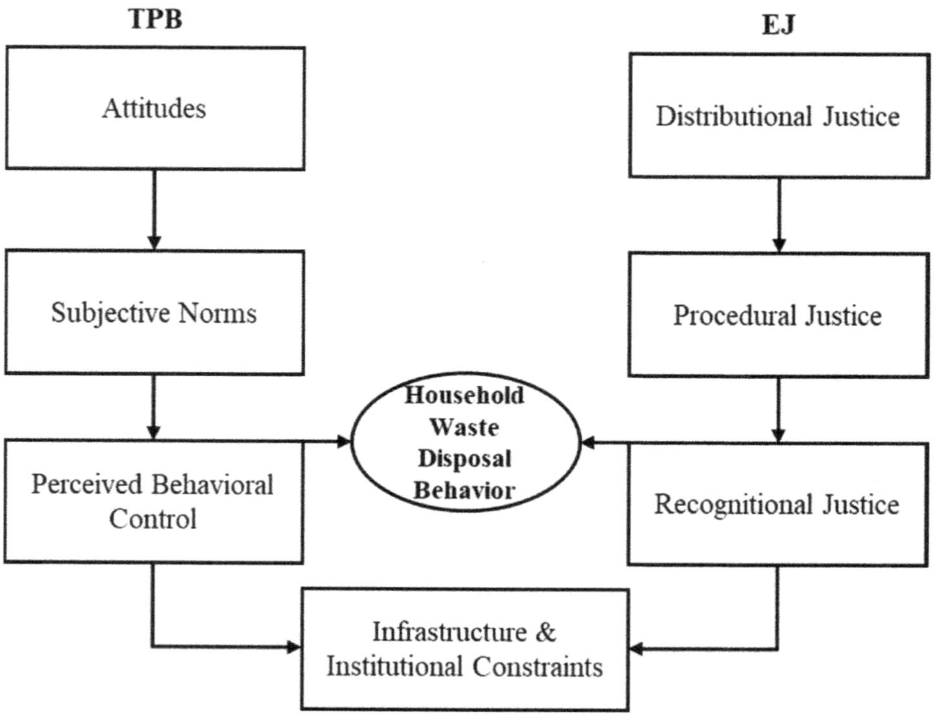

Figure 1. Conceptual framework integrating TPB and EJ constructs to interpret household waste disposal patterns. Arrows indicate theorized influence, not tested relationships.

This conceptual model synthesizes the TPB with the EJ framework. Within this model, behavioral intention is influenced by attitudes, subjective norms, and perceived behavioral control, as outlined in the TPB. However, the actualization of this intention is either constrained or facilitated by infrastructural and institutional conditions, as described in the EJ framework. In environments with limited resources, perceived behavioral control is redefined as contingent upon material and governance systems rather than solely on individual volition. This integration enables the TPB to incorporate both individual-level determinants of behavior and the structural constraints emphasized by EJ. While TPB explains how attitudes, social norms, and perceived behavioral control shape behavioral intentions, it assumes that individuals can act on those intentions. In low-resource settings, however, this capacity is often constrained by limited infrastructure, spatial exclusion, and unequal access to services, dimensions that are central to the EJ framework. By combining these perspectives, the TPB–EJ model offers a more comprehensive explanation of informal waste disposal practices, framing them not as irrational or negligent behaviors, but as adaptive responses to systemic inequities. This theoretical synthesis enhances explanatory power and provides a stronger foundation for designing context-sensitive and equity-oriented waste management policies in marginalized communities.

Table 1. Conceptual mapping of integrated TPB–EJ constructs to study variables and relevant SDGs.

Variable	Measurement/Coding	Variable Type	TPB Construct	EJ Dimension	SDG Link(s)	Operationalization
Formal Service Use	1 = public/private collection; 0 = none	Binary	Perceived Behavioral Control	Distributional Justice	SDG 11.6, 12.5	Indicates access to formal bin-emptying services; higher values reflect greater control over proper disposal.
Bin Proximity	1 = public bin ≤ 1 km; 0 = >1 km	Binary	Perceived Behavioral Control	Distributional Justice	SDG 11.6	Physical accessibility affects perceived ease of behavior execution.
Disposal Methods (5 Types)	Binary: formal bin, river, roadside, open space, burial	Binary (DV)	Behavior (Outcome)	—	SDG 6.3, 12.5	Actual behaviors used to test intention vs. structural constraint.
Satisfaction with Waste Services	Likert scale (1 = Strongly Agree→5 = Strongly Disagree)	Ordinal	Attitude (Direct)	Procedural Justice	SDG 11.6	Captures evaluative judgments about the effectiveness, reliability, and responsiveness of municipal waste services.
Environmental Concern	1 = Concerned; 0 = Not Concerned	Binary	Attitude (Proxy)	Recognitional Justice	SDG 12.8	Indicates general environmental values limited as a binary proxy.
Perceived Barriers to Disposal	Likert scale (1–5): cost, distance, lack of bins, frequency	Ordinal	Perceived Behavioral Control (Direct)	Distributional Justice	SDG 11.6	Measures perceived obstacles to compliant waste disposal
Recycling/Separation Practices	Multiple response: yes/no	Nominal	Subjective Norm (Proxy)	Procedural Justice	SDG 12.5	Indicates prevailing household norms regarding the reuse, separation, and recycling of waste.
Household Income	1 = <ZAR 10k; 2 = ZAR 10–30k; 3 = ZAR 30–50k	Ordinal	Attitude (Proxy)	Distributional Justice	SDG 10.2	Income shapes the affordability of alternatives and correlates with values.
Education Level	1 = none→ 5 = graduate	Ordinal	Attitude (Proxy)	Recognitional Justice	SDG 4.7, 12.8	Education influences environmental values and the ability to act; it also reflects cultural exclusion.
Municipality	Categorical: Makhado (ref), Musina, etc.	Nominal	—	Distributional and Procedural Justice	SDG 11.6, 16.7	Reveals spatial patterns in service inequality and decision-making inclusion.
Residence Typology	Rural (ref), Farm, Township, City centre	Nominal	—	Distributional and Procedural Justice	SDG 10.2, 11.6	Settlement type influences exposure to dumping and the likelihood of exclusion from planning.
Household Size and Tenure	Size: 1–2→9+; Tenure: <1 yr→≥20 yrs	Ordinal	—	Procedural and Recognitional Justice	SDG 16.7	Long residence increases local social capital; large households face waste management burdens.
Demographics (Age, Gender, Occupation)	Standard categorical	Nominal	Subjective Norm (Proxy)	Recognitional Justice	SDG 5.4, 10.2	Reflects social roles and expectations influencing household disposal decisions.

3. Materials and Methods

3.1. Study Area and Design

This study was conducted in the Vhembe District of Limpopo Province, South Africa, encompassing four municipalities: Makhado, Musina, Collins Chabane, and Thulamela. These areas are characterized by diverse settlement types, including rural villages, farms, and peri-urban towns. A cross-sectional household survey was conducted to investigate the socio-demographic and infrastructural determinants influencing waste disposal behavior.

A multistage sampling approach was employed, beginning with the purposive selection of residential clusters to capture variations in access to waste services. The study area was first stratified by municipality (Makhado, Musina, Collins Chabane, and Thulamela), followed by further stratification by settlement typology: rural villages, farms, townships, and city centers to ensure proportional representation of marginalized communities. Within each cluster, households were selected using systematic random sampling based on local household registers or guided by community leaders in areas lacking formal records. The four municipalities collectively comprise approximately 436,959 households [40]. Using Krejcie and Morgan's formula [41] for known populations, a minimum of 385 households was required for statistical validity. Hence, a total of 400 households were surveyed (approximately 100 per municipality), with one case excluded due to missing data, resulting in a final sample of 399 households, with a response rate of 99.75%.

3.2. Survey Instrument and Data Collection

Data collection was conducted from May to July 2024 using a structured questionnaire, which was administered through face-to-face interviews with an adult respondent (≥18 years) in each household. The instrument was initially developed in English and subsequently translated into Tshivenda and Xitsonga by trained, bilingual fieldworkers. A pilot test involving 20 households not included in the study sample was performed to enhance item clarity. The instrument included Likert-scale items to measure TPB constructs such as attitudes and perceived behavioral control. Binary items captured variables like environmental concern and access to services. Subjective norms were proxied through reported reuse and recycling practices.

3.3. Data Analysis

Survey responses were systematically coded, cleaned, and analyzed using SPSS (Version 27) and R (Version 4.2). One incomplete questionnaire was excluded, resulting in a final analytical sample of 399 households in the study. Item-level missing data across the remaining cases were minimal (<1%) and were addressed through a listwise deletion.

Descriptive statistics, including frequencies and percentages, were initially computed for all key variables, including household demographics, socioeconomic characteristics, waste production patterns, and disposal behaviors. These summaries, presented in Table 2, provide an empirical foundation for subsequent inferential analysis. To investigate the determinants of household waste disposal behavior, a series of five binary logistic regression models were estimated, each with one of the following binary dependent variables (1 = use, 0 = non-use): (a) use of formal bin collection, (b) dumping in rivers, valleys, or lakes, (c) roadside dumping, (d) disposal in open public spaces, and (e) burial of waste within the household compound. The independent variables included municipality, residential location type, gender, age group, education level, employment status, household income, household size, and length of residence. The service-related variables included access to waste collection (public/private/none) and the presence of a bin within 1 km. All variables were selected and operationalized based on the integrated TPB–EJ conceptual framework detailed in Table 1, ensuring theoretical coherence in translating constructs such as attitudes, perceived control, and justice dimensions into measurable variables relevant to multivariate analysis. Categorical variables were dummy-coded using the following reference categories: Makhado (municipality), rural village (settlement type), female (gender), no formal education (education), unemployed (employment), and income below ZAR 10,000/month (household income).

Results are reported as adjusted odds ratios (AORs) with 95% confidence intervals (CIs), and a significance threshold of $\alpha = 0.05$ was employed. No interaction terms were tested, and model diagnostics were not performed at this stage. The sampling weights were omitted. This analytical approach aligns with the study's explanatory aim of assessing how socio-demographic, spatial, and service-related variables collectively influence the observed variability in household waste disposal outcomes, particularly in contexts characterized by infrastructural inequity.

Table 2. Demographic and socioeconomic characteristics of respondents (n = 399).

Variables	Frequency (%)
Municipality	
Makhado	99 (24.8%)
Musina	100 (25.1%)
Collins Chabane	100 (25.1%)
Thulamela Municipality	100 (25.1%)
Location of Residence	
Rural	143 (35.8%)
Farm	108 (27.1%)
City Center	61 (15.3%)
Township	87 (21.8%)
Gender	
Female	245 (61.4%)
Male	154 (38.6%)
Age	
18–29 years	111 (27.8%)
30–39 years	113 (28.3%)
40–49 years	93 (23.3%)
50–59 years	59 (14.8%)
60–69 years	16 (4.0%)
≥70 years	7 (1.8%)
Number of Household Members	
1–2	20 (5.0%)
3–4	152 (38.1%)
5–6	189 (47.4%)
7–8	25 (6.3%)
≥9	13 (3.3%)
Number of Years Residing in the Area	
Less than a year	2 (0.5%)
1–5 years	43 (10.8%)
5–10 years	81 (20.3%)
10–15 years	44 (11.0%)
15–20 years	90 (22.6%)
20 years or more	139 (34.8%)
Highest Qualification	
Never Schooled	15 (3.8%)
Primary	81 (20.3%)
Secondary	142 (35.6%)
Undergraduate	74 (18.5%)
University Graduate	86 (21.6%)
Other	1 (0.3%)

Table 2. *Cont.*

Variables	Frequency (%)
Occupation	
Unemployed	132 (33.1%)
Self-employed	148 (37.1%)
Government Employee	68 (17.0%)
Farmworker	37 (9.3%)
Other	14 (3.5%)
Average Monthly Household Income	
<ZAR 10,000	172 (43.1%)
ZAR 10,000–30,000	174 (43.6%)
ZAR 30,000–50,000	53 (13.3%)

3.4. Demographic and Socioeconomic Profile of Respondents

Among the 399 valid responses analyzed, households were nearly evenly distributed across the four municipalities: Makhado (24.8%), Musina (25.1%), Collins Chabane (25.1%), and Thulamela (25.1%), as shown in Table 2. This proportional allocation ensured a balanced geographical representation, thereby enhancing the robustness of cross-municipal comparisons. Regarding settlement typology, most respondents resided in rural areas (35.8%) or on farms (27.1%), followed by those living in townships (21.8%) and city centers (15.3%).

The demographic composition was predominantly female (61.4%), with the majority aged between 18 and 39 years (56.1%), indicating a relatively young population structure. Household sizes were typically large, with nearly half (47.4%) reporting 5–6 members. Additionally, 34.8% had resided in their current location for over two decades, suggesting strong community ties and potential implications for long-term waste-behavior patterns.

The educational attainment was modest, with 35.6% having completed secondary education and 21.6% holding university degrees. A smaller fraction (3.8%) reported that they had never attended school. Employment status varied: 33.1% were unemployed, 37.1% were self-employed, and 17.0% were employed in the public sector. Income distribution revealed financial vulnerability, with 43.1% earning less than ZAR 10,000 per month. These socioeconomic characteristics provide an essential context for understanding differential waste disposal practices and access to waste services across demographic groups.

3.5. Ethical Considerations

The study protocol was approved by the Durban University of Technology Research Ethics Committee (Ethics Clearance Number: IREC 294/22). Informed consent was obtained from all the participants. The data were anonymized and securely stored on an encrypted server. Additionally, community permission was obtained from municipal authorities in all participating areas.

3.6. Methodological Limitations

This study employed an exploratory and interpretive design. Although logistic regression was used to identify associations between socio-demographic and service access factors and disposal behaviors, the objective was not to statistically validate a formal theory. Instead, this study seeks to understand the structural influences on household-level practices. However, the regression analysis does not fully capture the underlying reasons for hazardous disposal behaviors. Future research should incorporate qualitative methods to provide deeper insight into behavioral motivations and structural adaptations.

4. Results

4.1. Waste Production Behavior

4.1.1. Waste Collection Services

The data indicate that waste collection services are primarily administered by private entities, with 81% of respondents utilizing these services. Conversely, only 14.3% of respondents relied on public waste collection services, suggesting a diminished level of government participation in this sector. A minor segment (4.8%) relies on alternative forms of waste collection.

4.1.2. Availability of Public Bins Within 1 km from House

The data revealed that a substantial proportion of respondents (78.2%) lacked access to public waste bins within a 1-kilometer radius of their residences. Conversely, only 21.8% of respondents reported the availability of public bins within this distance. This disparity highlights a notable deficiency in waste management infrastructure that may pose challenges to effective waste disposal. The absence of proximate public bins could exacerbate improper waste disposal practices, such as illegal dumping or littering, and underscores the necessity for enhanced waste management planning and service delivery in the region to address this issue.

4.1.3. How Long Does It Take to Get There?

Among respondents residing within one kilometer of public waste disposal bins, the convenience of waste disposal exhibited considerable variation. Specifically, 37.9% of the individuals had a bin located directly outside their residence, facilitating immediate and effortless waste disposal. Conversely, 28.7% are required to walk between 5 and 10 min, while 11.5% must walk 11–15 min to access the nearest bin. A smaller segment of the population experienced even longer travel times, with 5.7% needing 16–20 min and another 5.7% requiring 21–25 min. Additionally, 10.3% of respondents reported a walking time of 25 min to reach a public bin.

These data underscore that, although some residents enjoy convenient access to waste disposal facilities, a significant proportion must traverse considerable distances, potentially discouraging proper waste disposal and contributing to littering or illegal dumping. Enhancing the distribution of public bins could improve waste management efficiency and promote better disposal practices in affected areas.

4.1.4. Type of Solid Waste That Comes from the Household

Households were surveyed regarding the types and frequency of solid waste generated. The waste stream in this rural district is notably heterogeneous, reflecting the diverse household activities. Figure 2 illustrates that households produce various types of solid waste at different frequencies. Clothing and garden or yard waste were among the most frequently discarded items, with 65.2% and 61.9% of households, respectively, indicating that these waste types occasionally originate from their homes. This suggests that clothing disposal may occur seasonally or as part of decluttering efforts, whereas yard waste is likely influenced by gardening activities and seasonal changes.

Glass bottles (57.6%), diapers (54.1%), and paper or carton waste (45.4%) are also commonly discarded, although their frequencies vary among households. Notably, plastic bags (39.8%), food waste (36.3%), and aluminum tins/cans (50.9%) were reported to "sometimes" originate from households, highlighting their presence but perhaps also the influence of waste reduction efforts, such as reusable bags or composting initiatives. Conversely, a smaller percentage of households reported that certain waste types never originated from their homes. For instance, 12.3% of households did not generate glass waste, whereas 9.5% reported never discarding clothing or diapers. This variation likely reflects differences in household composition, lifestyle choices, and consumption habits; for example, homes without infants would not generate diaper waste, whereas households that prioritize recycling may have lower glass or aluminum waste output.

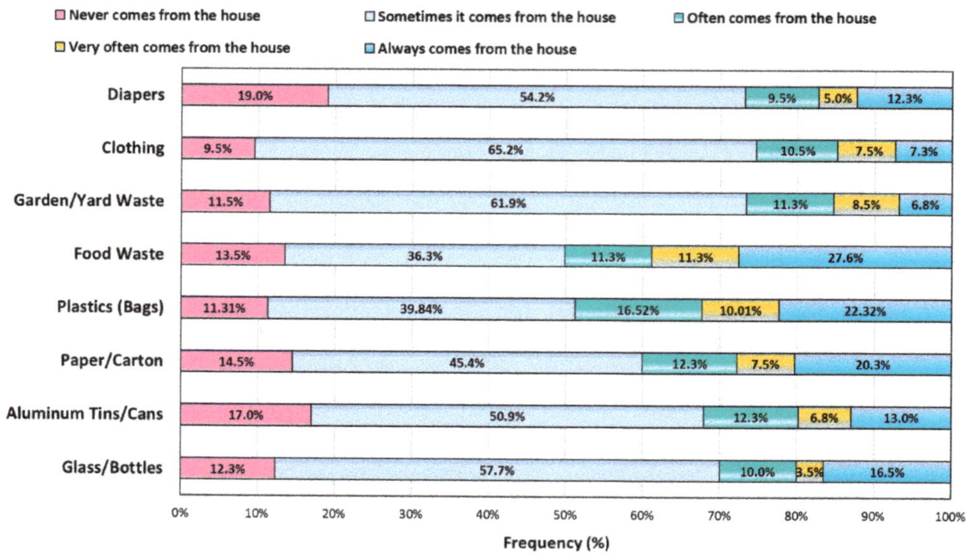

Figure 2. Types of solid waste commonly generated by households.

4.2. Waste Disposal Behaviors

The data indicate that most households predominantly utilize informal waste disposal methods, with only 8% reporting the use of bins for waste disposal. This suggests a substantial deficiency in access to formal waste collection services or a general inclination towards alternative disposal methods. The overwhelming 92% of households not utilizing bins implies a potential inadequacy in waste management infrastructure in many regions.

Alarmingly, 23.8% of households acknowledged waste disposal in valleys, lakesides, or rivers, posing significant environmental and health hazards. Similarly, 22.6% of respondents disposed of waste on roadsides or streets, exacerbating urban pollution and sanitation issues. Although these practices are not adopted by the majority, the fact that nearly one in four households engages in them underscores the necessity of improved waste management interventions.

Open dumping in public spaces is relatively uncommon, with only 3% of households employing this method, whereas 97% refrain from it. However, a significant 41.4% of households bury their waste within their compounds, a practice that, despite its convenience, can result in soil contamination and long-term environmental repercussions.

The findings, summarized in Table 3, reveal a pronounced reliance on informal and occasionally hazardous waste disposal methods. The prevalent lack of bin usage and the high incidence of dumping in natural areas and on roadsides highlight serious deficiencies in waste management.

Table 3. Patterns of waste disposal in the community.

Variables	Frequency (%)
	Waste Disposal: In the Bin
No	367 (92%)
Yes	32 (8%)
	Waste Disposal: Dumped in the Valley/Lakeside/River
No	304 (76.2%)
Yes	95 (23.8%)
	Waste Disposal: The Road or Street Side
No	309 (77.4%)
Yes	90 (22.4%)
	Waste Disposal: In an Open Space
No	387 (97%)
Yes	12 (3%)
	Waste Disposal: In a Hole in Own Compound
No	234 (58.6%)
Yes	165 (41.4%)

4.3. How Often Is the Waste Container Emptied by the Local Municipality?

Figure 3 shows the frequency with which waste containers are serviced by local municipal authorities. The data revealed substantial disparities in waste collection services, with nearly half of the respondents (48.1%) indicating that their waste was never collected by municipal services. This finding suggests a significant deficiency in formal waste management infrastructure, which may contribute to increased illegal dumping, environmental hazards, and sanitation challenges. Among those who received waste collection services, 21.1% of households had their waste containers emptied once a week, representing the most common frequency for those receiving the service. Furthermore, 15.3% of respondents reported waste collection occurring twice a week, while 11.8% had their waste collected three times per week. Less frequent waste collection is reported by a small percentage of households, with 3% experiencing waste collection once a month and a very small number of households receiving service twice a month (0.3%) or thrice a month (0.5%).

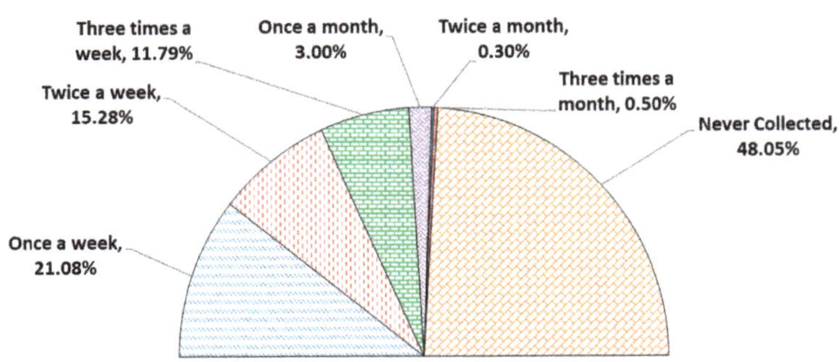

Figure 3. Frequency of municipal waste collection as reported by households.

4.4. Unadjusted Odds Ratios for Predictors of Household Waste Disposal Methods

Binary logistic regression models assessed associations among socio-demographic, economic, and spatial variables and five household waste disposal practices: (1) disposal in formal bins, (2) dumping in valleys, rivers, or lakesides, (3) disposal on roadsides or street edges, (4) dumping in open spaces, and (5) burying waste in household compound holes. Each model used a single predictor, yielding crude ORs with 95% CIs and *p*-values for each variable. Table 4 presents the outcomes of simple binary logistic regression models.

Table 4. Results of binary logistic regression of housing characteristics, socio-demographic factors, and environmental health risk factors regarding waste disposal methods.

Demographic Variables	Waste Disposal Methods									
	Disposal in the Bin		Dumped in the Valley/Lakeside/River		On the Road or Street Side		In an Open Space		In a Hole in the Own Compound	
	OR (95%CI)	*p*-Value	OR (95%CI)	*p*-Value	OR (95%CI)	*p*-Value	OR (95%CI)	*p*-Value	OR (95%CI)	*p*-Value
Municipality										
Makhado	Reference		Reference		Reference		Reference		Reference	
Musina	0.2 (0.03–0.62)	0.010	156 (44–552)	<0.001	2.8 (0.72–10.8)	0.139	0.5 (0.04–5.5)	0.563	0.06 (0.001–0.03)	<0.001
Collins Chabane	1.2 (0.52–2.60)	0.705	1.0 (0.2–5.0)	0.990	15.1 (4.4–51)	<0.001	4.2 (0.9–20.4)	0.073	0.2 (0.1–0.4)	<0.001
Thulamela	0.1 (0.03–0.62)	0.010	2.0 (0.5–8.4)	0.322	28.4 (8.2–96)	<0.001	0.5 (0.04–5.5)	0.563	0.2 (0.1–0.4)	<0.001
Location of Residence										
Rural	Reference		Reference		Reference		Reference		Reference	
Farm	0.3 (0.04–2.95)	0.318	5.4 (3.1–9.6)	<0.001	0.1 (0.01–0.27)	<0.001	0.1 (0.02–1.1)	0.063	0.7 (0.4–1.2)	0.184
City Center	1.2 (0.21–6.61)	0.852	0.7 (0.3–1.6)	0.376	2.2 (1.1–4.1)	0.019	0.2 (0.03–2.0)	0.191	0.7 (0.4–1.4)	0.341
Township	14 (4.68–42)	<0.001	0.1 (0.02–0.5)	0.003	1.8 (1.0–3.3)	0.041	0.2 (0.2–1.4)	0.099	0.5 (0.3–0.9)	0.017
Gender										
Female	Reference		Reference		Reference		Reference		Reference	
Male	0.8 (0.38–1.75)	0.609	1.2 (0.8–1.9)	0.421	1.1 (0.7–1.8)	0.578	0.8 (0.2–2.7)	0.704	0.9 (0.6–1.3)	0.575
Age										
18–29 years	Reference		Reference		Reference		Reference		Reference	
30–39 years	1.2 (0.35–4.01)	0.781	0.4 (0.2–1.0)	0.063	0.8 (0.4–1.6)	0.578	0.4 (0.07–2.0)	0.256	1.7 (0.97–2.8)	0.062
40–49 years	1.7 (0.53–5.63)	0.366	3.1 (1.6–6.0)	0.001	1.1 (0.6–2.1)	0.678	0.2 (0.03–2.0)	0.184	0.4 (0.2–0.7)	0.001
50–59 years	3.8 (1.21–12)	0.022	4.1 (1.9–8.4)	<0.001	0.7 (0.3–1.6)	0.399	0.7 (0.1–3.9)	0.729	0.2 (0.07–0.4)	<0.001
60–69 years	7.1 (1.67–29)	0.008	2.3 (0.7–7.6)	0.153	0.7 (0.2–2.7)	0.625	1.4 (0.2–12.9)	0.759	1.4 (0.03–0.6)	0.012
≥70 years	3.5 (0.35–35.2)	0.282	2.1 (0.4–11.5)	0.407	N/A		3.5 (0.4–35.2)	0.282	0.7 (0.2–3.4)	0.698
Number of Household Members										
1–2	Reference		Reference		Reference		Reference		Reference	
3–4	0.6 (0.11–2.83)	0.489	0.5 (0.2–1.2)	0.097	9.9 (1.3–75.9)	0.028	N/A		0.6 (0.2–1.8)	0.408
5–6	0.7 (0.15–3.42)	0.680	0.1 (0.05–0.4)	0.000	3.9 (0.5–29.9)	0.195	0.1 (0.05–1.4)	0.080	3.2 (1.2–8.6)	0.023
7–8	1.2 (0.18–8.16)	0.832	0.03 (0.003–0.3)	0.002	3.6 (0.4–35.3)	0.268	0.3 (0.04–3.0)	0.324	2.2 (0.6–7.4)	0.224
≥9	4 (0.61–26.1)	0.148	0.1 (0.03–0.9)	0.032	1.6 (0.09–27.8)	0.753	3 (0.3–28.8)	0.341	1.5 (0.3–6.3)	0.615
Number of Years Residing in the Area										
Less than a year	Reference		Reference		Reference		Reference		Reference	
1–5 years	1.0 (0.31–3.2)	0.992	2.1 (0.1–35.6)	0.616	0.05 (0.002–1.1)	0.057	N/A		0.1 (0.05–0.3)	<0.001
5–10 years	0.5 (0.16–1.60)	0.245	1.8 (0.1–29.7)	0.684	0.2 (0.01–3.5)	0.278	N/A		0.07 (0.03–0.2)	<0.001
10–15 years	1.5 (0.54–4.30)	0.420	0.9 (0.01–3.4)	0.258	0.3 (0.02–5.1)	0.402	0.2 (0.03–1.7)	0.138	0.4 (0.2–0.9)	0.016
15–20 years	0.6 (0.20–1.66)	0.302	0.01 (0.003–0.3)	0.010	0.6 (0.04–9.6)	0.703	1.2 (0.3–4.7)	0.796	0.8 (0.5–1.4)	0.534
20 years or more	N/A		0.04 (0.002–0.7)	0.027	0.3 (0.02–4.5)	0.367	N/A		N/A	

Table 4. *Cont.*

Demographic Variables	Disposal in the Bin		Dumped in the Valley/Lakeside/River		On the Road or Street Side		In an Open Space		In a Hole in the Own Compound	
	OR (95%CI)	***p*-Value**	**OR (95%CI)**	***p*-Value**	**OR (95%CI)**	***p*-Value**	**OR (95%CI)**	***p*-Value**	**OR (95%CI)**	***p*-Value**
Highest Qualification										
Never Schooled	Reference		Reference		Reference		Reference		Reference	
Primary	0.3 (0.06–2.04)	0.236	3.6 (1.1–11.3)	0.031	0.9 (0.4–2.0)	0.749	0.7 (0.1–4.3)	0.701	0.2 (0.02–1.3)	0.084
Secondary	0.6 (0.12–2.98)	0.532	0.1 (0.04–0.4)	<0.001	N/A		0.8 (0.2–3.7)	0.776	6.5 (1.4–29.9)	0.016
Undergraduate	0.6 (0.1–3.16)	0.523	0.05 (0.01–0.2)	<0.001	N/A		1.2 (0.2–5.9)	0.851	6.9 (1.4–32.6)	0.015
University Graduate	0.6 (0.11–3.08)	0.519	0.01 (0.001–0.1)	<0.001	N/A		N/A		9.9 (2.1–46.8)	0.004
Other	N/A		N/A		N/A		N/A		N/A	
Occupation										
Unemployed	Reference		Reference		Reference		Reference		Reference	
Self-employed	0.6 (0.26–14.3)	0.254	11.9 (5.8–24.4)	<0.001	0.6 (0.4–1.1)	0.098	0.2 (0.02–1.9)	0.175	0.3 (0.2–0.5)	<0.001
Government Employee	0.5 (0.17–1.67)	0.276	2.4 (0.9–5.9)	0.066	0.6 (0.3–1.2)	0.176	1.5 (0.3–6.8)	0.617	1.0 (0.6–1.9)	0.919
Farmworker	N/A		0.3 (0.04–2.7)	0.310	0.1 (0.03–0.6)	0.008	1.8 (0.3–10.4)	0.496	5.3 (2.1–13.6)	<0.001
Other	3.4 (0.93–12.19)	0.064	N/A		0.95 (0.3–3.2)	0.939	5.3 (0.9–32.2)	0.068	0.4 (0.1–1.4)	0.151
Average Monthly Household Income										
<ZAR 10,000	Reference		Reference		Reference		Reference		Reference	
ZAR 10,000–ZAR 30,000	0.4 (0.2–0.88)	0.024	16.2 (7.2–36.7)	<0.001	0.7 (0.4–1.3)	0.248	0.2 (0.04–0.9)	0.048	0.4 (0.3–0.6)	<0.001
ZAR 30,000–ZAR 50,000	0.3 (0.06–1.24)	0.095	11.1 (4.3–28.8)	<0.001	3.6 (1.9–6.9)	<0.001	0.3 (0.04–2.8)	0.322	0.1 (0.04–0.2)	<0.001

N/A: not available (no data), OR: odds ratio, 95%CI: 95% Confidence Interval.

The following are the key findings.

4.4.1. Geographical Differences

The binary logistic regression analysis shows that, compared to Makhado, Musina residents had a lower propensity to dispose of waste in bins (OR = 0.2, p = 0.010). However, they showed a higher likelihood of discarding waste in valleys, lakesides, or rivers (OR = 156, $p < 0.001$). The probability of these residents disposing of waste in a hole within their compound was reduced (OR = 0.06, $p < 0.001$). In contrast, Collins Chabane Municipality showed increased likelihood of waste disposal on roadsides (OR = 15.1, $p < 0.001$) and decreased likelihood of using a hole within their compound (OR = 0.2, $p < 0.001$). Similarly, Thulamela Municipality residents were more likely to dispose of waste on roadsides (OR = 28.4, $p < 0.001$) and less likely to use a hole within their compound (OR = 0.2, $p < 0.001$).

4.4.2. Type of Residential Location

Residents of rural areas demonstrated a significantly higher propensity to dispose of waste in valleys, lakesides, or rivers (OR = 5.4, $p < 0.001$), while exhibiting a markedly lower likelihood of depositing waste along roadsides (OR = 0.1, $p < 0.001$). Conversely, individuals residing in townships were predominantly inclined to utilize bins for waste disposal (OR = 14, $p < 0.001$) and were less prone to discard waste in valleys (OR = 0.1, $p < 0.003$). However, township residents also showed a modestly increased tendency to dispose of waste on roadsides (OR = 1.8, p = 0.041). Notably, gender did not present significant correlations with waste disposal practices, as the odds ratios comparing males to females did not reveal substantial differences across the various disposal categories.

4.4.3. Gender

In the adjusted analysis, sex did not significantly affect disposal behavior. The odds ratios for male-headed households, compared to female-headed households, did not demonstrate statistical differences for any disposal method. This indicates that when controlling for other variables, gender does not determine whether a household engages in dumping or utilizes bins. Although women often manage daily household waste, the ultimate disposal decisions, particularly for large-scale dumping, may involve both genders. Our data suggest that gender parity in behavior may also reflect the influence of community norms and infrastructure limitations, which may supersede individual gender tendencies.

4.4.4. Age

Analysis of age-related waste disposal behaviors revealed that individuals within the older age brackets, specifically those aged 50–59 years and 60–69 years, exhibited a markedly higher propensity to utilize waste bins (OR = 3.8, $p = 0.022$ and OR = 7.1, $p = 0.008$, respectively). Conversely, individuals aged 40–49 and 50–59 years demonstrate a significantly increased likelihood of disposing of waste in valleys, lakesides, or rivers (OR = 3.1, $p < 0.001$, and OR = 4.1, $p < 0.001$, respectively). Furthermore, those within the 40–49, 50–59, and 60–69 age groups were significantly less inclined to dispose of waste in a hole within their compound ($p < 0.05$).

4.4.5. Household Size and Length of Residence

Households comprising three to four members exhibited a significantly higher propensity to dispose of waste along roadsides (OR = 9.9, $p = 0.028$). In contrast, larger households consisting of five to six members demonstrated a markedly reduced likelihood of depositing waste in valleys (OR = 0.1, $p < 0.001$) and an increased tendency to utilize a hole within their compound for waste disposal (OR = 3.2, $p = 0.023$). Residents with a tenure of less than one year in the area were used as the reference group. Those residing in the area for one to five years and five to ten years exhibited a significantly diminished likelihood of employing a hole in their compound for waste disposal ($p < 0.001$). Similarly, individuals who have resided in the area for 15–20 years are significantly less inclined to dispose of waste in valleys (OR = 0.01, $p < 0.010$).

4.4.6. Education

Individuals without formal education were the reference group in this analysis. In comparison, those with a primary education exhibited a significantly higher likelihood of disposing of waste in valleys (OR = 3.6, $p = 0.031$). Conversely, individuals who had completed secondary education, undergraduate studies, or held university degrees demonstrated a markedly lower propensity to engage in such waste disposal practices ($p < 0.001$). Notably, university graduates were particularly inclined to utilize a hole within their compound for waste disposal (OR = 9.9, $p = 0.004$).

4.4.7. Employment Status

Unemployed individuals were used as the reference category in this analysis. The findings indicate that self-employed individuals exhibit a significantly higher propensity to dispose of waste in valleys (OR = 11.9, $p < 0.001$) and are less inclined to utilize a hole within their compound for waste disposal (OR = 0.3, $p < 0.001$). Conversely, farmworkers demonstrated a markedly higher likelihood of using a hole within their compound for waste disposal (OR = 5.3, $p < 0.001$) but were significantly less likely to dispose of waste on roadsides (OR = 0.1, $p = 0.008$).

4.4.8. Household Income

Households with an income of less than 10,000 serve as the reference category for income. Those with earnings between 10,000 and 30,000 were significantly more likely to dispose of waste in valleys (OR = 16.2, $p < 0.001$) and less likely to utilize a hole within their compound for waste disposal (OR = 0.4, $p < 0.001$). Similarly, households with an income ranging from 30,000 to 50,000 were significantly more inclined to dispose of waste in valleys (OR = 11.1, $p < 0.001$) and on roadsides (OR = 3.6, $p < 0.001$), while they were significantly less likely to use a hole within their compound (OR = 0.1, $p < 0.001$). The findings suggest that waste disposal behavior is strongly influenced by demographic, socioeconomic, and environmental factors.

There are significant differences in waste management practices among municipalities, with Musina and Thulamela exhibiting a high tendency for improper waste disposal. The location of residence is also a critical factor, with farms and rural areas showing a high propensity to dump waste in valleys. Additionally, age, education, and household income influence waste disposal habits; older individuals were more likely to use formal bins, while university graduates, despite greater environmental awareness, often resorted to burial within their compounds. Middle-income households were more likely to engage in river dumping. These findings suggest that higher socioeconomic status or educational attainment does not automatically translate into more environmentally sound practices, particularly in contexts characterized by inadequate waste infrastructure.

4.5. Multivariate Analysis of Predictors of Household Waste Disposal Practices

The multivariate analysis of waste disposal patterns is presented in Table 5. Results indicate Musina Municipality residents are significantly less likely to dispose of waste in bins compared to Makhado (OR = 0.1, 95% CI: 0.02–0.95, $p = 0.044$), suggesting deficient formal waste collection. The incidence of waste dumping in natural water bodies was high, with residents 141.6 times more likely to engage in this practice (OR = 141.6; 95% CI: 39.5–508.5, $p < 0.001$), highlighting critical environmental and health concerns. Waste management challenges exist in Collins Chabane Municipality. Residents are significantly less likely to use bins (OR = 0.2, 95% CI: 0.07–0.9, $p = 0.029$), indicating limited collection services. While dumping in rivers does not differ significantly from Makhado (OR = 0.6, 95% CI: 0.08–4.2, $p = 0.589$), roadside dumping is concerning, with residents 18.2 times more likely to dispose of waste this way (OR = 18.2, 95% CI: 5.3–62.9, $p < 0.001$).

The Thulamela Municipality exhibits similar challenges. Although no significant difference was observed in bin usage (OR = 0.4, 95% CI: 0.06–2.9, $p = 0.361$), the most concerning finding was that residents were 30 times more likely to dispose of waste on roadsides than those in Makhado (OR = 30.0, 95% CI: 8.8–101.9, $p < 0.001$). This suggests that roadside dumping is a prevalent practice, likely due to the inadequate waste collection infrastructure.

The location of residence plays a crucial role in determining waste disposal behavior. Farm residents were 5.1 times more likely to dump waste in rivers than their rural counterparts (OR = 5.1, 95% CI: 2.8–9.4, $p < 0.001$), raising environmental concerns regarding water contamination. However, they were significantly less likely to dispose of waste on roadsides (OR = 0.6, 95% CI: 0.01–0.2, $p < 0.001$), possibly due to the rural nature of their living environment.

In urban areas, particularly city centers, residents exhibited an increased tendency to dispose of waste on roadsides (OR = 1.9, 95% CI: 1.0–3.7, $p < 0.001$), reflecting common urban waste management issues such as illegal dumping due to overflowing bins or irregular waste collection. Township residents displayed a mixed pattern of waste disposal. They were more likely to use bins (OR = 3.9, 95% CI: 0.8–19.4, $p = 0.092$), suggesting better access to formal waste collection. However, they were also significantly less likely to dump waste in rivers (OR = 0.1, 95% CI: 0.02–0.4, $p = 0.003$), indicating more environmentally conscious practices than rural or farm areas.

Table 5. Adjusted odds ratios from multivariate logistic regression models examining predictors of household waste disposal methods.

Demographic Variables	Waste Disposal Methods									
	Disposal in the Bin		Dumped in the Valley/Lakeside/River		On the Road or Street Side		In an Open Space		In a Hole in the Own Compound	
	OR (95%CI)	*p*-Value	OR (95%CI)	*p*-Value	OR (95%CI)	*p*-Value	OR (95%CI)	*p*-Value	OR (95%CI)	*p*-Value
Municipality										
Makhado	Reference		Reference		Reference		Reference		Reference	
Musina	0.1 (0.02–0.95)	0.044	141.6 (39–508)	<0.001	3.0 (0.8–11.8)	0.116	0.4 (0.03–4.5)	0.420	0.004 (0.01–0.02)	<0.001
Collins Chabane	0.2 (0.07–0.9)	0.029	0.6 (0.08–4.2)	0.589	18.2 (5.3–62.9)	<0.001	2.5 (0.4–14.0)	0.309	0.2 (0.09–0.5)	<0.001
Thulamela	0.4 (0.06–2.9)	0.361	1.8 (0.4–7.5)	0.420	30.0 (8.8–102)	<0.001	0.5 (0.04–5.2)	0.526	0.1 (0.1–0.2)	<0.001
Location of Residence										
Rural	Reference		Reference		Reference		Reference		Reference	
Farm	0.5 (0.05–5.4)	0.562	5.1 (2.8–9.4)	<0.001	0.6 (0.01–0.2)	<0.001	0.2 (0.02–1.6)	0.120	0.8 (0.4–1.3)	0.344
City Center	1.2 (0.1–10.5)	0.890	0.6 (0.3–1.6)	0.327	1.9 (1.0–3.7)	<0.001	0.4 (0.04–3.1)	0.349	0.7 (0.4–1.4)	0.379
Township	3.9 (0.8–19.4)	0.092	0.1 (0.02–0.4)	0.003	1.8 (0.9–3.6)	0.071	0.1 (0.01–1.5)	0.103	1.0 (0.5–1.9)	0.977
Highest Qualification										
Never Schooled	Reference		Reference		Reference		Reference		Reference	
Primary	1.6 (0.16–15.5)	0.687	2.7 (0.7–10.1)	0.135	0.3 (0.1–1.5)	0.145	0.4 (0.1–2.8)	0.342	0.1 (0.02–1.3)	0.081
Secondary	1.2 (0.16–8.87)	0.864	0.7 (0.02–0.27)	<0.001	1.2 (0.3–4.5)	0.833	0.7 (0.1–3.3)	0.615	7.5 (1.4–40.0)	0.019
Undergraduate	3.7 (0.37–38.0)	0.266	0.3 (0.01–0.14)	<0.001	1.7 (0.4–6.7)	0.480	1.0 (0.2–5.8)	0.974	6.4 (1.2–34.98)	0.034
University Graduate	1.0 (0.13–8.49)	0.970	0.011 (0.001–0.06)	<0.001	1.3 (0.3–5.1)	0.736	N/A		11.7 (2.1–64.9)	0.005
Other	N/A		N/A		N/A		N/A		N/A	
Average Monthly Household Income										
ZAR 10,000>	Reference		Reference		Reference		Reference		Reference	
ZAR 10,000–30,000	0.8 (0.3–2.4)	0.737	19.0 (7.3–49.5)	<0.001	0.7 (0.4–1.2)	0.234	0.3 (0.06–0.7)	0.186	0.3 (0.2–0.5)	<0.001
ZAR 30,000–50,000	1.3 (0.2–9.6)	0.811	12.9 (4.4–38.5)	<0.001	3.6 (1.8–7.2)	<0.001	0.6 (0.07–5.4)	0.644	0.05 (0.02–0.1)	<0.001

N/A: not available (no data), OR: odds ratio, 95%CI: 95% confidence interval.

Household income levels have a significant impact on waste disposal practices. Individuals with an income range of ZAR 10,000 to ZAR 30,000 were 19 times more likely to dispose of waste in rivers (OR = 19.0, 95% CI: 7.3–49.5, $p < 0.001$), indicating that even middle-income households may face challenges in accessing formal waste disposal systems. For those earning between ZAR 30,000 and ZAR 50,000, the propensity for river dumping remained elevated (OR = 12.9, 95% CI: 4.4–38.5, $p < 0.001$), and these households were also 3.6 times more likely to dispose of waste on roadsides (OR = 3.6, 95% CI: 1.8–7.2, $p < 0.001$). Notably, higher-income households were significantly less likely to dispose of waste in a hole within their compound (OR = 0.05, 95% CI: 0.02–0.1, $p < 0.001$), suggesting a preference for alternative waste disposal methods, such as private waste collection services.

The findings underscore that waste disposal behavior is profoundly influenced by geographic location, municipal infrastructure, and income levels. Municipalities such as Musina and Collins Chabane face substantial environmental challenges due to prevalent dumping in rivers and on roadsides. Rural and agricultural areas lack adequate disposal infrastructure, resulting in increased dumping in open spaces and water bodies in these areas. Furthermore, middle-income households demonstrated suboptimal waste disposal behaviors, particularly with high incidences of river dumping.

4.6. Thematic Interpretation of Findings

4.6.1. Service Access and Perceived Behavioral Control (TPB Construct: Perceived Behavioral Control, EJ Dimension: Distributional Justice, SDGs: 11.6, 6.3)

The low percentage (8%) of households utilizing formal waste bins stands in stark contrast to the 92% of households engaging in informal disposal methods. This outcome reflects limited perceived behavioral control within the TPB framework, which does not arise solely from internal volition but is significantly influenced by objective infrastructural constraints. The lack of proximate public bins, with 78.2% of households lacking access within 1 km, exemplifies a critical form of distributional injustice: the material enablers for responsible waste disposal are inequitably allocated to the poor. Multivariate regression analysis confirmed that residents in underserved municipalities, such as Musina (OR = 0.1, $p = 0.044$), faced disproportionately constrained disposal options. These patterns undermine SDG 11.6, which pertains to sustainable waste management, and SDG 6.3, which concerns water quality, reinforcing the argument that pro-environmental behavior cannot be disentangled from equitable infrastructure provision.

4.6.2. Spatial Disparities and Environmental Injustice (TPB Construct: Subjective Norms, EJ Dimension: Spatial and Procedural Justice, SDGs: 10.2, 11.1)

The spatial distribution of waste management behavior highlights persistent inequities. Residents of urban centers and townships have greater access to formal waste management services; however, they also engage in practices such as roadside dumping, indicating a normalization of deviance from environmental standards in densely populated areas. Conversely, individuals residing in rural areas were over five times more likely to dispose of waste in rivers (OR = 5.1, $p < 0.001$), which serves as a clear indicator of ecological vulnerability exacerbated by geographic marginalization. These disparities undermine the objectives of SDG 10.2 (inclusion and equity) and 11.1 (universal access to basic services), while underscoring the spatial dimensions of environmental norms and their enforcement in Brazil.

4.6.3. Education, Income, and the Behavior–Infrastructure Paradox (TPB Construct: Attitudes, EJ Dimension: Recognitional Justice, SDGs: 4.7, 12.5)

Contrary to the expectations derived from the TPB, higher levels of educational attainment and income do not uniformly predict sustainable waste disposal practices. Although households with university-educated members were more inclined to bury waste within their compounds (OR = 9.9, $p = 0.004$), they did not consistently utilize waste bins, revealing a paradox where pro-environmental attitudes were present, yet infrastructural limitations persisted. Furthermore, individuals within the middle-income bracket (ZAR 10,000–50,000) demonstrated a significantly higher likelihood of engaging in river dumping (ORs > 11, $p < 0.001$), reflecting systemic neglect rather than a lack of environmental concern. This observation highlights the principle of recognitional justice, emphasizing the necessity of acknowledging socio-environmental barriers when interpreting behavioral outcomes. Additionally, it aligns with SDG 4.7, which focuses on education for sustainable development, and 12.5, which aims at waste reduction.

4.6.4. Gender, Age, and the Myth of Individual Choice (TPB Constructs: All, EJ Dimension: Recognitional and Procedural Justice, SDGs: 5.5, 10.2)

The study identified no significant differences in waste disposal practices based on gender, thereby challenging the assumption that women's traditional roles in household sanitation inherently lead to environmentally conscious behavior. Similarly, older adults exhibited contradictory patterns, with increased bin usage and a higher incidence of river dumping, suggesting that age-based assumptions of environmental responsibility may oversimplify complex realities. These findings question the notion of purely individual choice in environmental behavior, highlighting that structural and procedural inequities influence perceived norms and actual opportunities for pro-environmental behavior. They align with SDG 5.5 (gender equality in decision-making) and 10.2 (social inclusion).

4.6.5. Theoretical Synthesis and Framework Reflection

The findings substantiate the study's integrated TPB–EJ framework by demonstrating that perceived behavioral control extends beyond a mere cognitive belief and is materially constrained by infrastructural deficiencies and spatial exclusion. The correlations between informal disposal and variables such as settlement type, income, and service access affirm that this behavior is influenced by both individual agency and structural limitations. These results underscore the value of integrating psychological and justice-based models in the analysis of environmental action, particularly within underserved LMIC contexts (see Figure 1).

5. Discussion

This study examined the factors influencing household waste disposal practices in South Africa's Vhembe District by employing a conceptual framework that integrates the TPB and EJ. By operationalizing perceived behavioral control, subjective norms, and attitudes in conjunction with spatial, procedural, and recognitional inequalities, this study provides a nuanced understanding of informal disposal practices in regions lacking adequate infrastructure. These findings contribute to ongoing global discussions on sustainable waste governance, particularly in rural and peri-urban settings of LMICs and align with the localization agenda of the SDGs.

5.1. Spatial Inequality and Distributional Injustice (TPB: Perceived Behavioural Control, EJ: Distributional Justice, SDG 11.6, 6.3)

The analysis indicated that only 8% of households utilized formal waste bins, whereas over 78% reported a lack of access to bins within a 1 km radius. This spatial disparity was most evident in Musina and Collins Chabane, where it was associated with environmentally detrimental practices such as river dumping (OR = 141.6) and roadside disposal (OR = 18.2). From the perspective of the TPB, limited access constrains perceived behavioral control, thereby undermining the feasibility of normative intentions to engage in responsible waste disposal.

These patterns reflect global studies that demonstrate how geospatial marginality leads to infrastructure exclusion. For instance, in rural China, spatial remoteness predicts household reliance on illegal dumping [42]. Similarly, in the Kibera and Mukuru slums of Kenya, inadequate service coverage correlates with self-managed waste disposal practices [43]. This confirms the concept of distributional injustice as articulated in the EJ theory, where location determines exposure to risk and access to basic services [23].

5.2. Informal Disposal as Structural Adaptation (TPB: Limit of Intentionality, EJ: Structural Constraint, SDG 12.5, 13.1)

Despite established behavioral norms against waste dumping, the study revealed that over 90% of households engaged in informal waste management practices, with 41.4% opting to bury waste within their homes. These actions are indicative of constrained behavioral execution, rather than attitudinal indifference. According to EJ theory, such practices are rational adaptations to systemic neglect rather than deviant behaviors [26,42,44,45].

Similar findings have been reported in Bangladesh, where Tajkir-Uz-Zaman [46] observed that residents without access to municipal waste collection routinely resorted to burning or burying waste. Likewise, in the peri-urban areas of Indonesia, Amir et al. [18]

found that informal waste burning persists in the absence of formal infrastructure, despite high levels of environmental awareness. These cases highlight a significant contribution of this study: the TPB alone is inadequate to explain waste management behavior in resource-constrained contexts. The integration of the TPB and EJ frameworks advances behavioral theory by contextualizing "perceived control" within structural realities.

5.3. Socioeconomic Status and Conditional Agency (TPB: Attitudes and Norms, EJ: Recognitional and Procedural Justice, SDGs 10.2, 12.8)

Multivariate analyses revealed intricate socioeconomic gradients. Paradoxically, households with higher incomes were more inclined to engage in river and roadside dumping. This phenomenon may be attributed to increased waste generation rates that are not matched by proportional service provision, a pattern observed in the urbanizing regions of Nigeria [47] and Indonesia [18].

This finding appears to contradict assumptions of the TPB, which links higher socioeconomic status with pro-environmental behavior. However, the elevated river dumping among wealthier households (OR = 19.0) likely reflects persistent infrastructural gaps rather than behavioral deficits. Even in higher-income households, limited access to formal collection, especially in peri-urban and rural areas, can constrain environmentally responsible choices. In this context, perceived behavioral control must be understood not merely as a matter of individual agency but as fundamentally shaped by material and institutional limitations. Moreover, greater consumption among affluent households may amplify disposal pressures, further complicating the feasibility of sustainable behavior.

Education demonstrated a protective effect against environmentally detrimental behaviors, consistent with the findings of Estrada-Araoz et al. [48]. Nonetheless, even university graduates resorted to waste burial (OR = 9.9), highlighting the necessity for recognition of justice that values lived experiences to complement behavioral interventions. This evidence emphasizes that perceived control is not solely an internalized agency but also a function of external constraints, as reconceptualized by the framework, and that awareness of sustainable development (SDG 4.7) alone may be insufficient in the absence of enabling infrastructure.

Gender was not a statistically significant predictor of disposal practices. This finding aligns with the interpretation that structural and service-related constraints affect male- and female-headed households uniformly. In low-service contexts, waste disposal decisions are shaped less by individual demographic attributes and more by shared household adaptations to infrastructural exclusion. Similar patterns have been observed in other LMICs, where gendered behavioral distinctions diminish under widespread service deficits [49,50]

5.4. Settlement Type and Spatial Governance Exclusion (TPB: Subjective Norms (Proxy) EJ: Procedural and Recognitional Justice, SDGs 11.a, 16.7)

Disposal practices are influenced by residential typology. Farm residents were five times more likely to dispose of waste in rivers, whereas township residents exhibited a relatively higher likelihood of utilizing formal waste bins. These findings are consistent with the spatial disparities observed in Accra [51,52] and Dar es Salaam [53], where municipal planning frameworks inadequately serve informal and peri-urban settlements.

This exclusion exemplifies both procedural and recognitional injustice, as these communities are neither involved in planning processes nor represented in institutional priorities [23,42]. From the TPB perspective, social norms within such settlements are shaped not only by cultural values but also by collective adaptations to exclusion. Consequently, this study underscores the necessity of situating subjective norms within local political geography.

5.5. Policy Integration and SDG Alignment (TPB–EJ Interface, Policy Planning, SDGs 11.6, 12.8, 13.1)

This study highlights the necessity of effective waste governance extending beyond mere behavior change messaging, emphasizing the need to address structural and institutional inequities. The key recommendations are detailed in Table 6.

Table 6. Key policy recommendations.

Policy Recommendation	Relevant SDGs	TPB–EJ Construct Addressed	Key Supporting Finding(s)
Expand formal waste services to underserved municipalities (e.g., Musina, Collins Chabane)	SDGs 11.6, 12.5	Perceived Behavioral Control (TPB); Distributional Justice (EJ)	78.2% lacked access to bins within 1 km; formal bin use only 8%; OR for river dumping in Musina = 141.6
Integrate behavioral science into waste policy (e.g., public campaigns informed by TPB)	SDGs 12.8	Attitudes and Norms (TPB); Recognitional Justice (EJ)	Even university-educated households showed high burial rates (OR = 9.9); intent–action gap evident despite awareness
Decentralize waste planning to include rural and farm communities in governance processes	SDGs 11.a, 16.7	Procedural and Recognitional Justice (EJ)	Farm residents are 5× more likely to dump in rivers; township residents have better service access
Reframe environmental messaging to focus on equity and adaptive responses, not compliance	SDGs 12.8, 13.3	Attitudes and PBC (TPB); Recognitional Justice (EJ)	Informal dumping driven by structural exclusion, not behavioral negligence; 41.4% resorted to burial
Integrate waste governance with climate resilience planning	SDGs 13.1, 6.3, 15.1	Distributional Justice (EJ); Environmental Risk Mitigation	River dumping and open disposal contaminate water and soil ecosystems, heightening climate vulnerability
Embed SDG and environmental literacy into waste awareness programs	SDGs 4.7, 12.8	Attitudes (TPB); Recognitional Justice (EJ)	Households with higher education still rely on burial (OR = 9.9); awareness alone is insufficient without systemic support

5.6. Implementation Challenges and Feasibility Considerations

While Table 6 presents relevant policy recommendations, their implementation faces key challenges, especially in resource-constrained municipalities. Expanding formal waste services requires significant funding, yet many local governments rely heavily on conditional grants and lack sustainable revenue streams. Applying behavioral science to waste policy also demands cross-sector coordination among environmental, education, and communication agencies, which is often weak in fragmented municipal systems. Similarly, decentralizing waste planning is limited by poor community participation mechanisms and a lack of institutional capacity in rural areas.

Reframing environmental messaging to focus on equity assumes largely underdeveloped and culturally responsive communication strategies. Integrating waste management with climate resilience planning requires technical expertise and interdepartmental coordination capacities often absent at the local level. These challenges reflect broader patterns identified by [54], which noted that limited fiscal autonomy and administrative capacity hinder municipal waste reform in sub-Saharan Africa. Addressing these constraints will require targeted investments, institutional reform, and support from national and international actors.

6. Conclusions

This study investigated household waste disposal behaviors in the Vhembe District, South Africa, using an integrated framework that combines the TPB with EJ. By aligning behavioral intent with structural constraints, this study elucidates how socioeconomic status, spatial marginality, and infrastructural exclusion converge to influence waste practices in rural and peri-urban low-resource settings. Unlike most prior research, which applies TPB or EJ independently and primarily in urban, high-service environments, this study addresses a significant empirical gap by operationalizing the TPB–EJ framework in a rural African context, where material deprivation and governance asymmetries render conventional behavioral models insufficient.

The findings indicate that informal and environmentally hazardous disposal methods are not solely a result of individual negligence but rather rational responses to persistent service deficits. Key determinants, such as income, education, residential location, and service access, were found to influence both behavioral choices and environmental outcomes. Importantly, the study reconceptualizes perceived behavioral control as shaped not only by psychological agency but also by material and institutional constraints, thus refining the TPB for low-service contexts.

The integration of TPB and EJ proved analytically robust in linking micro-level behavioral explanations with macro-level spatial and governance inequality. This theoretical synthesis refines behavioral–environmental models and extends the empirical scope of EJ scholarship into under-researched rural African contexts. It also contributes to the localization of the Sustainable Development Goals (SDGs) by highlighting the roles of distributional, procedural, and recognitional justice. Importantly, the findings underscore recognitional justice by showing that cultural practices and local knowledge significantly shape waste behavior, yet remain absent from formal governance processes

At the policy level, the findings highlight the urgency of decentralized, equity-driven waste governance strategies that are both socially inclusive and contextually grounded. Interventions should prioritize infrastructural investment, community engagement, and the recognition of lived realities in environmental planning. These imperatives are not only aligned with SDGs 11, 12, and 13 but are also essential for achieving climate-resilient and socially just urban transitions in the Global South.

Author Contributions: Conceptualization, A.C.T. and S.H.; methodology, A.C.T.; software, A.C.T.; validation, A.C.T., S.H., and M.E.L.; formal analysis, A.C.T.; investigation, A.C.T.; resources, S.H.; writing—original draft preparation, A.C.T.; writing—review and editing, S.H. and M.E.L.; visualization, A.C.T. and S.H.; funding acquisition, A.C.T. All authors have read and agreed to the published version of the manuscript.

Funding: This work was supported by the Space Science grant no U371 from the Department of Science and Technology at the Durban University of Technology.

Institutional Review Board Statement: The study was conducted in alignment with the Declaration of Helsinki and approved by the Institutional Review Board (or Ethics Committee) of Durban University of Technology (Ethics Clearance Number: IREC 294/22 and 4 July 2023).

Informed Consent Statement: Informed consent was obtained from all subjects involved in the study.

Data Availability Statement: The data that support the findings of this study are available from the corresponding author, Aifani Confidence Tahulela, upon reasonable request.

Conflicts of Interest: The authors declare no conflicts of interest. The funders had no role in the design of the study; in the collection, analyses, or interpretation of data; in the writing of the manuscript; or in the decision to publish the results.

Abbreviations

The following abbreviations are used in this manuscript:

AAMC	Association of American Medical Colleges
CI	Confidence Interval
EJ	Environmental Justice
LMICs	Low- and Middle-Income Countries
MSW	Municipal Solid Waste
MWS	Municipal Waste Service
OECD	Organization for Economic Co-operation and Development
OR	Odds Ratio
SDGs	Sustainable Development Goals
TPB	Theory of Planned Behavior
US-EPA	United States Environmental Protection Agency

References

1. Kaza, S.; Yao, L.; Bhada-Tata, P.; Van Woerden, F. *What a Waste 2.0: A Global Snapshot of Solid Waste Management to 2050*; Wor Bank Publications: Washington, DC, USA, 2018; ISBN 1464813477.
2. Lenkiewicz, Z. *Global Waste Management Outlook 2024*; United Nations Environment Programme: Nairobi, Kenya, 2024.
3. Zhang, Z.; Chen, Z.; Zhang, J.; Liu, Y.; Chen, L.; Yang, M.; Osman, A.I.; Farghali, M.; Liu, E.; Hassan, D.; et al. Municipal Sol Waste Management Challenges in Developing Regions: A Comprehensive Review and Future Perspectives for Asia and Afri *Sci. Total Environ.* **2024**, *930*, 172794. [CrossRef]
4. Haque, S.E.; Nahar, N.; Chowdhury, N.N.; Gazi-Khan, L.; Sayanno, T.K.; Muktadir, M.A.; Haque, M.S. Identification of Recycli Potential of Construction and Demolition Waste: Challenges and Opportunities in the Greater Dhaka Area. *Environ. Mon Assess.* **2025**, *197*, 646. [CrossRef]
5. Wilson, D.C.; Rodic, L.; Scheinberg, A.; Velis, C.A.; Alabaster, G. Comparative Analysis of Solid Waste Management in 20 Citi *Waste Manag. Res. J. Sustain. Circ. Econ.* **2012**, *30*, 237–254. [CrossRef]
6. Haywood, L.K.; Kapwata, T.; Oelofse, S.; Breetzke, G.; Wright, C.Y. Waste Disposal Practices in Low-Income Settlements of Sou Africa. *Int. J. Environ. Res. Public Health* **2021**, *18*, 8176. [CrossRef]
7. United Nations. *The Sustainable Development Goals Report 2021*; United Nations Publications: New York, NY, USA, 202 ISBN 9789210056083.
8. Godfrey, L.; Oelofse, S. Historical Review of Waste Management and Recycling in South Africa. *Resources* **2017**, *6*, 57. [CrossRe
9. Maphanga, T.; Grangxabe, X.S.; Madonsela, B.S. A Meta-Analysis Review of Waste Generation and Collection in Urban Inform Settlements South Africa. *Discov. Environ.* **2025**, *3*, 63. [CrossRef]
10. Statistics South Africa. *General Household Survey 2021*; Statistics South Africa: Pretoria, South Africa, 2022.
11. DEA. *South Africa State of Waste Report—First Draft*; Department of Environmental Affairs: Pretoria, South Africa, 2018.
12. Ingwani, E.; Thynell, M.; Gurure, L.R.; Ekelund, N.G.A.; Gumbo, T.; Schubert, P.; Nel, V. The Impacts of Peri-Urban Expansic on Municipal and Ecosystem Services: Experiences from Makhado Biaba Town, South Africa. *Urban Forum* **2024**, *35*, 297–32 [CrossRef]
13. Murei, A.; Mogane, B.; Mothiba, D.P.; Mochware, O.T.W.; Sekgobela, J.M.; Mudau, M.; Musumuvhi, N.; Khabo-Mmekoa, C.M Moropeng, R.C.; Momba, M.N.B. Barriers to Water and Sanitation Safety Plans in Rural Areas of South Africa—A Case Study i the Vhembe District, Limpopo Province. *Water* **2022**, *14*, 1244. [CrossRef]
14. Department of Environmental Affairs. *National Waste Management Strategy 2020*; DEA: Pretoria, South Africa, 2020.
15. National Environmental Management. *National Environmental Management: Waste Act 59 of 2008*; DEA: Pretoria, South Africa, 200
16. Ajzen, I. The Theory of Planned Behavior. *Organ. Behav. Hum. Decis. Process* **1991**, *50*, 179–211. [CrossRef]
17. Moeini, B.; Ayubi, E.; Barati, M.; Bashirian, S.; Tapak, L.; Ezzati-Rastgar, K.; Hashemian, M. Effect of Household Interventions o Promoting Waste Segregation Behavior at Source: A Systematic Review. *Sustainability* **2023**, *15*, 16546. [CrossRef]
18. Amir, F.; Miru, A.S.; Sabara, E. Urban Household Behavior in Indonesia: Drivers of Zero Waste Participation. *arXiv* **202** arXiv:2505.17864. [CrossRef]
19. Pan, J.; Liu, P. Exploring Waste Separation Using an Extended Theory of Planned Behavior: A Comparison between Adults an Children. *Front. Psychol.* **2024**, *15*, 1337969. [CrossRef] [PubMed]
20. Hasan, M.Z.; Hasan, A.M.R.; Rabbani, M.G.; Selim, M.A.; Mahmood, S.S. Knowledge, Attitude, and Practice of Bangladeshi Urba Slum Dwellers towards COVID-19 Transmission-Prevention: A Cross-Sectional Study. *PLoS Glob. Public Health* **2022**, *2*, e000101 [CrossRef]
21. Knussen, C.; Yule, F. I'm Not in the Habit of Recycling. *Environ. Behav.* **2008**, *40*, 683–702. [CrossRef]

Boeing, G.; Lu, Y.; Pilgram, C. Local Inequities in the Relative Production of and Exposure to Vehicular Air Pollution in Los Angeles. *arXiv* **2023**, arXiv:2301.00440. [CrossRef]

Walker, G. *Environmental Justice: Concepts, Evidence and Politics*; Routledge: London, UK, 2012; ISBN 9780203610671.

EEA. *Delivering Justice in Sustainability Transitions*; EEA Report No 13/2023; European Environment Agency: Copenhagen, Denmark, 2023.

Carvalho, C.; Del Campo, A.G.; de Carvalho Cabral, D. Scales of Inequality: The Role of Spatial Extent in Environmental Justice Analysis. *Landsc. Urban Plan.* **2022**, *221*, 104369. [CrossRef]

OECD. *Environmental Justice*; Organisation for Economic Co-Operation and Development: Paris, France, 2024.

Emmanouil, C.; Chachami-Chalioti, S.E.; Kyzas, G.Z.; Kungolos, A. Application of the Theory of Planned Behavior to Predict Waste Source Separation. *Sci. Total Environ.* **2024**, *956*, 177356. [CrossRef]

Eremionkhale, G.E.; Sekhon, H.; Lazell, J.; Spiteri-Cornish, L. The Role of An Augmented Theory of Planned Behaviour (TPB) On Recycling Behaviours in Lagos Nigeria. In Proceedings of the 22nd European Conference on Knowledge Management, ECKM 2021, Coventry, UK, 2–3 September 2021; Academic Conferences International Limited: Reading, UK, 2021; pp. 905–914.

Babazadeh, T.; Ranjbaran, S.; Kouzekanani, K.; Abedi Nerbin, S.; Heizomi, H.; Ramazani, M.E. Determinants of Waste Separation Behavior Tabriz, Iran: An Application of the Theory of Planned Behavior at Health Center. *Front. Environ. Sci.* **2023**, *11*, 985095. [CrossRef]

Alam, S.S.; Haque, I.M.M.S.; Kokash, H.A.; Ahmed, S.; Ahsan, M.N. Drivers of Waste Separation Behavior in Urban Bangladesh: Leveraging Social Norms and Environmental Awareness for Circular Economy Success. *Circ. Econ. Sustain.* **2025**, *5*, 1631–1663. [CrossRef]

Liu, Z.; Mostafavi, A. Collision of Environmental Injustice and Sea Level Rise: Assessment of Risk Inequality in Flood-Induced Pollutant Dispersion from Toxic Sites in Texas. *arXiv* **2023**, arXiv:2301.00312. [CrossRef]

Jiang, Y.; Yang, Y. Environmental Justice in Greater Los Angeles: Impacts of Spatial and Ethnic Factors on Residents' Socioeconomic and Health Status. *Int. J. Environ. Res. Public Health* **2022**, *19*, 5311. [CrossRef]

US-EPA. *Environmental Justice Collaborative Problem-Solving (EJCPS) Project Selections*; U.S. Environmental Protection Agency: Washington, DC, USA, 2024.

AAMC. *Polling on Environmental Justice and Waste Governance: Impact on Disadvantaged Communities*; Association of American Medical Colleges: Washington, DC, USA, 2024.

Akenji, L.; Bengtsson, M.; Hotta, Y.; Kato, M.; Hengesbaugh, M. Policy Responses to Plastic Pollution in Asia. In *Plastic Waste and Recycling*; Elsevier: Amsterdam, The Netherlands, 2020; pp. 531–567.

Frantz Schneider, A.; Aanestad, M.; Carvalho, T.C. Exploring Barriers in the Transition toward an Established E-Waste Management System in Brazil: A Multiple-Case Study of the Formal Sector. *Environ. Dev. Sustain.* **2024**, 1–21. [CrossRef]

Urme, S.A.; Radia, M.A.; Alam, R.; Chowdhury, M.U.; Hasan, S.; Ahmed, S.; Sara, H.H.; Islam, M.S.; Jerin, D.T.; Hema, P.S.; et al. Dhaka Landfill Waste Practices: Addressing Urban Pollution and Health Hazards. *Build. Cities* **2021**, *2*, 700–716. [CrossRef]

Zulkipli, F.; Mohd Nopiah, Z.; Jamian, N.H.; Ahmad Basri, N.E.; Jack Kie, C. Sustainable Public Awareness on Solid Waste Management and Environmental Care Using Logistics Regression. *J. Kejuruter.* **2022**, *2*, 139–148. [CrossRef]

Jou, Y.T.; Mariñas, K.A.; Saflor, C.S.; Bernabe, D.A.; Casuncad, J.R.; Geronimo, K.; Mabbagu, J.; Sales, F.; Verceles, K.A. Assessing the Community Perception in San Jose, Occidental Mindoro, of Proper Waste Disposal: A Structural Equation Modeling Approach. *Sustainability* **2024**, *16*, 1087. [CrossRef]

Statistics South Africa. *Mbalo-Briefnew, the Missing Piece of the Puzzle*; Statistics South Africa: Pretoria, South Africa, 2024.

Krejcie, R.V.; Morgan, D.W. Determining Sample Size for Research Activities. *Educ. Psychol. Meas.* **1970**, *30*, 607–610. [CrossRef]

Chand Malav, L.; Yadav, K.K.; Gupta, N.; Kumar, S.; Sharma, G.K.; Krishnan, S.; Rezania, S.; Kamyab, H.; Pham, Q.B.; Yadav, S.; et al. A Review on Municipal Solid Waste as a Renewable Source for Waste-to-Energy Project in India: Current Practices, Challenges, and Future Opportunities. *J. Clean. Prod.* **2020**, *277*, 123227. [CrossRef]

Kirea, E.M.; Omwenga, J.Q. Determinants of Waste Management Programmes on Sustainable Environmental Conservation in Mukuru Slums in Kenya. *Int. J. Soc. Sci. Hum. Res.* **2023**, *1*, 137–150. [CrossRef]

Strayer, S.; Stoeffler, S.W. The Intersection of Racism and Poverty in the Environment: A Systematic Review for Social Work. *Soc. Dev. Issues* **2024**, *46*, 81–99. [CrossRef]

Blake, J. Overcoming the 'Value-action Gap' in Environmental Policy: Tensions between National Policy and Local Experience. *Local Environ.* **1999**, *4*, 257–278. [CrossRef]

Tajkir-Uz-Zaman, A.K.M. Bridging Cultures: Lessons from Japan for Improving Rural Household Waste Management in Bangladesh. *Int. J. Humanit. Soc. Sci. Educ.* **2024**, *11*, 1–14. [CrossRef]

Afon, A. A Survey of Operational Characteristics, Socioeconomic and Health Effects of Scavenging Activity in Lagos, Nigeria. *Waste Manag. Res. J. A Sustain. Circ. Econ.* **2012**, *30*, 664–671. [CrossRef]

48. Estrada-Araoz, E.G.; Gallegos Ramos, N.A.; Paredes Valverde, Y.; Quispe Herrera, R.; Mori Bazán, J. Examining the Relationsh Between Environmental Education and Pro-Environmental Behavior in Regular Basic Education Students: A Cross-Section Study. *Soc. Sci.* **2023**, *12*, 307. [CrossRef]
49. Ezeah, C.; Roberts, C.L. Analysis of Barriers and Success Factors Affecting the Adoption of Sustainable Management of Municip Solid Waste in Nigeria. *J. Environ. Manag.* **2012**, *103*, 9–14. [CrossRef]
50. Mwangi, W.W.; Kimani, E.; Okong', G. Retrieved From. *Int. J. Res. Sch. Commun.* **2021**, *4*, 9–14.
51. Lissah, S.Y.; Ayanore, M.A.; Krugu, J.K.; Aberese-Ako, M.; Ruiter, R.A.C. Managing Urban Solid Waste in Ghana: Perspectives ar Experiences of Municipal Waste Company Managers and Supervisors in an Urban Municipality. *PLoS ONE* **2021**, *16*, e02483 [CrossRef] [PubMed]
52. Mudu, P.; Nartey, B.A.; Kanhai, G.; Spadaro, J.V.; Fobil, J. *Solid Waste Management and Health in Accra, Ghana*; World Heal Organization: Geneva, Switzerland, 2022; ISBN 9789240024250.
53. Adebayo, K.; Lateefat, M.; Abimbola, M.; Abosede, A.; Afolabi, O.; Olabisi, M. Challenges of Waste Disposal and Management Peri-Urban Location around Ilorin Metropolis North Central Nigeria. *Am. J. Environ. Stud.* **2024**, *7*, 17–38. [CrossRef]
54. OECD. *Going Granular with Regional and Municipal Fiscal Data OECD Regional Development Studies OECD and EU Countries*; OEC Paris, France, 2024.

sustainability

ticle

valuating the Characteristics of Disaster Waste Management ι Practice: Case Studies from Queensland and New South Vales, Australia

vindi Caldera [1,*], Chamari Jayarathna [1,*] and Cheryl Desha [2]

1 School of Science, Technology and Engineering, University of the Sunshine Coast, Petrie, QLD 4502, Australia
2 School of Engineering and Built Environment, Griffith University, Nathan, QLD 4111, Australia; c.desha@griffith.edu.au
* Correspondence: scaldera@usc.edu.au (S.C.); cjayarathna@usc.edu.au (C.J.)

Abstract

Disaster waste management (DWM) has gained much attention due to the issues associated with the enormous amount of disaster waste generated by natural disasters. However, moving beyond ad hoc and champion-based take-up by practitioners, there is generally a lack of momentum towards mainstreaming such DWM practices to achieve resilient outcomes. This study aims to explore the characteristics of DWM practices, drawing on the lived experiences of industry practitioners and government decision-makers. An interpretive case study method was used to investigate how local government organisations could readily engage in effective DWM processes using the "Resilient disaster management framework" previously established by the research team. Insights include a lack of documented plans for DWM and at best a moderate focus on recovery processes for disaster waste. With these issues counterproductive to community resilience outcomes, there is an urgent need to better manage disaster waste, which we propose can be enabled and supported through DWM plans. The extended 'resilient DWM framework' proposed in this study provides a useful reference to prepare practical, agile, and comprehensive DWM plans.

Keywords: disaster waste management; disaster resilience; case study; Queensland; New South Wales; Australia

1. Introduction and Background Context

Climate change effects and weather-driven natural hazards such as flooding, bushfire, and cyclones are becoming more intense and more frequent. When built up towns and cities are exposed to natural hazards, damage to property and infrastructure results in what is termed 'disaster waste', which requires management alongside other disaster response and recovery efforts [1]. There is growing global recognition of the impact that disaster waste management (DWM) has on achieving several Sustainable Development Goals (SDGs) [2]. Many of the targets under these goals emphasise the importance of resilience—such as significantly reducing the number of people affected by disasters and the resulting economic losses (Target 11.5) and enhancing resilience and adaptive capacity to climate-related hazards and natural disasters in all countries (Targets 13.1 and 13.2) [3]. Resilience is generally defined as "a system's capacity to adapt to or respond to singular, unique and most often radically surprising events" [4]. Despite increasing international efforts, such as those led by the United Nations, to build resilient communities, academic research has paid limited attention to DWM from a resilience perspective. This gap has been highlighted in recent literature reviews on DWM [5,6].

According to the "National Waste Report, 2022", 181,000 tonnes of disaster waste were recorded for 2020–21 from bushfire and flood events in New South Wales (NSW), Victoria (VC), Western Australia (WA), and Queensland (QLD) [7]. This disaster waste significantly impacts the health and wellbeing of people and the natural environment. Exposure to hazardous waste such as asbestos causes severe illness, and the magnitude of this waste including plastics and chemicals can have adverse impacts on biodiversity, air pollution, and water quality [8]. On the other hand, the volume of disaster waste impacts support services and immediate response efforts due to road blocking [8]. Despite these issues, proper handling and management of disaster waste can reduce negative economic, environmental, and social impacts through turning waste into valuable resources and building a resilient community. Effective planning and coordination are crucial to minimising disruption during disaster response [9]. Previous research has established four critical steps to investigate the pre-disaster planning of waste management, including the following: (1) estimation of disaster waste generation, (2) determination of temporary storage sites, (3) setting of system boundaries, and (4) data collection and inventory analysis. We explain these steps in the following section [10]. DWM plays a critical role by focusing on separating waste and subsequent treatment options [11], debris management planning, partnerships between state and local emergency management services [12]. However, much of the existing literature on disaster waste and debris management consists of isolated case studies and planning guidelines. Within a global context, debris management planning guidelines—such as those by Solis et al. [13], the United States Environmental Protection Agency [14], the Federal Emergency Management Agency, USA [15]—outline various technical and managerial strategies for handling disaster waste. Disaster research often lacks continuity, with studies typically focused on specific events. Challenges such as the variability between disasters, limited time frames, and difficulties in accessing reliable data hinder the development of broader, quantitative, cross-event analyses.

Holistic DWM plans should build on previous event experiences and also encompass both technical aspects (collection and transportation; temporary debris storage; recycling; disposal; hazardous waste handling and disposal) and management aspects (communication strategies; contract management; organisational roles and responsibilities; record keeping; Memorandum of Understanding (MOU) establishment) [6]. These guidelines are typically rooted in the existing legislative, organisational, and financial frameworks relevant to the specific context [16,17]. Furthermore, there are recycling opportunities for the material sorted after disasters in the form of recycled aggregates and other materials [18,19].

Within a national Australian context, plans such as State Emergency Management Plans (SEMP) in South Australia provide a framework for managing emergencies, including waste management during and after disasters [20]. This plan includes four parts: (1) providing a strategic overview, (2) providing overarching details of the arrangement and structures to address emergencies, (3) guidelines and frameworks with best practices and (4) standards required of the various levels of detailed plans. In New South Wales, the Hunter Joint Organisation provides a framework for local councils to collaboratively utilise in the event of a disaster to coordinate waste management [21]. QLD government offers an information sheet with coordinated waste management solutions focusing on managing waste in the disaster recovery period [22]. This highlights that, historically, local governments typically manage the clean-up process professionally and to a high standard via the engagement of local waste contractors.

With the increasing issues associated with the overwhelming amount of disaster waste generated, researchers have attempted to explore disaster waste management (DWM) in recent years [6,23,24]. DWM refers to the practices that are "*responsible for handling, storing, collecting, and disposing of waste without posing a threat to the built environment as well as peoples' health and wellbeing*" [24]. DWM is also defined as "The management of debris is involved at the recovery stage as debris generated through damaged buildings need to be collected, transported, reused, recycled, land filled or disposed" [25]. It also involves different processes, people, laws, and standard procedures to be followed at different phases of the disaster management cycle: Prevention, Preparedness, Response, and Recovery (PPRR) [26]. DWM has been a challenging task because of the challenging factors such as financial resources, capacity of organisations and individuals [1], constraint of time across various disaster types, and the difficulties associated with data accessibility [8]. Zhang et al. [3] conducted a comprehensive review of recent developments in DWM, building upon the foundational work of Brown et al. [6]. Their study examined progress across nine key areas: planning, waste characteristics, treatment options, environmental impact, economic factors, social considerations, organisational structures, legal frameworks, and funding mechanisms.

Although there are numerous guidelines available for DWM nationally and internationally, these factors affect practitioners differently in executing their DWM plans properly. This study aimed to explore the characteristics of DWM practices, drawing on the lived experiences of industry practitioners and government decision-makers.

To achieve this research aim, two research questions were formulated using a Resilient Disaster Waste Management Framework (RDWMF), based on the authors' previous work [8], as follows:

RQ 1. How could the characteristics of the RDWMF, such as data management, training, resource management, communication, stakeholder management, and recovery process, be used in practice?

RQ 2. How do government authorities and industry practitioners aim to implement their DWM plan in practice?

The purpose of this study was to use the previously established framework to explore the characteristics of DWM practices from the ground level and offer unique case-based insights. The following sections provide an overview of the method, key findings, and a discussion of the proposed extended Resilient Disaster Waste Management Framework.

2. Method

2.1. Research Design

The purpose of this study is to explore the characteristics of DWM practices from an operational viewpoint based on the RDWMF proposed in the authors' previous work [23]. This framework was initially developed using twelve semi-structured interviews, including 3 industry practitioners and 9 government decision-makers (7 local government decision-makers and 2 state government decision-makers) [27]. The current study is an extension of the authors' previous work, and it involved two additional case studies to explore the characteristics of DWM practices in depth and understand the process of making a DWM plan in practice. Figure 1 illustrates the research design process of the current study.

2.2. Research Approach

The current study used a qualitative approach that involved an interpretive multiple case study method, which is a popular method to understand processes, problems, or programmes to engender improvements [28,29]. The number of participants is relatively small due to the complexity of managing and analysing the large volume of data [27,30]. The key principle for selecting an appropriate case study is the richness of information with respect to the topic of investigation [29,31]. The two cases were selected from Queensland and New South Wales, Australia, to understand the phenomenon of DWM in practice.

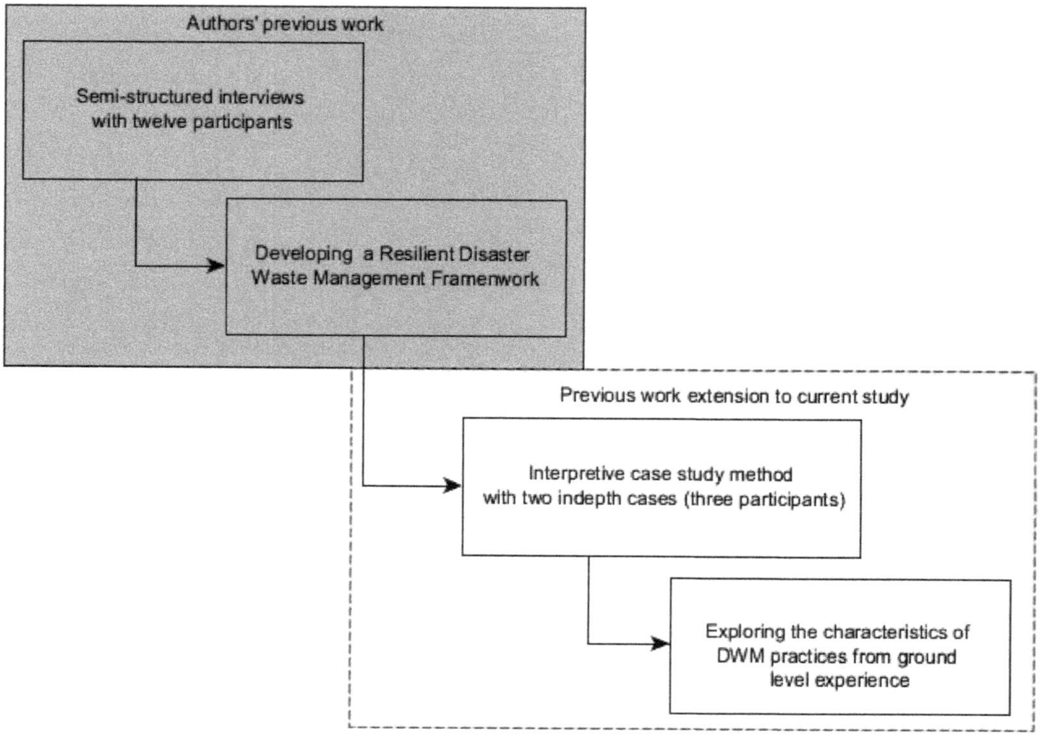

Figure 1. Research design process.

2.3. Case Selection

The case studies were selected based on purposive sampling approach [31], following the criteria of functional and suitability attributes which reflect the study purpose, accessibility to the key informants, and possibility of data collection within a limited timeframe [30]. Functional attributes were based on the geography of which communities that are highly vulnerable to disasters and generate a high volume of disaster waste (cases based in Queensland and New South Wales), accessibility to the key informants, and the ability to collect data within the desired time frame. Suitability attributes were based on involvement in disaster management activities, which the communities have experienced over a decade of natural disaster events. The aim was to draw on the lived experiences of the interview participants from the selected case studies and derive examples (current practices, practices they are aspiring to achieve, or possible plans to implement when aiming for better disaster management planning) aligned with the elements of the DWMF, as shown in Section 2.5. Although the selected entities did not have formal DWM plans at the time of the study, they were actively engaged in disaster waste-related activities. This allowed us to examine how elements of planning are approached in practice, particularly in informal or emerging forms, which aligns with the study's second research question.

2.4. Data Collection

Potential cases were contacted through the Queensland Office of the Inspector General Emergency Management's disaster management network. Semi-structured interviews were used to collect data for exploring the execution of DWM in practice. Interview participants comprised a manager of resource recovery, an operations manager, and senior staff involved in disaster management operations. An interview guide was developed based on the RDWM framework proposed by Caldera et al. [23], and it was approved by the Human Research Ethics Committee of the University of the Sunshine Coast (ethics approval number: A242194). Interviews were conducted online for 40–60 min and were audio recorded for transcription. This qualitative inquiry enabled deeper exploration and discovery, focusing on whether there was evidence to support the problem rather than on measuring the size of the problem [32].

2.5. Data Analysis

Audio data captured from three participants were manually transcribed before analysing. Transcripts were deeply analysed by the research team to explore the research questions. Thematic analysis was used employing deductive coding to examine interview data [33]. Before the analysis, a set of a priori codes were set up based on the RDWM framework.

According to Figure 2, stakeholders can be engaged across the six key focus areas—training, resource management, recovery processes, communication, data management, and stakeholder management—by aligning these efforts with the phases of the PPRR model (Prevention, Preparedness, Response, and Recovery). Since the PPRR model is widely adopted in many countries for their National Disaster Management Plans, integrating it into DWM (Disaster Waste Management) practices will help mainstream and strengthen resilience within the DWM framework [23].

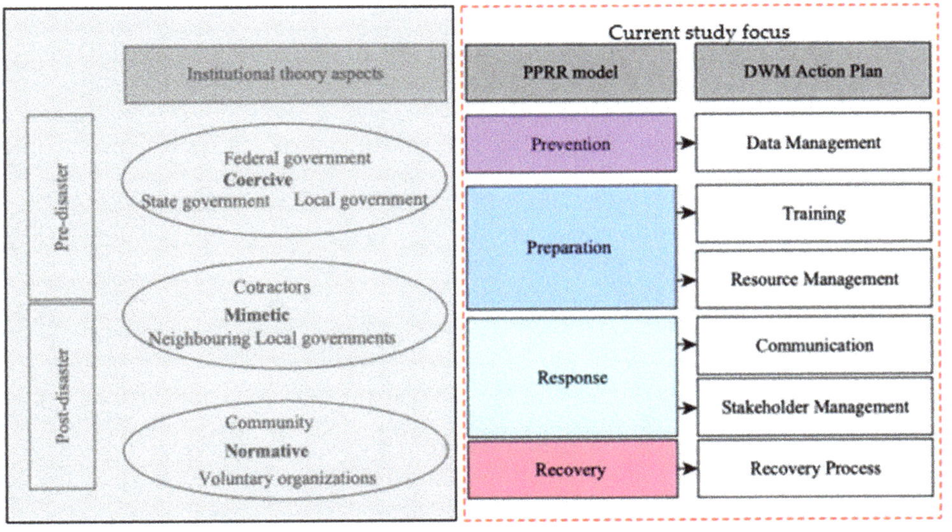

Figure 2. Priori codes extracted from the RDWM framework (source: [23]).

Building on the DWM action plan elements, the a priori codes including data management, training, resource management, communication, stakeholder management, and recovery process were used for the analysis. Other emergent codes were generated as in vivo codes to define the understanding and conceptualization of DWM and resilience.

A team of three researchers coded the interview data to corroborate the emergent themes to reduce bias in the synthesised data [34]. There was 80 per cent agreement, which is an acceptable inter-coder reliability rating [35].

3. Results

The following paragraphs summarise findings from the exploration of the two cases, referred to as "Case 1" and "Case 2".

3.1. Profile Summary of the Two Case Studies

Case 1 is a commercial waste business unit within a city council, located in the south-west of Greater Brisbane, covering an area of 1085 km^2 and serving a population of 229,208. It provides both domestic and commercial waste services to the cities of two Regional Councils in Queensland. Two participants were interviewed in this case. Participant 1 (P1) is the manager of resource recovery, with four and a half years' experience in the current role and 23 years of overall experience in waste management. Participant 2 (P2) is the operations manager, who has been in this role for three months but has a total of 26 years' experience in the waste management business.

Case 2 is a community-based initiative that has been actively working in recovery since the 2022 disaster that heavily impacted a city in NSW, located in the Northern Rivers region of New South Wales, Australia, within a 1290 km^2 area and with a population of 44,334 [20]. The role of this initiative at that time was essentially focused on cleanup and housing restoration, coordinating volunteers and other assistance to manage the waste generated from the event. Participant (P3) in this case is the executive director of this initiative, with 8 years of experience in disaster recovery, and P3 was formerly a councillor for a city council in NSW.

3.2. Understanding "DWM" and "Resilience" in the Context of DWM

As the foundational step, deeply understanding the concepts of disaster waste management and resilience was critical prior to evaluating the characteristics of DWM using the components in the RDWMF, as shown in Section 3.3. Therefore, the authors enquired about practitioners' understanding of the concept of DWM. Participants from Case 1 demonstrated a broad understanding of what DWM mean to them. P2 explained DWM as encompassing three broad phases, including pre-, during-, and post-disaster planning, with a focus on establishing a disaster management pad to handle, store, sort, and recover waste from both pre- and post-cleanup activities. *"we have a disaster management pad that we've just finished building at our facility. So, we know we can take all the products somewhere to handle it and manage it right"* [P2].

In addition, P1 explained DWM from a commercial viewpoint that there are government requirements to meet the expectations of communities when offering waste services [P1]. While P3 from Case 2 defined DWM as *"managing the significant volumes of waste that are generated after a disaster"*, P3 stressed that there is a missing piece of work in this definition. That is from the "**community level**" and the "**government level**", highlighting the importance of community awareness campaignsto educate people on t proper waste disposal and salvage practices. Furthermore, there should be targeted local government and state-level efforts to manage logistics and coordinate the removal of the quantum of waste that are generated [P3]. Based on the insights from both cases, DWM involves planning activities across pre-, during- and post-disaster phases, with the participation of community coordinated at the government level.

Resilience is another important concept that researchers always focus on with DWM. Despite the academic aspects of resilience, authors of this study expected to obtain the practitioners' viewpoint of resilience in DWM. P2 from Case 1 defined resilience from an operational aspect as *"resilience is always meant the ability to either withstand or show a toughness during an event, but it's how fast that you can recover from something from a disaster shows the resilience"* [P2]. Adding to the same conversation, P1 described different aspects of resilience in terms of residents, businesses, communities, personal resilience in how individuals deal with disasters, and facilities and equipment being designed to handle such events. Of these different aspects, P1 highlighted that the most interesting is personal resilience, giving an example; *"during the last flood I had to stand down one of my offices because they were becoming so traumatised because they kept seeing image upon image of people's lives being destroyed."* [P1]. This example illustrates the ability to confront the psychological or mental impact of disasters under personal resilience.

P3 from Case 2 defined resilience as *"it is about the ability to withstand significant shock, but still to return from that significant shock"* [P3]. In terms of DWM, P3 stated that they are embedding resilience into the types of materials used, the restoration building codes, and related measures, highlighting that lessons learnt from previous disasters suggest that disaster waste generation would decline as a result. This could be attributed to having more resilient disaster-educated communities [P3]. This can be identified as material resilience, where we think of material types that are being used in homes or offices to reduce waste in the event of disasters. According to insights from both cases, personal resilience and material resilience emerge as the main practical focuses in applyingresilience to DWM.

3.3. DWM Plan and Its Characteristics

In terms of the availability of a clearly defined DWM plan, interviewees in both cases responded "No". However, Case 1 has an overall plan to manage any disaster event, which is the **Business Continuity Plan (BCP)**. They believe that having a general plan is better than a specific one, emphasising *"it doesn't actually matter what the cause of the disruption was, usually your processes to deal with the disruption are generally the same"* [P1]. On the other hand, although they accept that having a specific DWM plan is useful, they do not put a lot of faith or time into very prescriptive plans, because every event is very different and requires a flexible response. Further, P1 stated, *"trying to hold people to very, very specific responses is almost impossible"* [P1].

The participant from Case 2 responded that they do not have a formal DWM plan because they have been funded to deliver a project over the next year, during which they will focus on building a plan specifically aimed at community awareness. However, interviewees from both cases described the components of a DWM plan in relation to what they currently have or what they have been doing. Table 1 shows evidence to support how extensively the components of a DWM plan have been used in practice.

Table 1. Participants' responses in relation to DWM plan components.

Characteristics of DWM Plan	Interviewee	Evidence (Participants' Responses)	Key Learnings from the Interview Findings
Data management	Case 1-P1	*"I suppose we use data around what to expect. What did we expect in the 2011 flood as opposed to the, you know, the 2021 flood and what we could expect for a flood of that nature again in relation to how much waste we will expect to come in".*	Be prepared by knowing what to expect based on previous disaster events
	Case 1-P2	*"we've just finished building a very large new resource recovery pad where you can store and sort and move, you get a chance at least at some level of resource recovery".*	Building infrastructure such as recovery pads to store, sort and recover disaster waste
	Case 2-P3	*"Having a handle on the data management that accurately tells that story in order to strategically plan for landfill facilities essentially and for managing hazardous waste as well".*	Data management to strategically plan for waste management facilities and landfill options

Table 1. *Cont.*

Characteristics of DWM Plan	Interviewee	Evidence (Participants' Responses)	Key Learnings from the Interview Findings
Training	Case 1-P2	*"We've identified that we could possibly do further training even though we've got a highly skilled workforce that operate".*	Training people to create a highly skilled workforce to activate during disasters
	Case 2-P3	*"we will be endeavouring to improve the literacy in the community around what needs to be thrown out and what maybe doesn't need to be thrown out".*	Educating communities to identify what waste types should be disposed of and retained
Resource management	Case 1-P1	*"I tend to use from a lot of things, which is a* ***people process and technology*** *summary".*	Having a range of resources such as people, process and technology summary accessible during disasters
	Case 1-P2	*"we need to think about how's this going to be paid for, how do we recover costs, especially if we've got to use external people like landfills. Bringing external companies to help with, you know, vehicles or equipment".*	Methods to utilise funding, manage costs, leverage external parties, vehicles, and equipment
	Case 1-P3	*"One of the things that we are doing is some asset mapping. And then, we deploy those resources accordingly, so we will be working to try to identify as much of that within the local community out front as we can, so that we can plan for more effective coordination as soon as possible".*	Asset mapping, effective resource deployment and coordination
Communication	Case 1-P2	*"That's clearly defined in our continuity plan in local government. So, from a communication viewpoint, it's clear who does what in what situation".*	Following a continuity plan
	Case 2-P3	*"It should be fast. It should be regular. It should be shareable and that way we've got a consistent communication platform rather than multiple sources of information in an already overloaded information environment".*	Critical need for a fast, regular, consistent, and a shareable communication platform
Stakeholder management	Case 1-P1	*"So whether that's the general public, whether that's businesses, whether that's the elected Members, whether that's higher up levels of government, you know you need to ensure that that communication is very clear up and around paths".*	Clear communication within and between stakeholders
	Case 2-P2	*"Common in disasters is there are always emergent community initiatives and community responses. So, linking in with those emergency groups and activities I think is really important and providing resources to those groups and activities".*	Connecting community initiatives with emergency response groups
Recovery process	Case 1-P1	*"we might have surge capacity where we hold on to different materials for different times. We know what the different landfills can take. We know what the different vendors can take".*	Prioritising and increasing capacity for resource recovery
	Case 2-P3	*"The first phase is the immediate removal of goods and belongings and furniture, chip board that's disintegrated. There are then subsequent and longer term waste phases that are associated with the rebuild and the restoration of properties, so people might not get around to the removal and replacement of the internal parts of the house until a number of months down the track".*	Using a phased approach depending on the quantity, quality and the capacity for recovery

3.3.1. Data Management

The interviewees from Case 1 clearly stated that data management plays a critical role at the prevention stage. Prevention data helps them identify areas based on risk levels (i.e., as high risk or low risk). Further, they considered data as historical knowledge. Substituting data management with historical records and understanding gives historical experience, so after every disaster event, debriefing those events to figure out what worked and what did not work helps capture this historical knowledge.

P3 from Case 2 stated that data management is important in other stages like preparation, response, and recovery. In their council, official government records track the amount of waste passing over bridges, arriving at the council's waste facility, and then how much was transported to other facilities.

3.3.2. Training

Participants from both cases responded that training is an essential part of the DWM plan. However, Case 1 considered training in a holistic view, highlighting that disaster management is not just about waste management—it is part of the whole process. For example, P2 stated, "*It's not just usually a waste disaster. It's a people disaster. It's a flood. It's you know, it's fires, it's holistic right, so back the waste is one portion, but it has to work inside everything else*". From a training point of view, Case 1 considers training to constantly look at what they can do better and how they can train their people to be prepared.

Based on the nature of Case 2, their focus on training is to make the community aware of the salvage and unnecessary waste creation. They provide clear advice, particularly on white goods, highlighting that "*you might not need to throw away your fridge. Your fridge might be able to be repaired. If we could prevent broad scale sending of fridges to landfill, that would be a really positive outcome*" [P3]. In this way, they will endeavour to improve literacy in the community around what needs to be thrown out and what may not need to be thrown out.

3.3.3. Resource Management

In terms of resource management, P1 from Case 1 categorised many aspects into **people, process, and technology**. From that point of view, they focus on who the key people to communicate with are, what processes are available, and what equipment or technology is available. In addition, P2 touched on funding and some challenges of putting information about resource management in the DWM plan. Although allocating funding is important from a business standpoint, in real situations, their priority is to complete the job first. On the other hand, though they accept that documenting required information is valuable, much of the necessary knowledge cannot be fully captured in written form. For example, P2 stated the importance of knowing what and how something can be recycled, "*we've got a whole heap of scrap metal and, let's recycle that*". They emphasised that it is not possible to document every action because the issues are complex and there is no quick, easy solution. However, they have an intuitive understanding of what can and cannot be done.

Case 2 uses information about available assets or equipment to plan for more effective coordination in DWM and build partnerships with other organisations involved in waste management, synchronizing responses with official government efforts. For example, P3 stated "*At a certain period of time following the disaster, what we know is that there is always more waste that comes out after the Council has completed its waste removal strategy. So we work to come in afterwards, and when the official government responses wind down. We embed that information and that planning into our procedures*". This example illustrates how information about certain processes can be used in DWM. Both cases highlight the importance of having information on people, processes, and technology for resource management.

3.3.4. Communication

P2 from Case 1 responded that a proper communication channel for responding to relevant parties involved is clearly defined in their BCP, and communication lines are set within the local government (see Table 1). For example, they have clearly established customer service, disaster management groups, and a chain of responsibility in relation to sign-offs for monetary purposes.

P3 from Case 2 responded that communication should be managed by local government as much as possible, emphasizing that local authorities need to quickly establishconsistent, comprehensive, and regular communication strategies that can then be shared and amplified by other stakeholders.

3.3.5. Stakeholder Management

Case 1 highlighted that their main stakeholders are internal stakeholders such as customer service, senior management groups, call centres, and a local disaster committee consisting of senior people, branch managers, general managers, Chief Executive Officers (CEOs), and politicians. P2 stated, "*these people already know the rules, they already know how to communicate, they already know who to deal with, so what they are doing is using experienced people in inexperienced roles to do the things*".

In addition to local government and councils, Case 2 highlighted the community and other business organisations as their stakeholders. P3 stated "*it is important to provide appropriate resources to assist community in their initiatives and PPE (Property, Plant and Equipment) for appropriate safe handling of hazardous*". For other business organisations, their approach is different from what it is for households and residents. P3 highlighted "*targeted planning for how to engage with those particular stakeholders is important as they have a significant number of businesses who are also managing waste removal and clean up*". From the insights of both cases, it is clear that different approaches are employed for different stakeholders in DWM.

3.3.6. Recovery Process

In terms of the recovery process for short-term and long-term DWM, P1 from Case 1 responded, "they haven't probably got as much detail in that as they probably wouldn't other than they utilise their normal processes to deal with stuff." P1 further stated that they probably need to do invest a bit more effort in this aspect, as it requires time, workshopping, and careful consideration.

P3 from Case 2 responded that there are definite phases for the recovery process for short-term and long-term DWM. For example, "*The first phase is immediate removal of easy to move belongings. There are then subsequent and longer-term waste phases that are associated with the rebuild and the restoration of properties. Simultaneously taking steps to manage environmental waste such as waste that has come down the river systems*" [P3]. This evidence indicates that Case 1 has not yet focused on the recovery process, while Case 2 has definite phases of the recovery process.

3.4. Making a DWM Plan in Practice

Following on the institutional theory aspect of the RDWMF [8], the authors of this paper attempted to figure out how government authorities (coercive), neighbouring local governments or organisations (mimetic), and community or voluntary organisations (normative) influence the development of a DWM plan. In terms of coercive influence, Case 1 highlighted that it is obvious to follow rules and regulations imposed by federal, state, or local government authorities when managing disaster waste. "*Anything you do, anything you store, would have to be done within licence, would have to be done within transport rules*" [P1]. Case 2 responded that they adhere to hazardous material handling policies and procedures by providing an example that *"we definitely embed a really strong adherence to safety in terms of how we would traverse the flood zone essentially and how we keep people safe in terms of managing that stuff"* [P3].

In terms of mimetic influence, P2 stated that they consider networking with other local government waste group a very important aspect of making a DWM plan, emphasizing the value of sharing information and networking. *"I will say for me and for most waste managers in the state, we all know each other across the board from Cairns to Tweed. So, we all share this stuff already and during events we're very quick to help each other out"* [P2].

In terms of normative influence, P3 from Case 2 stated that they definitely consider community practices in relation to regulation and policy, highlighting a story of creative repurposing *"we had a community member come to us with an idea for, like community art projects following the disaster. in terms of linking back to that preparedness stuff, there might be an opportunity for some, you know, creative repurposing of some of those materials that then ensures that that story about you know the waste that is generated following disasters It could be repurposed into a mural or a mosaic. It ensures that awareness is carried through to future community members"* [P3]. They also emphasised the role of voluntary organisations, highlighting that *"we refer to appropriate agencies like Fire and Rescue. So, if we identify hazard material, we refer up to appropriate government agencies, who can then respond accordingly"* [P3].

Evidence from both cases indicates the role of government authorities in setting up rules and regulations in managing disaster waste, sharing best practices of other neighbouring councils or organisations, and heightening community and voluntary organisations' participation, highlighting the applicability of institutional theory in making a DWM plan in practice.

4. Discussion

This section provides a discussion on the concept of DWM and resilience, components of the RDWMF, and key factors considered in making a DWM plan, based on the analysis and findings from the case studies above.

4.1. Concepts of DWM and Resilience

The understanding of DWM and resilience concepts in practice aligns closely with academic definitions found in the literature. Previous studies have described DWM based on the different phases of the disaster cycle—pre-, during-, and post-disaster [9,21]—which is evident in the practice, too. In addition, it is understood or defined based on the and council viewpoint (Case 1). Furthermore, missing pieces of the current definition of DWM are highlighted as "community level" and "government level" (P3-Case 2) in practice.

While resilience is generally defined as *"a system's capacity to adapt to or respond to singular, unique and most often radically surprising events"*, there are different aspects of resilience influencing DWM. A few of them are disaster resilient community [1,36], climate disaster resilience [37], circular economy resilience [38], urban resilience [39] and institutional resilience [40]. In addition to these aspects, case studies highlighted another two important aspects of resilience as "personal resilience" (Case 1) and "material resilience" (Case 2). Personal resilience is important when responding to clients or community or handing disaster waste during traumatic situations. Makwana [41] stated that individuals and communities experience a mental instability that might cause anxiety, depression or post-traumatic stress disorder because of disasters. It is important to support the development and strengthening of mental and emotional resilience to cope with difficult times. Additionally, crowds should be viewed as part of the solution rather than as part of the problem during mass emergencies and disasters [42,43]. Previous research indicates that strong social support received after a disaster is associated with increased psychological resilience [44]. It is about finding ways to navigate challenges, adapt to change, and, ultimately, recover and thrive. Material resilience is embedding resilience into types of material that are used by community or supporting workers, which is important in reducing waste in subsequent disasters. This concept has been incorporated by the Queensland Reconstruction Authority as resilient building design and construction to reduce damage caused by disasters [45].

4.2. DWM Plan and Its Components

Although previous research highlighted the importance of having a documented DWM plan [23], both case studies show that such formal documentation is often not maintained. The interview data highlighted how DWM planning can take place even in the absence of formal documentation. Activities such as preparing recovery infrastructure, coordinating stakeholders, conducting training, and implementing phased waste handling all represent key planning efforts. These actions illustrate how DWM planning can be developed and applied gradually, embedded in current business continuity plans, especially in practical, resource-limited, or evolving contexts.

Case 1 provided reasons for not having a specific DWM plan, explaining that it is very difficult to maintain multiple documents for different events. The plan must be reviewed annually and should not simply sit unused until the next disaster occurs, as every event is unique. They emphasized that relying heavily on a rigid, detailed plan is impractical because the response needs to be flexible to adapt to varying situations. These points indicate that DWM planning often unfolds organically through practice and experience. Therefore, they have an overall high-level plan that is a Business Continuity Plan, where they consider DWM as part of their business continuity plan. In the absence of a formal DWM plan, Case 2 follows a coordinated approach to work on DWM whereby they closely collaborate with the state government and volunteers. Even though both cases do not have a documented DWM plan, they discussed its components in relation to their current practice. Table 2 provides a summary that shows which components of the DWM plan have been used in practice and its importance.

Table 2. Summary of which components of the DWM plan have been used in practice.

Components of DWM Plan	Use of Components of DWM Plan in Practice		Level of Importance
	Case 1	Case 2	
Data management	Yes	Yes	High
Training	Yes	Yes	High
Resource management	Yes	Yes	High
Communication	Yes	Yes	High
Stakeholder management	Yes	Yes	High
Recovery process	No	Yes	Medium

Case 1 highlighted that data management plays an important role in the prevention stage, while Case 2 highlighted that it is important in other stages like preparation, response, and recovery. This includes capturing waste data to strategically plan for waste facilities, managing hazardous waste, and implementing responses accordingly. Oloruntoba et al. [46] highlighted that it is necessary to incorporate information gathered from previous events, as it can be used to lessen risks in preparation for future events. In terms of training, both cases considered it an essential part of DWM. Based on the evidence from both cases, in practice, training is provided for workers holistically covering everything in disaster management (including DWM), ensuring better preparation and for the community focusing on increasing awareness of avoiding unnecessary waste creation. Oloruntoba et al. [46] stated that skilled workers and volunteers are necessary for preparedness. In resource management, Case 1 discussed its importance from people, process, and technology summary, while Case 2 highlighted the importance of coordination and building partnerships among other stakeholders (local government and other business organisations). The literature discussed the role of humans and technology in DWM [47]. The researchers provided examples of using technology and processes in DWM, such as sensors for the sorting process, GPS for waste transportation process monitoring, a logistic digital system, waste mobile apps, and automatic compactor trucks [27]. The use of intelligent waste management systems, big data analysis, and automated processing technologies is becoming increasingly prevalent in disaster waste management [48].

Communication is well set up in practice, establishing a proper channel for how to contact and who to contact. Case 2 emphasised that the local government should take the lead in establishing a consistent, comprehensive, and regular communication strategy that can be shared with other stakeholders. However, Fatimah et al. [47] pointed out that different government and non-governmental organisations that collect the most waste information during the post-disaster phase lack communication and integration. In terms of stakeholder management, Case 1 mainly focused on internal stakeholders such as the senior management group, call centre, and the local disaster committee, whereas Case 2 focused on external stakeholders such as the community and other business organisations. In practice, it is evident that different approaches are applied to different stakeholders, and this indicates that targeted planning for how to engage with those stakeholders is important in stakeholder management. Elkady et al. [49] found a decision support tool for enhancing stakeholder interaction in disaster events. This includes initiatives such as enhancing the leadership skills of emergency managers and facilitating community participation in the decision-making process and disaster response activities.

For the recovery process, Case 1 did not provide much information about this component, but Case 2 discussed it in definite phases of short-term and long-term initiatives. The first phase is the immediate removal of easy-to-move belongings, which happens in short term. In the long term, rebuilding and the restoration of properties occur in subsequent phases, followed by steps to manage environmental waste. Tabata et al. [50] emphasised that immediate removal and treatment of waste generated during disasters are essential for the rapid recovery and reconstruction of the affected area and reduce the environmental impact of disaster waste. Resource recovery after a disaster not only involves restoring and rebuilding affected areas, focusing on rebuilding physical infrastructure, but also needs to focus on supporting the wellbeing of people and restoring the environment. Previous research has identified that competence of construction professionals (contractors and builders) and the government response and intervention in terms of legislation and policy, and government support and assistance, as common constraints to the resourcing of disaster recovery projects [51].

Based on the above discussion, the RDWMF can be extended with examples from the industry practice found by the current study (Figure 3). This framework will be a great source of reference to prepare a practical, agile, and a comprehensive but concise DWM plan, as it provides practical examples for each element of the DWM action plan. Although the previous RDWMF focuses on key elements that need to be included in the DWM plan, it lacks guidance on what to address in these key elements in practice. The extended framework proposed in this study will fill this gap by illustrating practical examples.

As indicated in Figure 3, data management is an essential element at the prevention stage of the DWM cycle, and it can be used to capture waste data and strategically plan waste facilities accordingly. It also helps manage hazardous waste and implement proper responses. Training can be given to increase awareness of unnecessary waste creation. Resource management can primarily be categorized into people, processes, and technology. In communication, well-consistent, comprehensive, and regular communication can be established under the local government leadership. Different approaches can be followed for managing different stakeholders. The recovery process can be executed through short-term and long-term actions to manage waste. Incorporating such examples in the DWM plans will support in achieving resilient outcomes through better managing disaster waste.

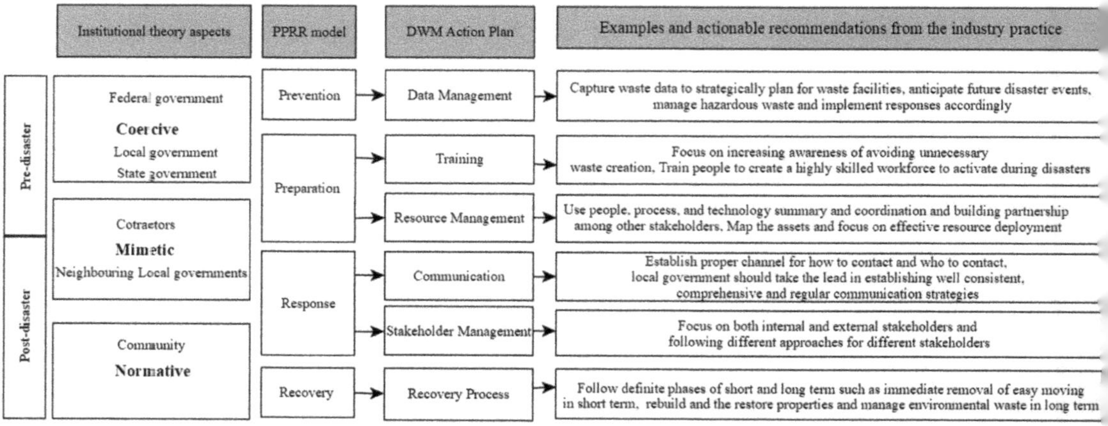

Figure 3. Extended resilient disaster waste management framework.

4.3. Driving Forces for the DWM Plan

In the literature, key factors that influence a DWM plan are largely discussed from the institutional theory aspect which are coercive, normative, and mimetic [52]. According to this theory, coercive elements represent rules and regulations, laws, and policies set up by federal, state, or local government bodies; normative elements represent norms or practices performed by a community or volunteer organisations. The mimetic element represents sharing best practices followed by the neighbouring local governments or other business organisations [23,53,54]. In practice, similar elements can be observed as driving forces behind the DWM plan, as shown in the findings. Therefore, all three isomorphisms of institutional theory are relevant and applicable to DWM practice.

5. Conclusions

The current study aimed to explore the characteristics of DWM practices, drawing on the lived experiences of industry practitioners and government decision-makers. An interpretive case study method was used to investigate how local government organisations could readily engage in effective DWM processes using the "Resilient disaster management framework" previously established by Caldera et al. [23]. Findings indicate that the components of the RDWM framework are relevant and used in practice. However, compared to the other five components of this framework (data management, training, resource management, communication, and stakeholder management), resource recovery received only moderate focus. Given that resource recovery is a major component of DWM, industry practitioners should place greater emphasis on this element, as it helps minimise the environmental burden of disaster waste and supports the achievement of resilient outcomes.

Furthermore, it was observed that there is a lack of documented DWM planning in practice. Within the investigated cases, it was evident that some industry practitioners are reluctant to maintain a formal document, as they believe that each event requires a different approach and that maintaining a single document may hinder flexibility during different events. This indicates that industry practitioners may not yet fully recognise the benefits of having a documented DWM plan, which can present measures in a concise yet adaptable way. Future efforts could focus on raising awareness among industry practitioners about the importance of having a documented DWM plan and on promoting good practices for its development and implementation.

This study contributes to knowledge by extending the current RDWM framework and offering rich insights from both local government and practitioner perspectives. While the previous literature has largely focused on the public sector, this paper offers a novel contribution through insights from both public and industry practitioner perspectives. The findings also highlight that social practices play a crucial role in effective disaster waste management, encompassing community engagement, collaboration, and the implementation of appropriate waste management strategies (to store, sort, and recover) during and after disasters. These practices help ensure efficient waste handling, recovery efforts, and reduce environmental and social impacts.

Although the RDWM framework was extended using a limited number of case studies in Australia, the richness of information in these cases should not be underestimated. Therefore, the extended RDWMF can serve as a valuable reference not only to educate industry practitioners but also to support the preparation of flexible, comprehensive, yet concise DWM plans. Future research could focus on further development and validation using a wider and more diverse range of cases, offering more generalisable insights intothe studied phenomena.

Author Contributions: Conceptualization, S.C. and C.J.; methodology, S.C. and C.J.; formal analysis, S.C. and S.C.; writing—original draft preparation, S.C. and C.J.; writing—review and editing, S.C. and C.J.; supervision, C.D.; project administration, S.C.; funding acquisition, S.C. All authors have read and agreed to the published version of the manuscript.

Funding: This research was funded by the University of the Sunshine Coast Spark Grant (Career interruption category, 2024).

Institutional Review Board Statement: The study was conducted in accordance with the Declaration of Helsinki, and approved by the Ethics Committee of the University of the Sunshine Coast (ethics approval number A242194 and date of approval 25 October 2024) for studies involving humans.

Informed Consent Statement: Informed consent was obtained from all subjects involved in the study.

Data Availability Statement: The data presented in this study are available on request from the corresponding author.

Conflicts of Interest: The authors declare no conflicts of interest.

Abbreviations

The following abbreviations are used in this manuscript:

BCP	Business Continuity Plan
CEO	Chief Executive Officer
DWM	Disaster Waste Management
MOU	Memorandum of Understanding
NSW	New South Wales
PPRR	Prevention, Preparedness, Response, Recovery
QLD	Queensland
RDWMF	Resilient Disaster Waste Management Framework
VC	Victoria
WA	Western Australia

References

1. Ahmed, I.; Johnson, T.; Fuller, S.; Guinto, M.B.; Sagapolutele, F. Strengthening capacity in disaster waste management in Vanua *Sustainability* **2022**, *14*, 13505. [CrossRef]
2. Ram, M.; Bracci, E. Waste Management, Waste Indicators and the Relationship with Sustainable Development Goals (SDG A Systematic Literature Review. *Sustainability* **2024**, *16*, 8486. [CrossRef]
3. Zhang, Y.-L.; Kim, Y.-C.; Cha, G.-W. Assessment of deep learning-based image analysis for disaster waste identification. *J. Cle Prod.* **2023**, *428*, 139351.
4. Kuhlicke, C. Resilience: A capacity and a myth: Findings from an in-depth case study in disaster management research. *N. Hazards* **2013**, *67*, 61–76.
5. Jalloul, H.; Choi, J.; Derrible, S.; Yesiller, N. Toward Sustainable Management of Disaster Debris: Three-Phase Post-Disaster Da Collection Planning. In *Construction Research Congress 2022*; ASCE: Reston, VA, USA, 2022.
6. Brown, C.O.; Hayes, J.L.; Milke, M.W. Planning to adapt: Identifying key decision drivers in disaster response planning. *Civ. Er Environ. Syst.* **2021**, *38*, 20–35.
7. The Department of Climate Change, Energy, the Environment and Water. In *National Waste Report 2022*; Blue Environment P Ltd.: Docklands, VIC, Australia, 2023.
8. Naderi, A.; Benis, K.Z.; Dowlati, M.; Seyedin, H.; Behnami, A.; Farzadkia, M. Identifying methods and challenges of was management in natural disasters. *J. Environ. Manag.* **2025**, *373*, 123514.
9. Gordon, K.; Dion, M. *Protection of 'Critical Infrastructure' and the Role of Investment Policies Relating to National Security*; Investme Division, Directorate for Financial and Enterprise Affairs, Organisation for Economic Cooperation and Development: Par France, 2008; p. 75116.
10. Tabata, T.; Zhang, O.; Yamanaka, Y.; Tsai, P. Estimating potential disaster waste generation for pre-disaster waste manageme *Clean Technol. Environ. Policy* **2016**, *18*, 1735–1744.
11. Asari, M.; Sakai, S.I.; Yoshioka, T.; Tojo, Y.; Tasaki, T.; Takigami, H.; Watanabe, K. Strategy for separation and treatment disaster waste: A manual for earthquake and tsunami disaster waste management in Japan. *J. Mater. Cycles Waste Manag.* **201** *15*, 290–299.
12. Crowley, J. Disaster debris management and the Maine floods of December 2023. *Int. J. Disaster Risk Reduct.* **2024**, *113*, 104911.
13. Solis, G.Y.; Hightower, H.C.; Sussex, J.; Kawaguchi, J. *Disaster Debris Management*; Disaster Preparedness Resources Cent University of British Columbia: Vancouver, BC, Canada, 1996.
14. EPA. *Planning for Natural Disaster Debris*; EPA: Washington, DC, USA, 2008.
15. Crowley, J. A measurement of the effectiveness and efficiency of pre-disaster debris management plans. *Waste Manag.* **201** *62*, 262–273.
16. Brown, C.; Milke, M.; Seville, E. Disaster waste management: A review article. *Waste Manag.* **2011**, *31*, 1085–1098.
17. Brown, C.O. Disaster Waste Management: A Systems Approach. Ph.D. Thesis, University of Canterbury, Christchurch, Ne Zeeland, 2012.
18. Wang, C.; Liu, J.; Lu, B.; Zhang, Y.; Ma, Z. Stiffness degradation and mechanical behavior of microfiber-modified high-toughne recycled aggregate concrete under constant load cycling. *Eng. Fract. Mech.* **2024**, *312*, 110608.
19. Ma, Z.; Wang, B.; Zhang, Z.; Zhang, Y.; Wang, C. New insights into the effects of silicate modulus, alkali content and modificatio on multi-properties of recycled brick powder-based geopolymer. *J. Build. Eng.* **2024**, *97*, 110989.
20. Green Industries South Australia. Disaster Waste Management Capability Plan. 2020. Available online: https:/ www.greenindustries.sa.gov.au/__media_downloads/190772/Disaster-Waste-Management-Capability-Plan.pdf (accessed c 20 May 2025).
21. Hunter Joint Organisation. *Regional Disaster Waste and Resource Recovery Management Plan*; Hunter Joint Organisation: Thornto Australia, 2024.
22. Queensland Government. *Managing Waste in Disaster Recovery Period*; Queensland Government: Brisbane, Australia, 2025.
23. Caldera, S.; Jayarathna, C.; Ferguson, S.; MCNeill, J.; Desha, C.l. A resilient disaster waste management framework fc communities: Experiences from Australian industry practitioners. *Clean. Waste Syst.* **2025**, *10*, 100197.
24. Pradhananga, P.; MElZomor; Kasabdji, G.S. Disaster waste management challenges in Nepal: Health impacts and the need fo safe practices. *Nat. Hazards Rev.* **2021**, *22*, 05021001.
25. Karunasena, G.; Amaratunga, D.; Haigh, R.; Lill, I. Post disaster waste management strategies in developing countries: Case c Sri Lanka. *Int. J. Strateg. Prop. Manag.* **2009**, *13*, 171–190.
26. Rogers, P. Development of resilient Australia: Enhancing the PPRR approach with anticipation, assessment and registration c risks. *Aust. J. Emerg. Manag.* **2011**, *26*, 54–58.
27. Holloway, I. *Basic Concepts for Qualitative Research*; Wiley: Hoboken, NJ, USA, 1997.
28. Klein, H.K.; Myers, M.D. A set of principles for conducting and evaluating interpretive field studies in information systems. *MI Q.* **1999**, *23*, 67–93.

Ponelis, S.R. Using interpretive qualitative case studies for exploratory research in doctoral studies: A case of information systems research in small and medium enterprises. *Int. J. Dr. Stud.* **2015**, *10*, 535.

Yin, R.K. *Case Study Research: Design and Methods*; SAGE: Thousand Oaks, CA, USA, 2009; Volume 5.

Patton, M.Q. *Qualitative Evaluation and Research Methods*; SAGE: Thousand Oaks, CA, USA, 1990.

Mertens, D.M. *Research and Evaluation in Education and Psychology: Integrating Diversity with Quantitative, Qualitative, and Mixed Methods*; Sage Publications: Thousand Oaks, CA, USA, 2019.

Braun, V.; Clarke, V. Using thematic analysis in psychology. *Qual. Res. Psychol.* **2006**, *3*, 77–101.

Eisenhardt, K.M. Building theories from case study research. *Acad. Manag. Rev.* **1989**, *14*, 532–550.

Lombard, M.; Snyder-Duch, J.; Bracken, C.C. Content analysis in mass communication: Assessment and reporting of intercoder reliability. *Hum. Commun. Res.* **2002**, *28*, 587–604.

Arifin, M.; Wunas, S.; Mushar, P.; Osman, W.W.; Lakatupa, G.; Mandasari, J. The Development of Mitigation-Based Infrastructure of Coastal Settlement and Fishery-Economic Resilience. *Civ. Eng. Archit.* **2022**, *10*, 692–702.

Adhikari, S.; Shrestha, D. Assessing Climate Disaster Resilience of Post-Earthquake Settlement: A Case Study of Bungamati, Lalitpur. In *Proceedings of 8th IOE Graduate Conference*; IOE: London, UK, 2020.

Pradhananga, P.; ElZomor, M. Revamping Sustainability Efforts Post-Disaster by Adopting Circular Economy Resilience Practices. *Sustainability* **2023**, *15*, 15870. [CrossRef]

Agustianingsih, D.P.; Shaw, R. Community disaster resilience using multi-hazard assessment during COVID-19: The case of Denpasar, Indonesia. *Nat. Hazards Res.* **2023**, *3*, 572–582.

Rahmafitria, F.; Sukmayadi, V.; Suryadi, K.; Rosyidie, A. Disaster management in Indonesian tourist destinations: How institutional roles and community resilience are mediated. *Worldw. Hosp. Tour. Themes* **2021**, *13*, 324–339.

Makwana, N. Disaster and its impact on mental health: A narrative review. *J. Fam. Med. Prim. Care* **2019**, *8*, 3090–3095.

Williams, R.; Drury, J. Psychosocial resilience and its influence on managing mass emergencies and disasters. *Psychiatry* **2009**, *8*, 293–296.

Paton, D.; Smith, L.; Violanti, J. Disaster response: Risk, vulnerability and resilience. *Disaster Prev. Manag. Int. J.* **2000**, *9*, 173–180.

Rodriguez-Llanes, J.M.; Vos, F.; Guha-Sapir, D. Measuring psychological resilience to disasters: Are evidence-based indicators an achievable goal? *Environ. Health* **2013**, *12*, 115. [PubMed]

Queensland Reconstruction Authority. State of Queensland 2025, Resilient Homes Building Guidance. 2025. Available online: https://www.qra.qld.gov.au/resilient-homes (accessed on 20 May 2025).

Oloruntoba, R.; Sridharan, R.; Davison, G. A proposed framework of key activities and processes in the preparedness and recovery phases of disaster management. *Disasters* **2018**, *42*, 541–570.

Fatimah, Y.A.; Murniningsih, R.; Setiawan, A.; Aman, M. A Smart Sustainable approach for waste management in post-natural disaster phase. In *IOP Conference Series: Materials Science and Engineering*; IOP Publishing: Bristol, UK, 2019.

Shah, S.A.; Seker, D.Z.; Hameed, S.; Draheim, D. The rising role of big data analytics and IoT in disaster management: Recent advances, taxonomy and prospects. *IEEE Access* **2019**, *7*, 54595–54614.

Elkady, S.; Mehryar, S.; Hernantes, J.; Labaka, L. Prioritizing stakeholder interactions in disaster management: A TOPSIS-based decision support tool for enhancing community resilience. *Prog. Disaster Sci.* **2024**, *22*, 100320.

Tabata, T.; Onishi, A.; Saeki, T.; Tsai, P. Earthquake disaster waste management reviews: Prediction, treatment, recycling, and prevention. *Int. J. Disaster Risk Reduct.* **2019**, *36*, 101119.

Chang, Y.; Wilkinson, S.; Potangaroa, R.; Seville, E. Managing resources in disaster recovery projects. *Eng. Constr. Archit. Manag.* **2012**, *19*, 557–580.

DiMaggio, P.J.; Powell, W.W. The iron cage revisited: Institutional isomorphism and collective rationality in organizational fields. *Am. Sociol. Rev.* **1983**, *48*, 147–160.

Tajima, R.; Tsuji, T.; Suzuki, K.; Muhamad, A.F. Conceptualizing disaster waste governance using network governance perspectives. *Int. J. Disaster Risk Reduct.* **2023**, *99*, 104137.

Jayarathna, B.; Wickramasinghe, C. Determinants of green supply chain practices of manufacturing companies in Sri Lanka. *Int. J. Product. Qual. Manag.* **2019**, *28*, 103–127.

sustainability

MDPI

Review

Zero-Waste Program Success: A Systems Approach to Indicator at the Micro, Meso, and Macro Levels

Ashpreet Kaur, Krista L. Thyberg and David J. Tonjes *

Waste Data and Analysis Center, Department of Technology and Society, Stony Brook University, Stony Brook, NY 11794-4404, USA; ashpreet.kaur@stonybrook.edu (A.K.); krista.thyberg@stonybrook.edu (K.L.T.)
* Correspondence: david.tonjes@stonybrook.edu

Abstract: Existing sustainability tracking frameworks primarily address water, energy, and material durability, with limited focus on factors driving zero-waste (ZW) initiatives. The current sustainability tracking tools especially lack comprehensive assessment mechanisms to evaluate organizational progress toward ZW, especially in higher education institutions (HEIs). This study reviews the global literature (2000–2023) on ZW indicators in organizations, focusing on HEIs, using databases such as Scopus and Web of Science, with an emphasis on waste reduction and sustainability practices. Institutional theory is used to explain the adoption of ZW programs in HEIs through normative, mimetic, and coercive isomorphisms. This study finds that ZW initiatives in HEIs are driven by shared professional norms (normative isomorphism), imitation of best practices (mimetic isomorphism), and compliance with regulatory policies (coercive isomorphism). Education, full-time staff, sustainability offices, and student advocacy play key roles, while collaboration within sustainability networks further enhances these efforts. This study emphasizes the complex interplay of these isomorphic forces across different levels of influence.

Keywords: zero waste; circular economy; indicators; systems theory; institutional theory; higher educational institutions

1. Introduction

Population growth, industrialization, urbanization, economic growth, and overconsumption are causing increased waste generation globally, particularly in the United States (US), where per-capita material consumption and waste production are the highest [1]. Due to the increasing waste, sustainable consumption and waste production have gained significant attention, which can be seen by the addition of Sustainable Development Goal (SDG) 12—"Responsible Consumption and Production" to the list of 17 SDGs in 2015, with a goal of achieving it by 2030 [2,3].

Some of the aims of this goal focus on implementing a 10-year sustainable policy framework and reporting systems to track progress toward responsible consumption and production, halve the global per capita food waste by 2030, reduce waste generation through prevention, reduction, recycling, and reuse by 2030, and ensure companies adopt sustainable practices and reporting frameworks [4]. Following these aims, communities, organizations, and institutions, including higher education institutions (HEIs), are implementing various zero waste (ZW) programs and frameworks as part of their sustainability goals to reduce their environmental impact and waste generation while increasing resource efficiency. ZW plans are strategic plans aimed at eliminating waste through resource conservation, waste reduction, and the promotion of recycling and reuse. Typically, ZW plans are structured to reduce waste sent to landfills and incinerators by diverting materials back into the economy through recycling, composting, and product redesign, with a focus on investing in upstream waste diversion efforts such as redesigning products to minimize waste, reducing consumption needs, and maximizing the reuse of existing materials [5]. They also focus on collaboration with various stakeholders, such as local governments, businesses, and communities, to foster sustainable waste management practices [6].

The literature on the factors promoting the circular economy (CE) usually divides these factors into micro, meso, and macro levels when studying organizations, mostly businesses, and industries [7–11]. However, there is limited literature that helps understand the organization and applicability of these frameworks and their reporting systems in promoting ZW in HEIs. The present sustainability frameworks predominantly focus on issues related to water, energy, and material durability, with less attention dedicated to identifying the factors that facilitate the achievement of ZW. Some studies have explored the economic dimension of waste, recognizing that waste incurs direct costs from production to disposal and indirect costs resulting from environmental damage and impacts on human health [12]. Others have linked waste to societal challenges, particularly the correlation between population growth and increased waste generation [13]. However, existing frameworks generally do not integrate the complex interactions of social, economic, political, and other underlying factors that influence ZW initiatives. Consequently, they do not offer a comprehensive understanding of activities that can effectively reduce waste at its source, including waste prevention through educational initiatives or hiring more employees. Therefore, while existing sustainability tracking systems like the Global Reporting Initiative (GRI) and Association for the Advancement of Sustainability in Higher Education (AASHE) contribute to monitoring the overall sustainability performance of HEIs, there is a need to develop assessment tools specifically tailored to gauge the progress in achieving ZW.

Additionally, it should be of interest to study the progress and motivations of HEIs because these institutions often mirror the governance and municipal structures of cities [14]. However, while HEIs are similar to cities, they also hold a unique status in society as education hubs. Through partnerships with local governments, businesses, and community stakeholders, HEIs conduct applied research that demonstrates sustainable practices in real-world contexts, further advancing their educational missions. For instance, initiatives such as implementing waste diversion programs, piloting renewable energy projects, or developing community-centric sustainability strategies highlight the role of HEIs as living laboratories for innovation and leadership in sustainability. Achieving any goal and making progress toward it in such institutions depend on the commitment of the entire organization, including its individual members. This collective sense of belonging, or the "social identity" shared by the organization and its members, differentiates the governance structure of HEIs from that of cities [15]. It also means that progress toward sustainability is not only dependent on the collectiveness or differences among the social groups inside an HEI but also on such differences from the physical and social structures in which the HEI is embedded. Similarly, HEI's sustainability policies will also be impacted by political and economic factors, inside and outside the institution. Therefore, studying these institutions is important due to their complex nature and their significant potential to influence society on a larger scale.

This paper will provide a categorization of zero-waste programs at colleges and universities using the concept of systems hierarchy. Such a categorization will help the conceptualization of a framework for ZW systems in organizations, as well as provide an opportunity to classify new indicators.

2. Conceptual and Theoretical Background—Systems and Institutional Theory

Systems theory was used to organize the factors being studied in a multi-level framework at the micro, meso, and macro levels. These levels draw from previous studies on CE indicators [10–12,16–18], which help contextualize how ZW practices are adopted within organizations, including HEIs.

At the micro level, indicators focus on internal processes within HEIs, such as operational efficiency, waste diversion efforts, and goal-setting for sustainability. This is influenced by CE studies where micro-level indicators assess the efficiency of products and processes within organizations [15,16]. In HEIs, these factors include initiatives like setting ZW goals or implementing waste reduction programs, which are under the direct control of the institution.

Meso-level indicators of the CE focus on industrial symbiosis, where organizations form networks to share resources, infrastructure, and by-products [9]. In organizational studies, the meso level refers to interactions between organizations, shaped by shared understandings and rules within a domain [19]. These interactions can involve the exchange of goods, services, and knowledge about norms and standards [20]. When applied to HEIs, the meso level involves collaboration between HEIs or between an HEI and other entities, like city governments or reuse centers, to achieve common goals. For instance, the Ohio State University collaborated with Rutgers University and the city of Columbus to launch a pilot food waste reduction campaign in the city of Columbus [21]. These universities are helping create educational material on food waste reduction to raise awareness and understand the food waste in the community, helping the city refine its food waste reduction strategies. The campaign is assessing existing efforts, identifying best practices, and creating a scalable model for other municipalities to follow. Similarly, Iowa State University's Polymer and Food Protection Consortium collaborated with industry partners to develop sustainable solutions for single-use plastics and agricultural by-products. Key projects included bio-based foam with HNI Corporation, plant-based adhesives with the American Packaging Corporation, and repurposing single-use plastics and fly ash with the city of Ames [22].

At the macro level, larger-scale indicators are examined, focusing on policies and practices at the regional, national, or international level. Macro-level factors are less within the control of individual HEIs but are influenced by broader regulatory environments and public policy on sustainability [18]. For HEIs, this might include compliance with state or national sustainability mandates or participating in international sustainability networks.

In this study, the HEI is the unit of analysis (Figure 1). The micro-level factors are the factors that occur inside the HEI and are under great direct control by the HEI, such as setting a ZW goal. Meso-level factors focus on collaborative efforts between two HEIs (e.g., HEI A and HEI B) or between an HEI and a local entity. These collaborative endeavors involve direct control by the HEI, albeit to a lesser extent compared to micro-level factors, as decisions made by other involved organizations can also impact the outcome. Lastly, at the macro level, larger-scale factors come into play at the municipal, state, or national scale, which are elements that are beyond the control of an HEI.

While systems theory provides a useful framework for categorizing these multi-level factors, it does not directly explain the organizational motivations for adopting ZW programs. To address that, this study draws on institutional theory (IT), which provides insights into why organizations, such as HEIs, adopt certain practices, including sustainability initiatives. IT posits that organizations face pressures from both internal and external sources that drive them to adopt similar practices to survive and gain legitimacy.

This process, known as isomorphism, can occur in three ways: coercive, mimetic, and normative [23].

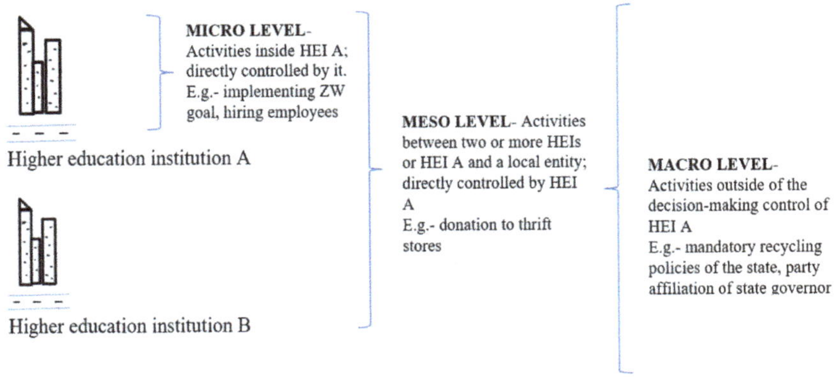

Figure 1. Micro, meso, and macro levels of a higher education institutional system.

Coercive isomorphism arises from formal pressures, such as regulations or governmental policies. HEIs might be compelled to adopt ZW programs to comply with state or federal environmental laws or to secure funding linked to sustainability performance [24]. Coercive isomorphism ensures that organizations meet external expectations, particularly from authorities, to avoid penalties or gain access to resources.

Mimetic isomorphism occurs when organizations imitate the successful practices of others, especially in situations of uncertainty. HEIs may adopt ZW programs by emulating other universities that have successfully implemented such initiatives. This imitation is often driven by the desire to improve performance and achieve similar success without necessarily testing the methods first [23]. For instance, HEIs participating in sustainability networks, such as the AASHE, may replicate successful practices from peer institutions [25,26].

Normative isomorphism stems from professional standards and expectations. HEIs often adopt practices like ZW initiatives due to pressures within the academic community or sustainability profession. For example, sustainability managers or faculty members, influenced by their professional networks or training, may advocate for the adoption of ZW policies within their institution as part of a broader commitment to sustainability goals [23]. In this context, HEIs may feel compelled to implement climate action plans or ZW programs to align with the norms of peer institutions and maintain their legitimacy within the academic and sustainability communities. These organizations do not have to "legally" adopt a climate action plan, but they do so because other HEIs are taking that step to maintain their social recognition and "legitimacy". According to IT, legitimacy refers to the extent to which an organization's actions, practices, or structures are seen as acceptable, appropriate, and valid by external stakeholders, including the broader society, customers, investors, regulators, and other organizations [23]. In that regard, when adopting climate action plans, members of HEIs feel their institutions are coming to be socially accepted and recognized by other HEIs since they are taking similar actions to those expected of a typical HEI in the US. Figure 2 provides a graphical representation of these three isomorphisms using the example of HEIs.

Isomorphisms thus account for why organizations might adopt ZW or other programmatic goals. Whether by coercion, mimesis, and/or normative reasons, organizations may choose to pursue ZW or other goals. Systems theory helps to structure the factors influencing ZW adoption in HEIs across multiple levels, from internal processes to external policies. Institutional theory, in turn, offers insights into why HEIs might adopt ZW programs, highlighting the role of coercive, mimetic, and normative pressures. Together, these frameworks provide a comprehensive understanding of how ZW initiatives in HEIs are influenced by both internal decision-making and external pressures.

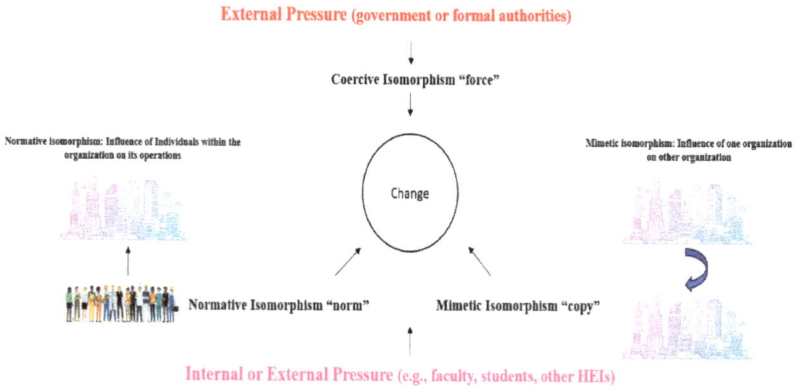

Figure 2. Institutional isomorphisms in a higher education institutional system.

3. Research Methodology

This study undertakes a comprehensive review of the existing literature to classify and analyze the current indicators used in assessing ZW initiatives in organizations, with a particular focus on HEIs.

The literature search was conducted using multiple academic databases, including Google Scholar, ProQuest, and Web of Science, from 2000 to 2023, to ensure a broad coverage of relevant studies. The search terms used on Google Scholar were ("zero waste implementation" OR "zero waste strategies" OR "waste reduction initiatives") AND ("higher education institutions" OR "colleges" OR "universities") AND ("landfill diversion" OR "composting" OR "source separation" OR "behavioral interventions" OR "recycling" OR "circular economy"). The key words provided a list of 455 studies, out of which the highest number of studies (73) was conducted in year 2023, as shown in Figure 3. Web of Science listed only 2 articles, and Proquest generated a list of 21 full-text and peer-reviewed articles with the same keywords, all of which were available on Google Scholar.

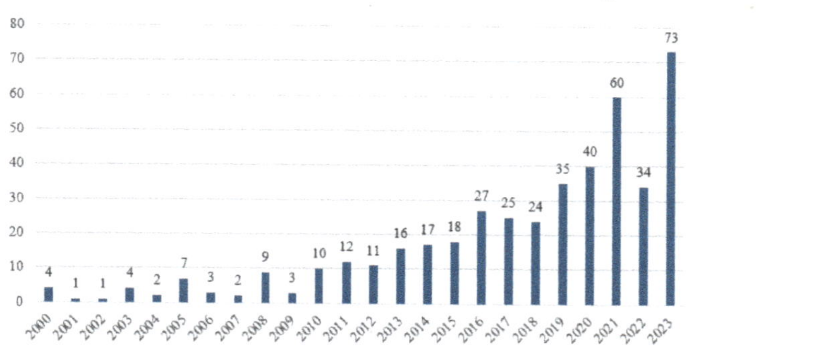

Figure 3. Trend in count of papers from Google Scholar search over 2000–2023.

Studies that focused on the general waste management process of an HEI, waste management plans, gray literature, web articles, and those irrelevant to factors or indicators of waste reduction, diversion, and minimization in cities, businesses, HEIs, and other organizational settings, as well as those published in languages other than English, were excluded from this study. The remaining studies were then reviewed to extract key themes related to waste reduction, diversion, and minimization in HEIs, municipalities, or businesses, ZW, the circular economy or sustainability indicators for HEIs, municipalities, or businesses, factors increasing success of program implementation at the micro, meso, or macro level, and factors affecting program implementation in relation to institutional theory.

The review covers studies conducted across all continents, although the majority of the literature is concentrated in the US, Canada, UK, Australia, and European nations, as can be seen in Table 1. It is important to note that the table only includes countries that were identifiable through VOSviewer (https://www.vosviewer.com); countries that could not be extracted are not represented. Nonetheless, the overall trend indicates that the United States accounts for the highest number of studies. This geographic distribution reflects the greater prevalence of sustainability policies and zero-waste initiatives in these regions.

Table 1. Count of papers per country.

US	106
Canada	30
UK	12
China	11
Australia	7
India	7
Turkey	6
Brazil	6
Ghana	5
Malaysia	5
Thailand	5
Saudi Arabia	5
Poland	4
New Zealand	4
Mexico	4
Sri Lanka	4
Philippines	3
Portugal	3
Sweden	3
Indonesia	2
Hungary	2
Italy	2
France	2
Kenya	2
Bangladesh	2
Germany	2
Tajikistan	2
Japan	2
Iran	2
Qatar	2

Table 1. *Cont.*

Colombia	2
Spain	2
Costa Rica	1
Finland	1
UAE	1
Nigeria	1
Lithuania	1
Norway	1
Singapore	1
Belgium	1
Cambodia	1
Estonia	1
Kuwait	1
Iraq	1
Nepal	1

Table 2 below summarizes the major studies included in this study on the basis of the major concept and whether occurring in micro-, meso-, or macro-level settings.

Table 2. Summary of studies included in this paper.

Major Concepts	Micro	Meso	Macro
ZW/sustainability plans	Tchobanoglous et al., 2022 [27]; Sherburne, 2019 [28]; Lidstone et al., 2015 [29]; Alexandra Wright, 2006 [30], McNamara, 2010 [31]; Locke and Latham, 1990 [32]; Lingard et al., 2001 [33]; Zhang et al., 2011 [34];		
Procurement policies	Ayarkwa,2011 [35]; UCLA, 2021 [36]; SB 54, 2022 [37]; ESF guidelines [38], Lisle, 2020 [39]		
Education and training	Zachariou, 2008 [40]; Ebrahimi and North, 2017 [41]; Brody and Ryu, 2006 [42]; Angelaki et al., 2024 [43]; Moqbel et al., 2020 [44]; Kaplowitz et al., 2009 [45]; Leal Filho, 2024 [46].		
Communication	Dahle and Neumayer, 2001 [47]; Mason, 2003 [48]; Wu, 2018 [49]		
Funding/Incentives	Dahle, 2001 [47]; Pitcher, 2010 [50]; Cheung, 2015 [51]		Allen, 2012 [52]
Presence of sustainability office/committee	Ebrahimi and North, 2017 [41]; Levy and Marans, 2012 [53]; Filho, 2015 [54]; Velazquez, 2006 [55]; Shi and Lai, 2013 [56]; Filho et al., 2019 [57]		

Table 2. *Cont.*

Major Concepts	Micro	Meso	Macro
Community engagement and collaboration		Kezar and Bernstein-Sierra, 2019 [58]	
Sustainability advocacy		Filho et al., 2019 [57]; Peer and Stoeglehner, 2013 [59]; Moore, 2005 [60]; Alshuwaikhat and Abubakar, 2008 [61]; Mosier and Ruxton, 2018 [62]; STARS (2019); Gardner, 2021 [63]	
Regulatory and legal frameworks			Hong (1999) [64]; Allen, 2017 [65]; Liyanage, 2019 [66]; Zaman and Lehmann, S., 2011 [67]; CalRecycle, 2023 [68]; USC, 2023 [69]; Latif, 2020 [70]; Schmidt and Rocconi, 2021 [71]; Henkels, 2022 [72]; NCSL 2020 [73]; Wilson, 2012 [74]; EPA, 2021 [75]
Infrastructure	Fuentes, 2022 [76]		Hao et al., 2020 [77]
Technological advancement			Murphy and Pincetl, 2013 [78]; Rubicon, 2023 [79]; Shreyas Madhav, 2022 [80]

4. Results

The existing literature underscores the significance of several factors across different levels of analysis, as delineated by systems theory. This review organizes those factors into micro, meso, and macro levels and institutional characteristics. At the micro level, a broad spectrum of the literature underscores the critical role played by factors such as education, funding, well-defined ZW plans, procurement policies, the employment of dedicated personnel, the establishment of sustainability offices, and the effective communication of initiatives. At the meso level, an emphasis is placed on active participation in sustainability networks and advocacy groups. At the macro level, the prevailing themes center on policy considerations at various governmental levels as pivotal elements in the progression toward ZW goals. Lastly, the institutional characteristics, such as being public, private, doctoral, or an associate (two-year) HEI, are kept in a separate category as they do not clearly fall under this study's definition of micro, meso, or macro level.

4.1. Micro-Level Factors

Micro-level factors encompass internal HEI dynamics such as ZW/waste management/sustainability plans, funding, full-time employees, and infrastructure—all critical for successful waste management [27–29]. An environmental plan or sustainability plan is deemed important for an HEI to establish tangible actions and timelines to move toward sustainability, ensure that all the activities conducted within the HEI align with the guidelines outlined in the sustainability plan, facilitate effective communication of key goals, and inspire individuals to commit to sustainability [30,81]. ZW goals are increasingly becoming a key component of HEIs' sustainability and environmental plans. However, academic research on these specific goals remains limited, with most studies focusing on broader sustainability or environmental initiatives, but similar conclusions for the presence of ZW plans in HEIs can be expected. According to AASHE, the tangible, measurable goals and objectives in a sustainability plan can allow HEIs to track their progress, document their successes, and manage the resources required to attain sustainability [82]. McNamara, in their study [31], found that HEIs achieve greater sustainability progress when they have a written sustainability plan that outlines specific goals, assigns clear individual responsibilities, includes a timeline, and establishes a process for measuring progress. The importance of having a ZW plan with clear, measurable, and aspirational objectives aligns with goal-setting theory, which suggests that human behavior is driven by purposeful goals and that setting specific, challenging goals improves performance [32]. For instance, when stakeholders set performance goals to reduce timber waste, they successfully reduced construction waste sent to landfills, whereas waste increased when no goals were set [33].

Furthermore, sustainability plans including sustainable purchasing policies promote the procurement of environmentally friendly products [28,34,39], enhancing waste management strategies, and can help an HEI achieve ZW goals. In 2005, the University of Southampton in the United Kingdom (UK) implemented an electronic material ordering system to centralize procurement and reduce upstream waste by prioritizing products with recycled content or which can be recycled or composted [34]. While this study highlights reduced operational and administrative costs, it does not quantify the waste reduction achieved. A qualitative study on barriers to implementing procurement programs in public universities found that decentralized procurement, where individual departments make their own purchases, complicates efforts to standardize purchasing policies [35]. Centralized procurement is thus crucial for effectively managing waste and streamlining sustainable purchasing practices. In some HEIs, sustainable purchasing is shaped by state environmental policies and ethics. For example, the University of California, Los Angeles's sustainability plan aims to phase out single-use plastics in line with California's ZW goals, supported by Senate Bill 54, which mandates that all packaging be reusable or compostable by 2032 [36,37]. Similarly, New York State's Executive Order 4 requires state agencies, including HEIs, to adhere to green procurement standards, prompting the State University of New York College of Environmental Science and Forestry to implement guidelines prohibiting the purchase of items like plastic containers, straws, and water bottles [38].

Educational programs and personnel training within HEIs play pivotal roles in instilling environmental consciousness that can drive ZW initiatives. A fundamental shift in mindset—achieved through education and awareness—can help prevent and reduce waste, leading to broader personal, cultural, and societal behavioral transformations [40,83]. Studies suggest that adding environmental education to the curriculum can help provide students with intellectual tools and techniques to address environmental problems [41,46] while instilling environmental consciousness. A study at Texas A&M showed that adding a course on sustainable development to the graduate curriculum brought positive changes in student behaviors, reducing their ecological footprint from an average of 19.5 acres (7.9 ha) (measured by the amount of land needed by these students to sustain their current lifestyle on Earth) to 16.8 acres (6.8 ha), a 14% reduction, after taking the course [42]. A study conducted with 52 students from a green university observed a significant increase in sustainability knowledge following an educational intervention. The participants' mean scores improved notably from the pre-test (mean = 3.40) to the post-test (mean = 4.08), indicating a substantial enhancement in their understanding of sustainability-related concepts. The effect size, calculated as Cohen's d = 1, demonstrates a large impact of the intervention, highlighting its effectiveness in improving sustainability knowledge among the students [43]. Training people, especially staff, about various aspects of ZW programs, such as recycling, composting, and even sustainable purchasing, is important to raise their awareness of correct waste management practices, as well as equip them with tools to manage waste correctly, and can lead to informed decisions made by these individuals to promote sustainability in HEIs [44,53]. A study conducted by Kaplowitz et al. [45]

found through a survey that staff members reported having the least knowledge about recycling (type of products that can be recycled, environmental benefits of recycling, different places on campus where products can be recycled, and types of waste reduction strategies on campus). Leal Filho et al. [46], in their review of food waste reduction in universities, emphasize the importance of training staff in food service areas to reduce food waste. These trainings can focus on key aspects such as sourcing fresh local food, adopting food preparation methods that minimize waste, and ensuring proper storage and preservation to reduce spoilage. These findings underscore the role of education in shaping waste-conscious behaviors, reinforcing that sustainability literacy is a crucial driver of ZW efforts within HEIs.

Beringer and Adomßent [84] explain that the research and development projects related to sustainability in HEIs are usually funded externally and aimed at institutional improvement for sustainability and advancing science. An example highlighting the importance of research as well as collaboration between state agencies and HEIs to reduce waste can be found in the recent selection of Iowa State University of Science and Technology by the Department of Energy to develop a closed-loop upcycling of single-use plastic films to convert them into biodegradable polymers [85]. The influence of research activity on sustainability was explored by Stafford [86], finding that more research activity by an HEI has a significant positive effect on the sustainability grade received by the HEI from the Sustainability Endowments Institute (SEI). From the perspective of IT, the SEI can be seen as a coercive force, applying external pressure on HEIs to enhance their sustainability performance. By assigning sustainability grades, the SEI establishes benchmarks and expectations that institutions must meet to maintain their reputation and competitive standing, thereby motivating HEIs to increase research efforts and adopt sustainable practices. In development projects, infrastructural changes can enhance zero-waste programs. For example, replacing multiple waste bins throughout a building with centralized waste and recycling collection units in lobbies and breakrooms [76] streamlines waste collection, making it more efficient and centralized.

Building on the importance of education and training, effective communication of waste reduction initiatives within an organization is equally vital. Semi-structured interviews with university stakeholders in the UK revealed that a lack of motivation among both the administration and students to reduce waste stems from ineffective communication regarding recycling processes and the importance of waste reduction [47]. A case study on the implementation of a ZW program at Massey University in New Zealand highlighted the significance of clear signage for recycling bins in promoting source separation behavior, which resulted in increased recycling and waste diversion, although the study did not quantify the diversion amount [48]. Additionally, an experimental study involving 20 university students in Canada found that using colored images improved the accuracy of identifying recyclables, garbage, and compostables by 6%, achieving a correct identification rate of 76%, compared to 70% with text-based messages [49].

Insufficient funding for ZW programs in HEIs can hinder program progress, often resulting in undercompensated employees and inadequate infrastructure to support effective program implementation. A semi-structured interview of 16 stakeholders (waste managers, student leaders, energy managers, and health and safety officers) from six HEIs in London revealed that, according to the participants, the biggest barrier to implementing energy-saving and waste-reduction measures in the institutions is the lack of financial resources [47]. Based on an environmental performance review of the University of Leeds, hiring more employees can result in better waste management programs [50]. The study recommends appointing an environmental officer responsible for implementing environmental activities on the campus [50]. A cost–benefit analysis of implementing an on-campus composting facility at a university in California found that it can divert 249 tons (226 tons) of food waste from a local commercial composting center and save USD 16,185 if food waste is managed at a dedicated facility, and a profit can start to be made in the ninth year [51].

Additionally, when students actively advocate for waste reduction, diversion, and other environmental initiatives at the grassroots level, they wield considerable influence in garnering support from university administrations (utilizing a bottom-up approach). For instance, Massey University's ZW program originated in response to student concerns about the inadequacy of recycling facilities on campus [48]. Phrophayak et al. [87] attribute Mahasarakham University's high sustainability scores to initiatives such as the creation of student clubs and associations. These groups play a key role in promoting sustainable activities on campus, including organizing recycling contests and other related projects.

The presence of a sustainability office and a dedicated committee plays a critical role in advancing sustainable practices on campus, and ZW. Filho [54] emphasizes the benefits of such offices in HEIs, noting their role in providing full-time staff, securing funding for sustainability initiatives, involving students and employees in decision-making, raising awareness, and managing sustainability data. Similarly, Levy and Marans [53] highlight that the presence of a sustainability office brings in the possibility of stakeholder collaboration by implementing environmental training in different departments, promoting sustainable purchasing, and organizing competitions among student residents, thereby enhancing student engagement and sustainability practices. Studies have also recommended the establishment of a sustainability committee, which can facilitate the task of creating and establishing campus-wide ZW and sustainability policies and targets, helping responsible authorities to implement such policies, and coordinating communication and outreach between various actors about progress on sustainability efforts [55–57]. Ebrahimi and North [41] further demonstrate the tangible benefits of a dedicated sustainability coordinator and committee, as seen in Western Kentucky University's decision to hire a full-time coordinator and establish a sustainability committee. This led to immediate actions, including the allocation of USD 30,000 for recycling infrastructure allowing the continued provision of necessary infrastructural facilities for waste management. These studies collectively illustrate that by providing structure, leadership, and resources, sustainability offices and committees may serve as key drivers in advancing ZW strategies within HEIs. Without such structures, the integration of sustainable practices risks becoming fragmented and unsystematic, limiting the potential for a long-term impact.

4.2. Meso-Level Factors

Meso-level factors encompass the broader institutional and community partnerships, collaborative networks, and inter-organizational relationships that support HEIs in advancing ZW programs through shared resources, knowledge exchange, and collective action. Based on anecdotal sources, by being a part of sustainability organizations, such as AASHE, HEIs can gain cutting-edge knowledge from experts around the world on ways to reduce and prevent waste, promote sustainability on a college campus, and hence contribute to ZW [25,26]. HEIs encompass a community of scholars, students, and teachers while also being integral components of broader societal communities at the city, regional, and state levels. By sharing their expertise and resources with a wider community, HEIs can facilitate the achievement of greater sustainability outcomes by leveraging the technical knowledge and intellectual capital within HEIs. Institutional theories have suggested that an organization may also collaborate with other organizations to solve a complex or uncertain problem [88]. In the study conducted by Kezar and Bernstein-Sierra [58], an influential relationship was observed between the HEIs affiliated with the Association of American

Universities and the established norms of this association concerning the enhancement of undergraduate teaching within the Science, Technology, Engineering, and Mathematics (STEM) disciplines. This influence was discerned through interconnected mechanisms, such as the desire to gain social recognition, competitions with other HEIs to improve their teachings, public acknowledgment, and the consequential opportunities for collaborative engagement with other HEIs, funding bodies, the National Research Council, and various STEM teaching associations, to advance their STEM teaching.

Studies have also emphasized the idea of HEIs advocating for sustainability, environmental policies, and climate change in communities, either by volunteering, creating community-based courses, or conducting participatory research focusing on and including local or regional communities [57,59]. HEIs can advance and advocate sustainability through the endorsement of legislation and public policies, active participation in sustainability campaigns, and involvement in decision-making with legislators [89]. According to AASHE, advocacy for sustainability includes efforts such as supporting or endorsing legislation, ordinances, and public policies that advance sustainability, participating in campaigns aiming to change public policy, and engaging in discussion with policymakers [22]. Advocating for sustainability by HEIs is considered important for several reasons. First, HEIs teach about sustainability in their courses and also seek to implement what they are teaching [60]. Second, advocating for sustainability at the local, state, or national level can be beneficial to HEIs via partnerships at local, state, and national levels. These could include partnerships with the governments providing research and development funding to HEIs to research in the field of sustainability and increase the visibility and applicability of their scientific research [90]. Additionally, NGOs can become involved with HEIs to help organize workshops and conferences on sustainability, or local communities can provide space or even research opportunities for students in HEIs to follow their academic goals [61,62]. Third, HEIs often embrace public service, and the idea of advocating for sustainability stems from the overarching objective of serving the public or the community in fostering greater sustainability [60]. By assuming this proactive role, HEIs, especially land grant HEIs that are historically meant to serve the community [62], fulfill their broader societal responsibility. The engagement of HEIs in communities beyond their academic and research roles can help bring a transformation in society and move it toward sustainability development [63].

4.3. Macro-Level Factors

Based on the literature review discussed below, macro-level factors that externally influence ZW programs in HEIs can be broadly divided into the following categories: geographic or regional effects, legal and regulatory factors, political factors affecting support of s specific policy, economic incentives, and program funding from the state or city, and technological advancement at the city, state or national level.

Various cities in the US have started adopting and integrating ZW strategies in existing waste management practices to reach sustainable development targets and promote a circular economy. In particular, ZW programs in the western US have advanced, which suggests that the region may constitute another macro-level variable that influences HEI ZW programs. Del Norte County of California prepared the first extensive ZW plan in the US in 2000 [91,92]. In 2002, San Francisco adopted a 75% waste diversion goal from landfills by 2010 and laid out a long-term ZW plan [92]. The city of Oakland passed a resolution in 2006 to adopted a strategic ZW plan [93]. Cities in other regions of the US have also adopted ZW initiatives. Austin created a ZW strategic plan in 2008 to reach the ambitious target of diverting 90% of waste from incinerators and landfills by 2040 [94]. Dallas released its ZW plan in 2013 to achieve 90% waste diversion by 2040 [95].

A previous study conducted a zonal comparison of hotels in the US on their behavior of adopting green practices using a survey (which included questions about their recycling, reuse, and waste reduction practices along with other questions on energy efficiency, education, and water conservation) [96]. They found that hotels in the Midwest adopted more green practices, followed by the West, South, and Northeast. In this study, because the hotels were divided based on zones, the zonal distribution was considered under macro-level factors as the effect of zones on the adoption of green practices depended on what state they were in and the green policies of the state, which cannot be controlled by hotels.

Regulatory and legal frameworks at the city, state, or national level play a pivotal role in shaping the effectiveness of ZW programs by enforcing measures such as mandatory recycling, source separation, and take-back policies, all of which drive waste reduction and sustainable practices in HEIs. A study performed in Asia showed that volume-based pricing or unit pricing of waste resulted in a sustained decrease in waste generation from 1.3 to 1.04 kg/capita/day in 1996 in Hong Kong and from 1.135 kg/cap/day in 1996 to 0.667 kg/cap/day in 2005 in Korea [64]. In Spain, a door-to-door waste collection system increased the source separation of waste and decreased contamination, resulting in reduced waste generation by 27% [65]. Other practices of enforcing regulations are holding producers and manufacturers liable for collecting their products back at the end of their usage, commonly known as extended producer responsibility (EPR), thus encouraging companies to manufacture innovative packaging and undertake the design of products that can last longer and be reused to reduce waste [66,67]. California passed a mandatory law to recycle in 2012 (AB 341) and compost in 2014 (AB 1826) in public and commercial facilities, including schools and HEIs that produce four cubic yards or more of commercial solid waste per week [68]. These facilities were required to provide an organic and recycling waste-collecting container next to the trash can to capture and divert recycling and organic waste [68]. The mandatory law has compelled HEIs in California to think about efforts to capture and reduce organic waste and recyclables. For example, the University of California San Diego adopted a ZW plan in 2019. In light of the ZW plan, they implemented a paper towel composting pilot program in 2022, and their waste hauler implemented an anaerobic waste digestion facility in 2021 (to digest food waste and convert it into biogas). Similarly, the University of Southern California (USC) adopted a ZW plan in 2019 to achieve ZW by 2028, and some efforts to reach ZW are being shaped by the mandatory regulations of the state [69]. In 2022, USC implemented 98 multi-stream recycling bins, which collect landfill trash, organics, mixed recyclables, and liquids separately to divert waste recyclables and compost from landfills. They have been putting on various ZW events to collect recyclables and organics during football games, tailgates, and other events on campus, diverting 18,479 pounds of waste from landfills in 2019 [69].

The effect of external stakeholders like government authorities and NGOs has been studied in the context of the adoption of environmental regulations and standards by companies, demonstrating that pressure from regulatory bodies plays a significant role in driving environmental initiatives. A quantitative study conducted on factors affecting the adoption of environmental management accounting (EMA analysis, management, and reduction in environmental costs in a mutually beneficial way for the company and the environment) by companies in Pakistan found that pressure from regulatory and authoritative bodies has the largest influence on EMA implementation in a firm [70]. Schmidt and Rocconi [71] measured the relationship between renewable energy targets of the state (targeted amount of energy that should come from renewable sources) and sustainability ratings of HEIs using a hierarchical linear model. Their results show that having renewable energy targets in the state increased the sustainability scores of HEIs.

The above examples of HEIs and non-academic firms implementing ZW or sustainability programs to collect and divert recyclables and composting from landfills to conform to state laws or regulatory standards show how external pressures can shape the waste reduction and prevention efforts of an HEI.

Political factors like party affiliation, support for a policy or policy preference [97,98], expenditure on pro-environmental policies [99], and the governor's party affiliation [99] have also been used to predict public and political support for environmental programs and policies. The political affiliation of an individual can predict their attitudes toward environmental policies. The literature shows that Democrats are more concerned about environmental and social issues than Republicans, who are mostly concerned about the economy and market [100]. Additionally, according to several studies [100–102], Democrats tend to allocate more funds toward environmental causes and exhibit greater concern for the environment compared to Republicans. Howe et al. [103] conducted a study on climate change opinions across different regions in the US using multilevel modeling and found that Republicans are less likely to believe in the anthropogenic effects of climate change. The literature suggests that one reason for Republicans' reluctance to support pro-environmental policies is the perceived financial burden on capitalist firms, thus prioritizing economic gains over environmental concerns [97]. However, studies have also shown that the state's political affiliation does not impact environmental policies. Schmidt and Rocconi [71] conducted a study using a hierarchical linear model and found that the composition of state legislatures (Republican or Democrat) did not have a significant impact on the sustainability ratings received by HEIs.

Incentives can be a motivating factor in increasing public participation in ZW programs. For instance, the city of Hernani in Spain offered a 40% discount on garbage collection fees if the residents segregated waste or composted it at home, which resulted in increased public participation in waste segregation [52]. The bottle bill incentive implemented in some states of the US, such as Oregon, Michigan, and California, has improved the collection of recyclables and reduced litter [72]. The governments of these states add a small fee (USD 0.05–0.10) during the purchase of products in plastic and glass bottles, aluminum cans, or any metal containers, which is refunded to the customers if they bring the empty containers back [73]. In Oregon, this bill resulted in a reduction in roadside litter from 40% before the bill was passed in 1971 to 6% in 1979 [72]. Incentivizing the recycling stream in such a way can reduce the quantity of waste going to landfills, increase the collection of recyclables, decrease money and time spent on collection and transport, and increase the conservation of virgin material [74]. In some cities, governmental and semi-governmental authorities may also decide to fund waste reduction programs in HEIs. For instance, the U.S. Environmental Protection Agency supports sustainability and waste reduction programs in HEIs, offering grants such as the recent USD 59,995 awarded to the University of Illinois, Chicago [75]. These incentives and funding programs, implemented at the state level, can seep into the operational mechanisms of HEIs and influence the waste reduction activities on HEI campuses. For example, the University of Oregon has various bottle redemption centers within a mile area, and it can be assumed that the students are dropping off bottles at participating grocery stores and redemption centers; in this way, they are reducing the waste on campus by diverting it. It is unknown by what percentage this reduction is occurring or if it is occurring at all, but the university is facilitating such reductions by setting up bottle collection centers in each building and collecting these bottles to generate around USD 35,000 to 45,000 annually from selling the deposited bottles and cans (Addison, D., personal communication, 10 May 2023) [104]. Even though the effect of bottle bills on recycling by students is not known, it can still be asserted that a macro-level policy such as a bottle bill has some effect on the waste reduction efforts in

HEIs (micro level) as the authorities are able to collect the bottles and cans and generate income from them. The above example shows an interplay between regulatory policies at the state level, funding or incentives from the government, and waste reduction at an HEI, and that macro-level policies (state, national, international) can shape waste governance at the micro level (HEI).

Apart from incentives, investment in improved waste collection, disposal, recycling infrastructure, advanced technologies, and innovations [105–107] can increase a city's waste diversion from landfills [108]. Such advancements in cities also affect the waste management of HEIs in those cities. For example, a mandatory waste separation policy in a city in China was implemented, which also focused on improving the waste-sorting infrastructure by providing separate bins for trash and recyclables and color coding the bins throughout the city, including its HEIs [77]. A survey conducted to understand the engagement and participation of students in waste separation before and after the implementation of the mandatory waste separation policy in the city found that the separation rate by college students improved from 10% to 18% after the city implemented the mandatory waste separation policy and provided the bins [77].

Technological advancements include tracking the flow of materials to landfills or other waste-processing facilities (recycling, composting, waste-to-energy). This can help understand the waste type, amount, and composition, provide improved data, and continuously monitor waste, which can improve the accountability of the cities in implementing vigilant waste management practices and policy changes [78]. Recently, companies like Rubicon have been leading the advancement in waste management by using artificial intelligence (AI) such as machine learning models to prioritize the waste collection area and time, avoid roads with parades or construction works, or even identify materials in the waste stream [79]. There are 11 cities in the US that have adopted Rubicons' technologies to optimize their waste management. However, the effectiveness of the partnership is not yet well-documented [79]. As technology advances, such systems can also be implemented in HEIs to optimize their waste management as the waste haulers collect waste from the city and HEIs. A study in India used machine learning to simulate and enhance the segregation of electronic waste from trash by designing a mobile robot arm that identifies and segregates electronic waste and can be attached to a trash-collecting truck [80]. The arm exhibited 96% accuracy in identifying and segregating electronic waste. The study proposes that using a similar machine learning concept can improve the collection and segregation of hazardous electronic waste with minimum harm to humans. As can be seen from the above studies, the application and acceptance of AI and machine learning in the field of waste management are very recent and mostly directed toward cities; however, these practices, such as using AI to segregate waste and investing in waste management infrastructure, have the potential be applied to HEIs as well.

In this study, being a public or private HEI (institutional control) and being a doctoral, master, baccalaureate, or associate HEI (institutional type) is not divided into micro, meso, or macro factors because the governance of the HEI can be subject to the discretion of the board of trustees, state government, accreditation agencies, and establishing stakeholders. Usually, public HEIs are funded by the state [78] and have some state governmental control [109]. Private HEIs usually govern themselves and hence have more autonomy over their governance [109], but sometimes private HEIs also receive funding from the state or federal government. Therefore, the institution's governance can be characterized as discretionary, and the investment of funds can depend on the source of funding. Studies have explored the relationship between institutional type and control and sustainability efforts. Studies using Sustainability Tracking Assessment and Rating Systems' data have shown that doctoral-level [71,110] and public HEIs [69] have the highest overall sustainability scores. This may be because Carnegie doctoral research HEIs usually gain higher funding for research and foundation grants from private donors and public funding agencies as they are viewed as elite research institutions compared to master-, baccalaureate-, and associate-granting institutions [111]. Therefore, they may have higher scores on sustainability as there are greater chances for them to invest in sustainability efforts. A reason why public HEIs may achieve higher sustainability scores than private HEIs is that public HEIs are funded through the state by means of grants or sponsored research funds usually paid through state taxes, and hence, they are subject to public or state pressure to implement sustainable management practices [112]. On the other hand, private HEIs may have less external stakeholder pressure as they rely on grants, endowments, tuition, and philanthropic donations [112,113]. In summary, for a study of national or international waste diversion and minimization, where the unit of analysis is not HEIs but states or nations, the institutional type and control could be considered a micro-level category, which will mean that states or nations with a higher number of public HEIs have higher diversion compared to private HEIs based on the above discussion. But at the institutional level, such as in this study, these factors do not clearly fall under the micro-, meso-, or macro-level category and, therefore, are kept in the "other" factors category.

5. Discussion

There are some indicators found through the literature review above that can be clearly explained by IT. For instance, the effect of a full-time employee on ZW programs can be explained by the concept of normative isomorphism, wherein professionals who share a common educational background and hold comparable positions in diverse organizations tend to exhibit similar behaviors. This similarity in behavior often results in the adoption of practices that are considered standard and widely accepted as effective in their professional field. Therefore, due to normative isomorphism, having full-time employees can result in better ZW programs.

The impact of education and training can significantly increase the likelihood that staff and students will comply with acceptable waste reduction and prevention practices. This improvement can be attributed to normative isomorphism, where, after acquiring knowledge, staff members gain a better understanding of the importance of waste reduction and are motivated to adopt similar sustainable practices. Additionally, coercive pressures from the administration to implement training programs and educational resources, as well as from SEI, which encourage institutions to improve their performance for better benchmarking and reputation, also play a role. These pressures drive institutions to actively pursue and achieve their sustainability goals.

It also comes to light that HEIs conducting research activities are more likely to have better sustainability programs since they are more likely to secure external research funds [84,85]. Due to normative isomorphism, HEIs might want to become involved in sustainability-related research since conducting research is one of the various purposes of an HEI. Normative isomorphism drives HEIs to participate in sustainability-related research, as conducting research is a core mission of these institutions. This creates a mutualistic relationship where HEIs actively pursue sustainability research to attract funding, which in turn benefits the funders through impactful research outcomes. Hence, the pursuit of and investment in sustainability research by HEIs, encompassing ZW paradigms, can yield enhanced ZW programs.

The role of students and student organizations in advocating for zero-waste programs highlights the presence of normative isomorphism within HEIs. As HEIs educate their students about sustainability, these students become empowered advocates for the practical implementation of zero-waste principles. Consequently, HEIs feel a responsibility to respond to and support these advocacy efforts.

It can also be inferred from the results that a sustainability office serves as a central hub for all sustainability-related activities on campus, facilitating effective communication among stakeholders, supporting funding availability, and facilitating the employment of dedicated staff for sustainability projects. It creates an enabling environment for the implementation of sustainability initiatives, fostering widespread engagement and acceptance within the campus community. This stakeholder participation is driven by normative isomorphism, as individuals feel obligated to join sustainability committees due to their professional interests. Moreover, HEIs respond to stakeholder expectations through normative isomorphism by seeking support and validation while addressing their demands. After establishing a sustainability office and committees, these entities may also feel compelled to adopt widely accepted practices in their field, such as hiring sustainability officers and promoting funding for sustainability projects. This assumption is also supported by a qualitative study involving interviews with seven sustainability leaders from construction companies based in the UK, which revealed a significant driver for construction companies to embrace sustainability is their pursuit of winning more contracts, stemming from a combination of factors, including client demands for sustainable practices and social recognition [114]. The study highlights the effect of normative isomorphism, where construction firms, in their quest for social recognition and a competitive edge, gravitate toward adopting sustainable practices. Additionally, the presence of a sustainability office may be influenced by mimetic isomorphism, as HEIs might imitate successful practices from other institutions. Thus, the establishment and impact of a sustainability office and committee can be shaped by both normative and mimetic isomorphism.

HEIs engage in sustainability networks through the lenses of mimetic and normative isomorphism [57,58]. Mimetic isomorphism explains this involvement as HEIs seek to access shared resources to enhance their sustainability efforts by learning from the best practices of their peers. Normative isomorphism further influences this behavior, as HEIs feel it is the norm to engage in and advocate for sustainability through such platforms, aiming to be seen as legitimate among their peers. An important aspect to consider is that strong advocacy from HEIs may also lead to significant changes in local government policies, which in turn shape the sustainability landscape within these institutions. This influence can manifest as coercive isomorphism, compelling HEIs to comply with newly established sustainability standards and regulations. Thus, sustainability advocacy within HEIs is a complex process that involves elements of mimetic and normative isomorphism, potentially giving rise to coercive, normative, and mimetic isomorphism. This interplay ultimately drives systemic changes in the higher education sector.

At last, the regulatory policies pertaining to the types of waste reduction programs and amounts of investment in such programs at city, state, or national levels are informed by coercive isomorphism. For instance, if a city or state implements advanced technology and policies to enhance waste reduction, these developments will ultimately influence how HEIs manage their waste. HEIs may experience both coercive and normative isomorphism when government agencies or NGOs provide funding for waste management projects. To secure such funding, HEIs may be motivated to conduct research and advance ZW programs within their institutions. Additionally, engaging in these initiatives can enhance their public image and improve their rankings.

To summarize, the findings and discussion suggest that macro-level factors predominantly influence ZW programs through coercive isomorphism. This reflects macro-level forces such as market prices or state regulatory policies. In contrast, at the meso and micro levels, there is a complex interplay of normative and mimetic influences. In the context of research on collaboration occurring at the meso level, significant emphasis is placed on organizations collaborating to acquire distinctive resources or skills that are challenging for other entities to imitate. This underscores the impact of norms and mimesis in advancing sustainability within the broader community, stemming from collaborative efforts to attain social recognition and access to shared resources. Table 3 summarizes the explanatory factors for ZW programs, organized based on the level of scale (micro, meso, and macro) and type of isomorphism (coercive, mimetic, and normative), and aims to shows how specific factors at different scales reflect different and convoluted forms of isomorphism.

Table 3. Factors based on literature review divided by scale (micro, meso, and macro) and type of isomorphism.

Type of Isomorphism	Micro	Meso	Macro
Coercive	Sustainable purchasing policies in HEIs influenced by state laws; mandated education and training of staff and students.		Volume-based pricing; mandatory extended producer responsibility; door-to-door waste collection; expenditure on pro-environmental policies and the governor's party affiliation; bottle bill incentive; government investing in waste reduction programs at HEIs; implementing waste-sorting infrastructure and mandatory waste segregation; tracking material flow to landfills.
Normative	Promoting sustainable practices inside the organization due to social recognition; presence of sustainability officer, sustainability office, full-time employees in a sustainability department, and sustainability committee; presence of student organizations and their efforts; environmental education; sustainability training; research on sustainability-related projects.	Advocating for sustainability at city, state, national, or international level; participation in sustainability networks.	
Mimetic	Creation of sustainability office; hiring a sustainability officer (mimicking other successful programs).	Participation in sustainability networks.	

6. Recommendations and Conclusions

This study aimed to address the growing challenge of implementing ZW programs, particularly in the context of higher education institutions (HEIs), using the concept of systems hierarchy. Despite the global emphasis on sustainable development, particularly through initiatives like SDG 12, a significant gap exists in understanding how HEIs can effectively implement ZW frameworks and track their progress. So, this study sought to conceptualize a comprehensive framework that integrates various dimensions in which an HEI interacts and supports the classification of the ZW indicators accordingly.

There are large variables that interplay at the micro level in an HEI, but the effectiveness of ZW programs is higher when these institutions receive external, often authoritative command, such as mandatory laws to divert waste. The HEIs do not interact much at the meso level, and if they do, it is solely for advocating for their needs or fulfilling criteria to meet the requirements of reporting institutes for a better sustainability score and reputation. So, future studies should explore the interaction at the meso level in more depth. One thing is very clear: there is an interplay between normative, mimetic, and coercive isomorphism, which together shape the effectiveness of ZW programs. The findings emphasize the complex and multifaceted nature of policy adoption in an institution, underscoring the importance of both internal organizational factors and external regulatory influences on institutional behavior. The review of the literature and findings also suggest that future research should prioritize research and case studies on ZW programs and their progress in HEIs to enrich the existing body of literature and practical knowledge about successful ZW programs in HEIs. Since this study is a broad review aimed at advancing the field of ZW in HEIs, our findings provide a foundation for future research rather than definitive conclusions. While we sought to capture diverse perspectives, the applicability of our results may vary across different types of HEIs and geographic contexts. This highlights the need for more localized studies and case-specific analyses to further refine and strengthen ZW strategies in higher education.

Additionally, the review highlights several key recommendations for HEIs, including developing a ZW plan with clear targets and goals for waste reduction and diversion, investing in education and training programs focused on ZW practices, enhancing infrastructure to support ZW efforts, hiring dedicated staff to oversee and implement ZW initiatives, and engaging in professional sustainability and ZW networks to stay informed about the latest developments in the field. For policymakers, a key recommendation from the review is to implement and support coercive measures, such as mandatory recycling and composting laws, as well as other waste reduction initiatives like bottle bills and extended producer responsibility laws. For other stakeholders, such as cities and communities, it is crucial to invest in waste sorting and recycling infrastructure to efficiently recover recyclables, track material flow to landfills to analyze trends, and enact policies accordingly.

Author Contributions: Conceptualization, A.K.; methodology, A.K.; validation, A.K. and D.J.T.; formal analysis, A.K.; resources, A.K.; data curation, A.K.; writing—original draft preparation, A.K.; writing—review and editing, D.J.T. and K.L.T.; visualization, A.K.; supervision, D.J.T. All authors have read and agreed to the published version of the manuscript.

Funding: This research received no external funding.

Institutional Review Board Statement: Not applicable.

Informed Consent Statement: Not applicable.

Data Availability Statement: No new data were created or analyzed in this study. Data sharing is not applicable to this article.

Acknowledgments: The authors are thankful to the faculty of the Department of Sociology, Criminology and Law of the University of Florida. Comments and suggestions from four peer reviewers made this paper better and are gratefully acknowledged. This is Waste Data and Analysis Center contribution #67.

Conflicts of Interest: The authors declare no conflicts of interest.

Abbreviations

AASHE	Association for the Advancement of Sustainability in Higher Education
AI	Artificial Intelligence
CE	Circular Economy
EMA	Environmental Management Accounting
GRI	Global Reporting Initiative
HEIs	Higher Education Institutions
IT	Institutional Theory

SDG	Sustainable Development Goal
SEI	Sustainability Endowments Institute
STEM	Science, Technology, Engineering, and Mathematics
USC	University of Southern California
UK	United Kingdom
US	United States
ZW	Zero Waste

eferences

Global Waste Index 2019. Available online: https://sensoneo.com/global-waste-index-2019/ (accessed on 11 July 2022).

Sustainable Consumption and Production. United Nations Sustainable Development. Available online: https://www.un.org/sustainabledevelopment/sustainable-consumption-production/ (accessed on 17 July 2022).

Goal 12—Ensuring Sustainable Consumption and Production Patterns: An Essential Requirement for Sustainable Development. *United Nation Chronicle*, 4(LI). Available online: https://www.un.org/en/chronicle/article/goal-12-ensuring-sustainable-consumption-and-production-patterns-essential-requirement-sustainable (accessed on 10 December 2022).

Goal 12. Department of Economic and Social Welfare. Available online: https://sdgs.un.org/goals/goal12 (accessed on 18 July 2022).

Top 10 Zero Waste Campuses in 2024. Available online: https://www.postlandfill.org/top-10-campuses-2024#:~:text=Zero%20waste%20programs%20with%20deep,student%20run%20Goodrich%20Coffee%20Bar (accessed on 31 January 2025).

Filho, W.L.; Vargas, V.R.; Salvia, A.L.; Brandli, L.L.; Pallant, E.; Klavins, M.; Ray, S.; Moggi, S.; Maruna, M.; Conticelli, E.; et al. The role of higher education institutions in sustainability initiatives at the local level. *J. Clean. Prod.* **2019**, *233*, 1004–1015. [CrossRef]

Baratsas, S.G.; Pistikopoulos, E.N.; Avraamidou, S. A quantitative and holistic circular economy assessment framework at the micro level. *Comput. Chem. Eng.* **2022**, *160*, 107697. [CrossRef]

Kristensen, H.S.; Mosgaard, M.A. A review of micro level indicators for a circular economy—Moving away from the three dimensions of sustainability? *J. Clean. Prod.* **2020**, *243*, 118531. [CrossRef]

Ghisellini, P.; Cialani, C.; Ulgiati, S. A review on circular economy: The expected transition to a balanced interplay of environmental and economic systems. *J. Clean. Prod.* **2016**, *114*, 11–32. [CrossRef]

Veleva, V.; Bodkin, G.; Todorova, S. The need for better measurement and employee enegaement to advance a circular economy: Lesson deom biogen's "zero waste" journey. *J. Clean. Prod.* **2017**, *154*, 517–529. [CrossRef]

Alamerew, Y.A.; Brissaud, D. Circular economy assessment tool for end of life product recovery strategies. *J. Remanufacturing* **2018**, *9*, 169–185. [CrossRef]

Basiago, A.D. Economic, social, and environmental sustainability in development theory and urban planning practice. *Environmentalist* **1998**, *19*, 145–161. [CrossRef]

Brown, D.P. Garbage: How population, landmass, and development interact with culture in the production of waste. *Resour. Conserv. Recycl.* **2015**, *98*, 41–54. [CrossRef]

Amaral, A.R.; Rodrigues, E.; Gaspar, A.R.; Gomes, A. A review of empirical data of sustainability initiatives in university campus operations. *J. Clean. Prod.* **2020**, *250*, 119558. [CrossRef]

Stets, J.E.; Burke, P.J. Identity theory and social identity theory. *Soc. Psychol. Q.* **2000**, *63*, 224–237. [CrossRef]

Saidani, M.; Yannou, B.; Leroy, Y.; Cluzel, F.; Kendall, A. A taxonomy of circular economy indicators. *J. Clean. Prod.* **2019**, *207*, 542–559. [CrossRef]

De Pascale, A.; Arbolino, R.; Szopik-Depczyńska, K.; Limosani, M.; Ioppolo, G. A systematic review for measuring circular economy: The 61 indicators. *J. Clean. Prod.* **2021**, *281*, 124942. [CrossRef]

Kirchherr, J.; Reike, D.; Hekkert, M. Conceptualizing the circular economy: An analysis of 114 definitions. *Resour. Conserv. Recycl.* **2017**, *127*, 221–232. [CrossRef]

Fligstein, N.; McAdam, D. *A Theory of Fields*; Oxford University Press: Oxford, UK, 2012; pp. 46–50.

Zietsma, C.; Groenewegen, P.; Logue, D.M.; Hinings, C.R. Field or fields? Building the scaffolding for cumulation of research on institutional fields. *Acad. Manag. Ann.* **2017**, *11*, 391–450. [CrossRef]

City of Columbus and Ohio State launch Pilot Food Waste Reduction Campaign. 2024. Available online: https://news.osu.edu/city-of-columbus-and-ohio-state-launch-pilot-food-waste-reduction-campaign/ (accessed on 31 January 2025).

Iowa State University Consortium Partners with Industry to Solve Intractable Waste Problems. 2022. Available online: https://www.cals.iastate.edu/news/2022/iowa-state-university-consortium-partners-industry-solve-intractable-waste-problems (accessed on 31 January 2025).

DiMaggio, P.J.; Powell, W.W. The iron cage revisited: Institutional isomorphism and collective rationality in organizational fields. *Am. Sociol. Rev.* **1983**, *48*, 147–160. [CrossRef]

24. Larrán, M.; Herrera, J.; Andrades, F.J. Measuring the linkage between strategies on sustainability and institutional forces: A empirical study of Spanish universities. *J. Environ. Plan. Manag.* **2015**, *59*, 967–992. [CrossRef]
25. Rowe, D. Education for a Sustainable Future. *Science* **2007**, *317*, 323–324. [CrossRef]
26. Kerr, K.G.; Hart-Steffes, J.S. Sustainability, student affairs, and students. *New Dir. Stud. Serv.* **2012**, *137*, 7–17. [CrossRef]
27. Tchobanoglous, G.; Kreith, F.; Williams, M.E. Planning for municipal solid waste programs. In *Handbook of Solid Waste Manageme* Tchobanoglous, G., Kreith, F., Eds.; McGraw-Hill Companies, Inc.: New York, NY, USA, 2002.
28. Sherburne, S. Let's Get Sorted: The Path to Zero Waste at Pomona College. Pomona Senior Thesis, Claremont College. 20 Available online: https://scholarship.claremont.edu/pomona_theses/205 (accessed on 23 July 2021).
29. Lidstone, L.; Wright, T.; Sherren, K. Canadian STARS-rated campus sustainability plans: Priorities, plan creation and desi *Sustainability* **2015**, *7*, 725–746. [CrossRef]
30. Alexandra Wright, T.S. Giving "teeth" to an environmental policy: A Delphi Study at Dalhousie University. *J. Clean. Prod.* **20** *14*, 761–768. [CrossRef]
31. McNamara, K.H. Fostering sustainability in higher education: A mixed-methods study of transformative leadership and chan strategies. *Environ. Pract.* **2010**, *12*, 48–58. [CrossRef]
32. Locke, E.A.; Latham, G.P.A. *Theory of Goal Setting & Task Performance*; Prentice Hall, Inc.: Hoboken, NJ, USA, 1990.
33. Lingard, H.; Gilbert, G.; Graham, P. Improving solid waste reduction and recycling performance using goal setting and feedba *Constr. Manag. Econ.* **2001**, *19*, 809–817. [CrossRef]
34. Zhang, N.; Williams, I.; Kemp, S.; Smith, N. Greening academia: Developing sustainable waste management at Higher Educati Institutions. *Waste Manag.* **2011**, *31*, 1606–1616. [CrossRef] [PubMed]
35. Ayarkwa, J.; Agyekum, K.; Opoku, D.G.J.; Appiagyei, A.A. Barriers to the implementation of environmentally sustainab procurement in public universities. *Int. J. Procure. Manag.* **2020**, *13*, 24–41. [CrossRef]
36. University of California, Los Angeles. Sustainable Procurement Guidelines. 2021. Available online: https://www.ucop.ed procurement-services/for-ucstaff/sustainable-procurement/sustainableprocurementguidelines.pdf (accessed on 21 April 202
37. SB 54. Solid Waste: Reporting, Packaging, and Plastic Food Service Ware. California Legislative Information. 2022. Availab online: https://leginfo.legislature.ca.gov/faces/billTextClient.xhtml?bill_id=202120220SB54 (accessed on 21 April 2023).
38. ESF. Green Purchasing. Available online: https://www.esf.edu/au/documents/green-purchasing-guide-5-12-2022.pdf (access on 19 April 2023).
39. Lisle, I. UVM's Roadmap to Zero Waste: Effective Strategies for Enhancing Zero Waste Management Goals at the Universi of Vermont 2020. Environmental Studies Electronic Thesis Collection. 65. Available online: https://scholarworks.uvm.ed envstheses/65 (accessed on 5 May 2023).
40. Zachariou, A.; Kaila, M.; Katsikis, A. Sustainable school: Facts, objectives and prospects. *Sci. Technol. Top. Educ.* **2008**, *1*, 269–2
41. Ebrahimi, K.; North, L.A. Effective strategies for enhancing waste management at university campuses. *Int. J. Sustain. High. Ed* **2017**, *18*, 1123–1141. [CrossRef]
42. Brody, S.D.; Ryu, H.C. Measuring the educational impacts of a graduate course on sustainable development. *Environ. Educ. R* **2006**, *12*, 179–199. [CrossRef]
43. Angelaki, M.E.; Bersimis, F.; Karvounidis, T.; Douligeris, C. Towards more sustainable higher education institutions: Implementir the sustainable development goals and embedding sustainability into the information and computer technology curricula. *Ed Inf. Technol.* **2024**, *29*, 5079–5113. [CrossRef]
44. Moqbel, S.; Abu-Zurayk, R.; Bozeya, A.; Alsisan, R.; Al Bawab, A. Assessment of sustainable recycling at The University Jordan. *Int. J. Sustain. High. Educ.* **2020**, *21*, 1111–1129. [CrossRef]
45. Kaplowitz, M.D.; Yeboah, F.K.; Thorp, L.; Wilson, A.M. Garnering input for recycling communication strategies at a Big Te University. *Resour. Conserv. Recycl.* **2009**, *53*, 612–623. [CrossRef]
46. Leal Filho, W.; Ribeiro, P.C.C.; Setti, A.F.F.; Azam, F.M.S.; Abubakar, I.R.; Castillo-Apraiz, J.; Tamayo, U.; Özuyar, P.G.; Frizzo, Borsari, B. Toward food waste reduction at universities. *Environ. Dev. Sustain.* **2024**, *26*, 16585–16606. [CrossRef] [PubMed]
47. Dahle, M.; Neumayer, E. Overcoming barriers to campus greening. *Int. J. Sustain. High. Educ.* **2001**, *2*, 139–160. [CrossRef]
48. Mason, I.; Brooking, A.; Oberender, A.; Harford, J.; Horsley, P. Implementation of a zero waste program at a university campu *Resour. Conserv. Recycl.* **2003**, *38*, 257–269. [CrossRef]
49. Wu, D.W.L.; Lenkic, P.J.; DiGiacomo, A.; Cech, P.; Zhao, J.; Kingstone, A. How does the design of waste disposal signage influen waste disposal behavior? *J. Environ. Psychol.* **2018**, *58*, 77–85. [CrossRef]
50. Pitcher, K.F. What does a sustainability manager do? Sustainable development at the University of Leeds. *Int. J. Environ. Sustai Dev.* **2010**, *9*, 349–363. [CrossRef]
51. Cheung, L. *Feasibility and Cost Analysis of Composting Campus Dining Food Waste in Cal Poly's Compost Facility*; California Polytechn University: San Luis Obispo, CA, USA; University of Florida: Gainesville, FL, USA, 2015.
52. Allen, C. Hernani, Spain: Door-to-door collection as a strategy to reduce waste disposal. In *On the Road to Zero Waste: Success and Lessons from Around the World*; GAIA: Sheung Wan, Hong Kong, 2012; pp. 1–8.

- Levy, B.L.; Marans, R.W. Toward a campus culture of environmental sustainability: Recommendations for a large university. *Int. J. Sustain. High. Educ.* **2012**, *13*, 365–377. [CrossRef]
- Filho, W.L. Transformative approaches to sustainable development at universities. In *Working Across Disciplines*; Springer: Cham, Switzerland, 2015.
- Velazquez, L.; Munguia, N.; Platt, A.; Taddei, J. Sustainable university: What can be the matter? *J. Clean. Prod.* **2006**, *14*, 810–819. [CrossRef]
- Shi, H.; Lai, E. An alternative university sustainability rating framework with a structured criteria tree. *J. Clean. Prod.* **2013**, *61*, 59–69. [CrossRef]
- Filho, W.L.; Will, M.; Salvia, A.L.; Adomssent, M.; Grahl, A.; Spira, F. The role of green and Sustainability Offices in fostering sustainability efforts at higher education institutions. *J. Clean. Prod.* **2019**, *232*, 1394–1401. [CrossRef]
- Kezar, A.; Bernstein-Sierra, S. Examining processes of normative isomorphism and influence in scaled change among higher education intermediary organizations. *Aera Open* **2019**, *5*, 2332858419884905. [CrossRef]
- Peer, V.; Stoeglehner, G. Universities as change agents for sustainability—Framing the role of knowledge transfer and generation in regional development processes. *J. Clean. Prod.* **2013**, *44*, 85–95. [CrossRef]
- Moore, J. Policy, priorities and action: A case study of the University of British Columbia's engagement with sustainability. *High. Educ. Policy* **2005**, *18*, 179–197. [CrossRef]
- Alshuwaikhat, H.M.; Abubakar, I. An integrated approach to achieving campus sustainability: Assessment of the current campus environmental management practices. *J. Clean. Prod.* **2008**, *16*, 1777–1785. [CrossRef]
- Mosier, S.; Ruxton, M. Sustainability university–community partnerships: Lessons for practitioners and scholars from highly sustainable communities. *Environ. Plan. C Politics Space* **2018**, *36*, 479–495. [CrossRef]
- Gardner, C.J.; Thierry, A.; Rowlandson, W.; Steinberger, J.K. From Publications to Public Actions: The role of universities in facilitating academic advocacy and activism in the climate and ecological emergency. *Front. Sustain.* **2021**, *2*, 42. [CrossRef]
- Hong, S. The effects of unit pricing system upon household solid waste management: The Korean experience. *J. Environ. Manag.* **1999**, *57*, 1–10. [CrossRef]
- Allen, M. (Ed.) *The SAGE Encyclopedia of Communication Research Methods*; SAGE Publications: Washington, DC, USA, 2017.
- Liyanage, K.L.A.K.T.; Waidyasekara, K.G.A.S.; Mallawarachchi, H. Toward the adoption of zero waste concept in the construction industry: A review of existing strategies. In Proceedings of the International Conference on Industrial Engineering and Operations Management, Bangkok, Thailand, 5–7 March 2019; pp. 2414–2424.
- Zaman, A.U.; Lehmann, S. Challenges and opportunities in transforming a city into a "zero waste city". *Challenges* **2011**, *2*, 73–93. [CrossRef]
- CalRecycle. School Waste Reduction Laws. Available online: https://calrecycle.ca.gov/recycle/schools/ (accessed on 30 March 2023).
- USC. Waste Diversion, Office of Sustainability. Available online: https://green.usc.edu/programs/waste-diversion/ (accessed on 30 March 2023).
- Latif, B.; Mahmood, Z.; Tze San, O.; Mohd Said, R.; Bakhsh, A. Coercive, normative and mimetic pressures as drivers of environmental management accounting adoption. *Sustainability* **2020**, *12*, 4506. [CrossRef]
- Schmidt, A.; Rocconi, L. STARS and stripes: Political and institutional factors that affect higher education sustainability ratings. *arXiv* **2021**. [CrossRef]
- Henkels, M. Beverage Container Act (Bottle-Billl). Available online: https://www.oregonencyclopedia.org/articles/beverage_container_act_bottle_bill/ (accessed on 28 March 2022).
- NCSL. State Beverage Container Deposit Laws. National Conference of State Legislatures. 2020. Available online: https://www.ncsl.org/environment-and-natural-resources/state-beverage-container-deposit-laws (accessed on 28 March 2023).
- Wilson, S.; Hauck, Y.; Bremner, A.; Finn, J. Quality nursing care in Australian pediatric hospitals: A delphi approach to identifying indicators. *J. Clin. Nurs.* **2012**, *21*, 1594–1605. [CrossRef]
- EPA. EPA Awards More than $230,000 for Recycling and Food Waste Prevention Projects in Illinois, Michigan, and Ohio. 2021. Available online: https://www.epa.gov/newsreleases/epa-awards-more-230000-recycling-and-food-waste-prevention-projects-illinois-michigan (accessed on 22 June 2022).
- Fuentes, M.R. Zero Waste by Whom? UC Berkeley's Zero-Waste Initiative and Its Impact on Campus Facility Employees. 2022. Available online: https://nature.berkeley.edu/ (accessed on 6 May 2023).
- Hao, M.; Zhang, D.; Morse, S. Waste separation behaviour of college students under a mandatory policy in China: A case study of Zhengzhou city. *Int. J. Environ. Res. Public Health* **2020**, *17*, 8190. [CrossRef] [PubMed]
- Murphy, S.; Pincetl, S. Zero waste in Los Angeles: Is the emperor wearing any clothes? *Resour. Conserv. Recycl.* **2013**, *81*, 40–51. [CrossRef]
- Rubicon. Corporate Factsheet. Available online: https://www.rubicon.com/wp-content/uploads/rubicon-corporate-fact-sheet.pdf (accessed on 23 March 2023).

80. Shreyas Madhav, A.V.; Rajaraman, R.; Harini, S.; Kiliroor, C.C. Application of artificial intelligence to enhance collection of E-waste: A potential solution for household WEEE collection and segregation in India. *Waste Manag. Res.* **2022**, *40*, 1047–105 [CrossRef] [PubMed]
81. Wright, T. Definitions and frameworks for environmental sustainability in higher education. *High. Educ. Policy* **2002**, *15*, 105–12 [CrossRef]
82. STARS 2.1 Technical Manual, (n.d.). Available online: https://drive.google.com/file/d/1RcujP2wuYTSXiY0N_Dzb53Ebq8x FMZQ/view (accessed on 23 November 2023).
83. Moncada Garcia, S.Y. Strategy to Become a Zero-Waste Community at Varsity Courts Family Housing at the University of Calgar Master's Thesis, University of Calgary, Calgary, AB, Canada, 2021.
84. Beringer, A.; Adomßent, M. Sustainable university research and development: Inspecting sustainability in higher educatio research. *Environ. Educ. Res.* **2008**, *14*, 607–623. [CrossRef]
85. DOE. DOE Invests $13.4 Million to Combat Plastic Waste, Reduce Plastic Industry Emissions. 2022. Available online: http //www.energy.gov/eere/ammto/articles/us-department-energy-invests-134-million-combat-plastic-waste-reduce-plastic ~:text=Plastic%20Industry%20Emissions-,U.S.%20Department%20of%20Energy%20Invests%20$13.4%20Million%20to%20 Combat%20Plastic,emissions%20of%20single-use%20plastics (accessed on 5 May 2023).
86. Stafford, S.L. How green is your campus? An analysis of the factors that drive universities to embrace sustainability. *Contem Econ. Policy* **2011**, *29*, 337–356. Available online: https://ssrn.com/abstract=1720995 (accessed on 12 May 2023). [CrossRef]
87. Phrophayak, J.; Techarungruengsakul, R.; Khotdee, M.; Thuangchon, S.; Ngamsert, R.; Prasanchum, H.; Sivanpheng, C Kangrang, A. Enhancing Green University Practices through Effective Waste Management Strategies. *Sustainability* **2024**, *16*, 334 [CrossRef]
88. Emery, F.E.; Trist, E.L. The causal texture of organizational environments. *Hum. Relat.* **1965**, *18*, 21–32. [CrossRef]
89. STARS. STARS Technical Manual Version 2.2. 2019. Available online: https://stars.aashe.org/wp-content/uploads/2019/07 STARS-2.2-Technical-Manual.pdf (accessed on 23 November 2023).
90. Trencher, G.P.; Yarime, M.; Kharrazi, A. Co-creating sustainability: Cross-sector university collaborations for driving sustainab urban transformations. *J. Clean. Prod.* **2013**, *50*, 40–55. [CrossRef]
91. Connett, P. *The Zero Waste Solution: Untrashing the Planet One Community at a Time*; Chelsea Green Publishing: Chelsea, VT, US 2013.
92. US EPA. Zero Waste Case Study: San Francisco. 2013. Available online: https://www.epa.gov/transforming-waste-tool/zer waste-case-study-san-francisco (accessed on 16 December 2023).
93. Zero Waste. City of Oakland. Available online: https://www.oaklandca.gov/topics/zero-waste (accessed on 22 July 2022).
94. Zero Waste by 2040. AustinTexas.Gov. Available online: https://www.austintexas.gov/zerowaste (accessed on 15 July 2022).
95. City of Dallas' Local Solid Waste Management Plan. Texas Campaign for the Environment Fund. 2013. Available o line: https://dallascityhall.com/departments/sanitation/lswmp/Pages/default.aspx#:~:text=The%20LSWMP,%20also%2 known%20as,over%20the%20next%2050%20years (accessed on 10 July 2022).
96. Rahman, I.; Reynolds, D.; Svaren, S. How "green" are North American hotels? An exploration of low-cost adoption practices. *In J. Hosp. Manag.* **2012**, *31*, 720–727. [CrossRef]
97. Wan, C.; Shen, G.Q.; Choi, S. A review on political factors influencing public support for urban environmental policy. *Environ. Sc Policy* **2017**, *75*, 70–80. [CrossRef]
98. Wan, C.; Shen, G.Q.; Choi, S. Understanding public support for recycling policy: To unveil the political side of influence an implications. *Environ. Sci. Policy* **2018**, *82*, 30–43. [CrossRef]
99. Pacca, L.; Curzi, D.; Rausser, G.; Olper, A. The role of party affiliation, lobbying, and electoral incentives in decentralized US stat support of the environment. *J. Assoc. Environ. Resour. Econ.* **2021**, *8*, 617–653. [CrossRef]
100. Uyeki, E.S.; Holland, L.J. Diffusion of pro-environment attitudes? *Am. Behav. Sci.* **2000**, *43*, 646–662. [CrossRef]
101. Shriberg, M.P. *Sustainability in United States Higher Education: Organizational Factors Influencing Campus Environmental Performanc and Leadership*; University of Michigan: Ann Arbor, MI, USA, 2002.
102. Hull, J.D. Who's Greener–Democrats or Republicans? Environmental Protection. 2006. Available online: https://eponline.com articles/2006/01/01/whos-greener--democrats-or-republicans.aspx (accessed on 31 March 2023).
103. Howe, K.R. Two dogmas of educational research. *Educ. Res.* **1985**, *14*, 10–18. [CrossRef]
104. Addison, D. (University of Oregon, Eugene, OR, USA). Personal communication, 10 May 2023.
105. Lu, L.T.; Hsiao, T.Y.; Shang, N.C.; Yu, Y.H.; Ma, H.W. MSW management for waste minimization in Taiwan: The last two decade *Waste Manag.* **2006**, *26*, 661–667. [CrossRef]
106. Zaman, A.U. Identification of key assessment indicators of the zero waste management systems. *Ecol. Indic.* **2014**, *36*, 682–69 [CrossRef]
107. Yukalang, N.; Clarke, B.; Ross, K. Barriers to effective municipal solid waste management in a rapidly urbanizing area in Thailan *Int. J. Environ. Res. Public Health* **2017**, *14*, 1013. [CrossRef]

8. Nizar, M.; Munir, E.; Munawar, E. Implementation of zero waste concept in waste management of Banda Aceh City. *J. Phys. Conf. Ser.* **2018**, *1116*, 052045. [CrossRef]
9. Volkwein, J.F.; Malik, S.M. State Regulation and Administrative Flexibility at Public Universities. *Res. High. Educ.* **1997**, *38*, 17–42. [CrossRef]
0. Urbanski, M.; Filho, W.L. Measuring sustainability at universities by means of the Sustainability Tracking, Assessment and Rating System (STARS): Early findings from STARS data. *Environ. Dev. Sustain.* **2015**, *17*, 209–220. [CrossRef]
1. Devereux, E. Carnegie Classification of Institutions of Higher Education & Perceived Administrative Biases in Funding Review Processes. Ph.D. Thesis, West Chester University, West Chester, PA, USA, 2020. Available online: https://digitalcommons.wcupa.edu/all_doctoral/84 (accessed on 17 May 2023).
2. Filho, W.L.; Kovaleva, M.; Fritzen Gomes, B.; Fudjumdjum, H.; Emblen-Perry, K.; Platje, J.; Tuladhar, L.; Vasconcelos, C.R.; LeVasseur, T.J.; Minhas, A.; et al. Sustainability practices at private universities: A state-of-the art assessment. *Int. J. Sustain. Dev. World Ecol.* **2020**, *28*, 402–416. [CrossRef]
3. Tolbert, P.S. Institutional environments and resource dependence: Sources of administrative structure in institutions of higher education. *Adm. Sci. Q.* **1985**, *30*, 1–13. [CrossRef]
4. Opoku, A.; Fortune, C. The implementation of sustainable practices through leadership in construction organizations. In *Annual ARCOM Conference, Proceedings of the 27th Annual ARCOM Conference, Bristol, UK, 5–7 September 2011*; Egbu, C., Lou, E.C.W., Eds.; Association of Researchers in Construction Management: Reading, UK, 2011; pp. 1145–1154.

sustainability

MDPI

Article

Sustainability-Oriented Furnace Temperature Prediction for Municipal Solid Waste Incineration Using IWOA-SAGRU

Jinxiang Pian [1], Mayan Si [1], Ao Sun [2,*] and Jian Tang [3]

1 School of Electrical and Control Engineering, Shenyang Jianzhu University, Shenyang 110168, China; pianjx@sjzu.edu.cn (J.P.); simy@stu.sjzu.edu.cn (M.S.)
2 Arts and Information Engineering College, Dalian Polytechnic University, Dalian 116400, China
3 School of Information Science and Technology, Beijing University of Technology, Beijing 100124, China; freeflytang@bjut.edu.cn
* Correspondence: suna@caie.edu.cn

Abstract

Municipal solid waste incineration promotes sustainable development by reducing waste, recovering resources, and minimizing environmental impact, with furnace temperature control playing a key role in maximizing efficiency. Accurate real-time temperature prediction is crucial in developing countries to optimize incineration, re-duce emissions, and enhance energy recovery for global sustainability. To address this, we propose a method integrating an improved whale optimization algorithm (IWOA) with a self-attention gated recurrent unit (SAGRU). Using the maximal information coefficient (MIC) to identify key factors, we optimize SAGRU parameters with IWOA, enhancing prediction accuracy by capturing temporal dependencies. Experimental validation from an MSWI plant in China demonstrates that the proposed model significantly enhances prediction accuracy under complex conditions. When compared with the Elman and LSTM models, the error is reduced by 0.7146 and 0.4689, respectively, highlighting its strong potential for practical applications in waste incineration temperature control.

Keywords: furnace temperature prediction; municipal solid waste incineration (MSWI); improved whale optimization algorithm (IWOA); self-attention mechanism gated recurrent unit (SAGRU)

1. Introduction

Municipal solid waste incineration (MSWI) directly contributes to the achievement of sustainable development goals, such as "clean environment," "clean energy," and "sustainable cities," through reduction (decreasing the waste volume by 60–90%), resource recovery (converting thermal energy into electricity), and harmless treatment (eliminating pathogens at high temperatures). This process not only enables effective waste disposal but also facilitates resource recycling. However, to ensure that MSWI is both efficient and environmentally friendly, furnace temperature control plays a pivotal role. Accurate prediction and maintenance of the incineration temperature within the range of 850–1000 °C can maximize energy conversion efficiency (increasing power generation by 10–15% per 10,000 tons of waste), while simultaneously preventing dioxin formation at low temperatures and equipment damage at excessively high temperatures. Therefore, furnace temperature control is the key to ensuring a green, low-carbon, and harmless incineration process, and it significantly enhances the overall efficiency of waste-to-energy systems [1].

Although many studies have investigated combustion parameters in waste incineration, developed countries such as those in Europe and North America have already implemented waste classification systems. For example, in France, municipal solid waste (MSW) is primarily incinerated, and recent studies have provided comprehensive data on technological, environmental, and energy aspects across 90 MSW incineration plants, highlighting common air pollution control techniques, emissions, and energy recovery efficiency [2]. In these regions, public awareness of environmental protection is relatively strong, the calorific value of classified combustible components is relatively stable, and advanced temperature control technologies have been developed and supported by well-established expertise. In contrast, in developing countries, waste classification policies and regulations are still in the process of improvement and promotion [3]. As a result, the composition of municipal solid waste (MSW) in developing countries remains highly uncertain, with relatively low and fluctuating calorific values, making it difficult to directly apply the mature control technologies developed in developed countries [4]. Taking China as an example, waste incineration in many developing countries still relies heavily on operating engineers to anticipate changes in operating conditions based on experience, predict possible temperature fluctuations, and manually adjust temperature control. However, this experience-dependent approach cannot fully cope with the complex and dynamic nature of the incineration process, which often results in unstable temperature control [5].

Nevertheless, the heterogeneous and variable composition of solid waste leads to fluctuations in its physical and chemical properties, which in turn cause dynamic changes and uncertainties in furnace temperature. These factors make it difficult for traditional mathematical models to provide accurate descriptions of the process [6]. Consequently, developing advanced prediction models capable of handling such complexity has become a research priority. These models are designed to improve incineration efficiency, reduce pollutant emissions, and enhance resource recovery [7], which are all directly aligned with global sustainability objectives in terms of energy efficiency, circular economy, and environmental protection. Moreover, waste incineration can effectively reduce harmful factors associated with electricity generation and phosphate fertilizer production, thereby minimizing environmental impacts [8].

In the field of modeling for solid waste incineration processes, early studies primarily focused on mechanism-based approaches to establish temperature models. For example, Ref. [9] developed a comprehensive mathematical model by dividing the movement of solid waste into zones and constructing mass and energy balance equations for the fuel bed. Simulation results showed that the model was consistent with the operating characteristics of an incinerator. Similarly, Ref. [10] proposed different combustion modes of volatile matter based on its release and combustion processes inside the furnace, and constructed a combustion model using chemical reaction equations. However, the MSWI process is characterized by intricate physical and chemical reactions. Furnace temperature is influenced by multiple factors, such as air supply, fuel composition, and reaction rate. Due to the strong complexity and non-stationary nature of MSWI, furnace temperature exhibits dynamic and nonlinear characteristics, making it extremely difficult to establish accurate mechanism-based mathematical models.

With the rapid advancement of machine learning and deep learning technologies, data-driven modeling approaches have emerged as a prominent area of research in the field of solid waste incineration processes. For instance, references [8,11] employed back-propagation techniques to establish nonlinear relationships between key process parameters, such as energy consumption and pollutant emissions, and their influencing factors. Experimental results demonstrated that these methods significantly outperformed linear regression models. In reference [12], a T-S fuzzy neural network was used to model the temperature during the solid waste incineration process, with a parameter learning mechanism incorporated to enhance adaptability under varying conditions. Reference [13] introduced a novel time-span input neural network, considering the time factor in input variables, to predict trends in the main steam temperature during the incineration process. Furthermore, a temperature model for the incinerator was established in [14] using the Tikhonov regularization–least regression decision tree (TR-LRDT) algorithm. To improve the model's generalization ability, reference [15] developed an explainable Bayesian support vector regression (E-BSVR) model for generalized prediction and optimization of operating parameters, including PCDD/F emissions. However, these methods often face limitations when handling complex, multivariable, and high-dimensional data.

On the other hand, in order to cope with the strong nonlinearity and dynamic temporal features of the incineration process, reference [16] proposes a heterogeneous feature ensemble modeling method for furnace temperature prediction in municipal solid waste incineration. This approach enhances accuracy, generalization, stability, and training efficiency by combining multiple base models and employing a negative correlation learning strategy. In reference [17], a model for predicting gas concentration is proposed by combining convolutional neural networks (CNN) with long short-term memory (LSTM) networks. However, the accuracy of the model is influenced by manually set key parameters, such as the number of neurons and the learning rate. Therefore, despite some progress, challenges remain, including the lack of a systematic approach for selecting input variables and insufficient model generalization capability.

In recent years, research on complex process modeling in industrial applications has mainly evolved along three directions. First, in the traditional modeling stage, researchers employed methods such as multiple linear regression [18] and partial least squares [19], as well as shallow machine learning approaches including support vector machines [20] and backpropagation neural networks [21]. These methods perform reliably when dealing with linear or weakly nonlinear problems. However, they face limitations in capturing the dynamic temporal features and strong nonlinear relationships inherent in the complex MSWI process [22,23].

With the rapid development of deep learning, modeling techniques have entered the deep learning stage. Typical models include recurrent neural networks (RNN) and LSTM networks have significantly improved the prediction accuracy of nonlinear dynamic processes by deeply exploiting temporal dependencies. Among them, the gated recurrent unit (GRU) simplifies the structure of LSTM while maintaining comparable predictive performance, thereby improving computational efficiency. GRU has been widely applied in areas such as chemical reaction modeling and energy forecasting [24]. However, the hyperparameters of GRU are typically set empirically, which often leads to fluctuations in prediction results. To tackle this problem, heuristic methods like genetic algorithms and artificial bee colony optimization have been applied by researchers to tune GRU hyperparameters [25,26]. Experimental studies have shown that incorporating intelligent optimization algorithms can substantially improve the predictive accuracy of single models. Nevertheless, when dealing with long input sequences, GRU-based models may still suffer from information loss across time steps, reducing their predictive reliability. Consequently, as industrial processes become increasingly complex, the generalization ability of single models has become limited. This has given rise to the stage of intelligent optimization and algorithm integration, where hybrid models combining attention mechanisms, ensemble learning, and intelligent optimization algorithms—such as whale optimization and particle swarm optimization—have emerged as a new research trend [27]. However, there is still a lack of studies explicitly linking advanced modeling methods with sustainability-oriented objectives, such as enhancing energy efficiency, improving combustion stability, and supporting more efficient resource recovery in municipal solid waste incineration.

Based on the above analysis, the primary goal of this study is to develop an accurate and robust furnace temperature prediction model for municipal solid waste incineration (MSWI) that can effectively handle the nonlinear, dynamic, and complex characteristics of the process. To achieve this goal, we propose a model integrating a Self-Attention Gated Recurrent Unit (SAGRU) with an Improved Whale Optimization Algorithm (IWOA) to address the limitations of traditional temperature prediction methods. First, the Maximal Information Coefficient (MIC) is employed to evaluate the correlation between different features and temperature, thereby selecting appropriate input variables. Next, a temperature prediction model is constructed using GRU and an attention mechanism. The attention mechanism assigns adaptive weights across different time steps, which enhances the modeling accuracy. To overcome the shortcomings of the conventional Whale Optimization Algorithm (WOA), including slow convergence and a propensity to become trapped in local optima during the later stages of training, this study improves the algorithm by incorporating Latin hypercube sampling, an adaptive threshold strategy, and Gaussian perturbation. Finally, the improved IWOA is applied to optimize the key parameters of the SAGRU model, including the learning rate (η), the number of hidden nodes (n) and the dropout rate (p), thereby establishing a more accurate prediction model. Experimental validation is conducted using real operating data from a MSWI plant in China. The results demonstrate that the proposed furnace temperature prediction model achieves superior accuracy, generalization, and robustness compared with other networks, while also contributing to the broader sustainability agenda, reducing operational risks, and supporting the transition toward low-carbon waste management systems.

2. Description of the MSWI Process and Problem Formulation for Furnace Temperature Modeling

2.1. Process Description

As shown in Figure 1 [28], a typical MSWI process consists of several stages, including pretreatment, drying, combustion, and grate handling. First, a grab crane delivers the pretreated solid waste into the incinerator, where it is subjected to drying, combustion, and burnout. During this process, the combustible components of the waste are fully oxidized, and the organic matter is converted into carbon dioxide, water vapor, and other gases. To ensure efficient combustion, primary and secondary air is supplied through blowers, injected from the bottom of the grate and the middle of the furnace chamber. The incineration residues are discharged by a slag conveyor, while the high-temperature flue gas is directed into a waste heat boiler for energy recovery and subsequently treated by a flue gas purification system before being released into the atmosphere.

In this process, conventional temperature measurement instruments often exhibit slow response under dynamic operating conditions, which becomes a critical factor limiting the efficiency of temperature control. The temperature in MSWI processes typically fluctuates significantly, as both the composition of the waste and the combustion state are constantly changing. These factors make real-time monitoring of furnace temperature highly challenging. Traditional measurement methods suffer from considerable lag and cannot capture rapid or transient temperature variations with sufficient accuracy. Particularly under conditions of sharp or frequent fluctuations, sensors are unable to respond promptly. This time lag prevents the recorded data from fully reflecting the actual furnace temperature dynamics, reducing the effectiveness of real-time combustion regulation. Consequently, the accuracy of temperature control is compromised, which may even result in inadequate control strategies, thereby threatening the safety and stability of the entire incineration process.

Therefore, the development of an accurate furnace temperature model is a prerequisite for effective furnace temperature control.

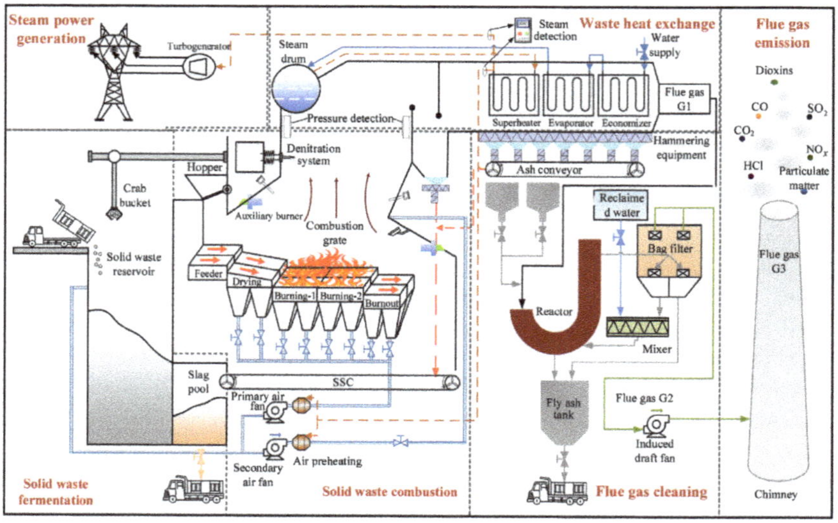

Figure 1. MSWI process flow diagram.

2.2. Problem Formulation for Furnace Temperature Prediction Modeling

The municipal solid waste incineration (MSWI) process is a complex thermochemical reaction involving multiple physical and chemical transformations. During this process, furnace temperature is influenced by various factors, which are usually characterized by complex nonlinear relationships. Moreover, temperature often fluctuates significantly under different operating conditions.

From a mechanistic perspective, the combustion of municipal solid waste involves complex physical and chemical transformations.

(1) The mechanistic model of the moisture evaporation process can be expressed as follows [29]:

$$m_w = \frac{dR_{air} + dR_{rad}}{R_{water}} \tag{1}$$

m_w represents the change in moisture mass of a single waste module, dR_{air} represents the change in the thermal energy of the waste module. dR_{rad} represents the change in radiative heat of the waste module, R_{water} represents the heat absorbed for the evaporation of each kilogram of water, which is taken as 2571.4 kJ/kg.

(2) After the moisture has evaporated to a certain extent, the waste undergoes pyrolysis reactions under high-temperature conditions.

The release of volatile matter corresponds to the pyrolysis process of the waste. The mechanistic model for the release of volatile matter can be expressed as follows [9]:

$$m_T = m_{vol} \int_0^T A_V e^{-\frac{E_V}{R y_V}} (1-\alpha)^n d\tau \tag{2}$$

m_T represents the amount of volatile matter released over a period T, m_{vol} represents the initial amount of volatile matter, A_V refers to the pre-exponential factor. E_V represents the activation energy, The volatile conversion occurs only when the activation energy of the molecules reaches a certain threshold. R is the universal gas constant, and y_v is the temperature of the waste in the material layer. α represents the conversion rate of volatile matter during the pyrolysis process. n represents the reaction order.

(3) The combustion performance can be classified into three states based on the ratio of oxygen to the mass of volatile matter: oxygen-limited combustion, incomplete combustion, and complete combustion [30].

Assuming that the ratio of oxygen mass in the air to the mass of volatile matter is less than k_1, the combustion is in an oxygen-limited state, and the products formed are CO, CO_2, and CH_4:

$$C_xH_yO_z + (\frac{3x}{4} - \frac{3y}{16} - \frac{z}{2})O_2 \rightarrow (\frac{x}{2} - \frac{y}{8})CO_2 + (\frac{x}{2} - \frac{y}{8})CO + \frac{y}{4}CH_4 \quad (3)$$

When the ratio of oxygen mass in the air to the mass of volatile matter is greater than k_1 but less than k_2, the combustion is in an incomplete state, and the products formed are CO, CH_4, CO_2 and H_2O:

$$\begin{cases} C_xH_yO_z + (\frac{3x}{4} - \frac{3y}{16} - \frac{z}{2})O_2 \rightarrow (\frac{x}{2} - \frac{y}{8})CO_2 + (\frac{x}{2} - \frac{y}{8})CO + \frac{y}{4}CH_4 \\ 2CO + CH_4 + 3O_2 \rightarrow 3CO_2 + 2H_2O \end{cases} \quad (4)$$

When the ratio of oxygen mass in the air to the mass of volatile matter is greater than k_2, the combustion is in a complete state, and the products formed areCO_2 and H_2O.

$$2CO + CH_4 + 3O_2 \rightarrow 3CO_2 + 2H_2O \quad (5)$$

The combustion process involves complex physical and chemical reactions. In practice, the furnace temperature is influenced by multiple operational parameters, which collectively determine the stability and efficiency of the incineration process. These parameters include primary air volume, primary air temperature, secondary air volume, secondary air temperature, main steam flow rate, grate temperature, left-side furnace temperature, furnace negative pressure, and inlet temperature of the combustion zone.

Specifically, primary air volume directly determines oxygen supply: a higher primary air volume enhances combustion reactions and raises the furnace temperature, whereas insufficient air supply leads to incomplete combustion and lowers the temperature. Similarly, primary air temperature plays a vital role in the combustion process; higher temperatures improve combustion efficiency, reduce heat loss, and increase furnace temperature, while lower temperatures may weaken combustion and reduce temperature. A moderate increase in secondary air volume enhances airflow and combustion uniformity, whereas excessive supply can lead to incomplete combustion and impair temperature regulation. Secondary air temperature also influences combustion efficiency: hot secondary air accelerates combustion and raises furnace temperature, while cooler air may slow down combustion and reduce temperature.

In addition, the main steam flow rate is closely related to heat output. Excessive steam flow may cause rapid heat loss in the furnace, lowering the temperature, while a moderate flow rate helps maintain temperature stability. Grate temperature reflects the combustion conditions at the furnace bottom: a higher grate temperature indicates efficient combustion, contributing to an overall increase in furnace temperature, whereas a lower grate temperature suggests incomplete combustion and reduced furnace temperature. The left-side furnace temperature is linked to airflow distribution; uneven temperature distribution may result in non-uniform combustion and hinder temperature rise, while uniform distribution contributes to stability. Furnace negative pressure regulates airflow and combustion efficiency: higher negative pressure increases oxygen supply and enhances combustion, thereby raising furnace temperature, whereas insufficient negative pressure may cause poor airflow and weaken combustion performance. Finally, the inlet temperature of the combustion zone, as the initial airflow temperature entering the furnace, has a direct effect on furnace heating. A higher inlet temperature accelerates combustion reactions and promotes an increase in furnace temperature.

In summary, the furnace temperature can be represented as being influenced by multiple variables, with complex nonlinear relationships existing between the temperature and these influencing factors, as shown in Equation (6).

$$Y(i) = F(x_1(i), x_2(i), x_3(i) \cdots x_9(i)) \tag{6}$$

In the equation, $Y(i)$ represents the furnace temperature, $F()$ indicates the unknown nonlinear relationship, and i refers to the sampling time. The input vector $x_1 \sim x_9$ consists of primary air volume, secondary air volume, main steam flow rate, grate speed, primary air temperature, secondary air temperature, left-side furnace temperature, furnace negative pressure, and inlet temperature of the combustion zone. Therefore, in essence, furnace temperature prediction is achieved by indirectly measuring these influencing variables and applying data-driven methods to establish a nonlinear model of the furnace temperature.

3. Furnace Temperature Prediction Model Based on IWOA-SAGRU

To address the challenges of accurately measuring furnace temperature and constructing reliable prediction models, this study proposes a data-driven IWOA-SAGRU furnace temperature prediction model for MSWI. As shown in Figure 2, the model consists of three modules: the MIC feature selection module, the SAGRU-based furnace temperature prediction module, and the IWOA-based parameter optimization module.

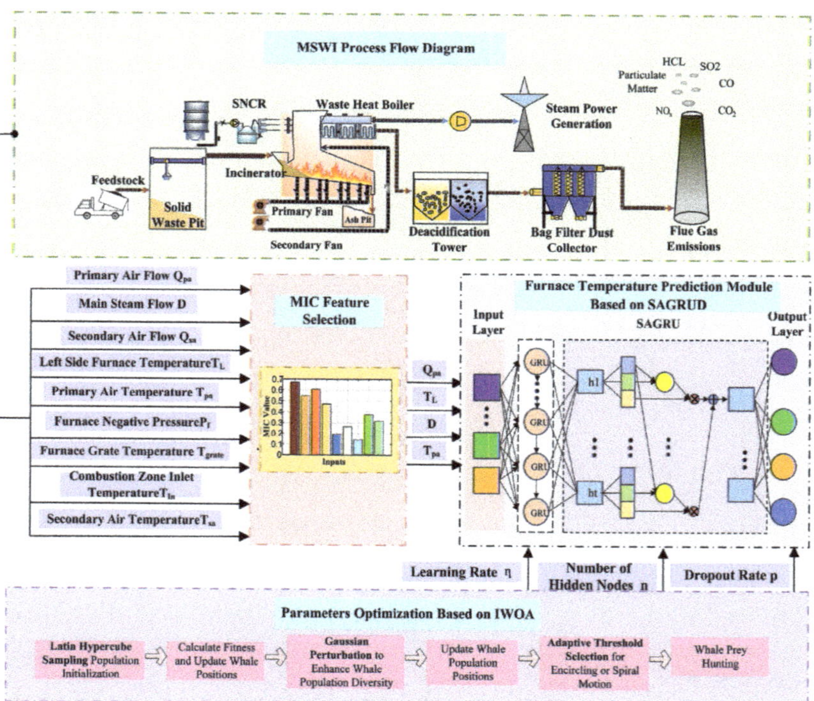

Figure 2. Furnace Temperature Prediction Modeling Strategy in MSWI.

In the MIC feature selection module, the maximal information coefficient (MIC) method is employed to screen the influencing factors of furnace temperature, balancing model accuracy with complexity reduction [31]. Based on expert knowledge, nine key variables—including the left-side temperature, combustion section inlet temperature, primary air flow, secondary air flow, primary air temperature, secondary air temperature, main steam flow, flue gas temperature, and furnace pressure—are identified as key temperature-influencing factors. These nine factors are further screened using MIC. Based on the MIC analysis results, four feature variables—Primary air flow, Main steam flow, Primary air temperature, and Left-side furnace temperature—are finally selected as inputs for the SAGRU-based furnace temperature prediction module, with the furnace temperature serving as the model output.

The SAGRU-based module integrates a self-attention (SA) mechanism with a gated recurrent unit (GRU) network to model the temporal dynamics of furnace temperature. In the IWOA-based parameter optimization module, the IWOA is used to optimize network parameters, including the learning rate (η), the number of hidden nodes (n), and the dropout rate (p), thereby improving the model's prediction performance.

3.1. Furnace Temperature Prediction Model Based on SAGRU

The Gated Recurrent Unit (GRU) effectively captures temporal dependencies of varying lengths in time-series data, rendering it well-suited for dynamic temperature forecasting. However, relying solely on GRU to extract sequence features is insufficient to address the complexity of furnace temperature prediction. To better capture the nonlinear relationships between measurable variables and furnace temperature, this study combines GRU with a SA mechanism and proposes the SAGRU model. In this model, the GRU processes the input furnace temperature time-series data, effectively memorizing and updating historical temperature information, while the self-attention mechanism adaptively assigns weights to different time steps, thereby highlighting critical operating variables and key points of temperature variation.

As shown in Figure 3, the structure of the model is provided. When the self-attention mechanism is combined with the GRU, four feature variables selected by the MIC analysis are used as inputs and processed by the GRU encoder to generate a series of hidden state $h_t = [h_1, h_2, h_3, \cdots h_t]$. The self-attention mechanism then assigns weights to these hidden states by generating queries (Q), keys (K), and values (V) to capture the critical features at different time steps within the sequence. The dot product between the queries and keys is computed to measure the similarity or relevance between the current state (query) and the historical states (keys), resulting in attention weights. These weights are then used to calculate a weighted average of the input sequence, which ultimately produces the predicted furnace temperature output from the model.

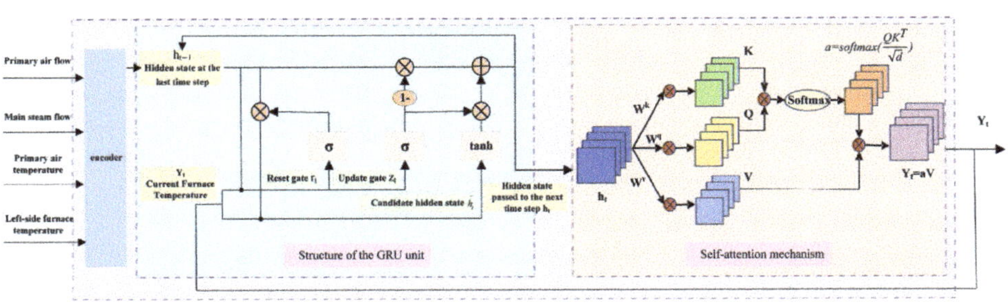

Figure 3. SAGRU Model Structure.

3.1.1. GRU-Based Furnace Temperature Prediction Model Structure

Since furnace temperature exhibits strong temporal dependence and dynamic characteristics, and is influenced by multiple factors such as feed rate, waste composition, and combustion air supply, the GRU, an improved architecture of the RNN [32], introduces a gating mechanism that effectively alleviates the vanishing gradient problem encountered by traditional RNNs when processing long sequences. This advantage makes GRU highly suitable for furnace temperature prediction, as it can efficiently handle temperature data with reduced risk of overfitting during training, while accurately capturing both short-term fluctuations and long-term trends in temperature variation. Compared with the LSTM [33] network, GRU has a simpler structure and fewer parameters. The GRU architecture contains two core gating mechanisms: the update gate and the reset gate. The update gate dynamically adjusts the balance between historical temperature memory and current operating information, which is particularly effective for handling abrupt temperature transitions caused by sudden changes in waste calorific value. The reset gate selectively discards irrelevant historical states, thereby filtering out sensor noise and improving the robustness of prediction results.

The reset gate r_t and update gate z_t are influenced by the previous hidden state h_{t-1} and the current furnace temperature Y_t. The calculation formulas for r_t and z_t are given in Equation (7) [34]:

$$\begin{aligned} r_t &= \sigma(w_{xr}Y_t + w_{hr}h_{t-1} + b_r) \\ z_t &= \sigma(w_{xr}Y_t + w_{hz}h_{t-1} + b_z) \end{aligned} \tag{7}$$

For the calculation of the candidate hidden state $\widetilde{h}_t$ the update relies on the current furnace temperature input together with part of the preceding hidden state. The reset gate r_t determines the extent to which historical furnace temperature information is retained:

$$\widetilde{h}_t = \tanh(w_{xh}Y_t + r_t \otimes (w_{hh}h_{t-1}) + b_h) \tag{8}$$

Finally, the update gate z_t updates the hidden state h_t at the current time step, which incorporates the furnace temperature information, as shown in Equation (9):

$$h_t = z_t \otimes h_{t-1} + (1 - z_t) \otimes \widetilde{h}_t \tag{9}$$

In the equation, σ represents the Sigmoid function; w_{xr}, w_{hr}, w_{hz}, w_{xh}, W_{hh} denotes the weight parameters; and b_r, b_h, b_z refers to the bias parameters.

3.1.2. Self-Learning of Influence Weight Coefficients Based on the Self-Attention Mechanism

When processing long-sequence incineration data, the GRU can effectively capture short-term dependencies in temperature fluctuations. However, it still faces limitations in modeling long-range dependencies, particularly under abnormal operating conditions that lead to significant furnace temperature variations. To address this issue, this study introduces the SA mechanism to enhance feature learning in long sequences and accurately capture complex long-range dependencies [35]. The core advantage of the self-attention mechanism lies in its ability to dynamically assign different weights to historical temperature data according to the current operating parameters. Specifically, the current operating state is encoded as a query, which is compared with historical states represented as keys to compute similarity scores. These scores are then normalized via a softmax function to produce attention weights, allowing the model to focus more on the most relevant past temperature data when predicting the current furnace temperature. In this way, the model can automatically learn the influence of each operating parameter on the present furnace temperature, effectively identifying and strengthening critical features while ignoring redundant or noisy information. Compared with traditional GRU networks, the self-attention mechanism is more suitable for handling complex data with significant temporal dependencies. In furnace temperature prediction tasks, it enables more precise weighting across different time steps, thereby not only capturing short-term fluctuations but also reinforcing long-term trends in temperature variation. This provides more accurate prediction performance under complex operating conditions. The specific implementation is as follows:

A series of hidden states ($h_t = [h_1, h_2, h_3, \cdots h_t]$) generated by the GRU model are linearly transformed to produce three matrices: the query (Q) matrix, key (K) matrix, and value (V) matrix. The query vectors are generated using the operating parameters at the current time step, and these operating conditions are compared with historical temperature data to determine their influence on the current temperature prediction. The specific equations are as follows [36]:

$$\mathbf{Q} = h_t \mathbf{W}^{\mathbf{q}} \tag{10}$$

The key matrix is constructed from the historical temperature data, where each time step of the historical sequence is mapped to a key vector. These key vectors are then compared with the query vector to calculate their similarity, which reflects the relevance of each historical moment to the current prediction.

$$\mathbf{K} = h_t \mathbf{W}^{\mathbf{k}} \tag{11}$$

The value matrix corresponds to the measured values or feature data of the historical temperature, aligned with the historical key vectors. After the attention weights are calculated, these weights are applied to the value matrix to compute a weighted average of the historical temperature data, thereby generating the result for the current temperature prediction.

$$\mathbf{V} = h_t \mathbf{W}^{\mathbf{v}} \tag{12}$$

In the equation: **Q**, **K**, **V** represent matrices obtained from the same input through different linear transformations; and $\mathbf{W}^{\mathbf{q}}$, $\mathbf{W}^{\mathbf{k}}$, $\mathbf{W}^{\mathbf{V}}$ are the learned weight matrices.

$$\mathbf{a} = \mathrm{softmax}(\frac{QK^T}{\sqrt{d}}) \tag{13}$$

$$Y_t = \mathbf{a}\mathbf{V} \tag{14}$$

Y_t represents the weighted temperature obtained from the historical temperature data, **a** denotes the attention weight, $\sqrt{d}$ indicates the vector dimension, and softmax is the normalization activation function.

3.2. IWOA-Based Parameter Optimization Algorithm

The IWOA optimization algorithm is employed to tune parameters of the GRU model, such as the learning rate, the number of hidden nodes (n) and the dropout rate (p), thereby reducing the tedious process of manual adjustment and improving parameter tuning efficiency. These parameters have a direct impact on the performance of the GRU model: the learning rate determines the step size of parameter updates during training, while the number of hidden nodes influences the model's representation and memory capacity, The dropout rate is used to prevent overfitting during training. By optimizing these parameters, IWOA enables the GRU to better adapt to complex datasets, thereby enhancing training efficiency and improving prediction accuracy.

The Whale Optimization Algorithm (WOA), inspired by the foraging behavior of humpback whales, is a swarm intelligence optimization method that mainly involves three stages: encircling prey, spiral hunting, and prey searching [37]. We selected WOA for this study because it offers strong global search capability, fast convergence, and relatively simple implementation with few parameters, making it well-suited for optimizing the parameters of our GRU-based furnace temperature prediction model. Compared with other algorithms such as Particle Swarm Optimization (PSO) or Genetic Algorithms (GA), WOA has demonstrated more stable and efficient performance in similar optimization tasks. WOA is adopted to optimize the parameters of the GRU model. First, a set of candidate initial parameters is randomly generated and used to train the GRU model. The training loss corresponding to each parameter configuration is then calculated and regarded as its fitness value. Next, whales simulate the behavior of encircling prey by approaching the current best solution and updating their positions according to the fitness of the candidate solutions. Guided by the global best solution, the whale individuals iteratively adjust their positions and gradually converge toward the global optimum through contraction behaviors. During the search process, the individuals update their positions based on training errors and fitness evaluations, and the population is updated accordingly. This iterative process continues until the predefined termination condition is satisfied. Finally, the optimized GRU parameters are output, ensuring that the GRU model achieves optimal performance in time-series prediction or other tasks. When new datasets or task requirements arise, the optimization process can be repeated to re-adjust the parameters.

(1) Encircling prey: In the WOA, the encircling phase simulates the process in which whales gradually approach their prey through a contraction behavior, and eventually converge to the global optimum in the later stage. This behavior is governed by the value of parameter A. When $|A| < 1$, the algorithm enters the encircling phase, and the whale individuals are updated according to the position of the current best solution. The mathematical model is given in Equation (15) [37].

$$\mathbf{L(t+1)} = \mathbf{L}^{*}(\mathbf{t}) - A\left|\mathbf{C}\mathbf{L}^{*}(\mathbf{t}) - \mathbf{L}(\mathbf{t})\right| \tag{15}$$

In the equation: t denotes the indicates how many iterations are performed. $\mathbf{L}(\mathbf{t})$, $\mathbf{L}^{*}(\mathbf{t})$ represent the position vectors of the whale individual and the prey, respectively; A denotes the coefficient vector, defined a $A = 2dr_1 - d$; C denotes the coefficient vector, $C = 2r_2$ (r is a random vector within [0, 1]), and d is a parameter linearly decreasing from 2 to 0.

$$d = 2 \times \frac{t_{\max} - t}{t_{\max}} \tag{16}$$

(2) Spiral hunting: In the WOA, during the spiral hunting phase, the process of whales moving around the prey along a spiral route is simulated. The corresponding mathematical model is given in Equation (17).

$$\mathbf{L(t+1)} = |\mathbf{L}^{*} - \mathbf{L}(\mathbf{t})| \cdot e^{bl}\cos(2\pi l) + \mathbf{L}^{*}(\mathbf{t}) \tag{17}$$

In the equation: b is used to define the logarithmic spiral shape, and l denotes a random number within the range $[-1, 1]$.

During the optimization process, the contraction-based encircling strategy and the spiral updating strategy of whale individuals are performed simultaneously. The selection between the two strategies is controlled by a probability p, and the position update is expressed as in Equation (18).

$$\mathbf{L(t+1)}\begin{cases} \mathbf{L}^*(\mathbf{t}) - \mathbf{A}\left|\mathbf{CL}^*(\mathbf{t}) - \mathbf{L}(\mathbf{t})\right|, p < 0.5 \\ \left|\mathbf{L}^* - \mathbf{L}(\mathbf{t})\right| \cdot e^{bl}\cos(2\pi l) + \mathbf{L}^*(\mathbf{t}), p \geq 0.5 \end{cases} \tag{18}$$

In the equation: p represents a random number between 0 and 1.

(3) Prey Search: In order to strengthen the global search ability and prevent convergence to local optima at an early stage, whale individuals randomly explore new prey positions within the search space. The corresponding mathematical model is expressed in Equation (19).

$$\mathbf{L(t+1)} = \mathbf{L_{rand}}(\mathbf{t}) - \mathbf{A} \cdot \left|\mathbf{CL_{rand}}(\mathbf{t}) - \mathbf{L}(\mathbf{t})\right| \tag{19}$$

In the equation: $\mathbf{L_{rand}}$ denotes the position vector of a randomly selected whale individual.

When $|A| \geq 1$, the whale does not move toward the current prey. Instead, it randomly selects another position $\mathbf{L_{rand}}$ in the search space and uses it as a reference point to perform global exploration.

To improve the traditional WOA's capability to avoid local optima and increase search efficiency, this study introduces three improvements to the algorithm.

(1) Population diversity enhancement based on Latin Hypercube Sampling (LHS).

During the population initialization stage, the LHS strategy is introduced to enhance population diversity. In the traditional WOA, the initial population is generated randomly, which may result in uneven distribution of individuals and overlaps among them [38]. Such issues can negatively affect the optimization performance of the algorithm [39]. Therefore, it is necessary to improve and optimize the population initialization method to strengthen the algorithm's search capability.

The specific implementation is as follows: let the initial population size of WOA be N, where each individual represents a solution and its position consists of v variables. For each variable $l_j (j = 1, 2, \cdots v)$, its domain $[l_j^{\min}, l_j^{\max}]$ is divided into M intervals, each with a length of $\left(l_j^{\max} - l_j^{\min}\right)/M$. Within each interval, a random point is selected as the sampling value of the variable, ensuring that each interval contains one sampling point. The N sampled points are then randomly recombined to form the individuals of the initial population. The position vector of an individual $\mathbf{L_{pos}}$ is represented as $\mathbf{L_{pos}} = (l_{i1}, l_{j=i2}, \cdots, l_{iv})$, where l_{ij} denotes the i-th variable value in the j-th sample combination. This sampling method provides a more uniform and efficient exploration of the search space, which not only improves search efficiency but also facilitates the identification of the optimal hyperparameter combination.

(2) Adaptive Threshold Selection Strategy

In the traditional WOA, during the prey-hunting phase, a random number within the range of [0, 1] is generated and compared with 0.5 to determine whether the whale chooses an encircling behavior or a spiral motion. However, this equal-probability selection strategy lacks reliability, as it does not allow whales to adaptively choose the most suitable movement pattern for the current individual. To address this limitation, this study introduces an adaptive threshold selection strategy to replace the fixed threshold in the conventional WOA. This method automatically adjusts the threshold according to the problem characteristics during the search process, thereby facilitating the identification of the optimal hyperparameter combination for the GRU model. The calculation formula for the adaptive threshold is given in Equation (20) [40].

$$p_a = 1 - \left[\frac{1}{\lambda + \xi} \cdot \left(\lambda \cdot \frac{t^{\lambda}}{t_{\max}{}^{\lambda}} + \xi \cdot \frac{t^{\xi}}{t_{\max}{}^{\xi}} \right) \right] \tag{20}$$

In the equation: t denotes the current iteration number; $t_{\max}$ represents the maximum number of iterations; and λ, ξ are the control parameters, with $\lambda = 2$, $\xi = 4$.

As shown in Equation (19), the adaptive selection threshold decreases gradually with the increase in iteration number. This indicates that, in the early stage of the algorithm, the adaptive probability threshold is relatively large, and whales tend to prioritize the encircling strategy. With the increase in iterations, the adaptive probability threshold gradually decreases, making it more likely for whales to adopt the spiral updating strategy. Accordingly, Equation (18) is updated to Equation (21).

$$\mathbf{L(t+1)} \begin{cases} \mathbf{L}^*(\mathbf{t}) - \mathbf{A}\left|\mathbf{CL}^*(\mathbf{t}) - \mathbf{L(t)}\right|, p < p_a \\ \left|\mathbf{L}^* - \mathbf{L(t)}\right| \cdot e^{bl} \cos(2\pi l) + \mathbf{L}^*(\mathbf{t}), p \geq p_a \end{cases} \tag{21}$$

(3) Population Resetting Based on Gaussian Disturbance

In the later stages of the search, swarm intelligence algorithms often encounter population aggregation, which reduces the diversity of the search population and makes the algorithm more prone to being trapped in local optima. To address this issue, Gaussian disturbance is introduced to implement population resetting. This strategy enhances diversity during the search process, prevents the algorithm from stagnating around local optima, improves global search efficiency, and ultimately enhances the quality of the optimization results.

The probability density function is given by [41]:

$$f(x) = \frac{1}{\sqrt{2\pi}\sigma} \exp\left(-\frac{(l-\mu)^2}{2\sigma^2} \right) \tag{22}$$

In Equation (17), when the value of σ is relatively large, the probability value at position l becomes lower, indicating that the curve is relatively flat and the probability distribution is more dispersed. In contrast, when the value of σ is smaller, the curve becomes steeper and the probability distribution is more concentrated. In the IWOA, the Gaussian perturbation applied to the optimal individual is formulated as Equation (23).

$$F(\mathrm{x}) = \mathbf{l}_*(\mathbf{t}) + \mathbf{l}_*(\mathbf{t}) \cdot normrnd(0,1) \tag{23}$$

In the equation, $normrnd(0,1)$ denotes a random number produced from a Gaussian distribution with a mean of 0 and variance of 1.

To evaluate the optimization capability of the improved IWOA, the objective function f_{object} and the fitness function $f_{fit}(IWOA)$ are defined as follows.

$$f_{object} = \frac{\sum\limits_{i=1}^{200} \sqrt{(b_i - \hat{b}_i)^2}}{200} \tag{24}$$

$$f_{fit}(IWOA) = \frac{1}{1 + f_{object}} \tag{25}$$

In the equation, b_i and $\hat{b}_i$ denote the measured temperature of the i-th sample and the temperature value calculated based on the optimized parameters in [42].

The correspondence between the IWOA and the parameter optimization problem is summarized in Table 1. In this framework, the position of a whale represents a potential feasible solution for the optimized parameters (η,n,p); the position of the prey represents the global optimal solution of the parameters (η,n,p).

Table 1. Correspondence between the IWOA and the Parameters Optimization Problem.

IWOA Parameter	Parameter Optimization Problem
Position of a whale	Potential feasible solution of the optimized parameters (η,n,p)
Positions of the whale population	Set of potential feasible solutions of the optimized parameters (η,n,p)
Movement of whales	Search process for the optimal solution of parameters (η,n,p)
Bubble-net attack	Update strategy for candidate solutions of parameters (η,n,p)
Encircling prey	Convergence process to the optimal solution of parameter (η,n,p)
Position of prey	Global optimal solution of parameter (η,n,p)

3.3. IWOA-SAGRU-Based Furnace Temperature Prediction: Algorithmic Workflow

The workflow of the IWOA-SAGRU-based furnace temperature prediction model is illustrated in Figure 4, The key steps are outlined as follows:

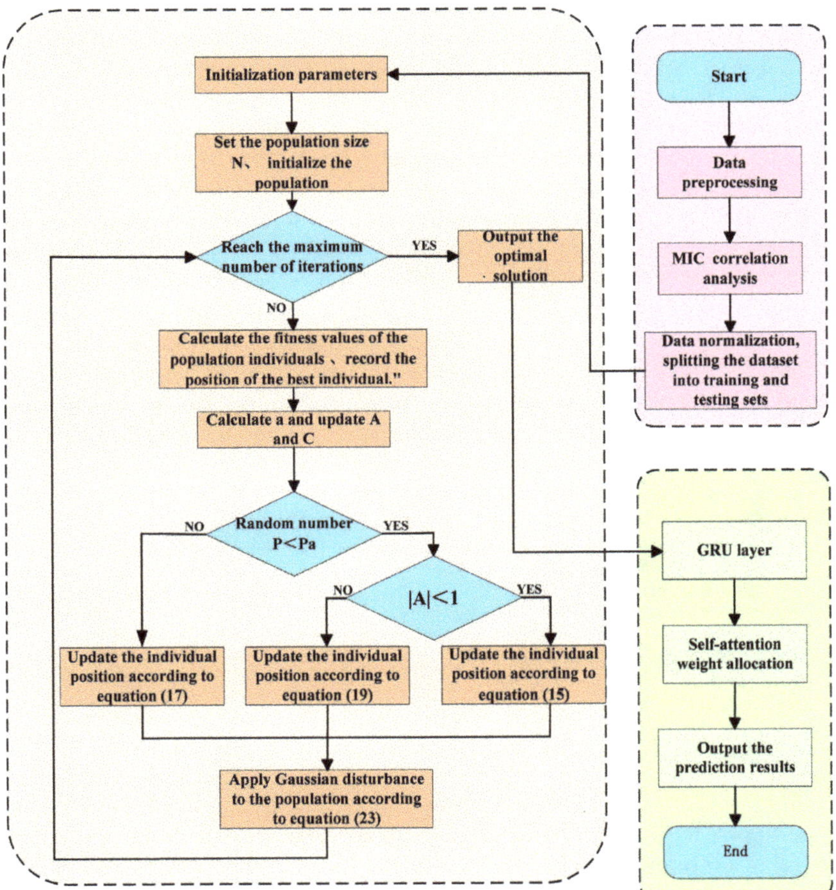

Figure 4. IWOA-SAGRU furnace temperature prediction model.

(1) According to MIC analysis results, the main variables influencing the furnace temperature are identified. The dataset is subsequently split into training and testing subsets, and the obtained data are normalized using the formula given in Equation (26).

$$s^* = \frac{s - s_{\min}}{s_{\max} - s_{\min}} \tag{26}$$

In the equation, s^* represents the normalized data, $s_{\max}$ and $s_{\min}$ denote the maximum and minimum values of the dataset, respectively.

(2) Construct the basic framework of the GRU-based network model and integrate the attention mechanism to improve the model's ability to generalize.

(3) Initialize the SAGRU model and employ the IWOA to update the learning rate (η), the number of hidden nodes (n) and dropout rate (p). The model parameters are iteratively updated until the error converges or the maximum number of iterations is reached, at which point the training process terminates.

(4) Output the prediction results and evaluate the model performance by comparing different evaluation metrics with those of other models.

4. Experimental Study

To verify the effectiveness of the IWOA-SAGRU model for furnace temperature prediction in MSWI, historical measured data from a municipal solid waste incineration plant in Beijing were used for case analysis. The municipal solid waste incineration (MSWI) plant adopts a moving grate combustion technology and mainly treats urban solid waste, with a daily treatment capacity of approximately 1600 tons. Electricity is primarily generated through waste incineration. The composition of the MSW treated in this plant is shown in Table 2 [43]. From the perspective of waste components, the moisture content is 48%, volatile matter is 33.31%, fixed carbon is 8.08%, and ash is 10.61%. In terms of elemental composition, carbon accounts for 65.2%, hydrogen for 8.09%, oxygen for 24.93%, nitrogen for 1.12%, and sulfur for 0.24%. These characteristics indicate a certain combustibility of the waste, but also highlight the necessity of pollutant control. At present, the operation of the plant relies mainly on manual control rather than full automation. From waste feeding and furnace combustion to waste heat recovery and power generation, the entire process depends on manual operation. Experienced operators perceive the changing operating conditions, predict temperature variations during combustion, and then adjust the manipulated variables to achieve temperature control.

Table 2. MSW properties.

Proximate (as a Received Basis, wt%)		**Ultimate (as a Dry Ash-Free Basis, wt%)**	
Moisture	48	C	65.2
Volatile matter	33.31	H	8.09
Fixed carbon	8.08	O	24.93
Ash	10.61	N	1.12
		S	0.24

The plant is equipped with a comprehensive monitoring and control system that can record furnace temperatures and related operating parameters in real time. A sampling interval of one second was adopted to continuously collect temperature data from key areas of the incinerator (from 08:00 to 24:00), covering the full production cycle. The core purpose of this design was to establish a raw temperature database with high temporal resolution. Considering the different requirements of data time scales in subsequent research applications, high-frequency sampling (1 s interval) was first used to preserve both the instantaneous fluctuations and the complete variation process of the temperature parameters. Then, targeted down-sampling was conducted according to specific research objectives (e.g., optimization of control algorithms and analysis of operating stability). This approach ensured the completeness and validity of the original data, while also providing flexible data support for studies at different levels, thus avoiding limitations in research scenarios caused by a fixed initial sampling interval.

In this experiment, 3000 sets of furnace temperature data and their related influencing factors were collected. After preprocessing, the dataset was divided into training and testing sets in a 4:1 ratio. The testing set was then used to evaluate the rationality of the proposed algorithm.

4.1. The MIC Feature Selection Module Algorithm

As previously mentioned, during the MSWI combustion process, there are numerous variables that are related to the furnace temperature. To guarantee accurate temperature prediction while simultaneously lowering model complexity, it is essential to select the relevant features associated with temperature. The MIC is a statistical method used to measure the degree of association between variables and can effectively capture the nonlinear relationships between them [44]. Compared to the Pearson linear correlation measure, MIC offers better stability and universality, enabling deeper analysis of the relationships between variables. In this study, a comparison between the Pearson and MIC analysis methods is performed. By calculating the correlation coefficients between different variables and the furnace temperature, variables with high correlation coefficients are selected as inputs, which helps reduce the number of input features and lower the model's dimensionality.

The correlation formula for the MIC analysis method is shown in Equation (27) [44]:

$$MIC(x,y) = \max_{a \cdot b < B} \frac{I(x,y)}{\log_2 \min(a,b)} \tag{27}$$

In the equation, a represents the number of horizontal grid divisions, b denotes the number of vertical grid divisions, and B refers to the variable, typically set to approximately 0.6 times the total data size. x is the feature variable, and y represents the furnace temperature.

The left-side temperature, combustion section inlet temperature, primary air flow, secondary air flow, primary air temperature, and secondary air temperature are identified as key temperature-influencing factors based on expert knowledge. The value ranges of each variable are presented in Table 3.

Table 3. Table of Temperature-Influencing Factors Variables.

Serial Number	Variable Name	Unit	Range of Variation
1	Primary air flow	m^3/s (N)	153.609–439.860
2	Primary air temperature	K	435.860–460.342
3	Main steam flow	kg/s	9191.778–9526.580
4	Left-side furnace temperature	K	1224.089–1301.078
5	Secondary air flow	m^3/s (N)	0.776–32.574
6	Inlet temperature of the combustion zone	K	461.757–502.531
7	Secondary air temperature	K	295.744–300.461

Table 3. *Cont.*

Serial Number	Variable Name	Unit	Range of Variation
8	Combustion grate temperature	K	451.638–598.742
9	Furnace negative pressure	Pa	−108.614–104.073

The correlation of the above-mentioned influencing variables was calculated to determine their effect on furnace temperature. The results were then compared with the Pearson correlation coefficient method, as detailed in Equation (28):

$$\rho = \frac{Cov(X, Y)}{\sigma_X \, \sigma_Y} \tag{28}$$

In the equation, $Cov(X, Y)$ represents the covariance between X and Y; σ_X denotes the standard deviation of X, σ_Y represents the standard deviation of Y. X refers to the influencing factor variable, while Y represents the furnace temperature.

To enhance the understanding of the influence of the above-mentioned variables on furnace temperature, visualizations were created. The results are shown in Figures 5 and 6.

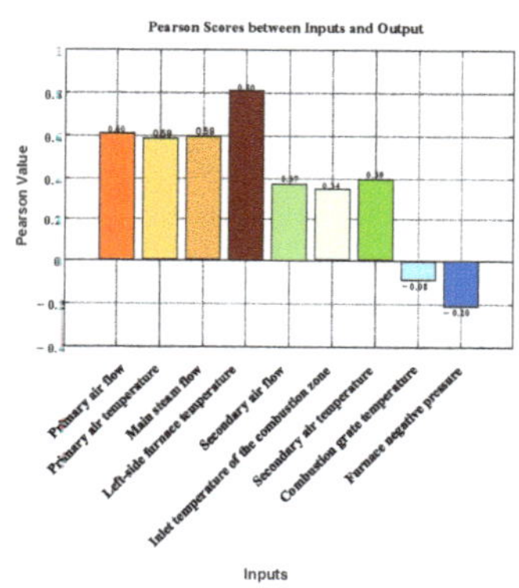

Figure 5. Pearson analysis was used to evaluate the correlation of the influencing factor variables.

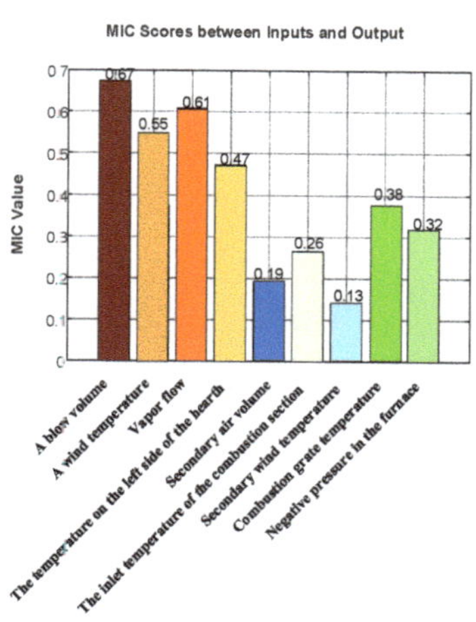

Figure 6. MIC analysis was used to evaluate the correlation of the influencing factor variables.

Figures 5 and 6 clearly reflect the correlation between the influencing variables and the output variable, furnace temperature. However, there are differences in the correlation among the variables.

In Table 4, variables showing a correlation above 0.5 with the output variable are presented, in order, are the furnace left-side temperature, primary air flow, main steam flow, and primary air temperature.

Table 4. Analysis of Pearson correlation.

Variable Name	Correlation Coefficient
Left-side furnace temperature	0.80
Primary air flow	0.60
Main steam flow	0.59
Primary air temperature	0.58
Secondary air temperature	0.39
Secondary air flow	0.37
Inlet temperature of the combustion zone	0.34
Furnace negative pressure	−0.20
Combustion grate temperature	−0.08

In Table 5, the four variables with the highest correlation identified by the MIC analysis method are, in descending order, the primary air volume, the main steam flow, the primary air temperature, and the left-side furnace temperature.

Table 5. Analysis of MIC correlation.

Variable Name	Correlation Coefficient
Primary air flow	0.67
Main steam flow	0.61
Primary air temperature	0.55
Left-side furnace temperature	0.47
Combustion grate temperature	0.38
Furnace negative pressure	0.32
Inlet temperature of the combustion zone	0.26
Secondary air flow	0.19
Secondary air temperature	0.13

There are differences in the correlation results of feature variables obtained by the two analysis methods. To verify the superiority of the MIC over the Pearson correlation coefficient in detecting nonlinear relationships, this study conducts a comparative analysis of the primary air temperature and the grate temperature in the combustion section using scatter plots combined with fitted curves, as shown in Figure 7.

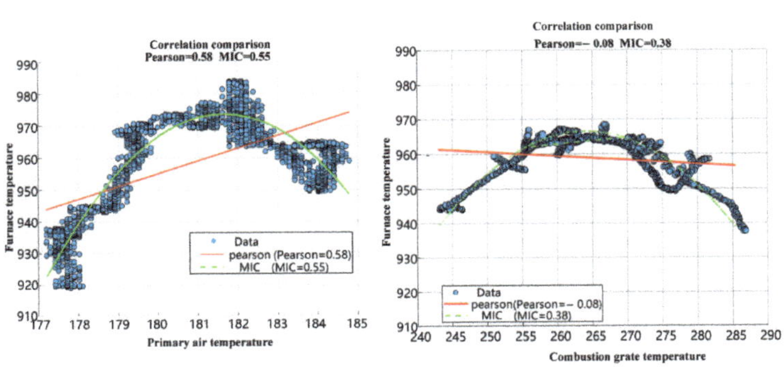

Figure 7. Fitted Plot of Feature Variables.

In Figure 7, the Pearson correlation coefficient can only capture linear relationships, and its fitted curve deviates significantly from the actual data distribution. In contrast, the MIC, through dynamic binning and mutual information maximization, successfully identifies the quadratic relationship embedded in the data. Its nonlinear fitted curve shows a markedly higher consistency with the data points. This finding confirms that, for complex nonlinear relationships, MIC provides a more comprehensive quantification of statistical dependence between variables. Therefore, four feature variables—primary air flow, main steam flow, primary air temperature, and left-side furnace temperature—are selected as inputs, with the furnace temperature as the output.

4.2. Performance Evaluation of the IWOA

To validate the optimization capability of the IWOA, this study selects the Genetic Algorithm (GA), Particle Swarm Optimization (PSO), and the standard WOA as comparison benchmarks. Six test functions with different characteristics are used for evaluation, as listed in Table 6. The algorithm is configured with a maximum of 500 iterations and a population size of 30. To eliminate errors caused by randomness, each of the four algorithms is independently executed multiple times on the six test functions, and the average optimal value is taken as the optimization result. The comparative results of different algorithms are summarized in Table 7.

Table 6. Test Functions.

Test Functions	Range
$F_1(x) = \sum_{i=1}^{n} x_i^2$	$[-100, 100]$
$F_2(x) = \sum_{i=1}^{n} \lvert x_i \rvert + \prod_{i=1}^{n} \lvert x_i \rvert$	$[-10, 10]$
$F_3(x) = \sum_{i=1}^{n} \left(\sum_{j-1}^{i} x_j \right)^2$	$[-100, 100]$
$F_4(x) = \sum_{i=1}^{n} i x_i^4 + random[0,1)$	$[-1.28, 1.28]$
$F_5(x) = \left[x_i^2 - 10\cos(2\pi x_i) + 10 \right]$	$[-5.12, 5.12]$
$F_6(x) = -20\exp\left(-0.2\sqrt{\frac{1}{n}\sum_{i=1}^{n} x_i^2} \right) - \exp\left[\frac{1}{n}\sum_{i=1}^{n} \cos(2\pi x_i) \right] + 20 + e$	$[-32, 32]$

Table 7. Results Comparison of Optimization Algorithms.

Test Functions	Evaluation Metrics	GA	PSO	WOA	IWOA
F_1	Mean	3.62×10^3	3.52×10^{-2}	7.20×10^{-10}	1.46×10^{-19}
	Best	1.45×10^3	1.41×10^{-3}	3.32×10^{-13}	1.17×10^{-24}
F_2	Mean	2.11×10^1	3.20×10^1	5.16×10^{-8}	1.74×10^{-13}
	Best	1.39×10^1	8.14×10^{-2}	5.12×10^{-9}	2.24×10^{-15}
F_3	Mean	3.48×10^3	1.00×10^3	8.98×10^{-10}	4.16×10^{-20}
	Best	1.77×10^3	1.75×10^{-3}	1.86×10^{-12}	1.40×10^{-22}
F_4	Mean	1.43×10^0	5.17×10^0	1.48×10^{-2}	7.48×10^{-4}
	Best	4.08×10^{-1}	6.53×10^{-2}	3.29×10^{-3}	1.42×10^{-4}
F_5	Mean	9.18×10^1	1.27×10^2	4.61×10^{-1}	0.00×10^0
	Best	6.48×10^1	6.92×10^1	6.78×10^{-11}	0.00×10^0
F_6	Mean	1.13×10^1	2.03×10^0	3.93×10^0	1.49×10^{-11}
	Best	9.19×10^0	2.32×10^{-2}	8.06×10^{-7}	1.35×10^{-12}

As shown in Table 7, when the average of the optimal values is used as the evaluation metric, the optimization results obtained by the IWOA outperform those of the other three algorithms across all six test functions. In particular, for the F(5) test function, both the average value and the optimal value found by the IWOA are equal to zero. Overall, the results indicate that the average values achieved by IWOA in multiple optimization runs are consistently smaller than those of the other algorithms, demonstrating that IWOA possesses superior global search capability.

4.3. Model Evaluation Metrics

To evaluate the performance of the IWOA-SAGRU temperature prediction model, this study employs the mean absolute error S_{MAE}, root mean square error S_{RMSE}, mean absolute percentage error S_{MAPE}, and coefficient of determination S_{R^2} as performance metrics. The corresponding formulas are given in Equations (29)–(32).

$$S_{\mathrm{MAE}} = \frac{1}{n_s}\sum_{i=1}^{n}\left|(\hat{y}_i - y_i)\right| \tag{29}$$

$$S_{\mathrm{RMSE}} = \sqrt{\frac{1}{n_s}\sum_{i=1}^{n}\left(y_i - \hat{y}_i\right)^2} \tag{30}$$

$$S_{\mathrm{MAPE}} = \frac{1}{n_s}\sum_{i=1}^{n}\left|\frac{y_i - \hat{y}_i}{y_i}\right| \times 100\% \tag{31}$$

$$S_{\mathrm{R}^2} = 1 - \frac{\sum_{i=1}^{n}(\hat{y}_i - y_i)^2}{\sum_{i=1}^{n}(y_i - \overline{y_i})^2} \tag{32}$$

In the equations, n_s denotes the number of samples; $\hat{y}_i$ and y_i represent the predicted temperature value and measured temperature value, respectively; $\overline{y_i}$ is the mean of measured temperature values. Parameters S_{MAE}, S_{RMSE} and S_{MAPE} reflect the error between the prediction and the measured values, where a smaller error indicates higher prediction accuracy of the model. Moreover, the closer S_{R^2} is to 1, the better the model fitting performance.

4.4. Experimental Analysis

The optimal parameter settings of the IWOA-SAGRU model are presented in Table 8. To compare the characteristics of the SAGRU prediction model, a GRU prediction model was also established. The key parameters of the GRU model were determined using IWOA. The calculation results and error comparisons are shown in Figure 8 and Table 9.

Table 8. Setting of model hyperparameters.

Hyperparameters	Initial Value	Optimized Value
Learning rate	0.001	0.039
Number of hidden nodes	20	32
dropout rate	0.13	0.25

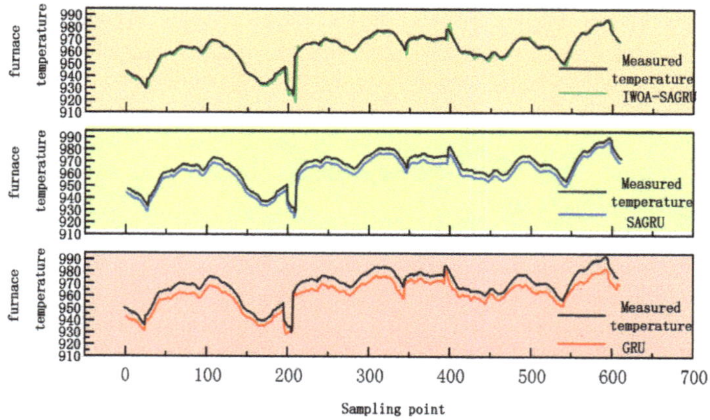

Figure 8. Comparison of Prediction Results of the Models.

Table 9. Comparison of model errors.

Model	S_{RMSE}	S_{MAE}	S_{MAPE}	S_{R^2}
GRU	2.1312	1.4367	0.1492	0.9717
SAGRU	1.8627	1.0022	0.1044	0.9789
IOWA-SAGRU	1.7275	0.8855	0.0927	0.9815

As shown in Figure 8, the single network model exhibits larger deviations in prediction results when the temperature data fluctuate significantly. In contrast, the IWOA-optimized SAGRU model captures the variation trend of the temperature data more effectively, resulting in a predicted curve that nearly overlaps with the measured data.

As shown in Table 9, the IWOA-SAGRU model achieves higher prediction accuracy and demonstrates better forecasting performance. The comparative experimental results indicate that, compared with the SAGRU model, the IWOA-SAGRU model reduces the errors by 7.82%, 12.62%, and 13.18%, respectively. This confirms that integrating the attention mechanism with the optimization algorithm can effectively improve prediction accuracy. The attention mechanism improves the forecasting accuracy of the GRU model by assigning suitable weights according to the inherent correlations between different variables. Meanwhile, the introduction of the IWOA optimization algorithm allows the determination of optimal values for key network parameters, leading to higher prediction accuracy compared with the trial-and-error method.

To further investigate the impact of different optimization algorithms on the performance of the SAGRU model, A population size of 10 and a maximum of 200 iterations were used. Under the same conditions, the proposed model was compared with the PSO-SAGRU and WOA-SAGRU models. The corresponding prediction results and errors are presented in Figure 9 and Table 10.

As shown in Figure 9 and Table 10, the prediction curve of the proposed IWOA-SAGRU model is closer to the measured furnace temperature variation curve compared with models optimized by other algorithms. The error metrics of the IWOA-optimized model are reduced by 11.31% and 3.23%, and by 22.76% and 8.95%, respectively, relative to the other models. This indicates that the IWOA can better identify the optimal combination of key parameters (learning rate η, number of hidden nodes n and dropout rate p), thereby reducing the prediction error of the SAGRU model and improving its accuracy. These results verify the superiority of the IWOA-SAGRU model.

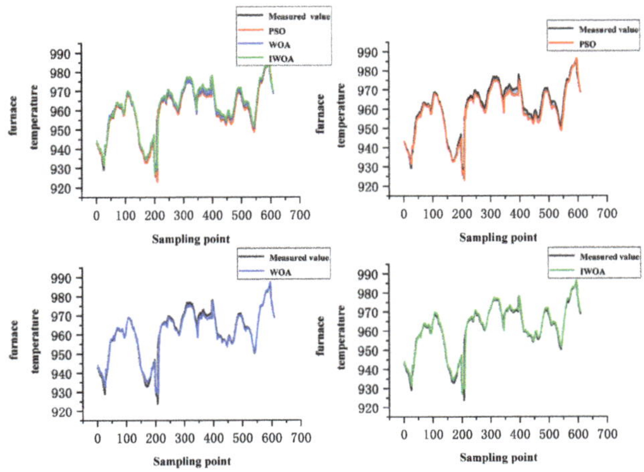

Figure 9. Comparison of Prediction Models Based on Different Algorithms.

Table 10. Comparison of algorithm errors.

Model	S_{RMSE}	S_{MAE}	S_{MAPE}	S_{R^2}
PSO	1.9229	1.0897	0.1138	0.9771
WOA	1.7833	0.9660	0.1010	0.9802
IWOA	1.7275	0.8855	0.0927	0.9815

To further assess the performance advantage of the IWOA-SAGRU model in temperature prediction, comparative tests were carried out using the BP neural network, Elman neural network, and LSTM network. To ensure fairness, the parameters of the BP, Elman, and LSTM models were standardized: each network contained 10 neurons, trained for 100 epochs, with a learning rate of 0.001 and a batch size of 64. The prediction outcomes of all models are illustrated in Figure 10 and summarized in Table 11.

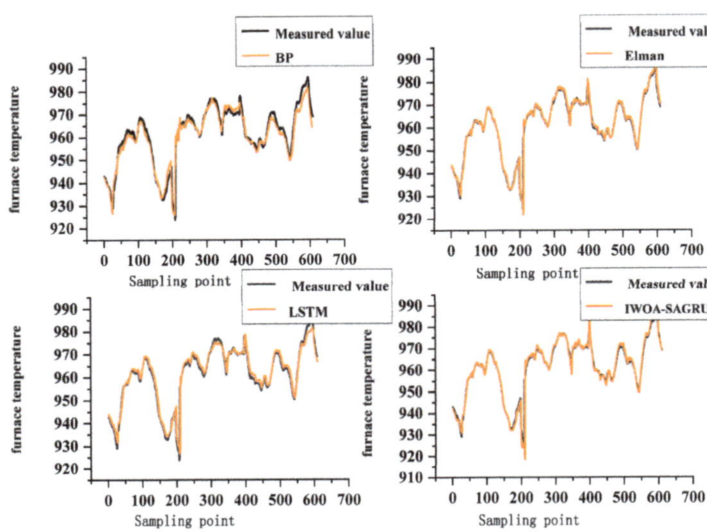

Figure 10. Comparison of Prediction Results of Different Models.

Table 11. Comparison of error metrics for forecasting models.

Model	S_{RMSE}	S_{MAE}	S_{MAPE}	S_{R^2}
BP	3.0976	1.9128	0.1981	0.9406
Elman	2.4421	1.5139	0.1576	0.9628
LSTM	2.1964	1.4533	0.1519	0.9701
IWOA-SAGRU	1.7275	0.8855	0.0927	0.9815

From Figure 10, Table 11, it can be observed that the IWOA-SAGRU model achieves the best prediction performance, with the lowest values of S_{RMSE}, S_{MAPE}, and S_{MAE}. In contrast, the BP and Elman neural network models, as classical deep learning algorithms, have limited ability to capture long-term temporal patterns. Although the LSTM network addresses the issue of long-term dependency to some extent, its complex structure results in low computational efficiency. Moreover, the absence of an attention mechanism limits its ability to focus on key features in complex tasks. By integrating the IWOA optimization algorithm, the IWOA-SAGRU model is able to identify optimal parameter combinations more effectively than the traditional gradient descent method. When compared with the Elman and LSTM models, the proposed model reduces error S_{RMSE} by 0.7146 and 0.4689, respectively, while errors S_{MAE} and S_{MAPE} are minimized to 0.8855 and 0.0927%. Furthermore, S_{R^2} is closer to 1, indicating superior fitting and generalization ability. Analysis of the prediction results indicates that the proposed furnace temperature prediction model attains high forecasting accuracy, enabling early detection of temperature anomalies and contributing to the reduction in harmful substance emissions.

5. Conclusions

This study proposes an IWOA-optimized SAGRU model for predicting the furnace temperature in municipal solid waste incineration (MSWI), aiming to overcome the limitations of conventional temperature measurement methods under complex operating conditions. By introducing a self-attention mechanism into the GRU network and optimizing key parameters such as learning rate, number of hidden nodes, and dropout rate using IWOA, the proposed approach significantly improves prediction accuracy, generalization capability, and computational efficiency. The results indicate that the model can more effectively capture both short-term dynamics and long-term trends of furnace temperature, thereby enabling stable and efficient combustion control.

From a sustainability perspective, this study has important practical significance. Accurate furnace temperature prediction not only helps optimize the municipal solid waste incineration process and improve energy recovery efficiency but also reduces reliance on fossil fuels and promotes the clean utilization of waste-derived thermal energy. In addition, by enabling more precise combustion control, harmful gas emissions can be effectively reduced, contributing to the achievement of emission reduction targets. These outcomes provide strong technical support for advancing low-carbon waste management and the development of a circular economy.

Nevertheless, this study has certain limitations. First, the model validation data are primarily derived from a specific MSWI plant in China, and variations in waste composition, equipment conditions, and operational modes across different regions may affect its generalizability and applicability. Second, potential equipment malfunctions or extreme environmental conditions during actual incineration processes have not been fully incorporated into the prediction framework. Future research could focus on expanding the collection and utilization of multi-source data to enhance the model's applicability and robustness. In addition, integrating equipment condition monitoring and environmental factors into the prediction system could improve the model's adaptability to complex operating conditions. Furthermore, exploring the combination of advanced techniques such as reinforcement learning and edge computing may further enhance the model's real-time performance and self-adaptive capabilities.

Overall, the proposed IWOA-SAGRU model not only offers a new technical approach for furnace temperature prediction and combustion optimization in MSWI processes but also provides theoretical support and practical value for achieving efficient energy utilization, emission reduction, and low-carbon waste management. It holds important reference significance for promoting intelligent and green urban solid waste management systems. Moreover, the findings of this study can provide data support and decision-making guidance for governments and management authorities in formulating emission reduction policies for waste incineration, optimizing energy utilization strategies, and developing low-carbon cities, thereby further advancing the achievement of sustainability goals.

Author Contributions: Methodology, J.P. and M.S.; Software, A.S.; Validation, A.S.; Investigation, M.S.; Resources, J.T.; Writing—original draft, M.S.; Writing—review & editing, J.T.; Supervision, J.P. All authors have read and agreed to the published version of the manuscript.

Funding: This research received no external funding.

Institutional Review Board Statement: Not applicable.

Informed Consent Statement: Not applicable.

Data Availability Statement: The original contributions presented in this study are included in the article. Further inquiries can be directed to the corresponding author(s).

Conflicts of Interest: The authors declare no conflicts of interest.

Abbreviations

Variable	Variable Name
n	Number of hidden nodes
η	Learning rate
p	Dropout rate
m_w	the change in moisture mass of a single waste module
dR_{air}	the change in the thermal energy of the waste module.
dR_{rad}	the change in radiative heat of the waste module
R_{water}	the heat absorbed for the evaporation of each kilogram of water
m_T	the amount of volatile matter released over a period T
m_{vol}	the initial amount of volatile matter
A_V	refers to the pre-exponential factor
E_V	the activation energy
R	the universal gas constant
Y	Furnace temperature
x	Variables affecting furnace temperature
h_t	Candidate hidden state
$\mathbf{Q}$	Query matrix
$\mathbf{K}$	Key matrix
$\mathbf{V}$	Value matrix
r_t	reset gate
z_t	update gate
h_{t-1}	Hidden state at the previous time step
σ	Sigmoid function
$\mathbf{W^q}$	Learned weight matrix
$\mathbf{W^k}$	Learned weight matrix

$\mathbf{W^v}$	Learned weight matrix
$\mathbf{a}$	Attention weight
T_W	Weighted temperature obtained from the historical temperature data
$\mathbf{L(t)}$	The position vectors of the whale individual
$\mathbf{L^*(t)}$	The prey, respectively
$\mathbf{A}$	The coefficient vector
$\mathbf{C}$	The coefficient vector
t	Number of iterations
N	The initial population size
$\mathbf{v}$	Number of population position vectors
M	Number of partitioned domain intervals
$\mathbf{L_{pos}}$	The position vector of an individual
p_a	The calculation formula for the adaptive threshold
b_i	Measured temperature of the i-th sample
f_{object}	Objective Function
f_{fit}	Fitness Function
s^*	The normalized data
s_{max}	The maximum values of the dataset
s_{min}	The minimum values of the dataset
n_s	The number of samples
$\hat{y}_i$	The predicted temperature value
y_i	The measured temperature value
$\overline{y_i}$	The mean of measured temperature values

References

1. Margallo, M.; Taddei, M.B.M.; Hernández-Pellón, A.; Aldaco, R.; Irabien, Á. Environmental Sustainability Assessment of the Management of Municipal Solid Waste Incineration Residues: A Review of the Current Situation. *Clean Technol. Environ. Poli* **2015**, *17*, 1333–1353. [CrossRef]
2. Beylot, A.; Hochar, A.; Michel, P.; Descat, M.; Ménard, Y.; Villeneuve, J. Municipal Solid Waste Incineration in France: A Overview of Air Pollution Control Techniques, Emissions, and Energy Efficiency. *J. Ind. Ecol.* **2018**, *22*, 1016–1026. [CrossRef]
3. Nguyen, T.D.; Kawai, K.; Nakakubo, T. Drivers and Constraints of Waste-to-Energy Incineration for Sustainable Municipal Sol Waste Management in Developing Countries. *J. Mater. Cycles Waste Manag.* **2021**, *23*, 1688–1697. [CrossRef]
4. Jian, T.; Heng, X.; Wen, Y.; Junfei, Q. Research Status and Prospects of Intelligent Optimization Control for Municipal Solid Was Incineration Process. *Zidonghua Xuebao* **2023**, *49*, 2019–2059.
5. Lu, J.-W.; Zhang, S.; Hai, J.; Lei, M. Status and Perspectives of Municipal Solid Waste Incineration in China: A Comparison wi Developed Regions. *Waste Manag.* **2017**, *69*, 170–186. [CrossRef]
6. Yan, A.; Wang, R. A Knowledge Transfer Online Stochastic Configuration Network-Based Prediction Model for Furna Temperature in Municipal Solid Waste Incineration Process | Request PDF. *ResearchGate* **2025**, *243*, 122733. [CrossRef]
7. Magnanelli, E.; Tranås, O.L.; Carlsson, P.; Mosby, J.; Becidan, M. Dynamic Modeling of Municipal Solid Waste Incineration. *Ener* **2020**, *209*, 118426. [CrossRef]
8. Nabavi-Pelesaraei, A.; Bayat, R.; Hosseinzadeh-Bandbafha, H.; Afrasyabi, H.; Chau, K. Modeling of Energy Consumption ar Environmental Life Cycle Assessment for Incineration and Landfill Systems of Municipal Solid Waste Management—A Ca Study in Tehran Metropolis of Iran. *J. Clean. Prod.* **2017**, *148*, 427–440. [CrossRef]
9. Lv, L.; Ding, Y.J. Modeling of an MSW Incinerator. *J. Tsinghua Univ. (Sci. Tech.)* **2006**, *11*, 1884–1887. [CrossRef]
10. Qin, Y.F.; Bai, Y. Modeling and Simulation of Combustion Process of MSW Volatile. *J. Chin. Soc. Power Eng.* **2010**, *30*, 444–449+45
11. Hu, Y.; Zhang, H.; Niu, Y.; Li, Z.; Li, Q.; Han, Y. Intelligent Control of Sludge and Municipal Solid Waste Incineration Usir Artificial Neural Network. *Energy Sources Part A Recovery Util. Environ. Eff.* **2025**, *47*, 2526671. [CrossRef]
12. Ding, H.; Qiao, J.; Huang, W.; Yu, T. Event-Triggered Online Learning Fuzzy-Neural Robust Control for Furnace Temperature Municipal Solid Waste Incineration Process. *IEEE Trans. Autom. Sci. Eng.* **2024**, *21*, 1201–1213. [CrossRef]
13. Hu, Q.; Long, J. A Novel Time-Span Input Neural Network for Accurate Municipal Solid Waste Incineration Boiler Stea Temperature Prediction. *J. Zhejiang Univ.-Sci. A* **2021**, *22*, 777–792. [CrossRef]
14. Wang, T.; Tang, J.; Xia, H. Intelligent Optimal Control of Furnace Temperature for the Municipal Solid Waste Incineration Proce Using Multi-Loop Controller and Particle Swarm Optimization. *Expert Syst. Appl.* **2024**, *257*, 125015. [CrossRef]

Xiong, S.; Peng, Y.; Lu, S.; Shang, F.; Li, X.; Yan, J.; Cen, K. Generalized Prediction and Optimal Operating Parameters of PCDD/F Emissions by Explainable Bayesian Support Vector Regression. *Waste Manag.* **2021**, *135*, 437–447. [CrossRef]

Yan, A.; Guo, J.; Wang, D. Heterogeneous Feature Ensemble Modeling with Stochastic Configuration Networks for Predicting Furnace Temperature of a Municipal Solid Waste Incineration Process. *Neural Comput. Appl.* **2022**, *34*, 15807–15819. [CrossRef]

CAI, M.; Wu, C. Prediction Method of Flue Gas Concentration of Waste Incinerator Based on CNN-LSTM. *Chem. Ind. Eng.* **2025**, *42*, 165–172. [CrossRef]

Zheng, J.; Ye, L.; Ge, Z. Laplacian Regularization of Linear Regression Model for Semi-Supervised Industrial Soft Sensor Development. *Expert Syst. Appl.* **2024**, *254*, 124459. [CrossRef]

Zhang, Z.; Yan, G. Instance Transfer Partial Least Squares for Semi-Supervised Adaptive Soft Sensor. *Chemom. Intell. Lab. Syst.* **2024**, *245*, 105062. [CrossRef]

Herceg, S.; Ujević Andrijić, Ž. Support Vector Machine-Based Soft Sensors in the Isomerisation Process. *Chem. Biochem. Eng. Q.* **2025**, *34*, 1825. [CrossRef]

Hua, H.; Xie, X.; Sun, J.; Qin, G.; Tang, C.; Zhang, Z.; Ding, Z.; Yue, W. Graphene Foam Chemical Sensor System Based on Principal Component Analysis and Backpropagation Neural Network. *Adv. Condens. Matter Phys.* **2018**, *2018*, 2361571. [CrossRef]

Curreri, F.; Patanè, L.; Xibilia, M.G. Soft Sensor Transferability: A Survey. *Appl. Sci.* **2021**, *11*, 7710. [CrossRef]

Alakent, B. Soft-Sensor Design via Task Transferred Just-in-Time-Learning Coupled Transductive Moving Window Learner. *J. Process Control.* **2021**, *101*, 52–67. [CrossRef]

Ławryńczuk, M.; Zarzycki, K. LSTM and GRU Type Recurrent Neural Networks in Model Predictive Control: A Review. *Neurocomputing* **2025**, *632*, 129712. [CrossRef]

Fan, Y.; Liu, X.; Zhang, C.; Li, C.; Li, X.; Wang, H. Dynamic Prediction of Boiler NOx Emission with Graph Convolutional Gated Recurrent Unit Model Optimized by Genetic Algorithm. *Energy* **2024**, *294*, 130957. [CrossRef]

Guo, J.; Li, J.; Sato, Y.; Yan, Z. A Gated Recurrent Unit Model with Fibonacci Attenuation Particle Swarm Optimization for Carbon Emission Prediction. *Processes* **2024**, *12*, 1063. [CrossRef]

Chen, W.; Hussain, W. Deep Learning for Financial Time Series Prediction: A State-of-the-Art Review of Standalone and Hybrid Models. *ResearchGate* **2025**, *139*, 187–224. [CrossRef]

Tian, H.; Tang, J.; Xiao, H.; Wang, T. Furnace Temperature Control Using IT2FBLS-based Reinforcement Learning PID for MSWI Process. *Acta Autom. Sin.* **2025**, *51*, 1626–1641. [CrossRef]

Báreková, A.; Franeková, Z. Composition Analysis of Municipal Solid Waste at a University Dormitory. *Acta Hortic. Regiotectuare* **2015**, *18*, 49–52. [CrossRef]

Wan, S.; Zhou, X.; Zhou, M.; Han, Y.; Chen, Y.; Geng, J.; Wang, T.; Xu, S.; Qiu, Z.; Hou, H. Hydration Characteristics and Modeling of Ternary System of Municipal Solid Wastes Incineration Fly Ash-Blast Furnace Slag-Cement. *Constr. Build. Mater.* **2018**, *180*, 154–166. [CrossRef]

Li, T.; Xia, T.; Zhang, H.; Liu, D.; Zhao, H.; Liu, Z. False Data Injection Attacks Detection Based on Stacking and MIC-DCXGB. *Sustainability* **2024**, *16*, 9692. [CrossRef]

Wang, Z.; Zhao, B. Short-Term Load Forecasting Method Based on GRU-Neural Network Model. *Autom. Electr. Power Syst.* **2019**, *43*, 53–58. [CrossRef]

Wang, Y.; Zhao, K. Short-Term Wind Power Prediction Using a Novel Model Based on Butterfly Optimization Algorithm-Variational Mode Decomposition-Long Short-Term Memory. *Appl. Energy* **2024**, *366*, 123313. [CrossRef]

Yu, H.; Xu, Y. Research on Gas Emission Prediction Model with Improved Pelican Algorithm to Optimize AM-GRU. *Control. Eng. China* **2024**. [CrossRef]

Wei, W.; Zhu, C.; Hu, L.; Liu, P. Application of a Transfer Learning Model Combining CNN and Self-Attention Mechanism in Wireless Signal Recognition. *Sensors* **2025**, *25*, 4202. [CrossRef]

Wu, X.; Chen, L.; Wei, T.; Fan, T. Sentiment Analysis of Chinese Short Text Based on Self-Attention and Bi-LSTM. *J. Chin. Inf. Process.* **2019**, *33*, 100–107.

Mirjalili, S.; Lewis, A. The Whale Optimization Algorithm. *Adv. Eng. Softw.* **2016**, *95*, 51–67. [CrossRef]

Si, Q.; Li, C. Indoor Robot Path Planning Using an Improved Whale Optimization Algorithm. *Sensors* **2023**, *23*, 3988. [CrossRef]

Gao, D.; Liang, H. Research on Test Data Generation for Improved Chimpanzee Optimization Algorithm. *Comput. Eng. Appl.* **2022**, *58*, 83–93. [CrossRef]

Yang, L.; Zhou, J.; Cao, C. Trajectory tracking of hydraulic manipulator optimized by improved whale genetic algorithm. *Transducer Microsyst. Technol.* **2025**, *44*, 94–98. [CrossRef]

Li, Q.; Wang, Z.; Du, H.; Lu, D. AdaptiveDensityPeakClusteringAlgorithmBasedonGaussianDistribution. *Comput. Eng.* **2025**, *51*, 137–148. [CrossRef]

He, H.; Meng, X. ET-RBF-PID-Based Control Method for Furnace Temperature of Municipal Waste Incineration Process. *Control. Theory Appl.* **2022**, *39*, 2262–2273. [CrossRef]

43. Chen, J.; Tang, J.; Xia, H.; Wang, T.; Gao, B. A Non-Manipulated Variable Analysis of Solid-Phase Combustion in the Furnace Municipal Solid-Waste Incineration Process Based on the Biorthogonal Numerical-Simulation Experiment. *Sustainability* **2023**, 14159. [CrossRef]
44. Bi, Y.; Zheng, B. Vibration Trend Prediction of Hydroelectric Units Based on MIC and BiGRU. *Shuili Xuebao* **2021**, *52*, 612–621+6 [CrossRef]

sustainability

MDPI

ticle

eal-Time Automatic Identification of Plastic Waste Streams for dvanced Waste Sorting Systems

bert Giel, Mateusz Fiedeń and Alicja Dąbrowska *

Faculty of Mechanical Engineering, Wroclaw University of Science and Technology, 50-371 Wroclaw, Poland; robert.giel@pwr.edu.pl (R.G.); mateusz.fieden@pwr.edu.pl (M.F.)
* Correspondence: alicja.dabrowska@pwr.edu.pl; Tel.: +48-71-320-23-91

Abstract: Despite the significant recycling potential, a massive generation of plastic waste is observed year after year. One of the causes of this phenomenon is the issue of ineffective waste stream sorting, primarily arising from the uncertainty in the composition of the waste stream. The recycling process cannot be carried out without the proper separation of different types of plastics from the waste stream. Current solutions in the field of automated waste stream identification rely on small-scale datasets that insufficiently reflect real-world conditions. For this reason, the article proposes a real-time identification model based on a CNN (convolutional neural network) and a newly constructed, self-built dataset. The model was evaluated in two stages. The first stage was based on the separated validation dataset, and the second was based on the developed test bench, a replica of the real system. The model was evaluated under laboratory conditions, with a strong emphasis on maximally reflecting real-world conditions. Once included in the sensor fusion, the proposed approach will provide full information on the characteristics of the waste stream, which will ultimately enable the efficient separation of plastic from the mixed stream. Improving this process will significantly support the United Nations' 2030 Agenda for Sustainable Development.

Keywords: waste sorting; plastic waste; recycling; circular economy; sustainability; automatic waste classification; waste identification; waste management

1. Introduction

Despite being in use for over a century, the global production and consumption of plastics are still increasing exponentially [1]. This ultimately contributes to the massive generation of plastic waste, which raises many environmental concerns associated with prolonged degradation time. However, it is possible to significantly reduce the negative environmental impacts by utilizing plastic's recycling potential, allowing for multiple recycling cycles while preserving its original value and functionality [2]. For recycling to be feasible, a crucial preliminary action is the efficient sorting of collected waste [3] to separate different types of plastics for further processing. Despite the availability of numerous advanced technologies designed to support this process [4,5], successful waste sorting remains a complex and demanding task due to unresolved issues related to the inherent uncertainty of the waste stream.

The solid waste stream indicates high heterogeneity and variability in composition [6]. These attributes are consistently identified as key factors contributing to the challenges in optimizing waste-sorting processes aimed at maximizing the recovery of recyclable materials [7]. The primary challenge lies in identifying the composition of the stream, which consists of a wide variety of materials. For this reason, numerous studies have been published in recent years reviewing this issue. A summary of such publications highlighting the key conclusions and limitations of existing developments is presented in Table 1.

Table 1. Summary of selected review publications in the area of automatic waste stream identification.

Ref.	Key Conclusions	Main Limitations of Current Developments
[3]	The predominant approach to waste identification involves processing images captured by visible light cameras utilizing a CNN model.	- unrealistic, generic datasets not specific to any particular geographical region - testing under laboratory conditions - rarely used sensor fusion
[8]	CNNs have been increasingly applied since 2018 and became the most frequently applied by 2021. Additionally, about a third of all studies focused on plastics.	- more than 90% of all studies were conducted on a laboratory scale - little research addressed material flow level analysis, and no studies were found at the process level
[9]	Deep learning has emerged as the leading computer vision algorithm for enabling efficient solid waste sorting.	- simplified operational conditions resulting in a disconnect between research outcomes and practical applications in real-world settings - lack of publicly accessible real-world datasets that are specifically aligned with industrial waste-sorting practices
[10]	In recent years, CNN has demonstrated outstanding performance in the field of image classification.	- there are few waste image datasets available for model training - lack of large-scale datasets
[11]	Integrating various sensors with suitable algorithms alongside industrial robots can achieve highly effective performance even in complex conditions that challenge human expertise.	- most experiments are conducted in controlled laboratory environments
[12]	In recent years, machine learning has gained growing significance in the advancement of sensor-based sorting systems.	- insufficient consideration of plastic waste heterogeneity - capabilities of multi-sensor solutions remain underutilized

In [3], more than fifty studies were reviewed to identify the physical enablers, datasets, and machine learning algorithms used for waste identification in indirect separation systems. A broader coverage is included in [8], where almost two hundred papers were analyzed over a 20-year period on the subject of optical sensors and machine learning algorithms for sensor-based material flow characterization. Similarly, ref. [10] reviewed over two hundred studies that used machine learning algorithms for municipal solid waste management, including automatic stream identification. In [9], an extensive review was presented addressing the development trajectory, current state, and future opportunities and challenges of utilizing computer vision for waste sorting. In another study [11], commonly utilized sensor-based technologies for solid waste sorting, including spectroscopic- and computer vision-based classification, were found. It should be noted that all these review publications from the past few years highlight similar limitations of existing solutions for automated waste stream identification. The most frequently reported limitations across all studies include testing under controlled laboratory conditions and challenges arising from using small-scale datasets, such as failure to account for waste diversity or insufficient generalization of models. On the one hand, this highlights the necessity of developing large-scale datasets that account for the specific characteristics of waste originating from different regions. On the other hand, laboratory conditions should be adjusted to replicate real-world scenarios strongly.

Object detection models have been intensively developed in recent years. In [11], the lines of development of object detection algorithms were presented, stating that currently, there are three mainstream series of object detectors: R-CNN, YOLO, and SSD. A comparative analysis demonstrated that YOLO outperforms both R-CNN and SSD due to its higher speed and feature extraction capabilities. Similar conclusions were drawn in [13], where YOLO, compared to R-CNN, demonstrated superior real-time performance with shorter inference times. Given that automatic waste stream identification requires the real-time, rapid detection of multiple objects within a single frame, accounting for variations in size and orientation, the YOLO series is one of the most commonly utilized frameworks [14]. Table 2 summarizes recent publications on automatic waste identification using YOLO.

Table 2. Summary of selected publications from recent years in the area of automatic waste identification based on CNN YOLO.

Ref.	Waste Type	Contaminants	Validation	Dataset	Classes
[15]	construction	included (without detection)	static	public and self-built	4 (bricks, wood, stones, plastics)
[16]	household	included (detected class)	static	self-built	10 (not recyclable, textile, leather, paper, metal, plastic, glass, wood, ceramic, food)
[13]	plastic	excluded	static	public	4 (PET, PE, PP, PS)
[17]	plastic	included	dynamic	self-built	2 (PET, PET-G)
[18]	river floating garbage	excluded	static	private and public	only detection
[19]	household	excluded	static	self-built	4 (recyclable, other, harmful, kitchen)
[20]	household	included (without detection)	static	self-built	9 (paper, plastic, snakeskin bags, packed waste, stone, sand, carton, foam, metal)

In [15], an improved YOLOv5 model was proposed considering the requirements of construction waste, i.e., small-object detection and inter-object occlusion. Household waste was subjected to automatic detection and classification in [16,19,20], where self-made datasets were used to train and validate YOLO models. Among studies of this type, the issue of detecting river floating waste can also be found [18]. Plastic waste is also the subject of research. However, it is typically limited to only four [13] or two types [17] of plastics. The analysis of the summary presented in Table 2 reveals three significant limitations. As already highlighted by review articles, a recurring constraint in many studies is that model testing is conducted under laboratory conditions. Most studies evaluate the model by partitioning a portion of the utilized dataset (what has been referred to as static validation), where waste remains stationary during testing. Only in [17] was dynamic validation performed, in which waste moved along a conveyor belt. However, the detection was limited to only PET and PET-G classes. Another limitation concerns the omission of contaminants present in the waste stream, which should serve as the background of the prepared samples rather than being the target of detection. This approach was implemented only in [15,20], where access to recordings from a real-world system was available. The final observed limitation is the low level of detail in the classification of plastic waste. Plastic is most commonly treated as a single class or, at best, divided into a maximum of four classes without distinguishing, for instance, different colors of PET, which is crucial from a recycling perspective.

Considering the identified limitations, this paper aims to develop a model for the automatic identification of plastic waste streams that takes into account the requirements of real-world waste-sorting facilities, such as the presence of contaminants and the transportation of waste on a conveyor belt with variable speed. The research contributes to the development of more sustainable waste management solutions by improving the efficiency of plastic waste sorting, thereby addressing the sustainability challenge of ecosystem degradation and its associated risks to human well-being. By enhancing the accuracy of automatic waste identification, the proposed approach supports circular economy principles, mitigates environmental pollution and reduces the unintended release of plastic waste into ecosystems. Due to the drawbacks of publicly available datasets, the model will incorporate a preparation process of the dataset. Additionally, the model will be based on CNNs, as they have achieved the best results in image classification in recent years. The main contributions of our paper are the following:

- We based the YOLO model training process on a self-built dataset consisting of over 4000 images of waste, including contamination.
- We have distinguished six classes reflecting the most commonly separated types of plastic, i.e., blue PET, white PET, green PET, HDPE, tetrapak, and, in addition to plastic, aluminum waste.
- We developed a set of recommendations for dataset creation aimed at automatic waste identification.
- We have applied a combination of three YOLO models to the multi-class waste classification issue.
- The model validation was conducted in a laboratory environment using a test setup and waste samples designed to closely replicate real-world conditions, with waste moving on a conveyor and being detected in real time by a camera.

2. Materials and Methods

The considered automatic waste stream identification is part of an ongoing project funded by the National Centre for Research and Development, entitled "Cyber-physical waste stream control system". The project aims to develop a cyber-physical control system for municipal waste sorting, i.e., a Digital Twin of a waste-sorting system. The structure of the Digital Twin consists of three fundamental components: the physical system, its digital replica, and a bidirectional link connecting the two [21]. A detailed description of the indicated elements and the Digital Twin framework for waste-sorting systems is presented in [22]. The foundation of the Digital Twin's operation is the automatic collection of data from the real-world system. Automatic identification of waste stream is therefore one of the key components of the data collection and control layer. The scope of this publication can thus be placed in the Digital Twin framework, as shown in Figure 1.

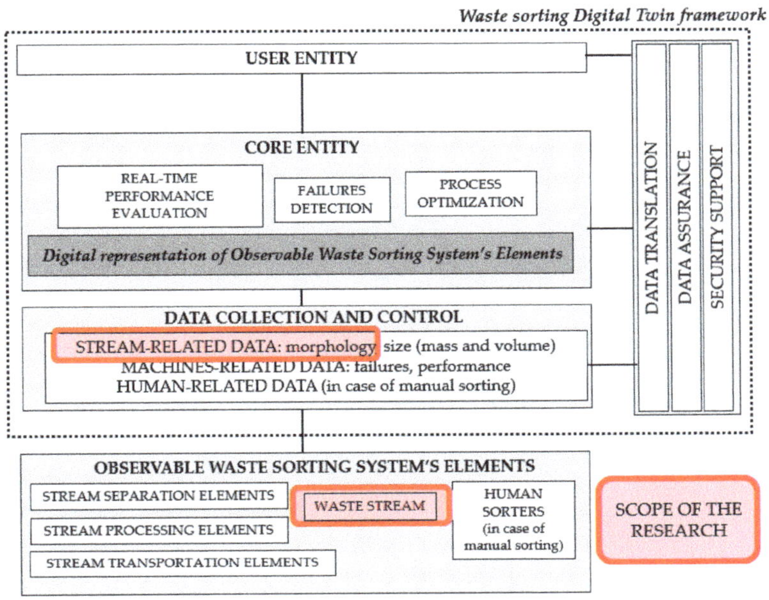

Figure 1. Scope of research in the context of the Digital Twin framework for waste-sorting systems.

Based on the results of the literature review, waste characterization, and stream monitoring requirements, the automatic identification of the stream was developed using CNN, specifically the YOLOv5 model.

2.1. Automatic Waste Identification Model Based on YOLOv5

The YOLO series are CNN-based frameworks commonly used in object detection and classification, including waste detection. Its three core components include the backbone, neck, and head. The backbone, a convolutional neural network, encodes image data into multi-scale feature maps. These maps are then refined by the neck, which consists of layers that enhance and integrate feature representations. Finally, the head module utilizes the processed features to predict object bounding boxes and class labels [23]. The general architecture of YOLOv5 is shown in Figure 2.

The automated waste stream identification model proposed in this paper is based on YOLOv5. Due to the complexity of detecting multiple waste classes and based on the results of our preliminary studies, we employed a combination of three YOLO models. Instead of burdening a single model with all the classes, the division into smaller models allows for focusing on a smaller set of classes in each case. This approach enables a more precise tailoring of the model to the specific characteristics of the waste. Such a strategy significantly improves detection accuracy and prevents situations where a single model might struggle to distinguish very similar classes. The use of three models allows for more accurate differentiation between classes, which is crucial in the context of waste sorting. Initially, the image is processed by two parallel models (1a and 1b). Each of them detects the same four waste classes (PET, HDPE, tetrapak, and Al), but these models are parameterized differently in terms of confidence score so that one focuses on PET, tetrapak, and Al, while the other focuses on the HDPE class. The other classes are ignored in both models. In the case of detecting the PET class, the output from model 1a is further processed by model 2, which assigns one of three colors to this class: blue, white, or green. The architecture of the proposed model is illustrated in Figure 3.

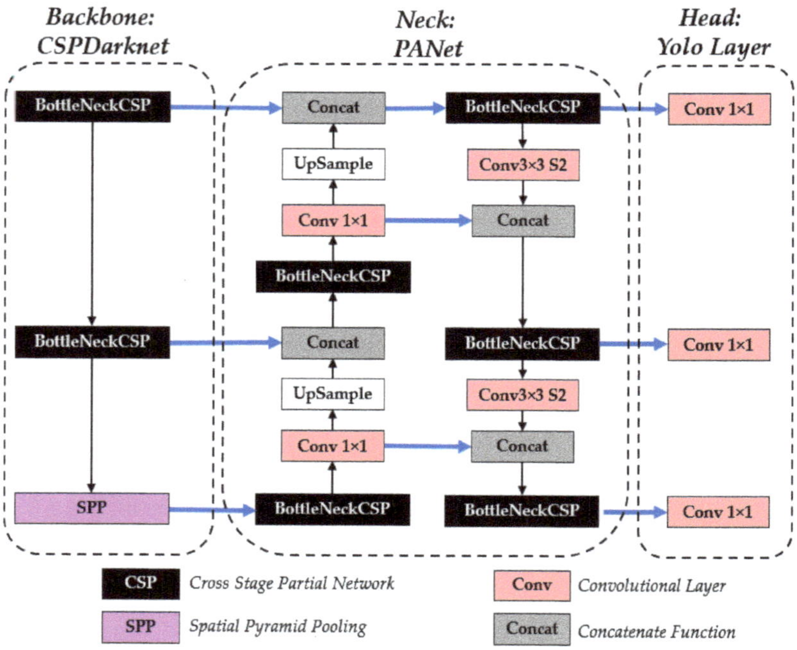

Figure 2. Architectural diagram of YOLOv5 [24–26].

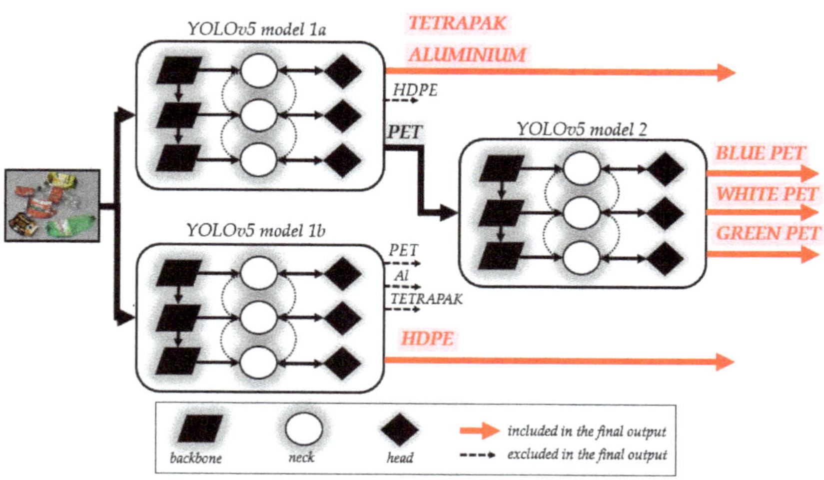

Figure 3. Architecture of the proposed model for automatic waste stream identification based on YOLOv5.

2.2. Dataset Preparation

A large-scale dataset, including many plastic waste images of different types, should be implemented in detection and classification models. There are several publicly available datasets with waste images, but they are insufficient for advanced sorting systems. Indeed, effective detection and classification of waste moving along the conveyor requires considering factors such as the piling of waste, the high irregularity of shapes, or the presence of contaminants that are not subjected to sorting. The development of a new complex dataset is the first major action needed when considering the problem of automatic waste identification. In the case of self-built datasets, the procedure for their creation is typically described only briefly in publications. Since the dataset significantly impacts both the learning efficiency and the subsequent performance of the model in a real-world system, this study presents the sequential steps required for the development of a self-built dataset. The main stages of dataset creation for automatic waste identification are presented in Figure 4.

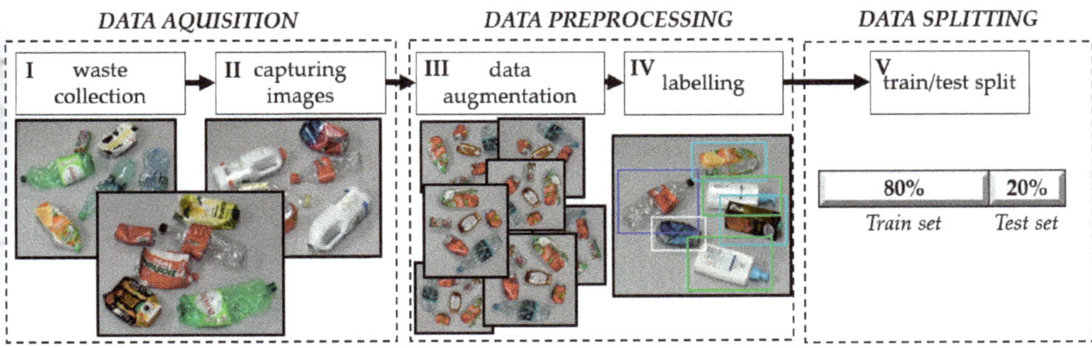

Figure 4. Main stages of dataset creation for automatic waste identification.

The first two stages are related to the data acquisition process. The input data for detection and classification models consist of images. Therefore, it is necessary to collect waste and subsequently capture images of it. The collected waste should include not only the target objects (Figure 5a) but also contaminants (Figure 5b) that typically appear in real-world sorting facilities due to errors in source separation. Such errors arise because people often misclassify waste, incorrectly assigning non-plastic items to plastic waste bins. Regarding contaminants, it is important to emphasize that their inclusion in a self-built dataset poses the greatest challenge. Contaminants exhibit significantly higher variability compared to the fractions intended for sorting. They may include textiles, wood, glass, paper, and many other materials. For this reason, a detailed representation of contaminants in a self-built dataset is realistically feasible only through the utilization of recorded waste streams from an actual sorting facility. In our study, contaminants were included to a limited extent, with an effort to construct a sample set that is as diverse as possible. We primarily gathered textiles, plastic bags, films, and paper with volumes comparable to those of the detected waste fractions. Additionally, it is necessary to collect multiple instances of the same type of waste to account for the high variation in shape, primarily resulting from different levels of compression.

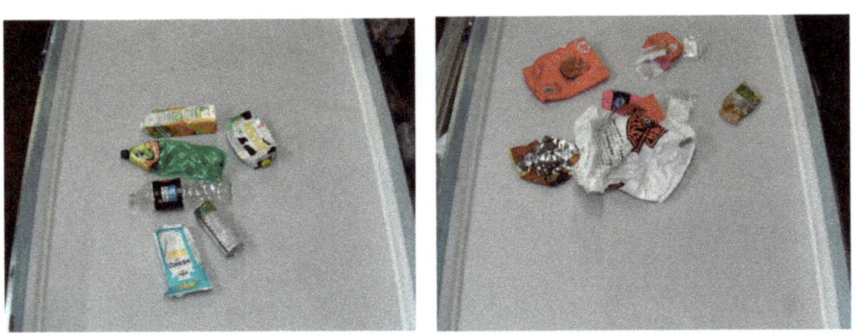

(a) (b)

Figure 5. Example of collected waste: (**a**) including only target objects; (**b**) including contaminants.

The waste classes considered include blue PET, white PET, green PET, HDPE, Tetrapak, and aluminum. For each class, waste was collected while ensuring a balanced number of instances across all categories. The collected waste was then photographed in small groups, positioned differently on the conveyor belt—both in the center and near the edges. Additionally, images were captured with overlapping waste items to reflect real-world conditions.

The collected images require preprocessing before being used for model training, which consists of two main steps: augmentation and labeling. Augmentation is optional; however, it is commonly performed to expand the dataset by applying transformations such as rotation or scaling. In contrast, labeling is a mandatory step for training the YOLO model, as it provides the necessary annotations for object detection. In the case of data augmentation for waste moving along a conveyor belt, geometric transformation is a key augmentation technique. In our dataset, we applied rotation. The final dataset, after augmentation, consisted of 2242 images used in models 1A and 1B. This dataset was further expanded with additional images for model 2, specifically for detecting blue PET, white PET, and green PET, resulting in a total of 4042 images. Each image underwent a labeling process involving the annotation of the detected classes. The summary of the applied labels is presented in Figure 6.

The final step in preparing the dataset for use in the YOLO model is to split it into two parts, the training and test sets, with the most commonly accepted ratio being 80/20.

As a summary, Table 3 presents a set of recommendations for each stage of the dataset creation process for waste identification.

Table 3. Summary of selected publications from recent years in the area of automatic waste.

Stage	Recommendations
I. Waste collection	- collecting waste with as much variation as possible - collecting the same waste many times with different shapes due to the level of compaction and with various features (e.g., no label and no cap) - consideration of subsequent dataset balancing (class equalization) - collection of not only waste intended for detection but also contaminants
II. Capturing images	- taking into account overlapping waste, varying positions on the conveyor, and changing lighting conditions
III. Data augmentation	- for the detection of waste, geometric augmentation (rotation and scaling) is of great importance
IV. Labeling	- the use of image annotation tools (e.g., Labelimg) to reduce labeling time
V. Train/test split	- rule of thumb gives an 80/20 ratio

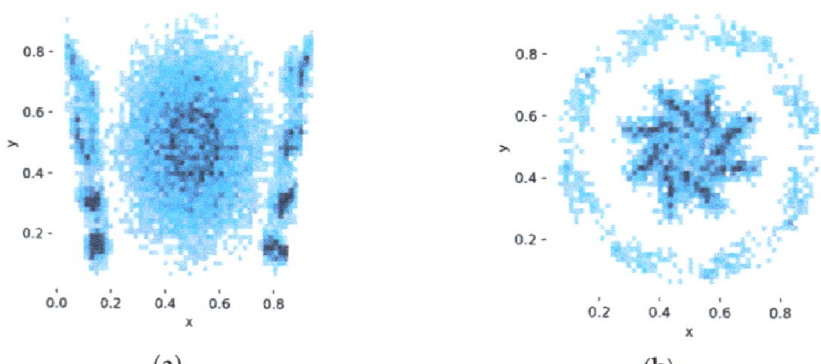

Figure 6. Distribution of label locations for (**a**) basic dataset (for model 1a and 1b) and (**b**) extended dataset (for model 2).

2.3. Experimental Environment

In advanced sorting systems, waste is transported between separating points (i.e., NIR separators, sorting robots, etc.). Effective sorting requires automatic identification at multiple stages of the sorting process (at the beginning—input stream; at the end—sorting residues; and before subsequent separation points). Regardless of the sorting stage, automatic identification using computer vision requires placing an RGB camera above the conveyor transporting the waste. To best replicate real-world conditions for the final model validation, a test bed representing a fragment of a real sorting system was constructed, namely, a manual waste-sorting cabin. The components of the developed test bed include the following:

- Conveyor belt (speed 0–0.5 m/s, 1.2 × 5 m);
- RGB camera positioned above the conveyor (4 K, 30 fps, 5–50 mm zoom lens);
- Computer;
- Input stream feeder;
- Output stream receiver;
- Two manual sorting stations (not considered in this paper);
- Hyperspectral camera (not in the scope of the study).

The scheme of the test bed and its real appearance are shown in Figures 7 and 8, respectively.

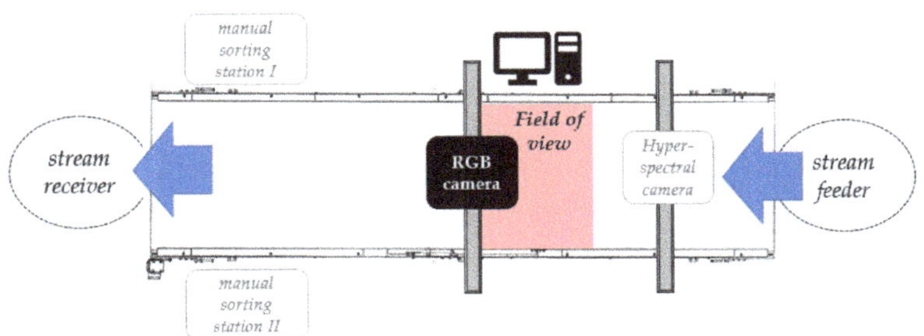

Figure 7. Scheme of the developed test bed.

Figure 8. The developed test bed.

2.4. Waste Sample Preparation for Model Validation

The primary limitation of previously applied approaches in automatic waste identification models is testing solutions under laboratory conditions based on a designated subset of the utilized dataset. The final validation of the model should be conducted under real-world conditions, considering the movement of waste along a conveyor. In the absence of access to a real sorting system, this was substituted by constructing a representative fragment of a real system and using waste samples that reflect the morphology of the Polish plastic waste stream.

In Poland, waste segregation is implemented at the source of generation. This means that waste is placed in designated containers for paper, glass, mixed waste, and plastics before collection. Metal waste is also disposed of in the plastic waste container; therefore, aluminum cans were included in the developed dataset.

For the final validation of the developed model, a 5 kg waste sample reflecting the morphology of Polish municipal waste was prepared based on the Institute of Environmental Protection—National Research Institute (IOS-PIB) report. The morphology of the prepared sample is shown in Figure 9.

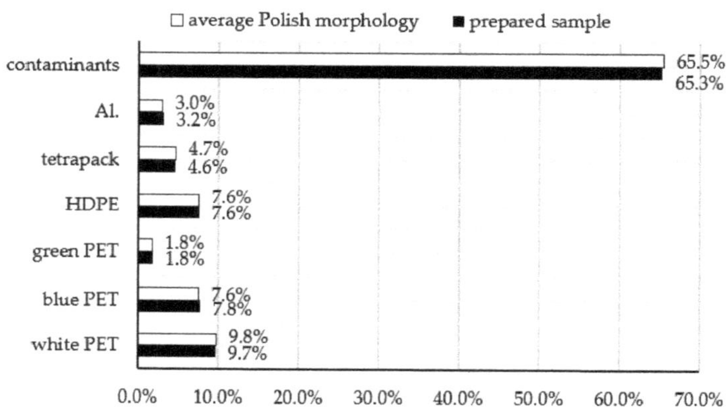

Figure 9. The morphology of the Polish plastic waste stream compared to the morphology of the prepared sample.

The presented figure confirms the significance of considering contaminants in the development of automatic identification models. In the average morphology of the Polish plastic waste stream, contaminants account for more than 50%.

3. Results

3.1. Model Performance Based on Validation Data Separated from the Self-Built Dataset

The proposed automatic waste stream identification model is based on three YOLOv5 models. The performance of each model was evaluated using precision, recall, and mean average precision (mAP), which are commonly used evaluation metrics for detection problems [27]. In the learning stage of the model, the indicated metrics represent a global perspective without considering each class separately.

Precision indicates the ratio of correct object detections (TP) to the total number of detections (both correct TP and incorrect FP), defined as follows:

$$\text{Precision} = \frac{\text{TP}}{(\text{TP} + \text{FP})} \tag{1}$$

Recall represents the ratio of correct object detections to the total number of actual objects, defined as follows:

$$\text{Recall} = \frac{\text{TP}}{(\text{TP} + \text{FN})} \tag{2}$$

Mean average precision (mAP) denotes the average performance of the detector across all recall values, which is equivalent to the area under the precision–recall curve. It can be mathematically expressed as Equation (4):

$$\text{AP} = \int_0^1 \text{P(R)dR}, \tag{3}$$

$$\text{mAP} = \frac{\sum_{i=1}^{n} \text{AP}_i}{n} \tag{4}$$

In addition, it measures the performance of an object detection model by averaging the area under the precision–recall curve across all classes. In this context, mAP_0.5 represents the average precision computed at a fixed IoU threshold of 0.5, whereas mAP_0.5:0.95 is a more rigorous and comprehensive metric, averaging precision over multiple IoU thresholds from 0.5 to 0.95 in increments of 0.05.

The progression of precision, recall, and mAP throughout the YOLOv5 training process for each of the three models used is presented in Figure 10.

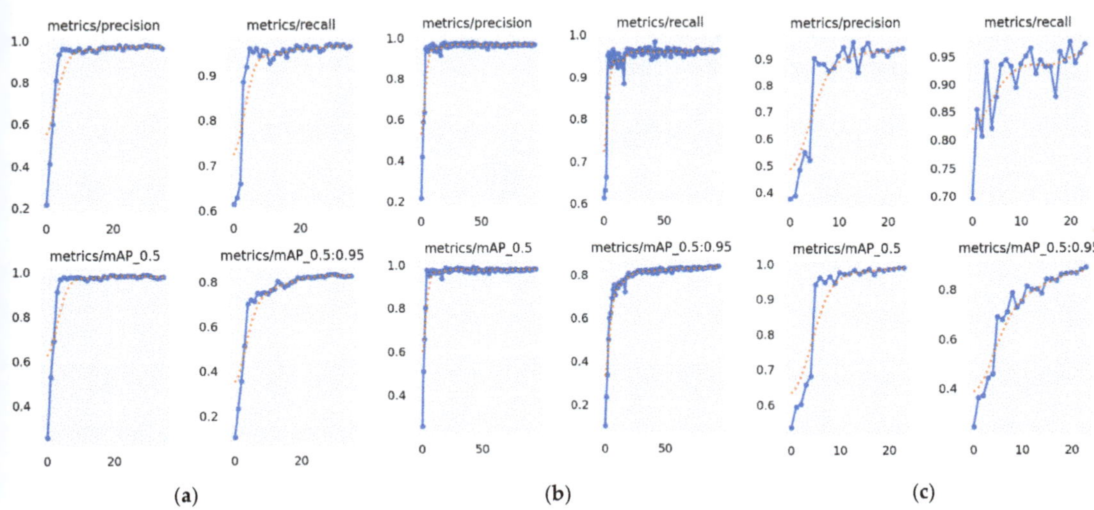

Figure 10. Progression of precision, recall, and mAP throughout the YOLOv5 training process (actual results with solid line and smoothed trend with dotted line) for (**a**) model 1a, (**b**) model 1b, and (**c**) model 2.

The comparison of the final performance metrics for each of the three models is presented in Table 4.

Table 4. Comparison of the final performance metrics of each model.

Model	Precision	Recall	mAP_0.5	mAP_0.5:0.95
1a	0.961	0.964	0.977	0.824
1b	0.967	0.963	0.979	0.837
2	0.936	0.972	0.988	0.899

Additionally, Figure 11 presents the confusion matrices generated by the trained YOLO models.

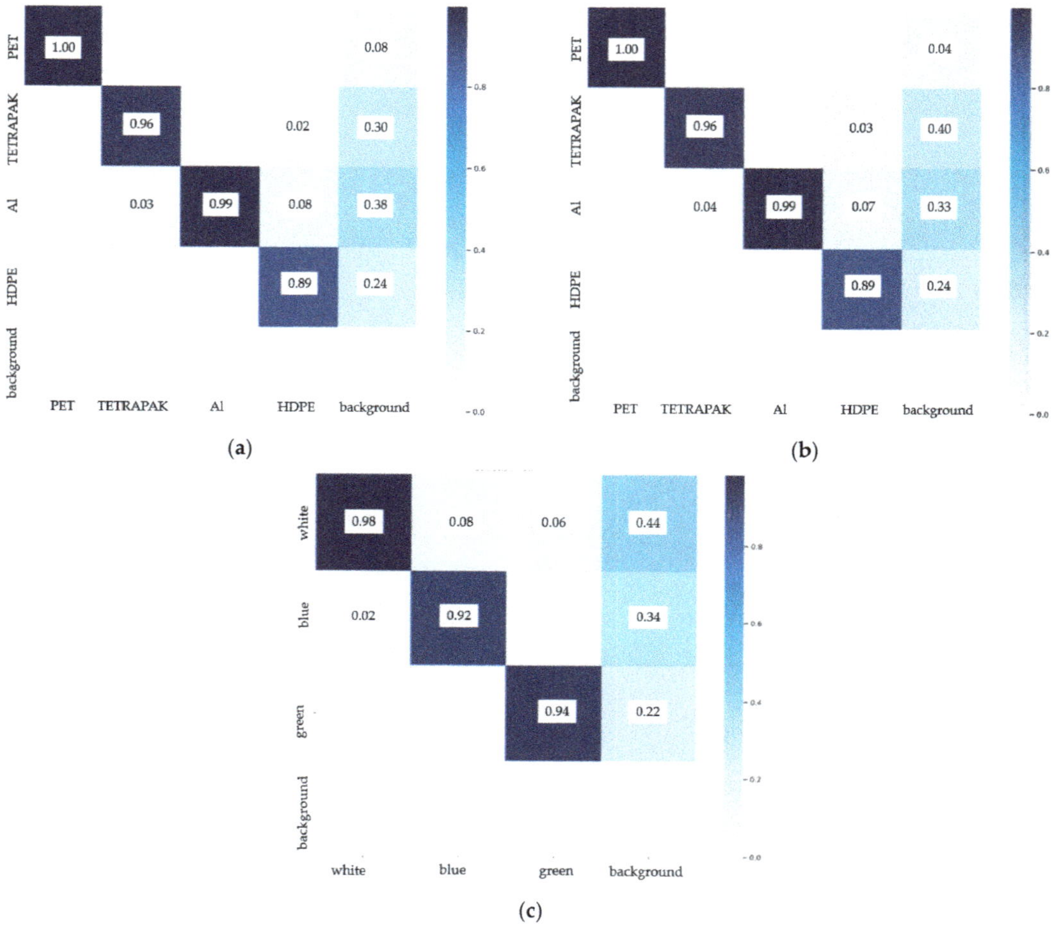

Figure 11. Confusion matrices generated by the trained YOLO models: (**a**) model 1a, (**b**) model 1b, and (**c**) model 2.

3.2. Model Performance Under Replicated Real-World Conditions

The lack of access to the real system was replaced by its replica in laboratory conditions using the developed test bed (described in Section 2.3). The mapping of real conditions mainly focused on using a conveyor with parameters consistent with those used in a real waste-sorting plant, using a waste stream with real-world morphology, and repeatedly loading the stream onto the conveyor to generate a random arrangement of waste with occlusion effects. The model was tested using waste samples with average Polish morphology (described in Section 2.4) moving at a speed of 0.2 m/s on the conveyor. The sample was fed onto the conveyor eight times consecutively. The visualization of the model's performance under conditions replicating a real-world sorting facility is presented in Figure 12.

Figure 12. The visualization of the model's performance under conditions replicating a real-world sorting facility (the red box shows detected HDPE, while the pink one – tetrapak).

The final results (model output) were analyzed to determine the accuracy, precision, recall, and F1-score for each detected class. These metrics are calculated based on four groups of results for each class separately:

- TPs—True Positives (e.g., HDPE detected, which is actually HDPE);
- FPs—False Positives (e.g., HDPE detected, which is actually a different fraction);
- TNs—True Negatives (e.g., a fraction other than HDPE detected, which is actually a fraction other than HDPE);
- FNs—False Negatives (e.g., HDPE not detected).

The comparison of the final performance metrics for each class is presented in Table 5.

Table 5. Comparison of the final performance metrics for each class.

Class	Accuracy	Precision	Recall	F1-Score
blue PET	0.96	0.99	0.90	0.94
green PET	0.98	0.99	0.85	0.91
white PET	0.94	1.00	0.84	0.91
tetrapak	0.97	0.89	0.90	0.89
Al	0.95	0.98	0.86	0.91
HDPE	0.95	0.96	0.76	0.84

4. Discussion

The primary performance measures of models 1a, 1b, and 2 show that model 2 is the best overall, achieving the highest recall and mAP values, making it the most accurate and generalizable. Model 1b offers a good balance between precision and recall, slightly outperforming 1a in mAP. The combination of high recall (1a), high precision (1b), and strong classification accuracy (2) suggests an optimized pipeline that balances detection and classification accuracy. Model 1a focuses on detecting Tetrapak, aluminum, and PET, while model 1b is trained primarily for HDPE detection. This approach improves the accuracy of HDPE detection, which is often more challenging. Since model 1b disregards all non-HDPE objects, and model 1a does not output HDPE, the system minimizes false detections and enhances classification confidence. Model 2 improves the accuracy of sorting PET based on color, which is crucial for recycling.

The confusion matrices indicate that the additional model trained for HDPE successfully eliminated the misclassification of HDPE as PET. It is also evident that model 1a struggles with detecting HDPE, often confusing it with Tetrapak and aluminum. Additionally, each model occasionally detects background objects due to contamination resembling the target fractions. Therefore, the proposed model effectively detects the assigned waste classes, but a significant challenge remains in misclassifying contaminants as sortable objects. This highlights the importance of introducing contaminants into self-built datasets for automatic waste identification, as real-world waste streams often contain more contaminants than recyclable materials—a situation particularly relevant in countries such as Poland. In further research, a more comprehensive examination of contaminants in the stream should be planned. Regarding misclassification problems, given that our dataset is balanced, future focus should be placed on post-processing techniques such as threshold adjustment and non-maximum suppression (NMS). Adjusting the confidence threshold could help reduce false positives by filtering out low-confidence detections while maintaining high recall. Additionally, fine-tuning the NMS parameters can mitigate cases where multiple overlapping detections are incorrectly assigned to different classes. These refinements would improve the model's precision and robustness, particularly in complex waste streams where object occlusion and visual similarities between classes pose additional challenges.

The final validation eliminated a key limitation present in most models available in the literature, which are typically not tested under real-world conditions. A binary classification accuracy ranging from 0.94 to 0.98 indicates a high level of correct detections among all identified objects. If an object is classified, the probability of assigning it to the wrong class remains low. The achieved precision values, ranging from 0.89 to 1.00, demonstrate high classification accuracy for each waste fraction. Once an object is detected, the model very rarely misclassifies it. The highest probability of such an error, at 0.11, was observed for the Tetrapak fraction. The recall values can also be considered high. The lowest recall was recorded for HDPE (0.76), while recall values ranged between 0.85 and 0.9 for the remaining fractions. These values indicate a low probability of failing to detect an object of a given class. PET classes (blue, green, and white) show strong overall performance, with particularly high precision, but there is room to improve recall, especially for green and white PET. Tetrapak and aluminum have a good balance of precision and recall, with minor improvements possible in precision for Tetrapak. HDPE is the weakest in terms of recall, which could mean that more HDPE objects are not being detected. Given the challenges associated with HDPE detection, a recall of 0.76 can be considered acceptable. However, it remains the lowest recall among all detected fractions, highlighting an area for improvement. Since the dataset is balanced, rebalancing techniques will not be applicable in this case. Therefore, potential corrective actions should focus on hyperparameter tuning to enhance HDPE detection performance.

Under conditions that replicate real-world scenarios, the results remain within acceptable limits, indicating a high potential for the practical applicability of the developed model in waste-sorting systems.

Potential Enhancements to Existing Solid Waste Management Technologies

The satisfactory results discussed above generate several possibilities for their application, thereby contributing to the advancement of existing solid waste management solutions. The developed model can be applied in the following areas:

- Real-time system monitoring—The developed model enables continuous real-time waste stream composition monitoring. This capability facilitates the detection of stream variations. By providing instant feedback, it supports decision making.
- Data collection layer of the waste-sorting system's Digital Twin—A fundamental characteristic of Digital Twins is the bidirectional data flow, which involves both data acquisition and system control. The proposed model is a foundational component for the development of the data collection layer within a Digital Twin framework, specifically in the context of waste composition.
- Waste-sorting system control—The automatic identification of waste stream composition enables dynamic process adjustments and improved operational control in sorting facilities.
- Sensor-based sorting—The integration of the proposed model with sensor-based sorting technologies enhances the efficiency of automated waste separation. By providing accurate and high-speed waste identification, the model supports optical sorters and robotic sorting systems operation.

5. Conclusions

In advanced waste sorting systems, the automatic identification of the waste stream is crucial to efficiently separate materials intended for recycling and contribute to advancing sustainability goals. It is primarily used by automated and robotic sorting components but also for monitoring the waste stream at various stages of processing, which enables the system to be controlled. The waste detection and classification model presented in this paper is one of the components of the municipal waste-sorting system's Digital Twin, which is a part of the ongoing project "Cyber-physical waste stream control system". The proposed model is based on the combination of three YOLOv5 models focused on classifying elements of the waste stream into the following classes: blue PET, white PET, green PET, HDPE, tetrapak, and aluminum. The model was trained using a self-built dataset, and its performance was validated in a replica of a real-world sorting system fragment.

The next step in the development of the data collection layer of the Digital Twin is the application of a lidar sensor for measuring the volume of transported waste. The developed model focuses only on the composition of the waste stream, i.e., detecting the number of pieces of waste needed to be sorted. From the point of view of the sorting system control and in the context of the requirements of the Digital Twins, knowledge only of the stream composition is insufficient. In fact, the composition of the stream must be supplemented with information about its size. This size can be partially estimated by mass using the detection results by assuming constant weights of the detected fractions. However, this does not take into account contaminants, so the size determined in this way may be severely underestimated. Therefore, the automatic detection of stream composition should be complemented by volume measurement. The final fusion of the RGB camera with the lidar sensor will enable the real-time collection of all key characteristics of the waste stream, including morphology, as well as volumetric and mass measurements.

Author Contributions: Conceptualization, R.G. and A.D.; methodology, R.G.; software, M.F.; validation, R.G., A.D. and M.F.; formal analysis, R.G.; investigation, A.D. and M.F.; resources, R.G. and A.D.; data curation, M.F.; writing—original draft preparation, A.D.; writing—review and editing, R.G.; visualization, A.D.; supervision, R.G.; project administration, R.G.; funding acquisition, R.G. All authors have read and agreed to the published version of the manuscript.

Funding: This research was funded by The National Centre for Research and Development, LEADER program, grant number 0212/L-13/2022. The APC was funded by the Faculty of Mechanical Engineering, Wroclaw University of Science and Technology.

Institutional Review Board Statement: Not applicable.

Informed Consent Statement: Not applicable.

Data Availability Statement: The datasets presented in this article are not readily available because the data are part of an ongoing research project.

Conflicts of Interest: The authors declare no conflicts of interest. The funders had no role in the design of the study; in the collection, analyses, or interpretation of the data; in the writing of the manuscript; or in the decision to publish the results.

References

1. Alhazmi, H.; Almansour, F.H.; Aldhafeeri, Z. Plastic waste management: A review of existing life cycle assessment studi *Sustainability* **2021**, *13*, 5340. [CrossRef]
2. Milios, L.; Davani, A.E.; Yu, Y. Sustainability impact assessment of increased plastic recycling and future pathways of plas waste management in Sweden. *Recycling* **2018**, *3*, 33. [CrossRef]
3. Arbeláez-Estrada, J.C.; Vallejo, P.; Aguilar, J.; Tabares-Betancur, M.S.; Ríos-Zapata, D.; Ruiz-Arenas, S.; Rendón-Vélez, E. Systematic Literature Review of Waste Identification in Automatic Separation Systems. *Recycling* **2023**, *8*, 86. [CrossRef]
4. Satav, A.G.; Kubade, S.; Amrutkar, C.; Arya, G.; Pawar, A. A state-of-the-art review on robotics in waste sorting: Scope a challenges. *Int. J. Interact. Des. Manuf.* **2023**, *17*, 2789–2806. [CrossRef]
5. Lubongo, C.; Alexandridis, P. Assessment of Performance and Challenges in Use of Commercial Automated Sorting Technolo for Plastic Waste. *Recycling* **2022**, *7*, 11. [CrossRef]
6. Tanguay-Rioux, F.; Legros, R.; Spreutels, L. On the limits of empirical partition coefficients for modeling material recovery facili unit operations in municipal solid waste management. *J. Clean. Prod.* **2021**, *293*, 126016. [CrossRef]
7. Tanguay-Rioux, F.; Spreutels, L.; Héroux, M.; Legros, R. Mixed modeling approach for mechanical sorting processes based physical properties of municipal solid waste. *Waste Manag.* **2022**, *144*, 533–542. [CrossRef]
8. Kroell, N.; Chen, X.; Greiff, K.; Feil, A. Optical sensors and machine learning algorithms in sensor-based material flow characte zation for mechanical recycling processes: A systematic literature review. *Waste Manag.* **2022**, *149*, 259–290. [CrossRef]
9. Lu, W.; Chen, J. Computer vision for solid waste sorting: A critical review of academic research. *Waste Manag.* **2022**, *142*, 29– [CrossRef]
10. Xia, W.; Jiang, Y.; Chen, X.; Zhao, R. Application of machine learning algorithms in municipal solid waste management: A mi review. *Waste Manag. Res.* **2022**, *40*, 609–624. [CrossRef]
11. Zhao, Y.; Li, J. Sensor-Based Technologies in Effective Solid Waste Sorting: Successful Applications, Sensor Combination, an Future Directions. *Environ. Sci. Technol.* **2022**, *56*, 17531–17544. [CrossRef]
12. Maier, G.; Gruna, R.; Langle, T.; Beyerer, J. A Survey of the State of the Art in Sensor-Based Sorting Technology and Researc *IEEE Access* **2024**, *12*, 6473–6493. [CrossRef]
13. Son, J.; Ahn, Y. AI-based plastic waste sorting method utilizing object detection models for enhanced classification. *Waste Mana* **2025**, *193*, 273–282. [CrossRef]
14. Ali, M.L.; Zhang, Z. The YOLO Framework: A Comprehensive Review of Evolution, Applications, and Benchmarks in Obje Detection. *Computers* **2024**, *13*, 336. [CrossRef]
15. Zhou, Q.; Liu, H.; Qiu, Y.; Zheng, W. Object Detection for Construction Waste Based on an Improved YOLOv5 Model. *Sustainabili* **2023**, *15*, 681. [CrossRef]
16. Ren, Y.; Li, Y.; Gao, X. An MRS-YOLO Model for High-Precision Waste Detection and Classification. *Sensors* **2024**, *24*, 433 [CrossRef]
17. Choi, J.; Lim, B.; Yoo, Y. Advancing Plastic Waste Classification and Recycling Efficiency: Integrating Image Sensors and Dee Learning Algorithms. *Appl. Sci.* **2023**, *13*, 10224. [CrossRef]
18. Yang, X.; Zhao, J.; Zhao, L.; Zhang, H.; Li, L.; Ji, Z.; Ganchev, I. Detection of River Floating Garbage Based on Improved YOLOv *Mathematics* **2022**, *10*, 4366. [CrossRef]
19. Fan, J.; Cui, L.; Fei, S. Waste Detection System Based on Data Augmentation and YOLO_EC. *Sensors* **2023**, *23*, 3646. [CrossRef]
20. Liu, S.; Chen, R.; Ye, M.; Luo, J.; Yang, D.; Dai, M. Methodology for Real-Time Detection of Domestic Waste. *Sensors* **2024**, *24*, 466 [CrossRef]
21. Werbińska-Wojciechowska, S.; Giel, R.; Winiarska, K. Digital Twin approach for operation and maintenance of transportatio system—systematic review. *Sensors* **2024**, *24*, 6069. [CrossRef]
22. Giel, R.; Dąbrowska, A. A Digital Twin framework for cyber-physical waste stream control system towards Reverse Logistics 4. *Logforum* **2024**, *20*, 297–306. [CrossRef]

Terven, J.; Córdova-Esparza, D.M.; Romero-González, J.A. A Comprehensive Review of YOLO Architectures in Computer Vision: From YOLOv1 to YOLOv8 and YOLO-NAS. *Mach. Learn. Knowl. Extr.* **2023**, *5*, 1680–1716. [CrossRef]
Xu, R.; Lin, H.; Lu, K.; Cao, L.; Liu, Y. A forest fire detection system based on ensemble learning. *Forests* **2021**, *12*, 217. [CrossRef]
Katsamenis, I.; Karolou, E.E.; Davradou, A.; Protopapadakis, E.; Doulamis, A.; Doulamis, N.; Kalogeras, D. TraCon: A Novel Dataset for Real-Time Traffic Cones Detection Using Deep Learning. *Lect. Notes Netw. Syst.* **2023**, *556*, 382–391. [CrossRef]
Jia, X.; Tong, Y.; Qiao, H.; Li, M.; Tong, J.; Liang, B. Fast and accurate object detector for autonomous driving based on improved YOLOv5. *Sci. Rep.* **2023**, *13*, 1–13. [CrossRef]
Sirisha, U.; Praveen, S.P.; Srinivasu, P.N.; Barsocchi, P.; Bhoi, A.K. Statistical Analysis of Design Aspects of Various YOLO-Based Deep Learning Models for Object Detection. *Int. J. Comput. Intell. Syst.* **2023**, *16*, 1–29. [CrossRef]

sustainability

MDPI

Article

Data-Driven Optimization of Construction and Demolition Waste Management: Pattern Recognition and Anomaly Detection

Ana Lopes [1] and Carlos Afonso Teixeira [2,*]

[1] Department of Biology and Environmental, School of Life and Environmental Sciences, University of Trás-os-Montes e Alto Douro, Quinta de Prados, 5001-801 Vila Real, Portugal; al71526@alunos.utad.pt

[2] Centre for the Research and Technology of Agroenvironmental and Biological Sciences, Institute for Innovation, Capacity Building and Sustainability of Agri-Food Production (Inov4Agro), University of Trás-os-Montes e Alto Douro, Quinta de Prados, 5000-801 Vila Real, Portugal

* Correspondence: cafonso@utad.pt

Abstract: Construction and Demolition Waste (CDW) forecasting is essential for sustainable waste management and circular economy objectives. Traditional prediction models often face limitations when dealing with small datasets and extreme variability. This study introduces a robust statistical framework that employs the median and Median Absolute Deviation (MAD), applied to standardized CDW indicators: tons per day (t day^{-1}) and tons per square meter (t m^{-2}). The method enables the detection of statistical anomalies using a $\pm 2 \cdot$MAD threshold, increasing the model's resilience to outliers and enhancing its predictive reliability. The analysis is based on a dataset of 16 construction and rehabilitation projects, carried out under consistent technical methodologies, operational practices, and centralized data collection protocols. The results show that median-based predictions offer greater stability than mean-based estimators, particularly in skewed datasets. The framework successfully identifies projects with significant deviations, supporting targeted audits, performance monitoring, and iterative model refinement. Although the current model focuses on the duration and area as predictors, future enhancements should incorporate additional contextual variables. Furthermore, embedding the median–MAD logic within machine learning architectures (e.g., LSTM, ARIMAX) could improve scalability and support real-time CDW monitoring. These findings contribute to the development of data-driven forecasting tools that are aligned with operational efficiency and circularity goals in the construction sector.

Keywords: construction and demolition waste (CDW); anomaly detection; predictive modeling; circular economy (CE); waste generation indicators; statistical analysis

1. Introduction

Construction and Demolition Waste (CDW) management presents a significant environmental, economic, and operational challenge, primarily due to the construction sector's intensive resource use, the substantial volume and heterogeneity of CDW generation, and the associated greenhouse gas emissions. Construction activities account for approximately 40% of natural resource use and generate 30–40% of urban waste globally, making CDW one of the largest waste streams worldwide [1,2]. The sector is also responsible for around 39% of energy-related carbon dioxide emissions, underscoring its critical role in mitigating climate change and related environmental concerns [3]. In particular, inadequate CDW management can contaminate soil and water in urban settings, demonstrating the pressing need for innovative and sustainable waste strategies. Addressing these challenges requires urgent measures, including the application of circular economy (CE) principles and the adoption of more advanced waste management approaches.

The CE offers a transformative framework to transition the construction sector from a "take–make–dispose" model to one that is centered on waste reduction, reuse, and recycling. Its core objective is to optimize resource efficiency and lower environmental impacts by fostering material circularity and decreasing the reliance on virgin resources. Notably, successful CE initiatives in regions like the European Union, aligned with the Green Deal, highlight the potential for enhancing sustainability in the construction industry [4–6]. For instance, employing secondary aggregates and selective demolition has reduced landfill use and improved resource efficiency [7,8]. Additional examples include strategies for recycling CDW into secondary aggregates, thereby minimizing both the use of virgin materials and the amount of waste that is sent to landfills [9]. Policies and technologies that support material circularity are therefore crucial to achieving a more sustainable construction sector [7,8].

The project duration and logistical factors are also influential in CDW generation. Extended project timelines often lead to the degradation of stored materials—particularly cement and plasterboard—and increase packaging waste, which can comprise up to 13% of total CDW in long-term efforts [10]. Meanwhile, proximity to recycling facilities affects recovery rates: long transport distances drive up costs and emissions, discouraging reuse or recycling in remote locations [9].

Indicators such as tons per square meter (t m^{-2}), tons per capita per year (t $person^{-1}$ $year^{-1}$), and tons per day (t day^{-1}) are widely used to measure, parameterize, and compare CDW across various contexts [7]. These metrics capture differences based on the construction activity (new construction, demolition, refurbishment), the materials involved (e.g., concrete, wood, metals), and regional characteristics. For example, new construction projects typically generate 30–300 kg m^{-2} of waste, while renovations can produce 200–1200 kg m^{-2} [8,9]. In Sichuan, China, the amount of construction waste per unit of built area reached 0.550 t m^{-2}, demolition waste 1.3 t m^{-2}, and renovation waste 0.1 t m^{-2} [11]. Likewise, the authors of Ref. [12] found a CDW generation rate of 0.069 t m^{-2} for institutional non-residential projects, with specific values of 0.034 t m^{-2} for concrete/masonry and 0.028 t m^{-2} for wood [13].

Using the t $person^{-1}$ $year^{-1}$ indicator, Coelho and Brito [7] reported regional waste generation in Portugal ranging from 0.325 t $person^{-1}$ $year^{-1}$ to an anticipated 0.400 t $person^{-1}$ $year^{-1}$ over the next decade. Meanwhile, Germany accumulated 0.720 t $person^{-1}$ $year^{-1}$ and Ireland 0.162 t $person^{-1}$ $year^{-1}$ [7]. In Hong Kong, approximately 60 000 tons of CDW are produced daily, with 4400 tons being destined for landfills, which can hinge on the total project duration or working days [14]. These numbers highlight a pressing need for improved waste management.

Studies focusing on t m^{-2} underscore the importance of material types. For instance, Ref. [8] documented 0.018–0.040 t m^{-2} for new constructions and 0.401–0.840 t m^{-2} for demolitions; among these, concrete may constitute up to 85% of the waste [8]. In Hanoi, Vietnam, Ref. [15] reported 0.079 t m^{-2} for small construction sites and 1.030 t m^{-2} for larger ones, with soil, concrete, and brick dominating the waste stream. Beyond area-based metrics, temporal indicators such as t day^{-1} have proven valuable. Comparable approaches are found in [12,16], where effective working days were built into regression models, highlighting the critical role of project duration in forecasting CDW [12,14,16].

Numerous studies have employed such indicators to calibrate and predict CDW. For example, Ref. [17] constructed a multiple linear regression model incorporating dwelling numbers and total floor area, achieving R^2 values up to 0.99 in weight estimation but with potential volume deviations of ±38.48%. Similarly, Ref. [13] used regression models to estimate construction waste, discovering a mean generation rate of 0.151 m^3/m^2 in Porto Alegre, Brazil, and reporting R^2 values of 0.81 and 0.91—indicative of strong predictive accuracy.

Despite progress in modeling (e.g., Support Vector Machine Regression or GM-SVR [11]), discrepancies persist between the forecasted and actual on-site CDW generation. In one example, Ref. [17] cited average deviations of ±0.75% by weight and ±10.28% by volume, with extreme outliers reaching ±38.48% (volume) and ±15.97% (weight). Likewise, Ref. [12] noted up to ±35% deviation for 79% of the dataset and highlighted how site-specific factors heavily influence the outcomes. The waste generation rate method and lifetime analyses are often limited by static datasets, failing to reflect real-world complexities, as [18] points out. Additionally, even buildings with comparable features may vary sixfold in waste generation, as shown in [13]. Consequently, more robust, context-specific techniques that incorporate real-time data and advanced statistical methods are essential [19].

The Median Absolute Deviation (MAD) method is an increasingly effective strategy for addressing forecasting errors and on-site variability in Construction and Demolition Waste (CDW) management. This robust statistical approach mitigates the influence of outliers, improving anomaly detection and enhancing the predictive accuracy in highly variable datasets [18–22]. For instance, Ref. [20] highlights the importance of advanced modeling techniques to manage CDW recycling volatility, a challenge that MAD is particularly well-suited to address. Similarly, research in Egypt emphasizes how site-specific outliers significantly impact CDW generation, reinforcing the need for robust statistical tools [21]. Unlike traditional dispersion measures, MAD calculates variability around the median, making it highly resistant to extreme values—a crucial advantage in construction projects that are characterized by unpredictability [23]. When integrated into hybrid or non-linear models such as GM-SVR—which combines Gray Model residuals with Support Vector Regression—MAD further enhances the predictive reliability by establishing adaptive confidence bands, reducing the risk of forecasting errors [11,24,25]. Improved forecasting accuracy leads to more efficient waste management decisions, better aligning predictive models with real-world conditions [26]. Additionally, MAD aids in identifying key drivers of large deviations, enabling the development of context-specific CDW management strategies that optimize resource allocation and minimize the environmental impact.

Notwithstanding the prevalence of indicators such as t m^{-2} and t day^{-1}, discrepancies often remain between the projected and actual CDW generation on-site. While standard machine learning models and mean-based approaches capture broad waste patterns, they can falter when confronted with project-specific deviations stemming from material variability or local logistics. Consequently, this research integrates the median and MAD into predictive frameworks to systematically identify outliers, characterize waste generation patterns, and refine CDW projections. By improving the robustness and accuracy of these forecasts, the methodology contributes to more effective, data-driven CDW management practices that embrace CE principles and increase resource efficiency.

This study aims to develop a robust and data-efficient forecasting framework for CDW by applying the median and the Median Absolute Deviation (MAD) to two standardized indicators: average daily CDW generation (t day^{-1}) and CDW generation per intervention area (t m^{-2}). The methodology improves the predictive reliability by enabling systematic anomaly detection and refined deviation assessment. By optimizing forecasting performance in small-sample, high-variability environments, the model supports informed decision-making in waste management practices that are aligned with circular economy (CE) principles.

In contrast to conventional models that are reliant on large datasets, multivariate regressions, or opaque machine learning techniques, this study introduces a statistically grounded and data-efficient approach. It applies robust estimators to standardized indicators across a set of construction and rehabilitation projects that were executed under consistent technical, operational, and data governance protocols. The combined use of the median and MAD supports outlier-resilient forecasting and enables anomaly detection through adaptive confidence intervals. This framework increases the statistical robustness in heterogeneous datasets and enables systematic performance monitoring, especially in data-constrained settings. It addresses a gap in the literature by offering a replicable method for anomaly-based forecasting, which is particularly relevant for organizations seeking to improve benchmarking and internal waste management through resilient, evidence-based tools.

2. Methods

This section presents the methodological framework that was used to analyze and predict Construction and Demolition Waste (CDW) generation. It includes the dataset, the definition of standardized indicators, the statistical techniques applied, and the anomaly detection method based on robust estimators.

2.1. Data Collection

Data on Construction and Demolition Waste (CDW) generation were collected from 16 projects, including 7 new construction and 9 rehabilitation interventions, all executed within a single metropolitan region in Portugal between 2017 and 2022. All projects were carried out by the same construction company, which employs approximately 250 members of staff and operates under a centralized data management system. This system ensured the application of uniform methodologies for construction execution, waste recording, and data handling.

Standardized construction techniques, machinery, and monitoring protocols were used across the entire sample. All projects were managed by the same technical team, using identical equipment and operational models. This high level of methodological consistency minimizes procedural variability and enhances the comparability of results, ensuring that observed differences in CDW generation primarily reflect intrinsic project characteristics—such as the duration, scale, and typology—rather than inconsistencies in data acquisition or site operations. Additionally, each project was categorized according to the construction type—namely, low-rise residential units (typically single-family dwellings or buildings up to two stories) and low-to-mid-rise buildings (including small apartment blocks or mixed-use structures with more complex structural requirements)—to support the subsequent interpretation of CDW variability.

Despite the internal coherence of the dataset, its relatively small size and localized nature (n = 16) limit the statistical generalizability of the findings. However, this methodological rigor offers a strong foundation for reliable intra-sample comparisons. Future studies should aim to expand the dataset by incorporating a broader range of construction typologies, systems, and geographic contexts, thereby improving the external validity and representativeness of the proposed forecasting model.

The CDW classification followed the European Waste List (EWL), as established by Commission Decision 2000/532/EC. This framework categorizes waste into 20 main categories and assigns specific six-digit codes to each type. The classification process in this study involved identifying waste generated across key construction phases—including raw material supply, transport, manufacturing, and on-site installation. Hazardous waste was isolated in accordance with European handling requirements.

To support consistent data analysis, five key variables were defined: (1) duration (days), representing the total project execution time; (2) intervention area (m^2), indicating the built or affected surface; (3) CDW (t), capturing the mass of each waste type; (4) total CDW (t), denoting the overall quantity of waste that was generated per project. These parameters formed the quantitative basis for all subsequent analyses and indicator calculations.

2.2. Indicators

Two standardized indicators were established to allow for meaningful comparisons across projects of varying scales and typologies: (1) average daily waste generation (Equation (1)) and (2) waste generation per square meter of intervention area (Equation (2)).

These indicators normalize the waste output based on the project's duration and size, respectively, allowing for more consistent assessment of the CDW intensity across heterogeneous cases.

$$\mathrm{CDWd} = \frac{\mathrm{CDW}}{\mathrm{Duration}} \tag{1}$$

$$\mathrm{CDWa} = \frac{\mathrm{CDW}}{\mathrm{Intervention\ Area}} \tag{2}$$

where

CDWd (t day^{-1}): The average daily CDW generation.
CDWa (t m^{-2}): CDW per square meter of intervention area.
CDW (t): The total amount of waste produced.
Duration (days): The project's execution time.
Intervention Area (m^2): The built or affected area.

These indicators provided the analytical foundation for the evaluation of central tendencies, predictive modeling, and anomaly detection that are described in subsequent sections. Each equation was applied directly using the project-specific data that were compiled during the data collection phase. Specifically, Equation (1) was used to calculate the average daily CDW generation by dividing the total observed waste by the project duration, while Equation (2) derived the waste intensity per intervention area. These operations were performed systematically across all 16 projects using a standardized spreadsheet environment, ensuring uniform computation and comparability of results.

2.3. Statistical Analysis

To examine the distributional characteristics of the indicators, both the mean (Equation (3)) and the median (Equations (4) and (5)) were calculated. While the mean reflects the overall central tendency, it is highly sensitive to extreme values. In contrast, the median provides a more robust estimate, particularly in the presence of skewed data or outliers.

$$\mathrm{Mean} = \frac{\sum_{i=1}^{n} X_i}{n} \tag{3}$$

where

M (t day^{-1} and t m^{-2}): The mean of the observed values.
X_i: The observed value (t day^{-1} or t m^{-2}).
n: The total number of projects.

The median depends on whether the number of observations (n) is odd or even:

If n is odd, the following applies:

$$\mathrm{Median} = X_{\left(\frac{n+1}{2}\right)} \tag{4}$$

If n is even, the following applies:

$$\text{Median} = \frac{X_{(\frac{n}{2})} + X_{(\frac{n}{2}+1)}}{2} \tag{5}$$

where

Med. (t day^{-1} or t m^{-2}): The median of the observed values.
X_i: The observed value (t day^{-1} or t m^{-2}).
n: The number of projects.

A simplified predictive model was developed using robust estimators by multiplying the median value of each indicator by the corresponding project-specific variable. This generated two predicted values: CDW_Pd, representing the total waste, estimated from the average daily generation rate (t day^{-1}), which was multiplied by the project duration in days (Equation (6)), and CDW_Pa, representing the total waste, estimated from the waste intensity per square meter (t m^{-2}), which was multiplied by the intervention area (Equation (7)).

$$CDW_{Pd} = \text{Median} \times \text{Duration} \tag{6}$$

$$CDW_{Pa} = \text{Median} \times \text{Intervention Area} \tag{7}$$

where

CDW_{Pd} (t): The predicted CDW based on average daily generation.
CDW_{Pa} (t): The predicted CDW based on generation per unit of intervention area.
Median (t day^{-1} and t m^{-2}): The median of the calculated indicator values (construction and rehabilitation).
Duration (days): The duration of the project.
Intervention Area (m^2): The area of the project.

The reliability of the median-based forecasting model was evaluated by comparing the predicted and observed CDW values. Two complementary metrics were applied: absolute deviation (AD), which captures the direct difference between predicted and observed values, and relative deviation (RD), which expresses this difference as a percentage of the observed value (Equations (8) and (9)).

$$AD = CDW_P - CDW \tag{8}$$

$$RD = \frac{CDW_P - CDW}{CDW} \times 100 \tag{9}$$

where

Absolute Deviation (AD, t day^{-1} or t m^{-2}): The direct difference between the predicted (CDW_P) and observed values (CDW).
Relative Deviation (RD, %): The deviation expressed as a percentage of the observed value, enabling comparisons across projects of different scales.
CDW_P (t day^{-1} or t m^{-2}): The predicted waste generation, calculated using Equations (6) and (7).
CDW (t day^{-1} or t m^{-2}): The observed waste generation recorded for each project.

These metrics quantify discrepancies between predicted and observed values and serve as the basis for the anomaly detection procedures that are detailed in the following section. Equation (6) estimates the total CDW by multiplying the median daily generation rate (t day^{-1}) by the project duration (days), while Equation (7) applies the median per-area value (t m^{-2}) to the project's intervention area (m^2). Absolute and relative deviations were then calculated using Equations (8) and (9), respectively. All equations were applied consistently across the dataset of 16 projects using a standardized spreadsheet-based framework to maintain methodological uniformity and analytical comparability.

2.4. Detection and Analysis of Anomalies and Patterns

Statistical anomalies between predicted and observed CDW values were identified using a robust detection method based on the Median Absolute Deviation (MAD). Rather than relying on arbitrary thresholds (e.g., ±10%), this approach defines confidence intervals that are grounded in the dataset's internal variability.

The MAD was calculated as the median of the absolute deviations from the group-specific median, as shown in Equation (10).

$$\text{MAD} = \text{Median}(|\text{Xi} - \text{Med.}(\text{X})|) \quad (10)$$

where

MAD (t day^{-1} or t m^{-2}): The Median Absolute Deviation, calculated as the median of the absolute differences between each observed value and the median of the group.
X_i (t day^{-1} or t m^{-2}): The observed value for project *i*, either in absolute or relative terms depending on the analysis.

Based on this measure, upper and lower bounds for anomaly detection were computed using Equations (11) and (12). Values beyond these thresholds' limits were flagged as anomalies.

$$\text{Upper Limit} = \text{Median} + (\text{k} \times \text{MAD}) \quad (11)$$

$$\text{Lower Limit} = \text{Median} - (\text{k} \times \text{MAD}) \quad (12)$$

where

Upper Limit/Lower Limit (t day^{-1} or t m^{-2}): Thresholds defining the bounds of expected variability in CDW values.
Median (t day^{-1} or t m^{-2}): The median of the absolute or relative deviations, calculated separately for construction and rehabilitation projects.
k (–): A scale factor determining the width of the confidence interval; set to 2 in this study.
MAD (t day^{-1} or t m^{-2}): The Median Absolute Deviation, calculated as the median of the absolute differences between each observed value and the median of the group (Equation (10)).

For anomaly detection, the scale factor k = 2, which follows empirical recommendations in the literature for balancing sensitivity and specificity in anomaly detection applications [22–25]. Lower values (e.g., k < 1.5) may lead to excessive false positives, while higher values (e.g., k > 3) can mask meaningful outliers. A sensitivity analysis of the parameter k was conducted and is further discussed in Section 3.3.

Moreover, the MAD was selected over other robust alternatives, such as the Interquartile Range (IQR), due to its superior ability to quantify dispersion symmetrically around the median. While the IQR identifies the middle 50% of the data, it does not yield a central measure that is suitable for defining symmetric intervals. In contrast, the MAD provides a direct measure of median-centered variability, making it particularly well suited for datasets with irregular or non-normal distributions.

The statistical thresholds derived through Equations (10)–(12) were applied to both the absolute and relative deviations that were calculated for each project. The MAD values and confidence bounds were determined separately for construction and rehabilitation projects to account for structural differences between the intervention types. Projects with deviations falling outside the calculated Median ± 2·MAD limits were flagged as anomalous. All calculations were performed using a standardized computational framework, ensuring consistent and reproducible application across the dataset.

3. Results and Discussion

This section presents the main findings on Construction and Demolition Waste (CDW) generation across a sample of 16 projects, including 7 new construction and 9 rehabilitation interventions. The analysis focuses on identifying key patterns, statistical anomalies, and the influence of project-specific characteristics such as the intervention area, project duration, and construction typology. Our CDW categorization follows the European Waste List (EWL), ensuring standardization and comparability across the dataset. To account for variations in project scale and execution time, two normalized indicators—waste generation per day (t day^{-1}) and per square meter (t m^{-2})—were utilized. Additionally, the use of the median and Median Absolute Deviation (MAD) provided a robust statistical foundation for both prediction and anomaly detection, making the method well suited to handle the heterogeneous nature of the dataset.

Despite the relatively limited sample size (n = 16), as detailed in Section 2.1, the methodological consistency across all projects strengthens the internal validity, ensuring that observed variations in CDW generation are attributed to project characteristics rather than procedural inconsistencies or data recording errors.

3.1. CDW Generation Patterns

Table 1 presents an overview of each project's key characteristics, including the intervention type, duration, intervention area, and construction typology. These variables are essential for understanding the sources of variability in CDW production and provide a structured basis for cross-case comparison.

Table 1. Summary of project characteristics, including intervention type, duration (days), intervention area (m^2), and construction typology.

Type	Project	Duration (Days)	Intervention Area (m^2)	Construction Type
Construction	C1	487	3580	Low-to-mid-rise building
	C2	548	5400	Low-rise residential unit
	C3	731	1321	Low-to-mid-rise building
	C4	243	3662	Low-rise residential unit
	C5	365	2581	Low-rise residential unit
	C6	731	4100	Low-rise residential unit
	C7	731	1413	Low-to-mid-rise building
Rehabilitation	R1	731	573	Low-to-mid-rise building
	R2	365	290	Low-rise residential unit
	R3	365	7000	Low-rise residential unit
	R4	545	1388	Low-to-mid-rise building
	R5	455	20,000	Low-rise residential unit
	R6	365	15,000	Low-rise residential unit
	R7	183	4000	Low-rise residential unit
	R8	183	15,000	Low-rise residential unit
	R9	365	621	Low-rise residential unit

Table 2 presents the classification and quantities of CDW generated per project, categorized according to the European Waste List (EWL) codes. The total CDW per project was calculated as the sum of all waste categories reported, forming the observed variable that was used in both the modeling and anomaly detection processes. The table provides insights into the diversity of waste compositions and the recurrence of specific high-volume waste streams, such as excavation materials (EWL 17 05 04), structural demolition waste (EWL 17 01 07), and mixed construction waste (EWL 17 09 04).

The total CDW generated by each project displayed marked heterogeneity across the dataset. In construction projects, the values ranged from 3.94 tonnes (C6) to 5949.34 tonnes (C2). Although larger intervention areas generally resulted in greater waste volumes, this relationship was not strictly linear. For instance, Project C2, with an area of 5400 m^2, generated exceptionally high volumes due to substantial excavation activity, while C3, despite a long duration of 731 days, produced only 24.76 tonnes due to its limited demolition scope. Project C7, although relatively small in size (1413 m^2), generated 449.46 tonnes of CDW, with a significant proportion consisting of mixed demolition waste.

Table 2. Classification and quantities of CDW generated per project according to European Waste List (EWL) codes. Total CDW represents the sum of all waste categories per project.

Type	Project	EWL	EWL Code Production (t)	CDW (t)
Construction	C1	17 04 05	16.76	113.58
		17 08 02	44.85	
		17 09 04	51.97	
	C2	17 01 07	141.59	5949.34
		17 02 01	9.76	
		17 02 03	0.29	
		17 04 05	2.57	
		17 05 04	5734.25	
		17 08 02	0.62	
		17 09 04	57.72	
		15 01 01	1.41	
		15 01 02	0.68	
		15 01 05	0.39	
		19 12 01	0.08	
	C3	17 04 05	2.4	24.76
		17 08 02	10.65	
		17 09 04	11.71	
	C4	17 09 04	7.92	7.92
	C5	17 09 04	4.36	4.36
	C6	17 09 04	3.94	3.94
	C7	17 02 01	4.42	449.46
		17 06 04	4.36	
		17 08 02	6.96	
		17 09 04	433.72	

Table 2. *Cont.*

Type	Project	EWL	EWL Code Production (t)	CDW (t)
Rehabilitation	R1	17 01 07	2825.31	3692.56
		17 02 01	77.4	
		17 04 05	62.95	
		17 05 04	182.86	
		17 09 04	541.08	
		20 03 06	2.96	
	R2	17 01 07	501.2	884.28
		17 02 01	23.1	
		17 04 05	12.86	
		17 04 07	0.32	
		17 05 04	48.84	
	R3	17 04 05	0.27	11.68
		17 09 04	11.415	
	R4	17 01 03	10.38	37.03
	R5	17 01 07	13.42	163.95
	R6	17 01 07	58.48	79.57
	R7	17 01 03	4.08	29.14
	R8	17 01 07	17.24	17.68
	R9	17 01 03	17.12	23.29

Rehabilitation projects demonstrated even greater variability. Project R1 (573 m^2, 731 days) generated 3692.56 tonnes of CDW, primarily structural demolition debris and mixed waste, while R3, with a considerably larger area (7000 m^2), produced just 11.68 tonnes. Notably, some large-area projects, such as R5 and R6 (20,000 m^2 and 15,000 m^2, respectively), exhibited comparatively modest CDW outputs (163.95 and 79.57 tonnes), further emphasizing that area alone is not a sufficient predictor of the waste volume. Conversely, R2, with only 290 m^2 of intervention, produced 884.28 tonnes, suggesting a highly demolition-intensive scope.

Across both categories, recurring waste types included excavation material (EWL 17 05 04), structural debris (EWL 17 01 07), and mixed construction waste (EWL 17 09 04). The predominance of these streams across project types reinforces their relevance in the design of waste reduction and valorization strategies.

The construction typology appears to influence waste patterns as well. Projects classified as low-to-mid-rise buildings (e.g., C1, C3, R1, R4) tended to exhibit both higher CDW quantities and greater heterogeneity in waste composition. Low-rise residential units, in contrast, generally produced lower and more homogeneous waste volumes. While the construction type was not explicitly modeled as a predictive variable, the observed trends suggest its potential as a relevant contextual factor in future analyses.

Based on the total waste, duration, and intervention area as listed in Table 2, the standardized indicators defined in Equations (1) and (2) were calculated for all 16 projects. These include average daily CDW generation (t day^{-1}) and CDW generation per square meter of intervention area (t m^{-2}), which provide a normalized basis for comparing waste intensities across different project types and scales. The computed indicator values are summarized in Table 3.

Table 3. Average daily CDW generation (t day^{-1}) and CDW generation per intervention area (t m^{-2}) across construction and rehabilitation projects.

Type	Project	Duration (t Day $^{-1}$)	Intervention Area (t m^{-2})
Construction	C1	0.23	0.03
	C2	10.85	1.10
	C3	0.03	0.02
	C4	0.03	0.0022
	C5	0.01	0.0017
	C6	0.0054	0.0010
	C7	0.61	0.31
Rehabilitation	R1	5.05	6.44
	R2	2.42	3.04
	R3	0.03	0.0017
	R4	0.06	0.02
	R5	0.36	0.0082
	R6	0.21	0.0053
	R7	0.15	0.0073
	R8	0.09	0.0012
	R9	0.06	0.037

Table 3 provides a quantitative overview of the CDW intensity across all projects, based on the two standardized indicators. Among the construction projects, the daily CDW generation ranged from 0.0054 to 10.85 t day^{-1}, while waste per square meter varied between 0.0010 and 1.10 t m^{-2}. Rehabilitation projects exhibited similarly wide dispersion, with daily values ranging from 0.06 to 5.05 t day^{-1} and area-based values from 0.0012 to 6.44 t m^{-2}. These broad ranges reflect substantial differences in demolition methods, structural complexity, and the scope of material removal.

The project typology appears to significantly Influence the CDW Intensity. Projects classified as low-to-mid-rise buildings, such as C1, C3, R1, and R4, generally show higher values in at least one of the indicators, likely due to the structural complexity and material density that are typical of these constructions. In contrast, several low-rise residential units—including C6, R5, and R8—exhibited comparatively low CDW generation per area and per day, despite large intervention scales in some cases. This suggests that typology-related characteristics, such as construction techniques and material composition, may be more predictive of the CDW output than physical dimensions alone.

When compared to values reported in the literature, the observed construction indicators in this study (0.0017–0.31 t m^{-2} or 1.7–310 kg m^{-2}) generally fall within expected ranges. Exceptions at the upper end—most notably Project C2—can be attributed to intensive excavation activities [7,8,20]. The rehabilitation indicators (0.0053–6.44 t m^{-2} or 5.3–6440 kg m^{-2}) are likewise consistent with previous findings, particularly in cases involving structural refurbishment or partial demolition [7,8].

These results underscore the relevance of including contextual variables—such as the construction type, material typology, and site-specific operational conditions—into CDW forecasting models. They also highlight the limitations of relying solely on physical parameters such as the duration or intervention area. The use of standardized indicators not only improves comparability but also establishes a solid foundation for the statistical anomaly detection that is discussed in the following section.

3.2. Indicator Variability and Robustness Analysis

The statistical distribution of the standardized indicators presented in Table 3—average daily CDW generation (t day^{-1}) and CDW per square meter of intervention area (t m^{-2})—across the 16 projects was analyzed to assess the variability and determine the most appropriate measure of a central tendency for predictive applications. Given the heterogeneity in project duration, intervention area, and typology, these indicators enable consistent comparisons across interventions of different scales and scopes.

Figures 1 and 2 display the distribution of both indicators across construction and rehabilitation projects, respectively. Each data point corresponds to a value listed in Table 3. The dashed and solid lines represent the group-specific mean and median, respectively, enabling direct visual comparison between these estimators.

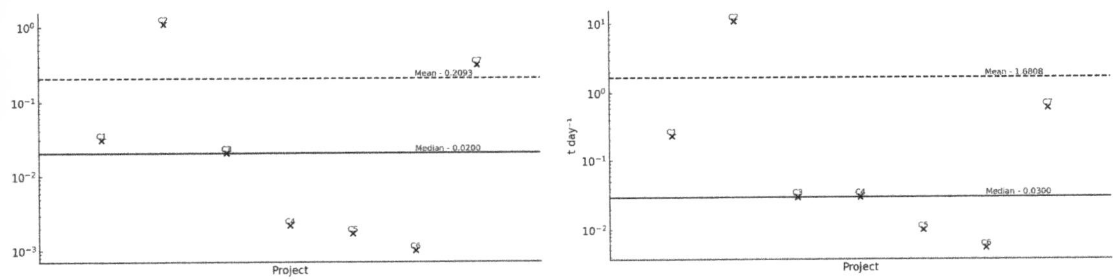

(a) (b)

Figure 1. Distribution of waste generation indicators for construction projects: (**a**) waste generation per intervention area (t m^{-2}); (**b**) average daily waste generation (t day^{-1}). Each data point represents a construction project (C1–C7). The solid line indicates the median; the dashed line shows the mean.

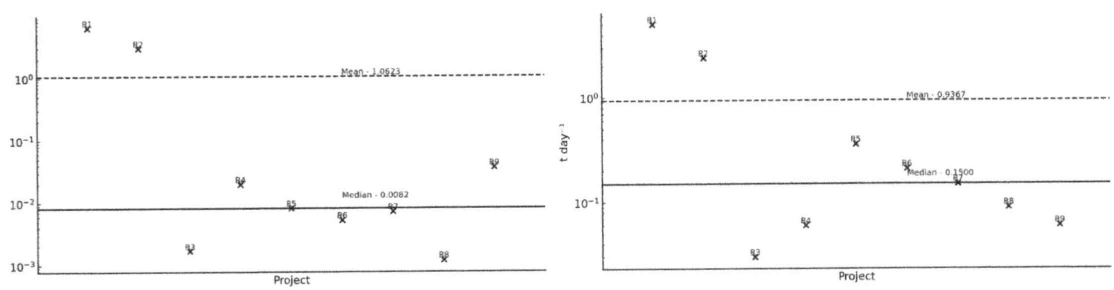

(a) (b)

Figure 2. Distribution of waste generation indicators for rehabilitation projects: (**a**) waste generation per intervention area (t m^{-2}); (**b**) average daily waste generation (t day^{-1}). Each data point represents a rehabilitation project (R1–R9). The solid line indicates the median; the dashed line shows the mean.

As Figures 1 and 2 illustrate, the mean is strongly influenced by high-output projects such as C2 and C7 (construction) and R1 and R2 (rehabilitation). These extreme values disproportionately elevate the mean, whereas the median remains stable and closer to the center of the distribution. This contrast highlights the robustness of the median and its superiority as a central tendency estimator for predictive purposes.

A quantitative analysis of the standardized indicators across all 16 projects reveals significant variability. The average daily CDW generation ranges from 0.0054 to 10.85 t day^{-1}, with a group mean of 1.31 t day^{-1} and a median of 0.12 t day^{-1}. For CDW generation per square meter, the values range from 0.0010 to 6.44 t m^{-2}, with a mean of 0.69 t m^{-2} and a median of 0.0141 t m^{-2} (Figures 1 and 2). These statistics highlight the dataset's asymmetry and reinforce the divergence between the mean and the more robust median.

This divergence arises because the arithmetic mean incorporates all values equally, making it highly sensitive to extreme values. In scenarios involving CDW forecasting—where variations in material intensity, structural typology, and site conditions are significant—such robustness of the median is critical to prevent systemic overestimation.

The construction typology was further examined as a potential explanatory factor for the observed variability. The analysis of the results shows that low-to-mid-rise buildings present substantially higher CDW generation per unit area. For instance, construction projects C2 and C7 generated 1.10 and 0.31 t m^{-2}, respectively, whereas low-rise projects such as C5 and C6 generated only 0.0017 and 0.0010 t m^{-2}. Similarly, rehabilitation projects R1 and R4—both low-to-mid-rise buildings—produced 6.44 and 0.02 t m^{-2}, respectively, while low-rise units like R6 and R8 yielded just 0.0053 and 0.0012 t m^{-2}. These differences reflect the greater material intensity and structural complexity of low-to-mid-rise buildings. Such typological distinctions underscore the relevance of considering the construction type in predictive modeling, particularly when estimating CDW across diverse project portfolios.

Based on these distributional characteristics, the median was selected as the central estimator for predictive modeling. As detailed in Section 2.3, Equations (6) and (7) were used to compute predicted CDW values (CDWpd and CDWpa) for each project by multiplying the group-specific median indicator by the project's duration or intervention area. The predicted values were then compared to the observed data (Table 3) using absolute deviation (AD; Equation (8)) and relative deviation (RD; Equation (9)). The full set of predicted values and their respective absolute and relative deviations from the observed data are presented in Table 4.

The deviation analysis reveals three main patterns. First, negative deviations—as observed in C2 and R1—correspond to cases in which the actual CDW generation significantly exceeded the predicted values. These discrepancies are likely associated with atypical interventions involving substantial excavation or extensive structural demolition. Second, positive deviations, such as those identified in C5 and R9, reflect overestimations by the model and may be linked to projects that are characterized by limited material removal or a high reuse efficiency. Third, in certain projects (e.g., C3 and R5), the predicted and observed values matched exactly. This does not reflect overfitting, since no calibration or parameter tuning was applied. Rather, this coincidence is expected in small samples when project characteristics naturally align with the group median—especially under consistent operational conditions and limited variability.

Although the construction typology was not included as an explicit modeling variable, it may partially account for some of the observed deviations. For instance, C2, categorized as a low-rise residential unit, exhibited an unusually high CDW generation rate (10.85 t day^{-1}), suggesting the presence of intensive excavation that was atypical of its typology. Similarly, R1, a low-to-mid-rise building with a relatively small footprint (573 m^2), generated an exceptionally high CDW per unit area (6.44 t m^{-2}), likely due to comprehensive structural demolition. In contrast, C5 and R9—both low-rise residential projects—produced considerably less CDW than predicted, which is indicative of less invasive interventions. These cases highlight the potential influence of construction type on CDW generation and support its consideration in future modeling efforts.

Table 4. Predicted values of CDW generation per day (CDWpd) and per intervention area (CDWpa), including absolute and relative deviations (%) from the observed values.

		Duration			Intervention Area		
Type	Project	CDWpd (t Day^{-1})	Absolute Deviation (t Day^{-1})	Relative Deviation (%)	CDWpa (t m^{-2})	Absolute Deviation (t m^{-2})	Relative Deviation (%)
Construction	C1	16.49	−97.09	−85.48	67.09	−46.49	−40.93
	C2	18.56	−5930.79	−99.69	101.2	−5848.15	−98.3
	C3	24.76	0.0	0.0	24.76	0.0	0.0
	C4	8.23	0.31	3.91	68.63	60.71	766.5
	C5	12.36	8.0	183.51	48.37	44.01	1009.38
	C6	24.76	20.82	528.32	76.84	72.9	1850.14
	C7	24.76	−424.7	−94.49	26.48	−422.98	−94.11
Rehabilitation	R1	116.4	−3576.16	−96.85	4.7	−3687.86	−99.87
	R2	58.12	−826.16	−93.43	2.38	−881.9	−99.73
	R3	58.12	46.44	397.57	57.38	45.7	391.23
	R4	86.78	49.76	134.39	11.38	−25.65	−69.27
	R5	72.45	−91.49	−55.81	163.95	0.0	0.0
	R6	58.12	−21.45	−26.96	122.96	43.39	54.53
	R7	29.14	0.0	0.0	32.79	3.65	12.52
	R8	29.14	11.47	64.87	122.96	105.28	595.67
	R9	58.12	34.83	149.52	5.09	−18.2	−78.15

The use of the median as a predictive baseline therefore provides a robust and resilient framework in the presence of outliers and skewed distributions. Its statistical stability enhances the reliability of CDW forecasting in heterogeneous datasets. The following section builds on this foundation by incorporating the Median Absolute Deviation (MAD) to formally establish thresholds for anomaly detection.

3.3. Statistical Detection of Anomalies and Patterns

Significant discrepancies between the predicted and observed CDW values were assessed using a robust anomaly detection method that combines the median with the Median Absolute Deviation (MAD). As detailed in Section 2.4, this method allows for the definition of upper and lower confidence limits that reflect the internal variability of the dataset and avoid the use of arbitrary thresholds—making it particularly suitable for small samples with high dispersion, as is common in construction-related studies.

Predicted CDW values were first calculated using Equations (6) and (7) by multiplying the group-specific median indicators by each project's duration or intervention area. Discrepancies between predicted and observed values were then quantified using absolute deviation (Equation (8)) and relative deviation (Equation (9)).

The anomaly detection procedure involved two sequential steps. First, the MAD was computed as the median of the absolute deviations from the group median (Equation (10)). Second, upper and lower statistical bounds were established using Equations (11) and (12), which define the thresholds as the median $\pm$ k $\times$ MAD, with k = 2, following established empirical precedent [23–25]. Observations falling outside these bounds were flagged as statistical anomalies.

The resulting MAD values and the corresponding statistical thresholds for both construction and rehabilitation projects—and for both indicators (CDW per day and per intervention area)—are presented in Table 5. These values provide the basis for identifying significant absolute and relative deviations in predicted CDW generation. An analysis of the values presented in Table 5 revealed several notable anomalies. In particular, Project C2 exhibited a considerable underestimation (−5848.15 t m^{-2}), while Project R2 showed a comparably large negative deviation (−5930.79 t day^{-1}). Several projects—including C1, C2, C7, R1, and R2—displayed substantial discrepancies across both temporal and spatial indicators.

Table 5. MAD values and statistical limits for anomaly detection. MAD and upper and lower thresholds (±2 × MAD) for absolute and relative deviations in predicted CDW per day (t day^{-1}) and per intervention area (t m^{-2}), by project type.

Type	Dimension	Deviation Type	MAD	Upper Limit	Lower Limit
Construction	Duration	Absolute Deviation	20.82	41.64	−41.64
		Relative Deviation (%)	94.49	188.98	−188.98
	Intervention Area	Absolute Deviation	60.71	121.42	−121.42
		Relative Deviation (%)	98.30	196.60	−196.60
Rehabilitation	Duration	Absolute Deviation	46.44	92.88	−92.88
		Relative Deviation (%)	93.43	186.86	−186.86
	Intervention Area	Absolute Deviation	43.39	86.78	−86.78
		Relative Deviation (%)	78.15	156.30	−156.30

Complementing the statistical thresholds outlined in Table 5, the results were also plotted by project type and indicator category to support visual interpretation. Figure 3 displays the anomaly detection results for construction projects, while Figure 4 focuses on rehabilitation projects. Each figure contains four panels: (a) absolute deviation in CDW per day, (b) relative deviation in CDW per day, (c) absolute deviation per intervention area, and (d) relative deviation per intervention area.

Figures 3 and 4 depict the deviation profiles for construction and rehabilitation projects, respectively, across four key dimensions: absolute and relative deviations in CDW generation per day (panels a and b) and per intervention area (panels c and d). The gray-shaded areas in each panel correspond to the statistical confidence bounds, calculated using the Median $\pm$ 2×MAD rule, with threshold values being summarized in Table 5.

In the absolute deviation plots (panels a and c), the vertical axis adopts a symmetric logarithmic (*symlog*) scale. This representation allows for balanced visualization of both negative and positive deviations while minimizing the influence of extreme values. Readability near zero is preserved, enabling a more faithful comparison across projects. In contrast, the relative deviation plots (panels b and d) are presented on a standard linear scale, as the distribution of values did not warrant transformation.

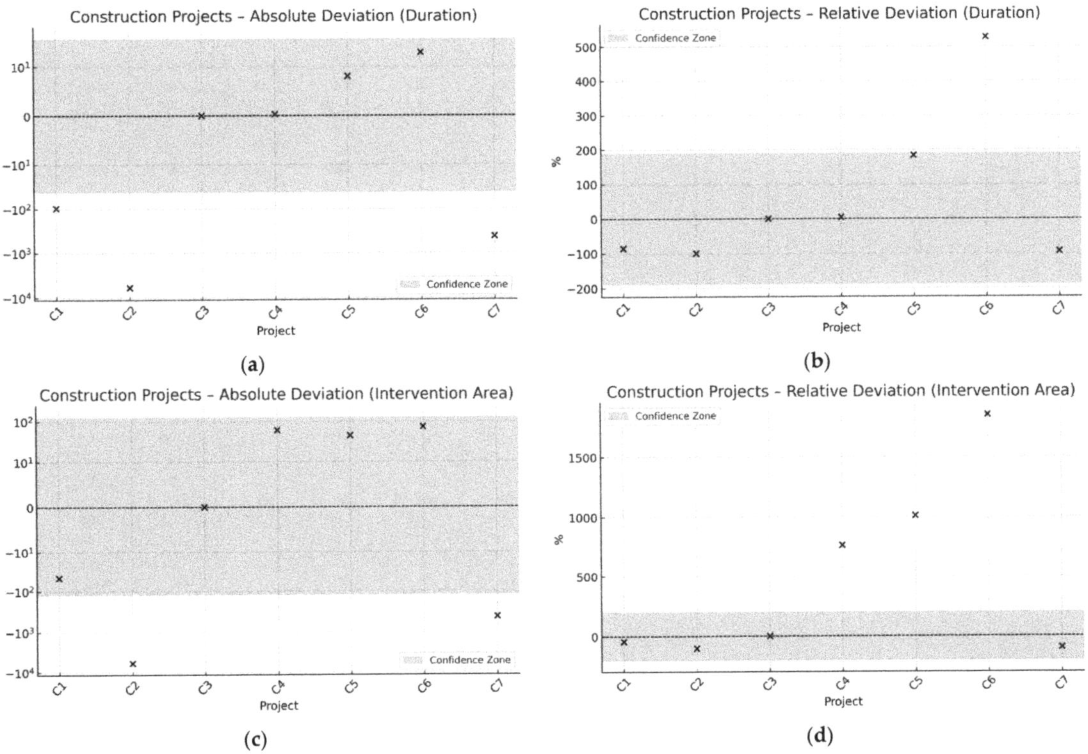

Figure 3. Anomaly detection in construction projects: (**a**) absolute deviation—CDW per day (t day^{-1}); (**b**) relative deviation—CDW per day (%); (**c**) absolute deviation—CDW per intervention area (t m^{-2}); (**d**) relative deviation—CDW per intervention area (%).

Projects C2, C7, R1, and R2 consistently fall outside the confidence bounds across all four indicators, indicating systemic deviations rather than isolated inconsistencies. These anomalies span both the spatial and temporal domains, suggesting persistent divergences from the expected CDW generation patterns. Project C2, for instance, a low-rise residential construction, recorded an observed output of 10.85 t day^{-1}, indicating substantial excavation or over-removal of materials. Similarly, R1, with a relatively small intervention area (573 m^2), exhibited a high waste intensity of 6.44 t m^{-2}, most likely resulting from extensive structural demolition associated with vertical rehabilitation.

Some projects showed deviations in only one metric. Project C6 exceeded the spatial confidence bounds (Figure 3c,d) but remained within the temporal limits (Figure 3a,b), suggesting localized inefficiencies in material handling. Conversely, C5 and R3 exhibited lower-than-expected values across indicators, which may reflect optimized demolition practices or material reuse strategies.

The robustness of the anomaly detection approach was evaluated through a sensitivity analysis of the scaling parameter k in the Median $\pm k \times$ MAD formulation, using values of 1.5, 2.0, and 3.0. Eight projects were flagged as outliers under k = 1.5 (C1, C2, C6, C7, R1, R2, R5, R8), while seven remained at k = 2.0. When the threshold was raised to k = 3.0, only four projects—C2, C7, R1, and R2—continued to fall outside the bounds. This persistence across all tested thresholds reinforces their classification as statistically robust anomalies, while the remaining cases appear to be sensitive to threshold variation and may reflect project-specific conditions rather than systematic discrepancies. The sensitivity analysis supports the choice of k = 2, as it provides the best balance between sensitivity and specificity, ensuring that meaningful anomalies are detected without generating excessive false positives.

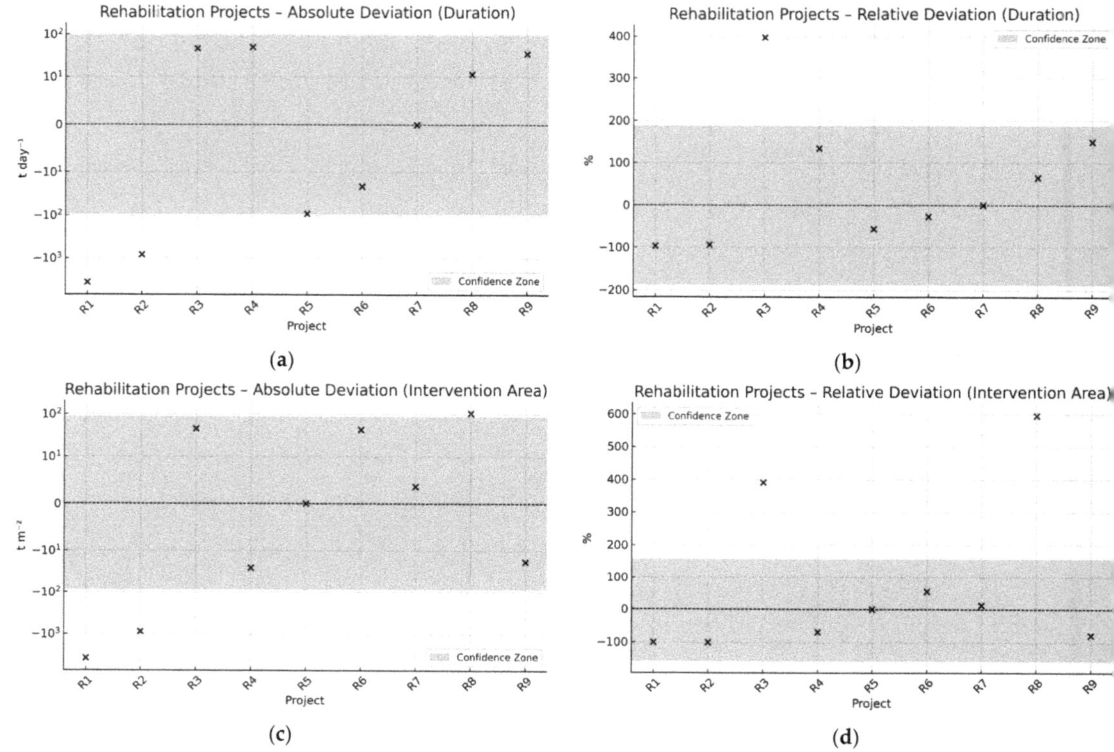

Figure 4. Anomaly detection in rehabilitation projects: (**a**) absolute deviation—CDW per day (t day^{-1}); (**b**) relative deviation—CDW per day (%); (**c**) absolute deviation—CDW per intervention area (t m^{-2}); (**d**) relative deviation—CDW per intervention area (%).

By jointly analyzing the absolute and relative deviations in both the temporal and spatial dimensions, this two-axis diagnostic framework provides a nuanced interpretation of CDW generation performance. Single-axis anomalies may point to targeted operational inefficiencies, while multi-axis deviations suggest deeper inconsistencies that merit comprehensive evaluation in project planning, execution, or reporting practices.

3.4. Addressing Variability and Systemic Challenges

The standardization of waste generation indicators—specifically, average daily production (t day^{-1}) and production per square meter (t m^{-2})—strengthens the analytical coherence and comparability of the proposed forecasting framework. This foundation is further enhanced through the application of robust statistical methods based on the median and the Median Absolute Deviation (MAD), which together mitigate the influence of extreme values and enable consistent cross-project comparisons. The robustness of the adopted methodology contributes not only to a more consistent interpretation of CDW patterns across diverse project typologies but also provides a solid foundation for operational planning and regulatory compliance [13,14,20,27,28].

Compared to traditional predictive approaches—such as linear regression or artificial neural networks—the MAD-based methodology exhibits increased reliability in anomaly detection. Conventional models generally require large datasets for effective calibration [14], while the use of non-parametric statistics ensures resilience, even in contexts characterized by high variability and limited data availability. This makes the framework particularly suitable for professionals and public authorities managing CDW in small-scale or regionally confined project settings. As shown in Sections 3.2 and 3.3, the median proves to be more stable than the mean in datasets with frequent outliers, further justifying its selection as the central estimator.

Despite its advantages, the current model formulation relies solely on two predictor variables: the project duration and intervention area. While these parameters are readily available and operationally relevant, they do not capture important contextual factors that can influence CDW generation. Variables such as the excavation depth, structural typology, construction techniques, or regional practices may significantly affect waste output, particularly in complex interventions. The absence of such variables is acknowledged as a limitation. To partially address this, the construction type was incorporated in Table 1 to distinguish between low-rise residential units and low-to-mid-rise buildings. Although not yet used as a formal predictor, this typological classification supports the interpretation of deviations and highlights pathways for future multivariate model refinement.

Beyond its forecasting accuracy, the MAD-based anomaly detection procedure provides a strategic advantage for targeted auditing and performance management. Projects exhibiting significant negative deviations—such as C2 and R2—can be prioritized for detailed review, while projects outperforming the predictions may reveal best practices that are worth generalizing. This dual functionality enhances decision-making and resource allocation. Integrating audit outcomes into the model's recalibration establishes a dynamic learning loop, progressively refining its predictive accuracy and responsiveness. As organizations continue to apply the model across multiple projects, the accumulation of internal data leads to more representative median and MAD values. This progressive refinement increases forecast precision, narrows the expected deviation ranges, and improves anomaly detection, thereby enabling earlier and more effective intervention at the project level.

The implications of improved CDW forecasting extend beyond operational efficiency. Although not directly quantified in this study, reductions in Construction and Demolition Waste are closely associated with greenhouse gas (GHG) mitigation. More precise waste predictions help minimize unnecessary extraction, transport, and disposal of construction materials—activities that are closely linked to CO_2 emissions—thereby aligning with broader environmental and climate mitigation goals. Future extensions of this research could incorporate Life Cycle Assessment (LCA) data to quantify these environmental benefits more explicitly and reinforce the model's contribution to sustainable construction practices [13,14,20,27–31].

Finally, incorporating advanced computational techniques may further improve the predictive capacity and scalability of the model. However, it is important to recognize that machine learning algorithms—such as LSTM networks or ARIMAX models—typically require large volumes of high-quality training data. This requirement may present a constraint, particularly for small or medium-sized construction enterprises with limited historical records. Nevertheless, in data-rich environments, such methods could complement the median–MAD approach by enabling more dynamic and adaptive forecasting systems. As audit processes become systematized and datasets grow, the model can evolve into a real-time monitoring system that is guided by key performance indicators (KPIs), enabling proactive waste management aligned with the principles of the circular economy—namely, material efficiency, emission reduction, and sustainable resource planning [29–33].

4. Conclusions

This study advances the predictive modeling of Construction and Demolition Waste (CDW) by applying standardized indicators—tons per day (t day^{-1}) and tons per intervention area (t m^{-2})—in conjunction with robust statistical estimators: the median and Median Absolute Deviation (MAD). This approach addresses key limitations of traditional forecasting methods by enhancing resilience to outliers and enabling statistically grounded anomaly detection.

The median demonstrated superior stability compared to the mean in datasets characterized by skewness and extreme values, offering a more reliable measure of central tendencies, as discussed in Sections 3.2 and 3.3. The MAD enabled the construction of adaptive confidence intervals and the detection of statistical anomalies using a threshold of $\pm 2 \cdot$MAD. This framework allowed for the classification of projects with significant deviations—both positive and negative—thus supporting targeted interventions and audit prioritization.

The findings also revealed that generalized models fail to adequately capture the variability that is present in empirical CDW data. The forecast accuracy improves when predictions are anchored in project-specific parameters—such as the duration and intervention area—and refined through robust statistical thresholds, resulting in closer alignment with actual waste generation patterns.

From an operational standpoint, the MAD-based framework supports systematic benchmarking, informed auditing, and real-time performance feedback. As datasets expand, the reliability of the median and MAD as estimators is expected to increase, enabling the development of more sensitive and adaptive forecasting models. This positions the methodology as a foundation for scalable CDW monitoring systems that are aligned with circular economy principles.

Despite the methodological contributions, this study presents limitations that should be acknowledged clearly. Most notably, the small dataset ($n = 16$), while internally coherent and standardized, limits the representativeness and external validity of the findings. The lack of broader regional and typological diversity constrains the generalizability. Furthermore, the current framework is based on only two predictors—the intervention area and project duration—whereas other potentially influential variables (e.g., construction techniques, labor productivity, and transport logistics) remain unaccounted for. Including such variables could enhance the model's predictive accuracy and contextual adaptability.

Future research should focus on consolidating and extending the applicability of the proposed framework by expanding the dataset and incorporating a broader range of construction typologies. The inclusion of additional contextual variables—such as structural systems, logistics constraints, or material typology—may further enhance the model's generalizability and predictive robustness. Embedding the median–MAD logic within advanced machine learning architectures (e.g., LSTM, ARIMAX) also emerges as a promising path toward adaptive, real-time forecasting. Such developments would not only improve the responsiveness and accuracy of CDW prediction tools but also reinforce the construction sector's alignment with broader goals of resource efficiency, regulatory compliance, and sustainable development.

Author Contributions: Conceptualization, C.A.T.; methodology, C.A.T.; A.L.; validation, C.A.T.; formal analysis, A.L.; investigation, A.L.; resources, A.L.; data curation, A.L.; writing—original draft preparation, A.L.; writing—review and editing, C.A.T.; visualization, C.A.T.; supervision, C.A.T.; project administration, C.A.T.; funding acquisition, C.A.T. All authors have read and agreed to the published version of the manuscript.

Funding: This work is supported by national funds from the FCT–Portuguese Foundation for Science and Technology, under the projects UID/04033: Centro de Investigação e de Tecnologias Agro-Ambienteis e Biológicas And LA/P/0126/2020 (https://doi.org/10.54499/LA/P/0126/2020, accessed on 15 January 2025).

Institutional Review Board Statement: Not applicable.

Informed Consent Statement: Not applicable.

Data Availability Statement: The datasets generated and/or analyzed during the current study are available within the article.

Conflicts of Interest: The authors declare no conflicts of interest.

eferences

Kaza, S.; Yao, L.; Bhada-Tata, P.; Van Woerden, F. *What a Waste 2.0: A Global Snapshot of Solid Waste Management to 2050*; Urban Development; World Bank: Washington, DC, USA, 2018; Volume 10986, p. 30317. [CrossRef]

World Bank. *Putting Waste to Work in a Circular Economy: Creating Good Jobs for Youth in the Waste Sector*; Knowledge Brief Series; Issue 24; World Bank: Washington, DC, USA, 2024. Available online: https://thedocs.worldbank.org/en/doc/f0ccf3267e94d9be1789dba8d7d01eab-0460012024/original/S4YE-Knowledge-Brief-Putting-Waste-to-Work-in-a-Circular-Economy-May-13-1.pdf (accessed on 10 January 2024).

Wu, H.; Zuo, J.; Yuan, H.; Zillante, G.; Wang, J. A review of performance assessment methods for construction and demolition waste management. *Resour. Conserv. Recycl.* **2019**, *150*, 104407. [CrossRef]

United Nations. Emissions Gap Report 2020. 2020. Available online: https://www.unep.org/emissions-gap-report-2020 (accessed on 10 January 2024).

European Commission. Circular Economy Action Plan. 2020. Available online: https://environment.ec.europa.eu/strategy/circular-economy-action-plan_en (accessed on 10 January 2024).

Ghisellini, P.; Cialani, C.; Ulgiati, S. A review on circular economy: The expected transition to a balanced interplay of environmental and economic systems. *J. Clean. Prod.* **2016**, *114*, 11–32. [CrossRef]

Yu, Y.; Yazan, D.M.; Junjan, V.; Iacob, M.-E. Circular economy in the construction industry: A review of decision support tools based on Information & Communication Technologies. *J. Clean. Prod.* **2022**, *349*, 131335. [CrossRef]

Coelho, A.; de Brito, J. Generation of construction and demolition waste in Portugal. *Waste Manag. Res.* **2011**, *29*, 739–750. [CrossRef] [PubMed]

Mália, M.; de Brito, J.; Pinheiro, M.; Bravo, M. Construction and demolition waste indicators. *Waste Manag. Res.* **2013**, *31*, 241–255. [CrossRef]

Ruiz, L.; Ramón, X.; Domingo, S. The circular economy in the construction and demolition waste sector—A review and an integrative model approach. *J. Clean. Prod.* **2020**, *248*, 119238. [CrossRef]

Lee, S.; Chang, H.; Lee, J. Construction and demolition waste management and its impacts on the environment and human health: Moving forward sustainability enhancement. *Sustain. Cities Soc.* **2024**, *115*, 105855. [CrossRef]

Hu, X.; Zhang, Y.; Li, Z.; Wang, T. Application of GM-SVR model for improved accuracy in CDW volume prediction. *Autom. Constr.* **2024**, *162*, 105380.

Bakchan, A.; Faust, K.M.; Leite, F. Seven-dimensional automated construction waste quantification and management framework: Integration with project and site planning. *Resour. Conserv. Recycl.* **2019**, *146*, 462–474. [CrossRef]

Teixeira, C.; Rodrigues, A.; Santos, P. Variation in construction and demolition waste generation: Analysis of influencing factors in different building types. *Resour. Conserv. Recycl.* **2019**, *145*, 52–60. [CrossRef]

Liang, H.; Liu, H.; Bian, X.; Wang, Y.; Dong, L. Supporting zero-waste building: A novel spatial explicit material flow analysis model for construction waste. *J. Clean. Prod.* **2024**, *456*, 142369. [CrossRef]

Hoang, N.H.; Ishigaki, T.; Kubota, R.; Tong, T.K.; Nguyen, T.T.; Nguyen, H.G.; Yamada, M.; Kawamoto, K. Waste generation, composition, and handling in building-related construction and demolition in Hanoi, Vietnam. *Waste Manag.* **2020**, *117*, 32–41. [CrossRef] [PubMed]

Domingo, N.; Batty, M. The impact of site-specific practices on construction waste generation: A regression analysis approach. *J. Clean. Prod.* **2021**, *292*, 126022. [CrossRef]

Villoria Sáez, P.; del Río Merino, M.; Porras-Amores, C.; San-Antonio González, A. Assessing the accumulation of construction waste generation during residential building construction works. *Resour. Conserv. Recycl.* **2014**, *93*, 67–74. [CrossRef]

Wu, Z.; Yu, A.T.W.; Shen, L. Factors influencing construction waste generation in building projects: An analysis of empirical data. *Waste Manag.* **2014**, *34*, 1683–1692. [CrossRef] [PubMed]

20. Lam, P.T.I.; Yu, A.T.W.; Wu, Z.; Poon, C.S. Methodology for upstream estimation of construction waste for new building proje *J. Clean. Prod.* **2019**, *230*, 1003–1012. [CrossRef]
21. Ding, Z.; Huang, X.; Wang, X.; Zuo, J. Assessment of promotional strategies for construction and demolition waste recycl products based on hybrid simulation system. *Environ. Impact Assess. Rev.* **2025**, *112*, 107814. [CrossRef]
22. Daoud, A.O.; Othman, A.A.E.; Ebohon, O.J.; Bayyati, A. Quantifying materials waste in the Egyptian construction industry: critical analysis of rates and factors. *Ain Shams Eng. J.* **2021**, *12*, 4275–4289. [CrossRef]
23. Leys, C.; Ley, C.; Klein, O.; Bernard, P.; Licata, L. Detecting outliers: Do not use standard deviation around the mean, use absol deviation around the median. *J. Exp. Soc. Psychol.* **2013**, *49*, 764–766. [CrossRef]
24. Nascimento, G.F.M.; Wurtz, F.; Kuo-Peng, P.; Delinchant, B.; Batistela, N.J. Outlier detection in buildings' power consumpti data using forecast error. *Energies* **2021**, *14*, 8325. [CrossRef]
25. Lu, W.; Long, W.; Yuan, L. A machine learning regression approach for pre-renovation construction waste auditing. *J. Clean. Pr* **2023**, *397*, 136596. [CrossRef]
26. Lyu, Y.; Xiong, N.; Liang, Y.; Lin, S.; Wang, J. Estimation of construction and demolition waste and assessment of disposal facil location: An integrated Material Flow Analysis and Geographic Information System approach. *Resour. Conserv. Recycl.* **20** *218*, 108226. [CrossRef]
27. Breysse, D.; Balayssac, J.-P. *Non-Destructive In Situ Strength Assessment of Concrete: Practical Application of the RILEM TC 249-IS Recommendations*; Springer: New York, NY, USA, 2021. [CrossRef]
28. Tam, V.W.Y.; Soomro, M.; Evangelista, A.C.J. Quality improvement of recycled concrete aggregate by removal of residual mort A comprehensive review of approaches adopted. *Constr. Build. Mater.* **2021**, *288*, 123066. [CrossRef]
29. Stephan, A.; Athanassiadis, A. Quantifying and mapping embodied environmental requirements of urban building stocks. *Bu Environ.* **2017**, *114*, 187–202. [CrossRef]
30. Na, S.; Heo, S.; Choi, W.; Kim, C.; Whang, S.W. Artificial intelligence (AI)-based technology adoption in the construction indust A cross-national perspective using the technology acceptance model. *Buildings* **2023**, *13*, 2518. [CrossRef]
31. Dodampegama, S.K.; Hou, L.; Asadi, E.; Setunge, S. Revolutionizing construction and demolition waste sorting: Insights fro artificial intelligence and robotic applications. *Resour. Conserv. Recycl.* **2024**, *202*, 107375. [CrossRef]
32. Tuyan, M.; Zhang, L.; Nehdi, M. Development of sustainable alkali-activated slag grout for preplaced aggregate concrete. *J. Cle Prod.* **2020**, *277*, 123488. [CrossRef]
33. Carpio, M.; Roldán-Fontana, J.; Pacheco-Torres, R.; Ordóñez, J. Construction waste estimation depending on urban plannir options in the design stage of residential buildings. *Constr. Build. Mater.* **2016**, *113*, 561–570. [CrossRef]

sustainability

MDPI

ticle

.nalysis of Solid Waste Treatment and Management in Typical :hinese Industrial Parks with the Goal of Sustainable)evelopment and Future Suggestions

ı Yu *, Sichen Chen and Zhe Tan

College of Materials Science and Engineering, Beijing University of Technology, No.100, Pingleyuan, Chaoyang District, Beijing 100124, China; chensien0@163.com (S.C.); zhe@bjut.edu.cn (Z.T.)

* Correspondence: yulu112266@bjut.edu.cn or yulu112266@163.com; Tel./Fax: +86-18810786623

Abstract: Solid waste disposal and management have become a global problem, which is particularly tricky in China with its large population and rapid urbanization. This study focused on the disposal status of multi-source solid waste as well as the park management of some typical cities of China. Firstly, the main technical methods for solid waste disposal were summarized as follows: landfill, incineration, anaerobic digestion and aerobic compost. Secondly, the network analysis method was applied to seek an optimized method for solid waste disposal and management. Thirdly, typical demonstration parks for solid waste disposal and management were analyzed to study their respective operating modes and strategies for synergistic development in terms of resources, environment and economy and to explore the sustainable development potential of the Guangdong–Hong Kong–Macao operating mode. The results showed that the collaborative disposal and recycling of solid waste are important for cities (especially megacities) to achieve resource conservation and environmental protection. The NIMBY effect and environmental pollution risks caused by decentralized construction could be reduced through the construction of circular industrial parks. Advanced technologies and the national policy for solid waste disposal and recycling in industrial parks of small–medium cities, large–medium cities and megacities were systematically analyzed so as to explore a self-operating management mode of industrial parks. Finally, reasonable suggestions, such as sharing, saving and cycling and propaganda education, as well as green and low-carbon solutions were put forward for solid waste disposal and management in typical industrial parks, effectively resolving the contradiction between economic development and environmental protection so as to help urban sustainable development.

Keywords: multi-source solid waste disposal; Guangdong–Hong Kong–Macao Greater Bay Area (GBA); industrial park management; network analysis; sustainable development

1. Introduction

With the acceleration of urbanization at the beginning of the 21st century, the urban population and resource consumption increased rapidly in China. Subsequently, the problems of human activity and the random stacking of waste have increased and caused more and more serious water and air pollution, endangering the urban environment and affecting the lives and health of residents. If this continues, it will undoubtedly limit the sustainable development of cities.

Municipal solid waste mainly includes general industrial solid waste (non-hazardous industrial solid waste), hazardous industrial solid waste, medical solid waste and urban solid waste. General industrial solid waste refers to any industrial solid waste other than hazardous waste produced by industrial production activities [1]. Industrial hazardous waste refers to toxic, flammable, corrosive, infectious, highly chemically reactive, or other hazardous waste produced from industrial production [2]. Medical waste, produced by medical and health institutions in medical treatment, prevention, health care and other related activities, is directly or indirectly infectious or toxic and includes other types of hazardous waste. It also includes other waste managed and disposed of in accordance with the *Management Regulations on Medical Waste* [3]. Urban solid waste includes solid waste discharged from urban daily life or from services provided to urban daily life, as well as solid waste that is regarded as urban house refuse under laws and administrative rules and regulations [4]. The annual output of all kinds of solid waste increased by 6.34%, with general industrial waste accounting for 77.07% of municipal solid waste (Figure 1 and Table S1), which could be ascribed to the sharp rise in the urban population with the continuous development of China's urbanization level.

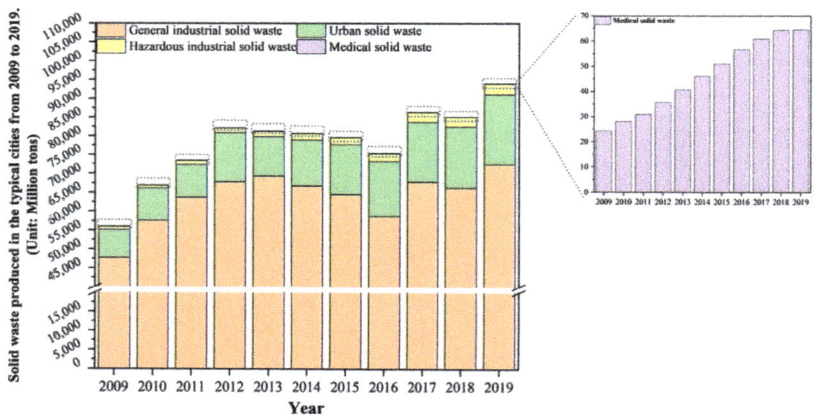

Figure 1. The amount of solid waste generation in major cities from 2009 to 2019 [5,6] (unit: million tons).

In recent years, the production of solid waste has increased year by year around the world. In 2020 alone, 2.24 billion tons of solid waste was generated by human beings, amounting to 0.79 kg per person per day. The annual waste generation level is expected to increase by 73% from 2020 to 2050, finally reaching 3.88 billion tons [7]. If so, it will have a serious impact on the urban environment and the daily lives of residents. Therefore, reasonable treatment and effective management of solid waste are particularly important, and their practice will contribute to the sustainable development of the environment, economy and society.

The development of solid waste treatment and disposal has experienced more than half a century of exploration and has made certain achievements in China. Although China has made great strides in establishing a comprehensive regulatory framework for the green development of industrial parks covering different legislative levels, less than 5% of industrial parks are green-certified nationwide. There are still some problems that need to be solved, not only at the technical level but also involving the management of solid waste industrial parks. An industrial park is designed and established according to the requirements of cleaner production, the concept of a circular economy and the principles of industrial ecology, which is the concrete embodiment of a circular economy at a certain regional level in the industrial field. It connects different factories or enterprises through logistics or energy flow transmission to form an industrial symbiotic combination of sharing resources and exchanging by-products, making the waste or by-products of one factory become the raw materials or energy of another factory, simulating the natural system and establishing a "producer–consumer–decomposer" chain in the industrial system. However, the present problem of solid waste industrial parks is that the real effect of a single treatment technology is limited, while the combination of multiple composite technologies could be too complex. In addition, a lack of proper disposal or management modes has damaged the environment and caused a waste of resources.

In this study, the current situation of solid waste disposal and management in typical industrial parks in China is discussed by analyzing real cases of solid waste disposal and management in typical industrial parks. Network analysis is applied to research the multi-source solid waste disposal of the demonstration park, as well as solid waste management in typical industrial parks of China, following different sizes of cities. We aim to provide possible suggestions for multi-solid waste disposal and management, with a final goal of a green and low-carbon solid waste disposal model to contribute to sustainable urban development.

2. Materials and Methods

Due to its characteristics of bioavailability and mobility, solid waste is difficult to decompose and usually exists for a long time in the environment [8]. It not only occupies land space [9] but also damages the water–soil environment and endangers the lives and health of human beings [10]. What is worse, volatile pollutants formed from the chemical degradation of waste could cause air pollution [11]. At present, solid waste treatment technologies [12,13] mainly include landfill, incineration and biological treatment technologies (aerobic composting and anaerobic digestion).

2.1. Solid Waste Disposal Technology of Landfill

Generally, the most common method of solid waste treatment is landfill [14,15], especially for unsorted waste treatment [16], which has economic advantages [17]. However, landfills take up a large amount of land resources and cause secondary pollution [18]. Moreover, biochemical degradation produces a large amount of toxic leachates in the process of landfill, which can further be aggravated by precipitation [19], surface runoff [20], and groundwater runoff [21,22]. Therefore, landfill technology for solid waste disposal needs improvement for solving the problems of leachates and secondary pollution.

2.2. Solid Waste Disposal Technology of Incineration

Incineration is a common method for solid waste disposal around the world [23–25] because of its dual ability to reduce the volume of solid waste [26] and recover energy [27]. However, solid waste incineration produces three types of residues, i.e., bottom ash, fly ash and air pollutants [28]. Among the air pollutants, flue gas is produced in the process of incineration in the form of a secondary pollutant [29]. Furthermore, heavy metal compounds in solid waste whose volatilization point is lower than the combustion temperature could also enter the flue gas and contribute to air pollution [30]. What is worse, incomplete incineration can produce flue gas that contains highly toxic dioxins [31,32]. At present, only a few enterprises in China can achieve clean production and zero waste emission during the incineration process. Thus, incineration treatment is not the best choice for China's solid waste disposal industry.

2.3. Solid Waste Biological Disposal Technology

Biological disposal methods, including aerobic composting and anaerobic digestion, are technologies that promote resource utilization in the process of solid waste disposal. Compared with traditional landfill, incineration and other physicochemical treatment methods, biological treatment technologies can not only reduce organic solid waste but also turn waste into wealth (organic fertilizers, biogas and others). Biological treatments have the advantages of low cost, green environmental protection and realizing the reuse of resources [33]. However, current biological disposal technologies still have some limitations. For instance, volatile compounds are derivatives of biological treatment that could cause air pollution when entering the environment [34].

Aerobic composting is a method of chemical decomposition of solid waste for extracting useful components through fermentation, degradation and other ways [35,36]. Anaerobic digestion is also a green and environmentally friendly solid waste disposal method with low energy consumption in the absence of oxygen [37]. However, the process of biological disposal can be affected by multiple environmental factors, e.g., temperature, pH, moisture, substrate, carbon source, nitrogen source, C/N ratio, etc. [38,39]. In addition, there is a lack of systematic evaluation for its effectiveness and technical economics [40].

Furthermore, the industrialization promotion of biological disposal technology is another issue worthy of attention.

2.4. Network Analysis Method

In order to improve the efficiency of multi-source solid waste disposal and management [41], VOS viewer 1.6.18, a software tool for constructing and visualizing bibliometric networks, was applied to understand which factors were more important in the field of solid waste disposal and management (Table S2). Co-authorship networks, citation-based networks, and co-occurrence networks were created based on data download from Web of Science (WoS), Scopus, Dimensions, and Lens. Co-authorship networks and co-occurrence networks were created based on PubMed data [42]. Data collection was mainly based on published data (see Section 3.1 for details), as well as field research of the Guangdong–Hong Kong–Macao Greater Bay Area (GBA) solid waste disposal demonstration park. Generally, WoS is the most widely used database worldwide [43]. The keywords for screening and research were as follows: TS = ("solid waste") and/or ("solid waste treatment" or "solid waste disposal" or "solid waste management"). All the works retrieved from 2000 to 2023 included papers and journals but excluded minutes, abstracts, editorial materials, data files and letters [44,45]. Finally, 54,765 works were retrieved from the WoS database (Table S3) and visualized according to the keywords (Table S2) in the studies. Moreover, keywords that appeared at least 800 times were selected for clustering. Among them, 36 general keywords without valuable information were deleted from the selected dataset. The occurrence frequency of keywords in the literature was obtained in the most representative years (Figure S1). Then, 185 refined keywords were imported into the VOS viewer software to obtain 4 data clusters (Table S2).

3. Results and Discussion

3.1. The Demonstration Park of Multi-Source Solid Waste Disposal and Management

The network analysis map for solid waste disposal/management (Figure 2) comprises around 63% occurring keywords [46,47]. The first cluster (green) includes keywords such as management, impact, assessment, waste management, methodology, etc., focusing on the impact assessment of solid waste disposal/management methodology. The second cluster (red) mainly introduces cases involved in the centralized disposal of solid waste, with keywords including concentration, removal, mechanism and sample. The third category (blue) is dedicated to disposal technologies and by-products, containing keywords of anaerobic digestion, biogas, temperature, pyrolysis and others. The fourth cluster (yellow) mainly includes substances produced in the solid waste disposal process, such as heavy metal, fly ash, copper, sludge and others. The nodes of management, impact and concentration are significantly larger, which indicates that the centralized disposal and management of solid waste have been a hot topic in the field of solid waste disposal since then, which is reflected in solid waste management and disposal parks.

The construction of solid waste management demonstration parks is a key measure to solve the "Not in My Back Yard (NIMBY) effect" of solid waste disposal [48]. Constructing a venous industrial park (waste-free cities) with "solid waste park coordinated disposal management" at the core is an effective and fundamental method to solve the problem of solid waste surplus and resource shortage while practicing the concept of sustainable development. It is an innovative way to realize the harmonious coexistence of urban development and the ecological environment [49]. Based on the seventh census data and city size criteria (Figure 3), cities with a permanent population of less than 500,000 are small cities. Cities with a permanent urban population of more than 500,000 and less than 1 million are medium-sized cities, and those with an urban permanent population of more than 1 million and less than 5 million cities are big cities. Meanwhile, cities with a permanent urban population of more than 5 million are megacities. Based on data from the Statistical Yearbook, we determined the production volume of general industrial waste and hazardous waste in Chinese provinces from 2016 to 2022 (Figure 3 and Figure S2).

The construction characteristics of demonstration parks of different sizes in small- and medium-sized cities, large- and medium-sized cities and megacities were analyzed in order to promote the experience of solid waste demonstration park construction.

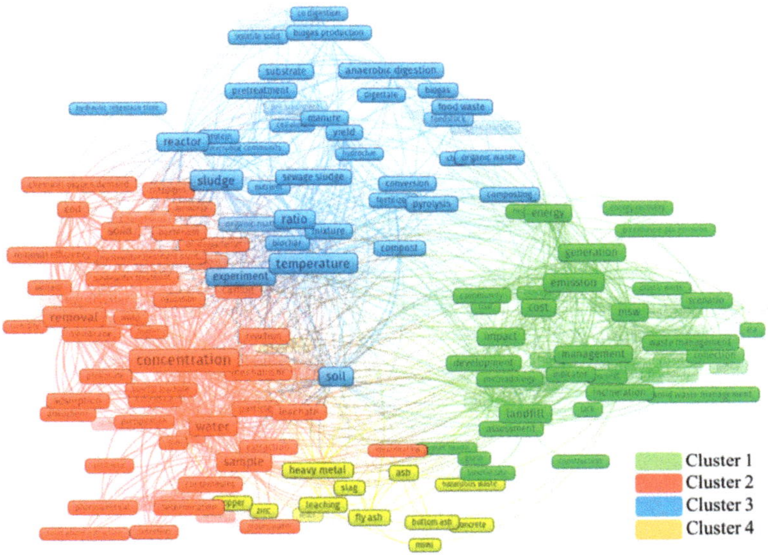

Figure 2. Network diagram of keyword screening and correlation analysis for solid waste disposal and management. The size of the rounded rectangle indicates how often each keyword appears in the dataset. The curves connect different rounded rectangles, and the thicker the curve is, the stronger the correlation is.

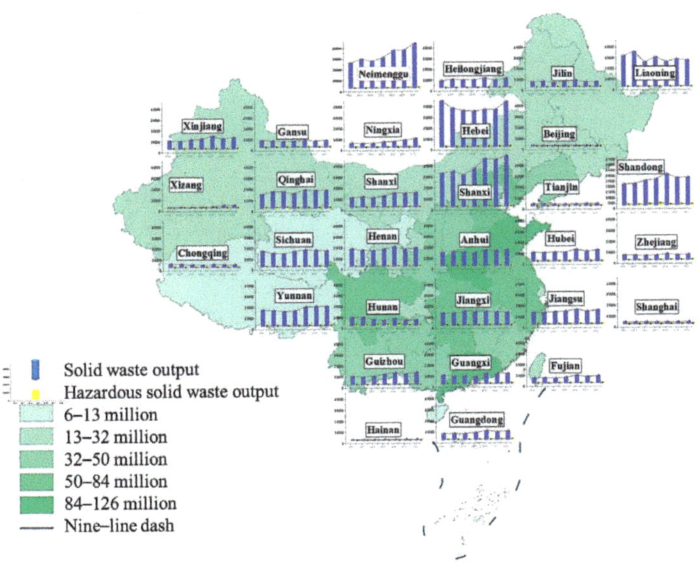

Figure 3. The production of solid waste in the main provinces in China. Among them, the population size of different provinces is distinguished by the shade of green color according to the seventh census data. The blue bar chart shows the output of total solid waste in each province, while the yellow bar chart shows the output of hazardous solid waste.

3.2. Solid Waste Management in Typical Industrial Parks of China

3.2.1. Multi-Source Solid Waste Collaborative Disposal Parks in Small–Medium Cities

Small–medium cities refer to cities with a population of less than one million. The number of small–medium cities in China has reached 2160, accounting for 56% of the total number of cities. Small–medium cities gather a huge population, resources, industries, environment and other development factors and have become an important support for China's economic and social development. Moreover, the improvement of their development quality has had complete and decisive significance for the modernization process of Chinese style. Due to the conflict between the growing demand for economic development and the shortage of land resources, the practice of "collaborative disposal" could optimize the economy of urban solid waste. Taking Zhangjiagang, a new port industrial city in Jiangsu Province, as an example [50,51], the average daily production of household waste was around 1800 tons [52]. In the solid waste collaborative treatment park, multiple solid wastes, such as domestic waste, kitchen waste, municipal sludge, sick and dead animals, local scrapped vehicles, general industrial solid waste and hazardous waste, were waiting to be treated [53]. The construction of a resource recycling chain in the solid waste disposal park mainly aimed to create "internal circulation with waste as the source" and "external circulation with products as the beginning" (Figure 4). Based on the characteristics of land projects and facilities, a water cycle chain (recycled water), gas cycle chain (landfill gas power generation), electric cycle chain (incineration power generation) and heat cycle chain (waste heat cycle) were constructed inside the park to achieve park management.

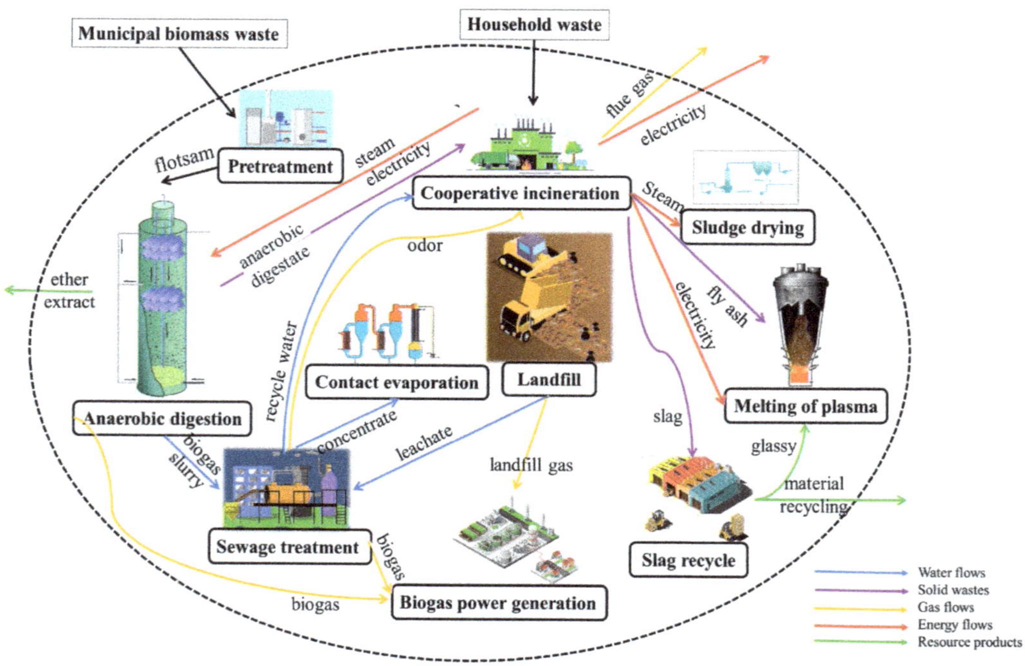

Figure 4. Schematic diagram of multi-source solid waste collaborative disposal park for small- and medium-sized cities (taking Zhangjiagang as an example). Black arrows indicate the solid waste entering the park; blue arrows indicate the direction of water flow; purple arrows indicate the solid waste; yellow arrows indicate the direction of gas flow; red arrows indicate the energy flows; green arrows indicate the direction of resource products. The content in the gray boxes indicates the main solid waste type; the content in the black boxes indicates the techniques for solid waste disposal; the outermost black dotted circle represents the system boundary.

In this way, household waste is transported to the incineration plant for power generation, while the generated electricity is sent to the charging station of logistics vehicles in the park. The heat generated is reused in the park through underground heating pipes, and the incineration residue is sent to a sanitary landfill for landfilling [54]. The wastewater generated in the processes of kitchen waste disposal, municipal sludge pretreatment and incineration power generation [55] as well as landfill leachate is centrally disposed to produce clean water for water recharge in the park [56]. The unusable residue generated in the disposal processes of the kitchen waste treatment plant, construction waste treatment plant and waste resource treatment enterprises should be sent to a sanitary landfill, while the biogas generated in the landfill process shall be used for power generation [57]. The electricity generated by incineration is sent to the power grid for surrounding residents to use [58]. Kitchen waste could be turned into microbial fertilizer for agriculture and soil condition adjustment [59]. Construction waste could be recycled or repurposed into building materials or other products through effective resource utilization. In general, this solid waste management constitutes the material and energy link between the project facilities in the park.

Due to the small size of Zhangjiagang city and the initial stage of waste classification, the output of solid waste in Zhangjiagang city was relatively small, centralized disposal had no economic benefit, and the environmental pollution of decentralized disposal was relatively large. With the application of the new sharing–saving–cycling mode, the solid waste disposal capacity of this park reached more than 900,000 tons per year, and the comprehensive utilization rate of solid waste was as high as 90%. In general, a collaborative disposal mode could help small–medium cities realize the goal of multi-source full green consumption.

3.2.2. Multi-Source Solid Waste Collaborative Disposal Parks in Large–Medium Cities

For a large–medium city with a permanent population of 500,000 to 5 million, the first thing to be solved in the process of solid waste treatment is the collaborative treatment of environmental pollution generated by cross-border solid waste. Taking the Yangtze River Economic Belt as an example (Figure 5), the basic idea was to promote urban–rural integration to build a cycle model of collaborative solid waste disposal [60]. Urban and rural areas help each other to deal with the surrounding solid waste and can jointly educate urban and rural residents to enhance their environmental awareness.

A comprehensive, integrated industrial park with the dual target of environmental governance and ecological protection was built for the coordinated treatment and disposal of solid waste in the Yangtze River Economic Belt, which includes three main parts (Figure 5): an incineration power generation system, a resource utilization system and a landfill system. The waste incineration power generation system is the energy center of the industrial park, which provides the energy supply for other processes internally [61]. Firstly, waste (especially the combustible portion) is sent to the incineration system for treatment. Secondly, fly ash and construction waste are used together to produce green recycled building materials [62]. Then, the remaining inert solid waste (including some hazardous waste) which cannot be treated by the current technologies is sent to the landfill plant for final disposal.

Solid waste pollution treated in a distributed organic ecological park is mainly used to treat organic waste that is not suitable for long-distance transportation in urban and rural areas [63]. For example, people could transform organic solid waste into ecological products and realize the goal of resource recycling and pollution reduction [64]. In detail, agricultural and forestry waste with a low moisture content could be converted into biochar, bio-oil and syngas through the pyrolysis process, while excrements of livestock with a high moisture content and fruit or vegetable residues could be converted into organic fertilizers and biogas through anaerobic digestion or aerobic composting [65,66]. This biogas could be used as vehicle energy for the environmental logistics chain externally and also provide energy for the pyrolysis system internally. Biochar could be further processed into carbon-based compounds or organic fertilizers to improve the soil environment quality. The remaining unusable part should be returned to the solid waste pollution comprehensive treatment industrial park for safe guaranteed treatment through the environmental logistics chain.

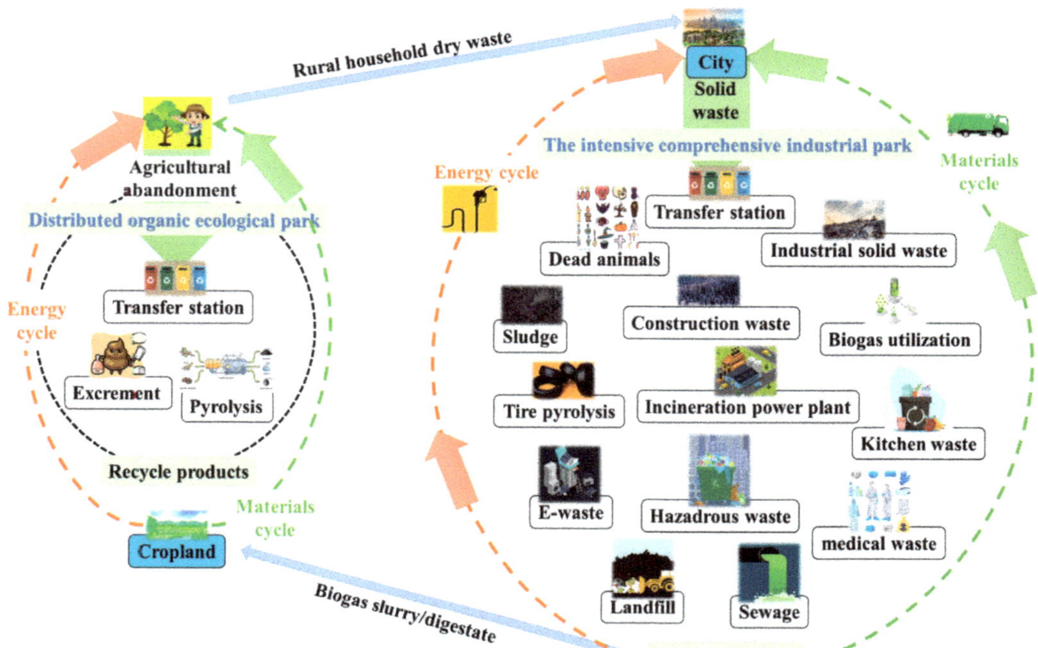

Figure 5. Schematic diagram of multi-source solid waste collaborative disposal and management park for large–medium cities (taking Yangtze River Economic Belt as an example). Orange dotted line indicates the energy flow; green dotted line indicates the material flow. The black text box indicates the solid waste disposal unit. The arrows represent the exchange transport directions.

Through the systematic coupling of pollution control and ecological construction in the intensive comprehensive industrial park, it was considered highly possible to achieve the goal of setting up a zero-waste city along the Yangtze River Economic Belt [67]. This approach could contribute to the construction of an ecological civilization in large–medium cities in China.

3.2.3. Solid Waste Disposal and Resource Recycling Industrial Parks in Megacities

Cities with a permanent population of more than five million are defined as megacities. The GBA covers only about one-fifth of the country's land area, but it supports one-third of its population and generates one-third of its economy. Therefore, a reasonable solid waste disposal mode for the GBA is essential to manage resources, energy and environmental factors in the whole process of multi-source solid waste recycling. In fact, Dongguan city has already set up a number of environmental protection projects for the disposal of municipal solid waste and solved the problems of industrial solid waste disposal by establishing a centralized disposal park. As a result, it handles all the tasks of solid waste disposal in the GBA [68,69]. In this industrial park, the coordinated treatment of kitchen waste and household waste, the sewage treatment system and other projects work together to achieve a circular process within the industrial park [70]. This comprehensive utilization treatment sometimes allows for the recycling and reuse of domestic waste for electricity generation. It could be an effective way in the long term to achieve the goal of resource utilization (Figure 6). Furthermore, hazardous waste is harmlessly treated by oxygen-enriched side blowing to eliminate environmental risks. In detail, the discarded materials enter the plasma reactor, where the organic materials react rapidly under the action of high temperature (>1200 °C), while the inorganic materials melt to form a molten slurry at an even higher temperature (>1450–1600 °C) [71]. Physicochemical treatment is quite important in the process of hazardous waste disposal to reduce or even eliminate the harm of liquid hazardous waste after heat treatment. After that, the material is then sent to the next process for final disposal [72]. The sludge with about 60% water content is dried by natural air to remove part of the adsorbed water on the surface of the sludge; then, it is dried by a rotary kiln with natural gas as the heat source [73]. When the temperature in the furnace is as high as 1200–1300 °C, the heat released by coke combustion melts the charge, where heat provided from the combustion of natural gas overheats the melt. Heavy metal salts in the sludge are decomposed into oxides at the high temperature and then reduced into elemental substances and other heavy metals by contacting with coke [74], which allows to achieve an innovative heavy metal removal technology.

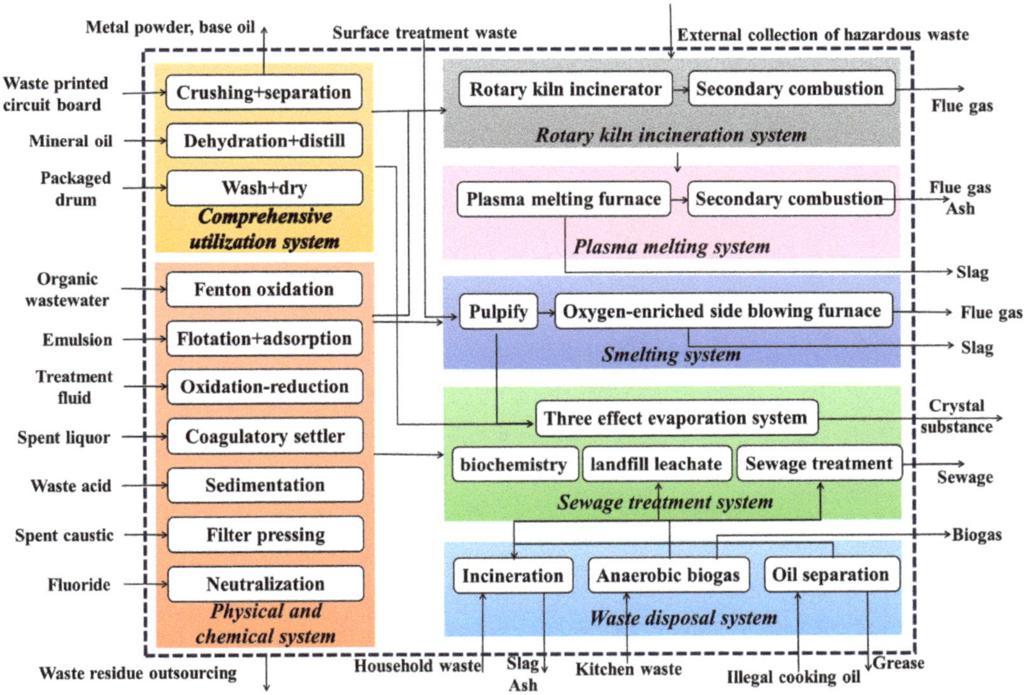

Figure 6. Solid waste disposal and resource recycling industrial park for megacities (taking the GBA as an example). Yellow-filled textbox indicates the comprehensive utilization system; orange-filled textbox indicates the physicochemical disposal system; gray-filled textbox indicates the rotary kiln incineration system; pink-filled textbox indicates the plasma melting system; light blue-filled textbox indicates the smelting system; green-filled textbox indicates the sewage treatment system; purplish blue textbox indicates the disposal system of household, kitchen and illegal cooking oil wastes.

The core process for kitchen waste treatment is "pretreatment + anaerobic digestion", which achieves a certain degree of clean production. For instance, biogas is produced continuously, and biogas residue can be recycled [75]. The sewage treatment center also applies biological filters for deodorization. In addition, an air curtain is set up at each gate of the workshop to prevent the leakage of toxic gases in the workshop, so as to ensure that all odors in the workshop are led to the deodorization equipment through the air duct after treatment.

After the implementation of the corresponding protective measures, the project site selection could meet the requirements of the site selection standards for the temporary storage and disposal of hazardous waste. The solid waste disposal park in Dongguan is a exemplary model in the GBA which can effectively dispose of solid waste to the greatest extent as well as solve the environmental problems caused by the non-standardized disposal of hazardous industrial waste. By accelerating innovative ways to meet human needs, integrating resource issues into the delivery of multilateral environmental agreements and identifying sustainable ways to use resources, the vision of green, low-carbon and sustainable development of the GBA could be realized with this system.

3.3. *Suggestions for Industrial Park Sustainable Development*

The demand for solid waste disposal in China has been growing rapidly during the past few years. Subsequently, the construction of solid waste treatment industrial parks has become a new trend, and their existence enables waste to be recycled and reused.

In order to promote efficient solid waste disposal and material recycling, different disposal options for solid wastes have been drawn up according to different city sizes (Figure 7). Small–medium cities should adopt collaborative disposal and intensive circulation for the rational use of land resources [76]. As for large–medium cities, urban–rural integration should be promoted to jointly educate urban and rural residents to enhance their environmental awareness [77]. Megacities should adopt the recycling mode to achieve green and low-carbon sustainable development [78].

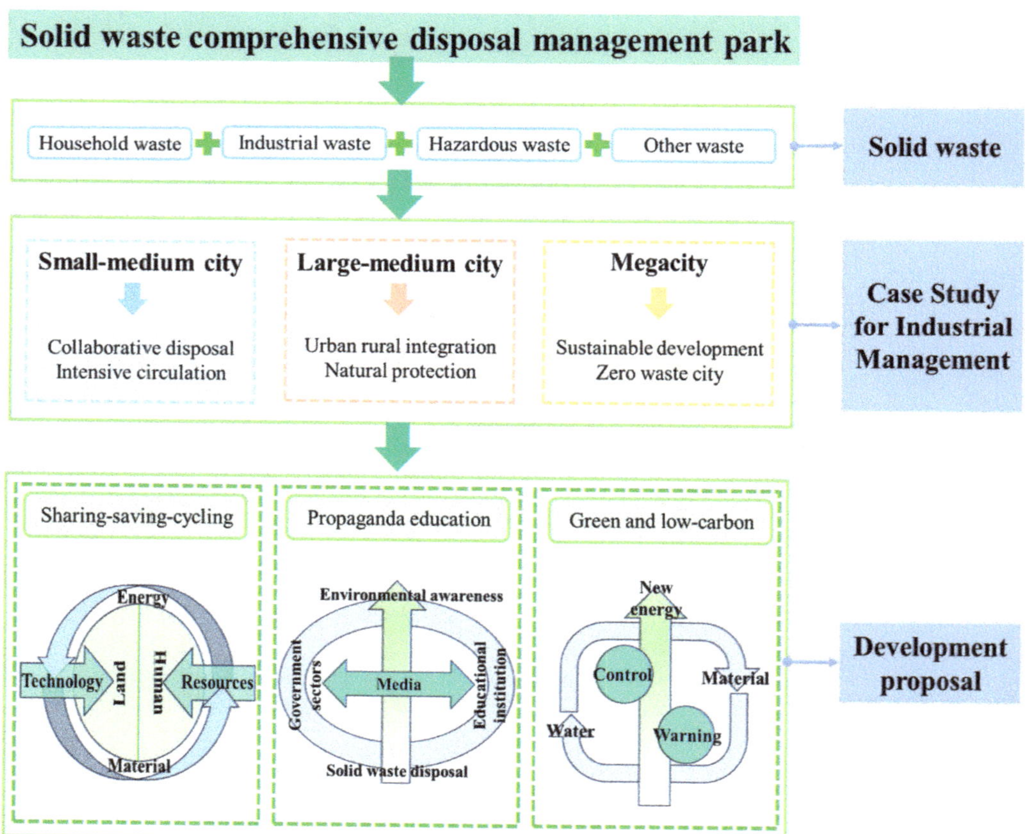

Figure 7. Concept map of construction and sustainable development of industrial parks for integrated disposal and effective management of solid waste.

Despite the differences based on city size, three common points of solid waste disposal and management should be considered [79]. Firstly, people should adhere to the construction concept of sharing–saving–recycling (Figure 7). The centralized construction mode in the park should be rationally selected and scientifically planned according to the characteristics of the wastes being processed and the technological characteristics of the project, with the aim of complementarily matching the technological and resource advantages [80]. Regardless of the city size, people should fully cooperate to share disposal technologies, as well as human resources and material resources, to improve the degree of waste resource reuse efficiency.

Secondly, government sectors and educational institutions at all levels should strengthen publicity and education on garbage classification and waste recycling (Figure 7) so as to promote public awareness of environmental protection [81]. In detail, relevant departments should take advantage of media resources to carry out science popularization activities on solid waste disposal technology, publicize information on solid waste classification [82] and help the public to understand the environmental protection effect of rational disposal of solid wastes.

Finally, green and low-carbon concepts should be implemented throughout the solid waste disposal process. This is precisely the key work that must be conducted in the field of solid waste disposal under the background of the "dual carbon target" [83]. It is necessary to develop new energy, strengthen water circulation and establish a material circulation system. A low-carbon and environmental protection demonstration park should be built in most cities according to the different urban scales. Further, the waste recycling park should incorporate and continuously innovate technologies related to solid waste treatment and environmental management [84]. In addition, research and development platforms should be provided for pollution control and environmental risk warning in the field of solid waste disposal and management.

3.4. Advancement of China's Solid Waste Industry Development

Developed countries such as the United States, Germany and Japan have established refined management models for the whole process of solid waste disposal. In the United States, data analysis was conducted on solid waste handling in order to understand the behavior and consumptive pattern of consumers and provide data support for the development of source reduction and recycling strategies. Germany's recycling of waste focuses on the research and development of waste treatment technology and the recycling of packaging plastics. Japan has made great efforts to develop a circular economy, formulated waste recycling rate indicators and paid attention to recycling and resource regeneration when sorting and disposing of garbage. In general, the utilization of solid waste resources is an important entry point and provides support for sustainable development. Based on foreign experience and China's national conditions, we put forward suggestions for the development of China's solid waste industry parks, namely to promote the industrialization process and the integration of an 'industrial chain, innovation chain, talent chain and capital chain'. Moreover, a scientific and creative resource highland should be created that strengthens the demonstrations and guidance, gathers talent advantages, builds a high-level cooperation platform and comprehensively promotes the high-quality development of the solid waste resource utilization industry.

4. Conclusions and Prospect

Generally, municipal solid wastes include industrial waste, hazardous waste, household waste, etc. The landfill, incineration and biological disposal technologies for different types of solid waste have become quite mature in China, although the combined techniques for multi-solid waste require more practice and improvement. In cities of different sizes, solid wastes could have distinct different disposal and management modes, as well as requiring different industrial demonstration parks. In this study, three typical industrial parks for multi-solid waste disposal and resource utilization were investigated in depth and systematically analyzed through statistical network analyses. After analyzing the characteristics of multi-source solid waste disposal and reuse models of small, medium and large cities in different dimensions, innovation points could be popularized and referenced for resource–environment synergy. Possible suggestions for solid waste disposal as well as industrial park management in China were provided, following three principles: sharing–saving–cycling, publicity and education and green and low-carbon. The above measures could not only promote solid waste disposal efficiency but also give opportunities for material recycling, energy saving and environmental protection, which would undoubtedly be beneficial for the sustainable development of urban resources and the environment.

Supplementary Materials: The following supporting information can be downloaded at: https://www.mdpi.com/article/10.3390/su16166731/s1, Figure S1: The occurrence frequency of keywords in the literature was obtained under the most representative years for solid waste management; Figure S2: The amount of general industrial waste and hazardous waste produced by typical provinces of China from 2016 to 2022; Table S1: The amount of solid waste generation in major cities from 2009 to 2019 (unit: million tons); Table S2: Network analysis keyword screening results and weight analysis; Table S3: List of partial literature for network analysis.

Author Contributions: Conceptualization, L.Y.; investigation, S.C.; data curation, L.Y.; data statistics, S.C.; writing—original draft preparation, S.C.; writing—review and editing, L.Y. and Z.T.; visualization, S.C.; supervision, L.Y.; project administration, L.Y.; funding acquisition, L.Y. All authors have read and agreed to the published version of the manuscript.

Funding: This research was funded by the National Key Research and Development Program of China (2023YFC3905005), the Beijing Natural Science Foundation (8232019), and The National Natural Science Foundation of China (42007126).

Institutional Review Board Statement: Not applicable.

Informed Consent Statement: Not applicable.

Data Availability Statement: Data are contained within the article and Supplementary Materials.

Acknowledgments: We sincerely thank the anonymous reviewers and editor for their constructive comments and suggestions.

Conflicts of Interest: The authors declare no conflicts of interest.

References

1. Rossit, D.G.; Nesmachnow, S. Waste bins location problem: A review of recent advances in the storage stage of the Municip Solid Waste reverse logistic chain. *J. Clean. Prod.* **2022**, *342*, 130793. [CrossRef]
2. Rodygin, K.S.; Lotsman, K.A.; Samoylenko, D.E.; Kuznetsov, V.M.; Ananikov, V.P. Towards Sustainable Carbon Return fro Waste to Industry via C2-Type Molecular Unit. *Int. J. Mol. Sci.* **2022**, *23*, 11828. [CrossRef]
3. Ministry of Ecology and Environment of the People's Republic of China, Regulations on the Administration of Medic Wastes. Available online: https://www.ajne.org/sites/default/files/document/laws/6874/regulations-on-the-administratio of-medical-wastes.pdf (accessed on 31 July 2024).
4. Ye, M.; Liang, X.; Lin, S.; Lin, H.; Deng, F. Efficiency Assessment of Hazardous Waste Disposal in EU Countries: A Three-Stag Super-Efficiency Data Envelopment Analysis Model. *Environ. Manag.* **2022**, *70*, 650–665. [CrossRef]
5. Ministry of Ecology and Environment of the People's Republic of China, Annual Report on the Prevention and Control Environmental Pollution by Solid Waste in Large and Medium Cities in 2020. Available online: https://www.mee.gov.cn/ywg gtfwyhxpgl/gtfw/202012/P020201228557295103367.pdf (accessed on 31 July 2024).
6. Ministry of Ecology and Environment of the People's Republic of China, Annual Report of China Ecological and Environment Statistics 2021. Available online: https://www.mee.gov.cn/hjzl/sthjzk/sthjtjnb/202301/t20230118_1013682.shtml (accessed o 31 July 2024).
7. World Bank Group. What a Waste 2.0: A Global Snapshot of Solid Waste Management to 2050 (English). Available o line: https://documents1.worldbank.org/curated/en/697271544470229584/pdf/What-a-Waste-2-0-A-Global-Snapshot-o Solid-Waste-Management-to-2050.pdf (accessed on 31 July 2024).
8. Anand, N.; Palani, S.G. A comprehensive investigation of toxicity and pollution potential of municipal solid waste landfi leachate. *Sci. Total Environ.* **2022**, *838*, 155891. [CrossRef]
9. Weligama, T.R.; Babel, S. Assessment of the environmental sustainability of municipal solid waste valorization by anaerobi digestion and by composting in Sri Lanka. *Environ. Technol.* **2022**, 1–14. [CrossRef]

Hubai, K.; Kováts, N.; Sainnokhoi, T.A.; Eck-Varanka, B.; Hoffer, A.; Tóth, Á.; Teke, G. Phytotoxicity of particulate matter from controlled burning of different plastic waste types. *Bull. Environ. Contam. Toxicol.* **2022**, *109*, 852–858. [CrossRef] [PubMed]

Namoun, A.; Tufail, A.; Khan, M.Y.; Alrehaili, A.; Syed, T.A.; BenRhouma, O. Solid Waste Generation and Disposal Using Machine Learning Approaches: A Survey of Solutions and Challenges. *Sustainability* **2022**, *14*, 13578. [CrossRef]

Ma, R.; Wang, J.; Liu, Y.; Wang, G.; Yang, Y.; Liu, Y.; Kong, Y.; Lin, J.; Li, Q.; Li, G.; et al. Dynamics of antibiotic resistance genes and bacterial community during pig manure, kitchen waste, and sewage sludge composting. *J. Environ. Manag.* **2023**, *345*, 118651. [CrossRef]

Liu, Y.; Zhang, Y.; Wang, M.; Wang, L.; Zheng, W.; Zeng, Q.; Wang, K. Comparison of the basic processes of aerobic, anaerobic, and aerobic-anaerobic coupling composting of Chinese medicinal herbal residues. *Bioresour. Technol.* **2023**, *379*, 128996. [CrossRef] [PubMed]

Al-Balushi, F.; Ibrahim, O.; Rajamohan, N. Sustainable treatment of landfill leachate: A review on methods. *Int. J. Environ. Sci. Technol.* **2024**. [CrossRef]

Guo, W.; Xi, B.; Huang, C.; Li, J.; Tang, Z.; Li, W.; Ma, C.; Wu, W. Solid waste management in China: Policy and driving factors in 2004–2019. *Resour. Conserv. Recycl.* **2021**, *173*, 105727. [CrossRef]

Abdel-Shafy, H.I.; Mansour, M.S. Solid waste issue: Sources, composition, disposal, recycling, and valorization. *Egypt. J. Pet.* **2018**, *27*, 1275–1290. [CrossRef]

Gu, B.; Tang, X.; Liu, L.; Li, Y.; Fujiwara, T.; Sun, H.; Gu, A.; Yao, Y.; Duan, R.; Song, J.; et al. The recyclable waste recycling potential towards zero waste cities—A comparison of three cities in China. *J. Clean. Prod.* **2021**, *295*, 126358. [CrossRef]

Zhu, Z.; Zhao, Y.; Zhu, Y.; Zhang, M.; Yu, Y.; Guo, Y.; Zhou, T. Efficient treatment of mature landfill leachate with a novel composite biological trickle reactor developed using refractory domestic waste and aged refuse. *J. Clean. Prod.* **2021**, *305*, 127194. [CrossRef]

Ma, S.; Zhou, C.; Pan, J.; Yang, G.; Sun, C.; Liu, Y.; Chen, X.; Zhao, Z. Leachate from municipal solid waste landfills in a global perspective: Characteristics, influential factors and environmental risks. *J. Clean. Prod.* **2022**, *333*, 130234. [CrossRef]

Bolan, N.; Hoang, S.A.; Tanveer, M.; Wang, L.; Bolan, S.; Sooriyakumar, P.; Robinson, B.; Wijesekara, H.; Wijesooriya, M.; Keerthanan, S.; et al. From mine to mind and mobiles—Lithium contamination and its risk management. *Environ. Pollut.* **2021**, *290*, 118067. [CrossRef] [PubMed]

Lipczynska-Kochany, E. Effect of climate change on humic substances and associated impacts on the quality of surface water and groundwater: A review. *Sci. Total Environ.* **2018**, *640–641*, 1548–1565. [CrossRef] [PubMed]

Huang, X.; Wei, X.; Liu, H.; Li, W.; Shi, D.; Qian, S.; Sun, W.; Yue, D.; Wang, X. Occurrence of per- and polyfluoroalkyl substances (PFAS) in municipal solid waste landfill leachates from western China. *Environ. Sci. Pollut. Res.* **2022**, *29*, 69588–69598. [CrossRef] [PubMed]

Ruan, X.; Wang, H.; Luo, Y.S.; Hou, C. Composite Failure Analysis of Municipal Solid Waste Landfill Cell Stability. *Iran. J. Sci. Technol. Trans. Civ. Eng.* **2022**, *46*, 2325–2343. [CrossRef]

Istrate, I.R.; Galvez-Martos, J.L.; Vázquez, D.; Guillén-Gosálbez, G.; Dufour, J. Prospective analysis of the optimal capacity, economics and carbon footprint of energy recovery from municipal solid waste incineration. *Resour. Conserv. Recycl.* **2023**, *193*, 106943. [CrossRef]

Wu, Q.; Wu, Z.; Wu, H.; Wang, Q.; Huang, N.; Zhou, J.; Li, X.; Shi, L.; Tian, S.; Li, M. Thermal treatment of municipal solid waste incineration fly ash and nanofiltration membrane concentrate co-processing: Speciation of chromium and its leachability. *J. Clean. Prod.* **2023**, *398*, 136619. [CrossRef]

Xiao, W.; Liu, T.; Tong, X. Assessing the carbon reduction potential of municipal solid waste management transition: Effects of incineration, technology and sorting in Chinese cities. *Resour. Conserv. Recycl.* **2023**, *188*, 106713. [CrossRef]

Han, X.; Chang, H.; Wang, C.; Tai, J.; Karellas, S.; Yan, J.; Song, L.; Bi, Z. Tracking the life-cycle greenhouse gas emissions of municipal solid waste incineration power plant: A case study in Shanghai. *J. Clean. Prod.* **2023**, *398*, 136635. [CrossRef]

Wei, J.; Li, H.; Liu, J.; Zhong, R. National and provincial dioxin emissions from municipal solid waste incineration in China. *Sci. Total Environ.* **2022**, *851*, 158128. [CrossRef] [PubMed]

Sorrentino, G.P.; Zanoletti, A.; Ducoli, S.; Zacco, A.; Iora, P.; Invernizzi, C.M.; Di Marcoberardino, G.; Depero, L.E.; Bontempi, E. Accelerated and natural carbonation of a municipal solid waste incineration (MSWI) fly ash mixture: Basic strategies for higher carbon dioxide sequestration and reliable mass quantification. *Environ. Res.* **2023**, *217*, 114805. [CrossRef]

Niu, Y.; Wen, L.; Guo, X. Co-disposal and reutilization of municipal solid waste and its hazardous incineration fly ash. *Environ. Int.* **2022**, *166*, 107346. [CrossRef] [PubMed]

Arunthathi, S.; Sivapriya, S.V.; Balaji, D. Assessing the compaction characteristics of modified soil with Incinerated municipal solid waste. *Mater. Today Proc.* **2022**, *62*, 629–632. [CrossRef]

Li, Y.; Jiang, X.; Lv, G.; Chen, Q.; Liu, X.; Yan, J.; Zhao, X.; Kong, L. Research on mechanism and effect of phosphorus-containing wastewater treatment by municipal solid waste incineration fly ash. *Int. J. Environ. Sci. Technol.* **2023**, *20*, 6479–6490. [CrossRef]

Nguyen, T.H.; Pham, Q.V.; Nguyen, T.P.; Vu, V.T.; Do, T.H.; Hoang, M.T.; Thu Thuy Thi, N.; Minh, T.B. Distribution characteristics and ecological risks of heavy metals in bottom ash, fly ash, and particulate matter released from municipal solid waste incinerators in northern Vietnam. *Environ. Geochem. Health* **2022**, *45*, 2579–2590. [CrossRef]

34. Salehiyoun, A.R.; Zilouei, H.; Safari, M.; Di Maria, F.; Samadi, S.H.; Norouzi, O. An investigation for improving dry anaerob digestion of municipal solid wastes by adding biochar derived from gasification of wood pellets. *Renew. Energy* **2022**, *186*, 1- [CrossRef]
35. Manikandan, S.; Vickram, S.; Sirohi, R.; Subbaiya, R.; Krishnan, R.Y.; Karmegam, N.; Sumathijones, C.; Rajagopal, R.; Chang, S.W Ravindran, B.; et al. Critical review of biochemical pathways to transformation of waste and biomass into bioenergy. *Bioreso Technol.* **2023**, *372*, 128679. [CrossRef] [PubMed]
36. Bernat, K.; Kulikowska, D.; Wojnowska-Baryła, I.; Kamińska, A. Can the biological stage of a mechanical–biological treatme plant that is designed for mixed municipal solid waste be successfully utilized for effective composting of selectively collect biowaste? *Waste Manag.* **2022**, *149*, 291–301. [CrossRef] [PubMed]
37. Saravanan, A.; Kumar, P.S.; Nhung, T.C.; Ramesh, B.; Srinivasan, S.; Rangasamy, G. A review on biological methodologies municipal solid waste management and landfilling: Resource and energy recovery. *Chemosphere* **2022**, *309*, 136630. [CrossRe [PubMed]
38. Le Pera, A.; Sellaro, M.; Bencivenni, E.; D'Amico, F. Environmental sustainability of an integrate anaerobic digestion-compostin treatment of food waste: Analysis of an Italian plant in the circular bioeconomy strategy. *Waste Manag.* **2022**, *139*, 341–3 [CrossRef] [PubMed]
39. Liang, J.; Luo, L.; Wong, J.W.; He, D. Recent advances in conductive materials amended anaerobic co-digestion of food waste ar municipal organic solid waste: Roles, mechanisms, and potential application. *Bioresour. Technol.* **2022**, *360*, 127613. [CrossRe [PubMed]
40. Mlaik, N.; Karray, F.; Sayadi, S.; Feki, F.; Khoufi, S. Semi-continuous anaerobic digestion of the organic fraction of municipal sol waste: Digester performance and microbial population dynamics. *J. Environ. Chem. Eng.* **2022**, *10*, 107941. [CrossRef]
41. Lei, G.; Guo, S.; Yuan, Z. Study on the Effect and Mechanism of Circular Economy Promotion Law on the Utilization Rate Industrial Solid Waste in Resource-Based Cities. *Sustainability* **2022**, *14*, 9878. [CrossRef]
42. Ndou, V.; Rampedi, I.T. Bibliometric Analysis of Municipal Solid Waste Management Research: Global and South African Trend *Sustainability* **2022**, *14*, 10229. [CrossRef]
43. Khan, A.H.; Kiyani, A.; Santiago-Herrera, M.; Ibánez, J.; Yousaf, S.; Iqbal, M.; Martel-Martín, S.; Barros, R. Sustainability phytoremediation: Post-harvest stratagems and economic opportunities for the produced metals contaminated biomass. *Environ. Manag.* **2023**, *326*, 116700. [CrossRef]
44. Ye, Q.; Umer, Q.; Zhou, R.; Asmi, A.; Asmi, F. How publications and patents are contributing to the development of municip solid waste management: Viewing the UN Sustainable Development Goals as ground zero. *J. Environ. Manag.* **2023**, *325*, 1164 [CrossRef]
45. Ansari, S.A.; Shakeel, A.; Sawarkar, R.; Maddalwar, S.; Khan, D.; Singh, L. Additive facilitated co-composting of lignocellulos biomass waste, approach towards minimizing greenhouse gas emissions: An up to date review. *Environ. Res.* **2023**, *224*, 11552 [CrossRef]
46. Li, J.; Suvarna, M.; Li, L.; Pan, L.; Pérez-Ramírez, J.; Ok, Y.S.; Wang, X. A review of computational modeling techniques for w waste valorization: Research trends and future perspectives. *J. Clean. Prod.* **2022**, *367*, 133025. [CrossRef]
47. Mondal, T.; Choudhury, M.; Kundu, D.; Dutta, D.; Samanta, P. Landfill: An eclectic review on structure, reactions and remediatio approach. *Waste Manag.* **2023**, *164*, 127–142. [CrossRef]
48. Hurlbert, M.; Bhardwaj, A.; Akbari, M. Best versus beneficial MP discourses: The significance of a change in discourse managin agricultural water quality in Canada. *J. Environ. Manag.* **2023**, *332*, 117289. [CrossRef]
49. Vlachokostas, C.; Achillas, C.; Michailidou, A.V.; Tsegas, G.; Moussiopoulos, N. Externalities of energy sources: The operation a municipal solid waste-to-energy incineration facility in the greater Thessaloniki area, Greece. *Waste Manag.* **2020**, *113*, 351–35 [CrossRef]
50. Zhao, Y.; Long, B. Study on Human Settlement Environment in Dai Traditional Villages of Different Regions. *IOP Conf. Ser. Eart Environ. Sci.* **2020**, *568*, 12035. [CrossRef]
51. Meng, L.; Huang, J.; Dong, J. Assessment of rural ecosystem health and type classification in Jiangsu province, China. *Sci. Tota Environ.* **2018**, *615*, 1218–1228. [CrossRef]
52. Liu, Y.S.; Wang, J.Y.; Long, H.L. Analysis of arable land loss and its impact on rural sustainability in Southern Jiangsu Province o China. *J. Environ. Manag.* **2010**, *91*, 646–653. [CrossRef]
53. Zhao, J.; Li, B.; Wei, X.; Zhang, Y.; Li, T. Slagging characteristics caused by alkali and alkaline earth metals during municipal soli waste and sewage sludge co-incineration. *Energy* **2020**, *202*, 117773. [CrossRef]
54. Chen, C.; Wen, Z.; Wang, Y.; Zhang, W.; Zhang, T. Multi-objective optimization of technology solutions in municipal solid wast treatment system coupled with pollutants cross-media metabolism issues. *Sci. Total Environ.* **2022**, *807*, 150664. [CrossRef]
55. Meng, M.; Wen, Z.; Luo, W.; Wang, S. Approaches and Policies to Promote Zero-Waste City Construction: China's Practices an Lessons. *Sustainability* **2021**, *13*, 13537. [CrossRef]
56. Li, Z.; Zhang, T.; Sun, Y.; Zheng, L.; Zhou, H.; Wang, H.; Chen, T.; Yan, J. Identifying the key policy drivers for behaviora improvement in waste source separation in the Yangtze Delta Region, China. *J. Clean. Prod.* **2022**, *351*, 131379. [CrossRef]
57. Wen, Z.; Xie, Y.; Chen, C.; Li, Y.; Xu, P. Water-energy-nutrient nexus: Multi-sectoral metabolism analysis and technical pat optimization for eco-towns. *J. Environ. Manag.* **2021**, *277*, 111395. [CrossRef]

Cheng, M.; Wen, Z.; Yang, S. The driving effect of technological innovation on green development: Dynamic efficiency spatial variation. *Environ. Sci. Pollut. Res.* **2022**, *29*, 84562–84580. [CrossRef]

Chen, C.; Wen, Z. Cross-media transfer of nitrogen pollution in the fast-urbanized Greater Bay Area of China: Trends and essential control paths. *J. Environ. Manag.* **2023**, *326*, 116796. [CrossRef]

Kosajan, V.; Wen, Z.; Fei, F.; Dinga, C.D.; Wang, Z.; Liu, P. Comprehensive assessment of cement kiln co-processing under MSW sustainable management requirements. *Resour. Conserv. Recycl.* **2021**, *174*, 105816. [CrossRef]

Zhang, R.; Tai, H.; Cheng, K.; Zhu, Y.; Hou, J. Carbon emission efficiency network formation mechanism and spatial correlation complexity analysis: Taking the Yangtze River Economic Belt as an example. *Sci. Total Environ.* **2022**, *841*, 156719. [CrossRef]

Tian, Y.; Pang, J. The role of internet development on green total-factor productivity—An empirical analysis based on 109 cities in Yangtze River economic belt. *J. Clean. Prod.* **2022**, *378*, 134415. [CrossRef]

Zhang, J.; Zhang, P.; Wang, R.; Liu, Y.; Lu, S. Identifying the coupling coordination relationship between urbanization and forest ecological security and its impact mechanism: Case study of the Yangtze River Economic Belt, China. *J. Environ. Manag.* **2023**, *342*, 118327. [CrossRef]

Xue, J.; Wang, Q.; Zhang, M. A review of non-point source water pollution modeling for the urban–rural transitional areas of China: Research status and prospect. *Sci. Total Environ.* **2022**, *826*, 154146. [CrossRef]

Atanasovska, I.; Choudhary, S.; Koh, L.; Ketikidis, P.H.; Solomon, A. Research gaps and future directions on social value stemming from circular economy practices in agri-food industrial parks: Insights from a systematic literature review. *J. Clean. Prod.* **2022**, *354*, 131753. [CrossRef]

Bîrgovan, A.L.; Lakatos, E.S.; Szilagyi, A.; Cioca, L.I.; Pacurariu, R.L.; Ciobanu, G.; Rada, E.C. How Should We Measure? A Review of Circular Cities Indicators. *Int. J. Environ. Res. Public Health* **2022**, *19*, 5177. [CrossRef]

Liu, W.; Wang, Y.; Xia, R.; Ding, X.; Xu, Z.; Li, G.; Nghiem, L.D.; Luo, W. Occurrence and fate of antibiotics in swine waste treatment: An industrial case. *Environ. Pollut.* **2023**, *331*, 121945. [CrossRef] [PubMed]

Coskun, A.; Metta, J.; Bakırlıoğlu, Y.; Çay, D.; Bachus, K. Make it a circular city: Experiences and challenges from European cities striving for sustainability through promoting circular making. *Resour. Conserv. Recycl.* **2022**, *185*, 106495. [CrossRef]

Lu, W.; Lou, J.; Webster, C.; Xue, F.; Bao, Z.; Chi, B. Estimating construction waste generation in the Greater Bay Area, China using machine learning. *Waste Manag.* **2021**, *134*, 78–88. [CrossRef] [PubMed]

Wu, M.; Wu, J.; Zang, C. A comprehensive evaluation of the eco-carrying capacity and green economy in the Guangdong-Hong Kong-Macao Greater Bay Area, China. *J. Clean. Prod.* **2021**, *281*, 124945. [CrossRef]

Luo, A.; Leipold, S. Chinese lessons on upscaling environmental policy concepts? A review of policy-oriented circular economy research. *J. Clean. Prod.* **2022**, *333*, 130047. [CrossRef]

Ghanem, H.; Kormann, M.; Gerhard, H.; Popovska, N. Processing of biomorphic porous TiO2 ceramics by chemical vapor infiltration and reaction (CVI-R) technique. *J. Eur. Ceram. Soc.* **2007**, *27*, 3433–3438. [CrossRef]

Chen, C.; Yao, Z.; Wen, Z.; Sheng, N. Impact of city characteristics on its phosphorus metabolism in the bay area: A comparative analysis of cities in the Greater Bay Area of China. *J. Clean. Prod.* **2021**, *286*, 124925. [CrossRef]

Liu, H.; Qiao, H.; Liu, S.; Wei, G.; Zhao, H.; Li, K.; Weng, F. Energy, environment and economy assessment of sewage sludge incineration technologies in China. *Energy* **2023**, *264*, 126294. [CrossRef]

Li, J.; Song, G.; Cai, M.; Bian, J.; Mohammed, B.S. Green environment and circular economy: A state-of-the-art analysis. *Sustain. Energy Technol. Assess.* **2022**, *52*, 102106. [CrossRef]

Zhu, Z.; Liu, W.; Ye, S.; Batista, L. Packaging design for the circular economy: A systematic review. *Sustain. Prod. Consum.* **2022**, *32*, 817–832. [CrossRef]

Chen, S.; Yu, L.; Zhang, C.; Wu, Y.; Li, T. Environmental impact assessment of multi-source solid waste based on a life cycle assessment, principal component analysis, and random forest algorithm. *J. Environ. Manag.* **2023**, *339*, 117942. [CrossRef]

Nayahi, N.T.; Ou, B.; Liu, Y.; Janjaroen, D. Municipal solid waste sanitary and open landfills: Contrasting sources of microplastics and its fate in their respective treatment systems. *J. Clean. Prod.* **2022**, *380*, 135095. [CrossRef]

Geng, X.; Song, N.; Zhao, Y.; Zhou, T. Waste plastic resource recovery from landfilled refuse: A novel waterless cleaning method and its cost-benefit analysis. *J. Environ. Manag.* **2022**, *306*, 114462. [CrossRef] [PubMed]

Li, Q.; Cui, H.; Li, Y.; Song, X.; Liu, W.; Wang, Y.; Hou, H.; Zhang, H.; Li, Y.; Wang, F.; et al. Challenges and engineering application of landfill leachate concentrate treatment. *Environ. Res.* **2023**, *231*, 116028. [CrossRef]

Fathinezhad, A.; Jafari, N.H.; Oldenburg, C.M.; Caldwell, M.D. Numerical investigation of air intrusion and aerobic reactions in municipal solid waste landfills. *Waste Manag.* **2022**, *147*, 60–72. [CrossRef]

Balaji, C.R.; de Azevedo, A.R.; Madurwar, M. Sustainable perspective of ancillary construction materials in infrastructure industry: An overview. *J. Clean. Prod.* **2022**, *365*, 132864. [CrossRef]

83. Choe, C.; Byun, M.; Lee, H.; Lim, H. Techno-economic and environmental assessments for sustainable bio-methanol producti as landfill gas valorization. *Waste Manag.* **2022**, *150*, 90–97. [CrossRef] [PubMed]
84. Medina-Mijangos, R.; Seguí-Amórtegui, L. Research Trends in the Economic Analysis of Municipal Solid Waste Manageme Systems: A Bibliometric Analysis from 1980 to 2019. *Sustainability* **2020**, *12*, 8509. [CrossRef]

sustainability

MDPI

ticle

'hermochemical Techniques for Disposal of Municipal Solid Vaste Based on the Intuitionistic Fuzzy Hypersoft Evaluation ;ased on the Distance from the Average Solution Technique

ana Muhammad Zulqarnain [1,2], Hongwei Wang [1,*], Imran Siddique [3,4], Rifaqat Ali [5], Hamza Naveed [6], Saalam i Virk [7] and Muhammad Irfan Ahamad [8,*]

1 School of Economics and Management, Tongji University, No. 1239 Siping Road, Yangpu District, Shanghai 200092, China; ranazulqarnain7777@gmail.com or ranazulqarnain@zjnu.edu.cn
2 Department of Mathematics, Saveetha School of Engineering, SIMATS Thandalam, Chennai 602105, India
3 Department of Mathematics, University of Sargodha, Sargodha 40100, Pakistan
4 Mathematics in Applied Sciences and Engineering Research Group, Scientific Research Center, Al-Ayen University, Nasiriyah 64001, Iraq
5 Department of Mathematics, Applied College in Mohayil Asir, King Khalid University, Abha 62521, Saudi Arabia
6 Department of Software Engineering, Faculty of Information Technology and Computer Science, University of Central Punjab, Lahore 54000, Pakistan
7 Department of Mathematics, University of Management and Technology, Sialkot 51310, Pakistan
8 College of Geography and Environmental Science, Henan University, Kaifeng 475004, China
* Correspondence: hwwang@tongji.edu.cn (H.W.); irfan@henu.edu.cn or mirfanahamad@hotmail.com (M.I.A.)

Abstract: The processing and disposal of municipal solid waste (MSW) are global problems, particularly in low- to middle-income states like Pakistan. These economic systems will need to tackle problems regarding municipal solid waste disposal to accomplish a sustainable future in waste management. Still, the determination of MSW procedures is frequently influenced by unstable, vague, and inadequately stated criteria. To deal with this issue, we designed an interactive model that uses intuitionistic fuzzy hypersoft sets (IFHSSs) to find the optimal thermochemical processing system for MSW. The main objective of this research is to define interactional operational laws for intuitionistic fuzzy hypersoft numbers and to use these laws to build interaction aggregation operators (AOs) and ordered AOs along with their basic characteristics. Based on developed operators, a novel Evaluation Based on the Distance from the Average Solution (EDAS) technique is proposed to integrate multiple attribute group decision making (MAGDM) issues. The suggested strategy is used to analyze five thermochemical treatment techniques for MSW, using a case study focusing on Pakistan's particular MSW administration problems to choose the most economical technique. Therefore, the new structure is assessed with established methodologies to illustrate its stability. The comparison of results proves that the implications of the stated approach will be more effective and capable than the existing approaches.

Keywords: intuitionistic fuzzy hypersoft set; interactive aggregation operators; EDAS method; multiple attribute group decision making; thermochemical techniques; municipal solid waste

1. Introduction

Several unexplored challenges exist for organizations, services, modern technologies, and infrastructure to move into MSW management. Pakistan discharges roughly 49.6 million tons of rubbish annually, with a yearly increase of over 2.4%. Pakistan lacks waste management infrastructure, like other low-income countries, causing significant environmental issues. A substantial amount of municipal waste can be disposed of through incineration, landfilling, or illegal dumping on empty plots of land, posing a significant risk to the overall well-being and safety of the public. According to the Pakistani administration's estimations, around 87,000 tons of solid garbage are produced weekly, mainly from large towns and cities. Karachi, the most populous city in Pakistan, produces almost 16,500 tons of municipal waste daily. The responsibility of collecting waste in almost every one of Pakistan's main areas falls within the authority of both local and regional governments. Roughly 60–70% of the solid waste produced in urban areas is gathered.

The world's growing population, economic growth, and rate of urbanization have all been responsible for an unprecedented increase in the utilization of resources and, by extension, the quantity of waste production [1]. As a result, waste-management-related issues with the environment are now receiving a lot of attention and causing controversy [2]. A significant obstacle for relevant organizations in developing countries is the adequate supervision of MSW [3–5]. The main objective of solid waste management is to handle a wide range of challenges caused by inadequate handling of waste, among which are the health of the community, the environment, the use of resources, aesthetics, land use, and finances [6]. Wang et al. [7] addressed the properties and worldwide utilization of bauxite residue (red mud) in cement-based materials. They highlighted its ability to mitigate energy use, emissions of carbon dioxide, and waste from factory formation.

Approximately 37% of MSW is dumped in landfills, whereas 33% of MSW is directed to open sites [8] as part of global waste disposal initiatives. Wu et al. [9] analyzed refurbishment techniques for end-of-life and used goods, concluding that reusing recycled materials delivers superior sustainable benefits. Because there is a shortage of proper disposal facilities in Pakistan [10], open-pit disposal is the preferred approach for managing MSW. Pan et al. [11] evaluated acidogenic fermentation for food waste administration, exposing that CO_2 splitting grows acetic acid fabrication by promoting acetate-producing organisms and limiting contending avenues. This waste, which amounts to 0.79 kg per person per day, negatively affects the globe's climate, health, and biodiversity. The study provides solutions tailored to specific countries, identifies significant issues, and presents feasible solutions to promote worldwide progress in sustainability in both the economy and the environment [12]. Sohoo et al. [13] performed an exhaustive study of the open disposal method for MSW in Pakistan, particularly in Karachi. Aslam et al. [14] examined the performance of the solid waste disposal system in Karachi, using materials flow analysis to explore the transit of solid waste. Iqbal et al. [15] explored global challenges with managing municipal waste, focusing on middle-income and low-income countries like Pakistan. The growing production of MSW in Asian metropolitan areas, combined with insufficient waste disposal systems and energy limitations [16], raise serious threats to people's health and the surroundings. Figure 1 shows the composition of MSW generated in Pakistan.

Unfortunately, urban areas in Pakistan are emerging as the most severe instances of negligence and mismanagement in handling MSW, causing a significant decline in the climate and the standard of community life. Han et al. [17] highlighted that using a tailored bacterial population derived from organically decomposed food debris improves the effectiveness of aerobic decomposition by facilitating the proliferation of hydrolyzed bacteria. Liu et al. [18] described the Multifaceted Anomaly Detection Framework, which merges Electrical Resistivity Tomography with sophisticated machine learning models to enhance the detection and surveillance of harmful substances immensely. The authors investigated the thermal and catalytic pyrolysis procedures for reducing waste into beneficial goods, emphasizing their structures, obstacles, and possibilities for waste-to-energy applications [19]. Wang et al. [20] described a breakthrough in situ chemical injection strategy exploiting an Fe-Mn oxide-activated persulfate structure for the sterilization of groundwater contamination with BPA. Meng et al. [21] explored biofilm formation on microplastics in wetlands, confirming that the plastisphere substantially affects the variety of bacteria and biological processes.

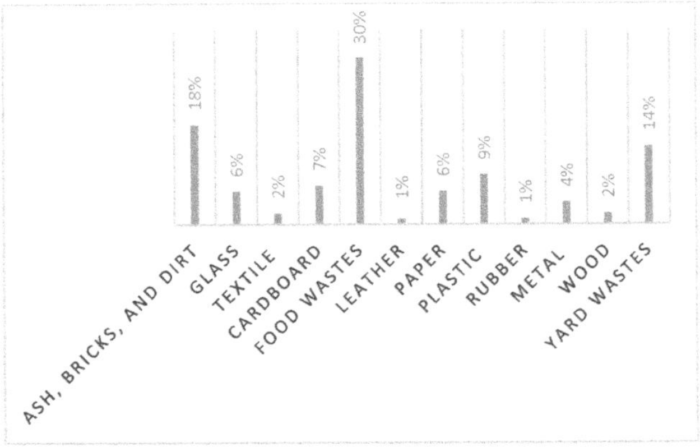

Figure 1. Composition of MSW in Pakistan (https://www.trade.gov/country-commercial-guides/pakistan-waste-management, accessed on 12 January 2024).

This research concentrates on one specific scenario of executing the stated approach, which examines five thermochemical techniques for regulating MSW: thermal gasification, incineration, pyrolysis, thermal depolymerization, and plasma arc gasification. Five experts with distinct acquisition experiences participated in the evaluation, all delivering quantitative observations. A novel hybrid approach is presented to boost the credibility of the findings.

This research is organized as follows. Section 1 illustrates the significance of merging factors of uncertainty and incomplete information into decision-making strategies. Section 2 contains a detailed examination of the fundamental concepts that form the basis for the remainder of the study. Section 3 describes the interactional operational laws for IFHSS regulating IFHSS and presents the IFHSIWA and IFHSIOWA operators, along with their essential characteristics. Section 4 describes the IFHSIWG and IFHSIOWG operators, clearly explaining their distinctive features. Section 5 proposes a novel EDAS model that employs the operators mentioned above to deal with the obstacles of MAGDM. Section 6 performs an empirical study to evaluate the feasibility of the proposed approach, intending to identify the optimum thermochemical treatment method for managing MSW. The proposed technique is assessed using a thorough comparison study with previous methodologies in Section 7.

1.1. Literature Review

The following subsection reviews relevant research on strategies used to treat MSW. It then discusses the interaction between AOs and the EDAS approach.

The disposal of MSW has been considered a top priority these days, as it has grown to be a critical problem globally and an essential component of environmental protection and ecologically sound resource protection. Arikan et al. [22] applied three MAGDM approaches, namely PROMETHEE, TOPSIS, and fuzzy TOPSIS, to find appropriate alternatives for solid waste treatment. Zhuang et al. [23] indicated that the integration of ZnO into a Co_3O_4 heterojunction configuration substantially improves its enzymatic efficacy for purifying peroxymonosulfate-based organic water. Mir et al. [24] implemented the TOPSIS approach to choose the most effective strategy to handle solid waste. Wang et al. [25] introduced a MAGDM strategy for determining a suitable process to regulate MSW.

Coban et al. [26] used three different multi-criteria decision-making (MCDM) strategies, specifically TOPSIS, PROMETHEE I, and PROMETHEE II, to assess and examine eight situations in solid waste treatment. Afrane et al. [27] analyzed the best waste-to-energy conversion system for use in Ghana. Akram et al. [28] investigated MAGDM problems using a linguistic Pythagorean fuzzy structure. Paul et al. [29] applied cubic Pythagorean fuzzy numbers to represent the fuzzy attributes of the criteria values associated with the five thermochemical techniques used to process MSW. Roy et al. [30] implemented the TOPSIS technique in a credibility structure to solve problems with MSW management. Rahimi et al. [31] presented a new structure that integrates Geographic Information System techniques with fuzzy MCDM processes to locate dump sites for MSW disposal. Rani et al. [32] developed the Pythagorean fuzzy stepwise weight assessment ratio analysis to solve MCDM issues in medical waste disposal systems. Garg and Rani [33] presented the MULTIMOORA method in the IFS structure to measure the feasibility of several ways of handling waste materials. Barakati and Rani [34] presented a novel parametric divergence measure and planned a twofold normalization-based multi-aggregation technique for interval-valued intuitionistic fuzzy sets (IVIFS) to deal with MCDM problems.

Zadeh [35] developed a framework for dealing with inconsistency by presenting the fuzzy set (FS) theory. Atanassov [36] introduced IFS to solve the drawbacks of FS. IFS is an FS extension integrating a membership degree (MD) with a non-membership degree (NMD). The structures mentioned above are insufficient when faced with the parametric representation of alternatives. Molodtsov [37] proposed the soft set (SS) as a conceptual structure to successfully tackle and resolve this problem. Several studies involve fuzzy set advancements with soft sets, especially fuzzy soft sets (FSS) [38] and intuitionistic fuzzy soft sets (IFSS) [39]. Smarandache [40] proposed the idea of hypersoft sets (HSS), which involves incorporating multiple multi-sub-parameters into the assigning function of an identified set of attributes. Saeed et al. [41] proposed a multi-fuzzy HSS structure that employs similarity and entropy measures to tackle MCDM challenges. Rahman et al. [42] proposed a mathematical structure for convexity and concavity in fuzzy HSS. Yolcu et al. [43] introduced essential operations for IFHSS. Zulqarnain et al. [44,45] developed the TOPSIS technique and aggregation operators (AOs) designed explicitly for IFHSS.

He et al. [46,47] proposed the intuitionistic fuzzy geometric interaction averaging operators and generalized intuitionistic fuzzy weighted geometric interaction and proposed the MADM model to resolve decision-making complications. Garg [48–50] presented the implementation of intuitionistic fuzzy Hamacher interactive weighted averaging, ordered weighted averaging, hybrid weighted averaging operators, interaction average, and geometric AOs in MCDM problems. Liu and Wang [51] introduced novel hammy mean AOs that integrate interaction in IFS. Garg [52] proposed interactive Einstein operational laws for IFS, which assisted in forming geometric interaction averaging AOs. Liu [53] enhanced the use of MADM techniques by implementing interaction AOs into the usual IFS framework. Garg and Arora [54] introduced prioritized interaction averaging AOs under IFSS. Several researchers have made major improvements in developing interaction AOs within different extensions of the FS theory [55–62].

Ghorabaee et al.'s [63] approach is considered a highly practical approach for numerically describing MAGDM difficulties that contain competing factors. Ghorabaee et al. [64] stated that the EDAS approach uses the average solution to calculate the importance of alternatives. It takes note of differences in the average solution by analyzing the positive distance from average (PDA) and negative distance from average (NDA) measures. Srivastava et al. [65] executed an investigation where they implemented the fuzzy EDAS approach to evaluate and rank autonomous maintenance characteristics from a technical perspective. Kahraman et al. [66] used the EDAS technique for solid waste management in an IF framework. Mishra et al. [67] implemented the EDAS method to investigate techniques for handling hospital waste, while Schitea et al. [68] engaged it to determine a suitable site within the IF context. Liu et al. [69] enlarged the EDAS technique to deal with Pythagorean fuzzy sets. Li et al. [70] constructed a novel EDAS method for Pythagorean fuzzy numbers to address decision-making obstacles. Zhang et al. [71] developed the EDAS method for picture fuzzy sets, while Darko and Liang [72] enhanced it by including Hamacher operators. Mishra et al. [73] explored the computation of attribute weights and used the EDAS method to evaluate the ranking of sustainable third-party reverse logistics providers.

1.2. Aim of the Proposed Study

The AOs in ref. [45] are insufficient for properly evaluating facts built into more secure categories that possess the exactitude required to produce precise outcomes. Let $\mathcal{U} = \{\delta_1, \delta_2\}$ be two experts with weights $\Omega_i = (0.6, 0.4)^T$ and let ℓ_1, ℓ_2 be the parameters with their sub-parameters, such as $\ell_1 = \{\ell_{11}, \ell_{12}\}$ and $\ell_2 = \{\ell_{21}\}$, where $\ddot{\mathcal{A}} = \ell_1 \times \ell_2 = \{\ell_{11}, \ell_{12}\} \times \{\ell_{21}\} = \{(\ell_{11}, \ell_{21}), (\ell_{12}, \ell_{21})\} = \{\check{c}_1, \check{c}_2\}$ with weights $\gamma_j = (0.4, 0.6)^T$. Let $\mathcal{Q}$ be an alternate, and then experts' evaluation $\mathcal{Q} = \begin{bmatrix} (0.7, 0.0) & (0.2, 0.5) \\ (0.3, 0.7) & (0.5, 0.3) \end{bmatrix}$ of the planned aspects in terms of IFHSNs. The obtained aggregated value using the IFHSWA [45] operator is $\langle 0.44717, 0.0 \rangle$. The validity of these AOs is unreliable, and only basic knowledge about the alternatives can be found. Thus, merging IFHSNs into the composition of interaction AOs offers an exciting and important way to investigate. The techniques described in [45] are inadequate for carefully examining information while observing previous theoretical structures and presenting obvious, practical implications.

This study is considered important for the following reasons.

1. The IFHSS concept has abundant applications and can represent complicated information in a versatile and detailed manner. This makes it an important DM method that considers the overall environment.
2. This research demonstrates the apparent convenience of interactional operations, which contributes to their specific variables. The study introduces interactive and interactive ordered AOs that have been optimized for IFHSS.
3. The EDAS approach is widely employed due to its accessibility and effectiveness in delivering more favorable outcomes in the DM procedure. The execution of the EDAS approach used in this study boosts the robustness and dependability of the DM system, solving an apparent drawback in present approaches.
4. Several researchers use various DM methodologies in different fuzzy structures for existing studies, such as TOPSIS, VIKOR, and EDAS. Still, there exists a lack of real-world studies on MSW management using the EDAS approach in an IFHSS context. This research handles this problem using the presented integrated model, which analyzes IFHSS information to identify the most effective thermochemical approach.

1.3. Research Problem and Contribution

This research analyzes five distinct alternatives and uses the intuitionistic fuzzy hypersoft EDAS technique for investigation. Municipalities must use appropriate solutions to mitigate the effects of trash disposal sites. The diverse composition of the collected waste, depicted in Figure 1, underscores the imperative for effective management to avert negative health impacts from infectious diseases and to mitigate the harmful consequences of climate change, particularly by controlling the release of methane and other deleterious gases. Moreover, the reality that 30% of MSW consists of food compounds stresses the importance of implementing waste-to-energy transformation.

Therefore, an intelligent selection of MSW disposal technologies is most important in managing the challenges caused by improperly handling MSW and encouraging waste-to-energy reprocessing. The EDAS approach is more effective for MAGDM than TOPSIS and VIKOR. This is because it concentrates on finding an average solution that takes into account both positive ideal and negative ideal solutions. The suggested strategy addresses significant challenges. How can one choose the optimum thermochemical treatment technique correctly in unreliable circumstances? How can we acquire data on the abilities of the MSW management group? How will the EDAS method be applied to create the MSW organizational structure to determine appropriate treatment processes?

This research presents multiple exciting developments, which can be briefly summarized as follows.

(1) Integrating interactional operational laws in the IFHSS structure assisted in designing and presenting IFHSS interaction AOs. The operators provided are IFHSIWA, IFHSIOWA, IFHSIWG, and IFHSIOWG operators. We also adequately presented some significant aspects of these operators, particularly idempotency, boundedness, monotonicity, homogeneity, and shift-invariance.
(2) The IFHSS structure uses the distinctive procedure EDAS to determine the most optimal alternative. This method delivers a practical solution for convoluted DM challenges and merges the prevailing DM theory through this novel strategy.
(3) The proposed approach investigates and chooses the thermochemical treatment strategy in MSW management.
(4) The approach executes detailed comparison and sensitivity evaluations, which are meticulously examined and evaluated. The research outcomes confirm the consistency and effectiveness of the proposed technique.

2. Preliminaries

This section recalls basic concepts, such as soft set (SS), fuzzy soft set (FSS), HSS, and IFHSS.

Definition 1 ([37]). *Let $\mathcal{U}$ and $\mathcal{E}$ be the universe of discourse and set of attributes, respectively. Let $\mathcal{P}(\mathcal{U})$ be the power set of $\mathcal{U}$ and $\mathcal{A} \subseteq \mathcal{E}$. A pair $(\mathcal{F}, \mathcal{A})$ is called an SS over $\mathcal{U}$. Its mapping is expressed as follows:*

$$\mathcal{F} : \mathcal{A} \rightarrow \mathcal{P}(\mathcal{U})$$

Also, it can be defined as follows:

$$(\mathcal{F}, \mathcal{A}) = \{\mathcal{F}(\urcorner) \in \mathcal{P}(\mathcal{U}) : \urcorner \in \mathcal{E}, \mathcal{F}(\urcorner) = \varnothing \ if \urcorner \notin \mathcal{A}\}$$

Definition 2 ([38]). *Let $\mathcal{U}$ be a universal set, let $\mathcal{E}$ be a set of attributes, and let $\mathcal{P}(\mathcal{U})$ be a fuzzy power set of $\mathcal{U}$. Let $\mathcal{A} \subseteq \mathcal{E}$, and then $(\mathcal{F}, \mathcal{A})$ is an FSS over $\mathcal{U}$; its mapping can be stated as*

$$\mathcal{F} : \mathcal{A} \rightarrow \mathcal{P}(\mathcal{U})$$

The previous studies have challenges in solving scenarios when an expert considers a parameter's sub-attributes. To solve this issue, Smarandache [40] introduced the idea of the hypersoft set as a solution.

Definition 3 ([40]). *Let $\mathcal{U}$ be a universal set, and let $\mathcal{P}(\mathcal{U})$ be a power set of $\mathcal{U}$ and $k = \{k_1, k_2, k_3, \ldots, k_n\}$, $(n \geq 1)$; K_i represents the set of attributes and their corresponding sub-attributes, such as $K_i \cap K_j = \varphi$, where $i \neq j$ for each $n \geq 1$ and $i, j \in \{1,2,3 \ldots n\}$. Assume $K_1 \times K_2 \times K_3 \times \ldots \times K_n = \dddot{A} = \{c_{1h} \times c_{2k} \times \cdots \times c_{nl}\}$ is a collection of sub-attributes, where $1 \leq h \leq \alpha$, $1 \leq k \leq \beta$, $1 \leq l \leq \gamma$, and $\alpha, \beta, \gamma \in \mathbb{N}$. Then, the pair $(\mathcal{F}, K_1 \times K_2 \times K_3 \times \ldots \times K_n) = (\mathcal{F}, \dddot{A})$ is known as HSS, defined as follows:*

$$\mathcal{F} : K_1 \times K_2 \times K_3 \times \ldots \times K_n = \dddot{A} \rightarrow \mathcal{P}(\mathcal{U})$$

It is also defined as

$$(\mathcal{F}, \dddot{A}) = \{\check{c}, \mathcal{F}_{\dddot{A}}(\check{c}) : \check{c} \in \dddot{A}, \mathcal{F}_{\dddot{A}}(\check{c}) \in \mathcal{P}(\mathcal{U})\}$$

Definition 4 ([40]). *Let $\mathcal{U}$ be a universal set, and let $\mathcal{P}(\mathcal{U})$ be a power set of $\mathcal{U}$ and $k = \{k_1, k_2, k_3, \ldots, k_n\}$, $(n \geq 1)$; K_i denotes the set of attributes and their corresponding sub-attributes, like $K_i \cap K_j = \varphi$, where $i \neq j$ for each $n \geq 1$ and $i, j \in \{1,2,3 \ldots n\}$. Assume $K_1 \times K_2 \times K_3 \times \ldots \times K_n = \dddot{A} = \{c_{1h} \times c_{2k} \times \cdots \times c_{nl}\}$ is a collection of sub-attributes, where $1 \leq h \leq \alpha$, $1 \leq k \leq \beta$, $1 \leq l \leq \gamma$, and $\alpha, \beta, \gamma \in \mathbb{N}$. $IFS^{\mathcal{U}}$ represents the intuitionistic fuzzy subsets of $\mathcal{U}$. Then, the pair $(\mathcal{F}, K_1 \times K_2 \times K_3 \times \ldots \times K_n) = (\mathcal{F}, \dddot{A})$ is known as IFHSS, defined as follows:*

$$\mathcal{F} : K_1 \times K_2 \times K_3 \times \ldots \times K_n = \dddot{A} \rightarrow IFS^{\mathcal{U}}$$

It is also defined as $(\mathcal{F}, \dddot{A}) = \{(\check{c}, \mathcal{F}_{\dddot{A}}(\check{c})) : \check{c} \in \dddot{A}, \mathcal{F}_{\dddot{A}}(\check{c}) \in IFS^{\mathcal{U}}\}$, where $\mathcal{F}_{\dddot{A}}(\check{c}) = \left\{\left\langle \delta, \mathcal{T}_{\mathcal{F}(\check{c})}(\delta), \mathcal{J}_{\mathcal{F}(\check{c})}(\delta) \right\rangle : \delta \in \mathcal{U}\right\}$, where $\mathcal{T}_{\mathcal{F}(\check{c})}(\delta)$ and $\mathcal{J}_{\mathcal{F}(\check{c})}(\delta)$ signify the membership and non-membership values of the sub-attributes, $\mathcal{T}_{\mathcal{F}(\check{c})}(\delta), \vartheta_{\mathcal{F}(\check{c})}(\delta) \in [0, 1]$, and $0 \leq \mathcal{T}_{\mathcal{F}(\check{c})}(\delta) + \mathcal{J}_{\mathcal{F}(\check{c})}(\delta) \leq 1$.

Example 1. *Let $\mathcal{U} = \{\delta_1, \delta_2\}$ and $\mathfrak{L} = \{\ell_1 =$ Teaching methdology, $\ell_2 =$ Subjects, $\ell_3 =$ Classes$\}$ represent the universe of discourse and attributes with their respective sub-attributes, e.g., teaching methodology = $L_1 = \{c_{11} =$ project base, $c_{12} =$ class discussion$\}$ Subjects = $L_2 = \{c_{21} =$ Mathematics, $c_{22} =$ Computer Science, $c_{23} =$ Statistics$\}$, and Classes = $L_3 = \{c_{31} =$ Masters,$\}$ $c_{32} =$ Doctorol, respectively. The group of attributes with succeeding conforming sub-attributes is specified as follows.*

Let $\dddot{A} = L_1 \times L_2 \times L_3$ be a set of attributes, such as $\dddot{A} = L_1 \times L_2 \times L_3 = \{c_{11}, c_{12}\} \times \{c_{21}, c_{22}, c_{23}\} \times \{c_{31}, c_{32}\} = \left\{\begin{array}{c} (c_{11}, c_{21}, c), (c_{11}, c, c_{32}), (c_{11}, c_{22}, c_{31}), (c_{11}, c_{22}, c_{32}), (c_{11}, c_{23}, c_{31}), (c_{11}, c_{23}, c_{32}), \\ (c_{12}, c_{21}, c_{31}), (c_{12}, c_{21}, c_{32}), (c_{12}, c_{22}, c_{31}), (c_{12}, c_{22}, c_{32}), (c_{12}, c_{23}, c_{31}), (c_{12}, c_{23}, c_{32}), \end{array}\right\}$ $\dddot{A} = \{\check{c}_1, \check{c}_2, \check{c}_3, \check{c}_4, \check{c}_5, \check{c}_6, \check{c}_7, \check{c}_8, \check{c}_9, \check{c}_{10}, \check{c}_{11}, \check{c}_{12}\}$

Then, the IFHSS over $\mathcal{U}$ is known as

$$(\mathcal{F}, \dddot{A}) = \left\{\begin{array}{l} (\check{c}_1, (\delta_1, (0.6, 0.3)), (\delta_2, (0.3, 0.5))), (\check{c}_2, (\delta_1, (0.2, 0.7)), (\delta_2, (0.1, 0.5))), (\check{c}_3, (\delta_1, (0.2, 0.8)), (\delta_2, (0.3, 0.4))), \\ (\check{c}_4, (\delta_1, (0.2, 0.5)), (\delta_2, (0.1, 0.6))), (\check{c}_5, (\delta_1, (0.4, 0.3)), (\delta_2, (0.3, 0.5))), (\check{c}_6, (\delta_1, (0.2, 0.4)), (\delta_2, (0.1, 0.5))), \\ (\check{c}_7, (\delta_1, (0.2, 0.6)), (\delta_2, (0.4, 0.2))), (\check{c}_8, (\delta_1, (0.2, 0.5)), (\delta_2, (0.3, 0.1))), (\check{c}_9, (\delta_1, (0.4, 0.2)), (\delta_2, (0.3, 0.5))), \\ (\check{c}_{10}, (\delta_1, (0.1, 0.4)), (\delta_2, (0.7, 0.2))), (\check{c}_{11}, (\delta_1, (0.4, 0.5)), (\delta_2, (0.2, 0.5))), (\check{c}_5, (\delta_1, (0.1, 0.2)), (\delta_2, (0.2, 0.7))) \end{array}\right\}$$

Remark 1. *If $\mathcal{T}_{\mathcal{F}(\check{c})}(\delta) + \mathcal{J}_{\mathcal{F}(\check{c})}(\delta) \leq 1$ is held, no one attribute has a sub-attribute. Then, IFHSS is condensed to IFSS [39].*

IFHSN $\mathcal{F}_{\delta_i}(\check{c}_j) = \left\{\left(\mathcal{T}_{\mathcal{F}(\check{c}_j)}(\delta_i), \mathcal{J}_{\mathcal{F}(\check{c}_j)}(\delta_i)\right) \mid \delta_i \in \mathcal{U}\right\}$ can be written as $\mathfrak{J}_{\check{c}_{ij}} = \left\langle \mathcal{T}_{\mathcal{F}(\check{c}_{ij})}, \mathcal{J}_{\mathcal{F}(\check{c}_{ij})} \right\rangle$. Alternative ranking can be found using score functions, as follows:

$$S\left(\mathfrak{J}_{\check{c}_{ij}}\right) = \mathcal{T}_{\mathcal{F}(\check{c}_{ij})} - \mathcal{J}_{\mathcal{F}(\check{c}_{ij})}, \ S\left(\mathfrak{J}_{\check{c}_{ij}}\right) \in [-1,1] \quad (1)$$

where $S\left(\mathfrak{J}_{\check{c}_{ij}}\right) \in [-1,\ 1]$. It is not easy to differentiate the IFSNs and which alternative is most appropriate, in some cases, using the score function. For example, let $\mathfrak{J}_{\check{c}_{11}} = \langle 0.4, 0.7\rangle$ and $\mathfrak{J}_{\check{c}_{12}} = \langle 0.5, 0.8\rangle$, and then $S(\mathfrak{J}_{\check{c}_{11}}) = -0.3$ and $S(\mathfrak{J}_{\check{c}_{12}}) = -0.3$. Therefore, it is impossible to decide the best alternative. To overcome the above situation, an accuracy function has been developed, such as

$$H\left(\mathfrak{J}_{\check{c}_{ij}}\right) = \mathcal{T}_{\mathcal{F}(\check{c}_{ij})} + \mathcal{J}_{\mathcal{F}(\check{c}_{ij})},\ H\left(\mathfrak{J}_{\check{c}_{ij}}\right) \in [0,1] \tag{2}$$

Let $\mathfrak{J}_{\check{c}_{ij}}$ and $\mathfrak{T}_{\check{c}_{ij}}$ be two distinct IFHSNs; then, comparison laws are defined as follows:

1. If $S\left(\mathfrak{J}_{\check{c}_{ij}}\right) > S\left(\mathfrak{T}_{\check{c}_{ij}}\right)$, then $\mathfrak{J}_{\check{c}_{ij}} > \mathfrak{T}_{\check{c}_{ij}}$.
2. If $S\left(\mathfrak{J}_{\check{c}_{ij}}\right) = S\left(\mathfrak{T}_{\check{c}_{ij}}\right)$, then.
   - If $H\left(\mathfrak{J}_{\check{c}_{ij}}\right) > H\left(\mathfrak{T}_{\check{c}_{ij}}\right)$, then $\mathfrak{J}_{\check{c}_{ij}} > \mathfrak{T}_{\check{c}_{ij}}$.
   - If $H\left(\mathfrak{J}_{\check{c}_{ij}}\right) = H\left(\mathfrak{T}_{\check{c}_{ij}}\right)$, then $\mathfrak{J}_{\check{c}_{ij}} = \mathfrak{T}_{\check{c}_{ij}}$.

Definition 5 (**[44]**). *Let* $\mathfrak{J}_{\check{c}_k} = (\mathcal{T}_{\check{c}_k}, \mathcal{J}_{\check{c}_k}), \mathfrak{J}_{\check{c}_{11}} = (\mathcal{T}_{\check{c}_{11}}, \mathcal{J}_{\check{c}_{11}})$, *and* $\mathfrak{J}_{\check{c}_{12}} = (\mathcal{T}_{\check{c}_{12}}, \mathcal{J}_{\check{c}_{12}})$ *be three IFHSNs and* $\alpha > 0$. *Then, the algebraic operational laws for IFHSNs are defined as follows:*

1. $\mathfrak{J}_{\check{c}_{11}} \oplus \mathfrak{J}_{\check{c}_{12}} = \langle \mathcal{T}_{\check{c}_{11}} + \mathcal{T}_{\check{c}_{12}} - \mathcal{T}_{\check{c}_{11}}\mathcal{T}_{\check{c}_{12}}, \mathcal{J}_{\check{c}_{11}}\mathcal{J}_{\check{c}_{12}}\rangle$.
2. $\mathfrak{J}_{\check{c}_{11}} \otimes \mathfrak{J}_{c_{12}} = \langle \mathcal{T}_{\check{c}_{11}}\mathcal{T}_{\check{c}_{12}}, \mathcal{J}_{\check{c}_{11}} + \mathcal{J}_{\check{c}_{12}} - \mathcal{J}_{\check{c}_{11}}\mathcal{J}_{\check{c}_{12}}\rangle$.
3. $\alpha\mathfrak{J}_{\check{c}_k} = \left\langle \left[1 - (1 - \mathcal{T}_{\check{c}_k})^{\alpha}, (\mathcal{J}_{\check{c}_k})^{\alpha}\right]\right\rangle$.
4. $\mathfrak{J}^{\alpha}_{\check{c}_k} = \left\langle \left[(\mathcal{T}_{\check{c}_k})^{\alpha}, 1 - (1 - \mathcal{J}_{\check{c}_k})^{\alpha}\right]\right\rangle$.

A comprehensive evaluation of the aggregation operators under the IFHSS framework clearly shows that in particular situations, these aggregation operators generate undesirable outcomes. We promote the interaction AOs designed for IFHSNs to communicate and deal with such issues.

3. Interactive Aggregation Operators for Intuitionistic Fuzzy Hypersoft Sets

This section will construct a couple of interactive average AOs, such as IFHSIWA and IFHSIOWA operators, with their essential properties.

Definition 6. *Let* $\mathfrak{J}_{\check{c}_k} = (\mathcal{T}_{\check{c}_k}, \mathcal{J}_{\check{c}_k})$, $\mathfrak{J}_{\check{c}_{11}} = (\mathcal{T}_{\check{c}_{11}}, \mathcal{J}_{\check{c}_{11}})$, *and* $\mathfrak{J}_{\check{c}_{12}} = (\mathcal{T}_{\check{c}_{12}}, \mathcal{J}_{\check{c}_{12}})$ *be the three IFHSNs and* $\alpha > 0$. *Then, the interactional operational laws for IFHSNs are defined as follows:*

1. $\mathfrak{J}_{\check{c}_{11}} \oplus \mathfrak{J}_{\check{c}_{12}} = \langle 1 - (1 - \mathcal{T}_{\check{c}_{11}})(1 - \mathcal{T}_{\check{c}_{12}}), (1 - \mathcal{T}_{\check{c}_{11}})(1 - \mathcal{T}_{\check{c}_{12}}) - (1 - \mathcal{T}_{\check{c}_{11}} - \mathcal{J}_{\check{c}_{11}})(1 - \mathcal{T}_{\check{c}_{12}} - \mathcal{J}_{\check{c}_{12}})\rangle$.
2. $\mathfrak{J}_{\check{c}_{11}} \otimes \mathfrak{J}_{\check{c}_{12}} = \left\langle (1 - \mathcal{J}_{\check{c}_{11}})(1 - \mathcal{J}_{\check{c}_{12}}) - \left(1 - \mathcal{T}_{\check{c}_{11}} - \mathcal{J}_{\check{c}_{11}}\right)(1 - \mathcal{T}_{\check{c}_{12}} - \mathcal{J}_{\check{c}_{12}}), 1 - (1 - \mathcal{J}_{\check{c}_{11}})(1 - \mathcal{J}_{\check{c}_{12}})\right\rangle$.
3. $\alpha\mathfrak{J}_{\check{c}_k} = \left\langle 1 - (1 - \mathcal{T}_{\check{c}_k})^{\alpha}, (1 - \mathcal{T}_{\check{c}_k})^{\alpha} - (1 - \mathcal{T}_{\check{c}_k} - \mathcal{J}_{\check{c}_k})^{\alpha}\right\rangle$.
4. $\mathfrak{J}^{\alpha}_{\check{c}_k} = \left\langle (1 - \mathcal{J}_{\check{c}_k})^{\alpha} - (1 - \mathcal{T}_{\check{c}_k} - \mathcal{J}_{\check{c}_k})^{\alpha}, 1 - (1 - \mathcal{J}_{\check{c}_k})^{\alpha}\right\rangle$.

Definition 7. *Let* $\mathfrak{J}_{\check{c}_{ij}} = \left(\mathcal{T}_{\check{c}_{ij}}, \mathcal{J}_{\check{c}_{ij}}\right)$ *be an IFHSN;* Ω_i *and* γ_j *represent the weights of experts and multi-sub-attributes along with stated conditions* $\Omega_i > 0$, $\sum_{i=1}^{n} \Omega_i = 1$, $\gamma_j > 0$, *and* $\sum_{j=1}^{m} \gamma_j = 1$, *where* $i = 1, 2, \ldots, n$ *and* $j = 1, 2, \ldots, m$. *Then,*

$$IFHSIWA\ (\mathfrak{J}_{\check{c}_{11}}, \mathfrak{J}_{\check{c}_{12}}, \ldots, \mathfrak{J}_{\check{c}_{nm}}) = \oplus_{j=1}^{m}\gamma_j\left(\oplus_{i=1}^{n}\Omega_i\mathfrak{J}_{\check{c}_{ij}}\right)$$

Theorem 1. *Let $\mathfrak{J}_{\check{c}_{ij}} = \left(\mathcal{T}_{\check{c}_{ij}}, \mathcal{J}_{\check{c}_{ij}}\right)$ be a family of IFHSNs. Then, the aggregated value of the IFHSIWA operator is also an IFHSN and*

$$IFHSIWA\left(\mathfrak{J}_{\check{c}_{11}}, \mathfrak{J}_{\check{c}_{12}}, \ldots, \mathfrak{J}_{\check{c}_{nm}}\right) = \left\langle \begin{array}{c} 1-\prod_{j=1}^{m}\left(\prod_{i=1}^{n}\left(1-\mathcal{T}_{\check{c}_{ij}}\right)^{\Omega_i}\right)^{\gamma_j}, \\ \prod_{j=1}^{m}\left(\prod_{i=1}^{n}\left(1-\mathcal{T}_{\check{c}_{ij}}\right)^{\Omega_i}\right)^{\gamma_j} - \prod_{j=1}^{m}\left(\prod_{i=1}^{n}\left(1-\mathcal{T}_{\check{c}_{ij}}-\mathcal{J}_{\check{c}_{ij}}\right)^{\Omega_i}\right)^{\gamma_j} \end{array} \right\rangle \tag{3}$$

where Ω_i and γ_j represent the experts' and the sub-attribute weights within certain circumstances, $\Omega_i > 0$, $\sum_{i=1}^{n} \Omega_i = 1$, $\gamma_j > 0$, $\sum_{j=1}^{m} \gamma_j = 1\ \forall\, i = 1, 2, \ldots, n$, and $j = 1, 2, \ldots, m$.

Proof. The IFHSIWA operator can be proved using mathematical induction.

Let $n = 1$; we obtain $\Omega_1 = 1$. Then,

$$IFHSIWA\left(\mathfrak{J}_{\check{c}_{11}}, \mathfrak{J}_{\check{c}_{12}}, \ldots, \mathfrak{J}_{\check{c}_{nm}}\right) = \oplus_{j=1}^{m} \gamma_j \mathfrak{J}_{\check{c}_{ij}}$$

$$IFHSIWA\left(\mathfrak{J}_{\check{c}_{11}}, \mathfrak{J}_{\check{c}_{12}}, \ldots, \mathfrak{J}_{\check{c}_{nm}}\right) = \left\langle 1-\prod_{j=1}^{m}\left(1-\mathcal{T}_{\check{c}_{ij}}\right)^{\gamma_j}, \prod_{j=1}^{m}\left(1-\mathcal{T}_{\check{c}_{ij}}\right)^{\gamma_j} - \prod_{j=1}^{m}\left(1-\mathcal{T}_{\check{c}_{ij}}-\mathcal{J}_{\check{c}_{ij}}\right)^{\gamma_j} \right\rangle$$

$$= \left\langle 1-\prod_{j=1}^{m}\left(\prod_{i=1}^{1}\left(1-\mathcal{T}_{\check{c}_{1j}}\right)^{\Omega_i}\right)^{\gamma_j}, \prod_{j=1}^{m}\left(\prod_{i=1}^{1}\left(1-\mathcal{T}_{\check{c}_{1j}}\right)^{\Omega_i}\right)^{\gamma_j} - \prod_{j=1}^{m}\left(\prod_{i=1}^{1}\left(1-\mathcal{T}_{\check{c}_{1j}}-\mathcal{J}_{\check{c}_{1j}}\right)^{\Omega_i}\right)^{\gamma_j} \right\rangle$$

For $m = 1$, we obtain $\gamma_1 = 1$. Then,

$$IFHSIWA\left(\mathfrak{J}_{\check{c}_{11}}, \mathfrak{J}_{\check{c}_{12}}, \ldots, \mathfrak{J}_{\check{c}_{nm}}\right) = \oplus_{i=1}^{n} \Omega_i \mathfrak{J}_{\check{c}_{ij}}$$

$$= \left\langle 1-\left(\prod_{i=1}^{n}\left(1-\mathcal{T}_{\check{c}_{i1}}\right)^{\Omega_i}\right), \prod_{i=1}^{n}\left(1-\mathcal{T}_{\check{c}_{i1}}\right)^{\Omega_i} - \prod_{i=1}^{n}\left(1-\mathcal{T}_{\check{c}_{i1}}-\mathcal{J}_{\check{c}_{i1}}\right)^{\Omega_i} \right\rangle$$

$$= \left\langle 1-\prod_{j=1}^{1}\left(\prod_{i=1}^{n}\left(1-\mathcal{T}_{\check{c}_{ij}}\right)^{\Omega_i}\right)^{\gamma_j}, \prod_{j=1}^{1}\left(\prod_{i=1}^{n}\left(1-\mathcal{T}_{\check{c}_{ij}}\right)^{\Omega_i}\right)^{\gamma_j} - \prod_{j=1}^{1}\left(\prod_{i=1}^{n}\left(1-\mathcal{T}_{\check{c}_{ij}}-\mathcal{J}_{\check{c}_{ij}}\right)^{\Omega_i}\right)^{\gamma_j} \right\rangle$$

So, Equation (3) holds for $n = 1$ and $m = 1$.

For $n = 2$ and $m = 1$. Then,

$$IFHSIWA\left(\mathfrak{J}_{\check{c}_{11}}, \mathfrak{J}_{\check{c}_{21}}\right) = \gamma_1\left(\Omega_1 \mathfrak{J}_{\check{c}_{11}}\right) \oplus \gamma_1\left(\Omega_2 \mathfrak{J}_{\check{c}_{21}}\right)$$

$$\gamma_1\left(\Omega_1 \mathfrak{J}_{\check{c}_{11}}\right) = \left\langle 1-\left(\left(1-\mathcal{T}_{\check{c}_{11}}\right)^{\Omega_1}\right)^{\gamma_1}, \left(\left(1-\mathcal{T}_{\check{c}_{11}}\right)^{\Omega_1}\right)^{\gamma_1} - \left(\left(1-\mathcal{T}_{\check{c}_{11}}-\mathcal{J}_{\check{c}_{11}}\right)^{\Omega_1}\right)^{\gamma_1} \right\rangle$$

And

$$\gamma_1\left(\Omega_2 \mathfrak{J}_{\check{c}_{21}}\right) = \left\langle 1-\left(\left(1-\mathcal{T}_{\check{c}_{21}}\right)^{\Omega_2}\right)^{\gamma_1}, \left(\left(1-\mathcal{T}_{\check{c}_{21}}\right)^{\Omega_2}\right)^{\gamma_1} - \left(\left(1-\mathcal{T}_{\check{c}_{21}}-\mathcal{J}_{\check{c}_{21}}\right)^{\Omega_2}\right)^{\gamma_1} \right\rangle$$

$$IFHSIWA\left(\mathfrak{J}_{\check{c}_{11}}, \mathfrak{J}_{\check{c}_{21}}\right) = \gamma_1\left(\Omega_1 \mathfrak{J}_{\check{c}_{11}}\right) \oplus \gamma_1\left(\Omega_2 \mathfrak{J}_{\check{c}_{21}}\right)$$

$$= \left\langle 1-\prod_{j=1}^{1}\left(\prod_{i=1}^{2}\left(1-\mathcal{T}_{\check{c}_{ij}}\right)^{\Omega_i}\right)^{\gamma_j}, \prod_{j=1}^{1}\left(\prod_{i=1}^{2}\left(1-\mathcal{T}_{\check{c}_{ij}}\right)^{\Omega_i}\right)^{\gamma_j} - \prod_{j=1}^{1}\left(\prod_{i=1}^{2}\left(1-\mathcal{T}_{\check{c}_{ij}}-\mathcal{J}_{\check{c}_{ij}}\right)^{\Omega_i}\right)^{\gamma_j} \right\rangle$$

So, Equation (3) holds for $n = 2$ and $m = 1$.

Consider that Equation (3) holds for $n = \beta_1$, $m = \beta_2$. Then,

$$IFHSIWA\left(\mathfrak{J}_{\check{c}_{11}}, \mathfrak{J}_{\check{c}_{12}}, \ldots, \mathfrak{J}_{\check{c}_{\beta_1\beta_2}}\right) = \oplus_{j=1}^{\beta_2} \gamma_j\left(\oplus_{i=1}^{\beta_1} \Omega_i \mathfrak{J}_{\check{c}_{ij}}\right)$$

$$= \left\langle 1-\prod_{j=1}^{\beta_2}\left(\prod_{i=1}^{\beta_1}\left(1-\mathcal{T}_{\check{c}_{ij}}\right)^{\Omega_i}\right)^{\gamma_j}, \prod_{j=1}^{\beta_2}\left(\prod_{i=1}^{\beta_1}\left(1-\mathcal{T}_{\check{c}_{ij}}\right)^{\Omega_i}\right)^{\gamma_j} - \prod_{j=1}^{\beta_2}\left(\prod_{i=1}^{\beta_1}\left(1-\mathcal{T}_{\check{c}_{ij}}-\mathcal{J}_{\check{c}_{ij}}\right)^{\Omega_i}\right)^{\gamma_j} \right\rangle$$

For $n = \beta_1 + 1$ and $m = \beta_2 + 1$, we have

$$IFHSIWA\left(\mathfrak{J}_{\check{c}_{11}}, \mathfrak{J}_{\check{c}_{12}}, \ldots, \mathfrak{J}_{\check{c}_{(\beta_1+1)(\beta_2+1)}}\right) = IFHSIWA\left(\mathfrak{J}_{\check{c}_{11}}, \mathfrak{J}_{\check{c}_{12}}, \ldots, \mathfrak{J}_{\check{c}_{\beta_1\beta_2}}\right) \oplus \Upsilon_{(\beta_2+1)}\left(\Omega_{(\beta_1+1)}\mathfrak{J}_{\check{c}_{(\beta_1+1)(\beta_2+1)}}\right)$$

$$\Upsilon_{(\beta_2+1)}\left(\Omega_{(\beta_1+1)}\mathfrak{J}_{\check{c}_{(\beta_1+1)(\beta_2+1)}}\right) = \left\langle \begin{array}{c} 1-\left(\left(1-\mathcal{T}_{\check{c}_{(\beta_1+1)(\beta_2+1)}}\right)^{\Omega_{(\beta_2+1)}}\right)^{\Upsilon_{(\beta_2+1)}}, \left(\left(1-\mathcal{T}_{\check{c}_{(\beta_1+1)(\beta_2+1)}}\right)^{\Omega_{(\beta_1+1)}}\right)^{\Upsilon_{(\beta_2+1)}} \\ -\left(\left(1-\mathcal{T}_{\check{c}_{(\beta_1+1)(\beta_2+1)}}-\mathcal{J}_{\check{c}_{(\beta_1+1)(\beta_2+1)}}\right)^{\Omega_{(\beta_1+1)}}\right)^{\Upsilon_{(\beta_2+1)}} \end{array} \right\rangle$$

For simplicity,
$M_1 = 1-\prod_{j=1}^{\beta_2}\left(\prod_{i=1}^{\beta_1}\left(1-\mathcal{T}_{\check{c}_{ij}}\right)^{\Omega_i}\right)^{\Upsilon_j}$, $N_1 = \prod_{j=1}^{\beta_2}\left(\prod_{i=1}^{\beta_1}\left(1-\mathcal{T}_{\check{c}_{ij}}\right)^{\Omega_i}\right)^{\Upsilon_j}$, $O_1 = \prod_{j=1}^{\beta_2}\left(\prod_{i=1}^{\beta_1}\left(1-\mathcal{T}_{\check{c}_{ij}}-\mathcal{J}_{\check{c}_{ij}}\right)^{\Omega_i}\right)^{\Upsilon_j}$, $M_2 = 1-\left(\left(1-\mathcal{T}_{\check{c}_{(\beta_1+1)(\beta_2+1)}}\right)^{\Omega_{(\beta_2+1)}}\right)^{\Upsilon_{(\beta_2+1)}}$, $N_2 = \left(\left(1-\mathcal{T}_{\check{c}_{(\beta_1+1)(\beta_2+1)}}\right)^{\Omega_{(\beta_1+1)}}\right)^{\Upsilon_{(\beta_2+1)}}$, $O_2 = \left(\left(1-\mathcal{T}_{\check{c}_{(\beta_1+1)(\beta_2+1)}}-\mathcal{J}_{\check{c}_{(\beta_1+1)(\beta_2+1)}}\right)^{\Omega_{(\beta_1+1)}}\right)^{\Upsilon_{(\beta_2+1)}}$.
Then,

$$\begin{gathered} IFHSIWA\left(\mathfrak{J}_{\check{c}_{11}}, \mathfrak{J}_{\check{c}_{12}}, \ldots, \mathfrak{J}_{\check{c}_{(\beta_1+1)(\beta_2+1)}}\right) = IFHSIWA\left(\mathfrak{J}_{\check{c}_{11}}, \mathfrak{J}_{\check{c}_{12}}, \ldots, \mathfrak{J}_{\check{c}_{\beta_1\beta_2}}\right) \oplus \Upsilon_{(\beta_2+1)}\left(\Omega_{(\beta_1+1)}\mathfrak{J}_{\check{c}_{(\beta_1+1)(\beta_2+1)}}\right) \\ = \langle M_1, N_1 - O_1\rangle \oplus \langle M_2, N_2 - O_2\rangle = \langle M_1M_2, N_1N_2 - O_1O_2\rangle \\ = \left\langle 1-\prod_{j=1}^{\beta_2+1}\left(\prod_{i=1}^{\beta_1+1}\left(1-\mathcal{T}_{\check{c}_{ij}}\right)^{\Omega_i}\right)^{\Upsilon_j}, \prod_{j=1}^{\beta_2+1}\left(\prod_{i=1}^{\beta_1+1}\left(1-\mathcal{T}_{\check{c}_{ij}}\right)^{\Omega_i}\right)^{\Upsilon_j} - \prod_{j=1}^{\beta_2+1}\left(\prod_{i=1}^{\beta_1+1}\left(1-\mathcal{T}_{\check{c}_{ij}}-\mathcal{J}_{\check{c}_{ij}}\right)^{\Omega_i}\right)^{\Upsilon_j}\right\rangle \end{gathered}$$

It is verified that Equation (3) holds for $n = \beta_1 + 1$, $m = \beta_2 + 1$. So, we can say that Equation (3) holds $\forall\, n, m$. □

Proposition 1. *Let* $\mathfrak{J}_{\check{c}_{ij}} = \left(\mathcal{T}_{\check{c}_{ij}}, \mathcal{J}_{\check{c}_{ij}}\right)$ *be a family of IFHSNs. Also, let* Ω_i *and* Υ_j *be the weights of experts and sub-attributes, respectively, such as* $\Omega_i > 0, \sum_{i=1}^{n}\Omega_i = 1$ *and* $\Upsilon_j > 0, \sum_{j=1}^{m}\Upsilon_j = 1$.

3.1. Idempotency

If $\mathfrak{J}_{\check{c}_{ij}} = \mathfrak{J}_{\check{c}_\delta} = \left(\mathcal{T}_{\check{c}_\delta}, \mathcal{J}_{\check{c}_\delta}\right)$, then

$$IFHSIWA\left(\mathfrak{J}_{\check{c}_{11}}, \mathfrak{J}_{\check{c}_{12}}, \ldots, \mathfrak{J}_{\check{c}_{nm}}\right) = \mathfrak{J}_{\check{c}_\delta}$$

Proof. As we know, $\mathfrak{J}_{\check{c}_{ij}} = \mathfrak{J}_{\check{c}_\delta} = \left(\mathcal{T}_{\check{c}_\delta}, \mathcal{J}_{\check{c}_\delta}\right)$ is a collection of *IFHSNs*. Then, by using Equation (3),

$$\begin{gathered} IFHSIWA\left(\mathfrak{J}_{\check{c}_{11}}, \mathfrak{J}_{\check{c}_{12}}, \ldots, \mathfrak{J}_{\check{c}_{nm}}\right) = IFHSIWA\left(\mathfrak{J}_{\check{c}_k}, \mathfrak{J}_{\check{c}_k}, \ldots, \mathfrak{J}_{\check{c}_\delta}\right) \\ = \left\langle 1-\prod_{j=1}^{m}\left(\prod_{i=1}^{n}\left(1-\mathcal{T}_{\check{c}_\delta}\right)^{\Omega_i}\right)^{\Upsilon_j}, \prod_{j=1}^{m}\left(\prod_{i=1}^{n}\left(1-\mathcal{T}_{\check{c}_\delta}\right)^{\Omega_i}\right)^{\Upsilon_j} - \prod_{j=1}^{m}\left(\prod_{i=1}^{n}\left(1-\mathcal{T}_{\check{c}_\delta}-\mathcal{J}_{\check{c}_\delta}\right)^{\Omega_i}\right)^{\Upsilon_j}\right\rangle \\ = \left\langle 1-\left(\left(1-\mathcal{T}_{\check{c}_\delta}\right)^{\sum_{i=1}^{n}\Omega_i}\right)^{\sum_{j=1}^{m}\Upsilon_j}, \left(\left(1-\mathcal{T}_{\check{c}_\delta}\right)^{\sum_{i=1}^{n}\Omega_i}\right)^{\sum_{j=1}^{m}\Upsilon_j} - \left(\left(1-\mathcal{T}_{\check{c}_\delta}-\mathcal{J}_{\check{c}_\delta}\right)^{\sum_{i=1}^{n}\Omega_i}\right)^{\sum_{j=1}^{m}\Upsilon_j}\right\rangle \end{gathered}$$

As $\sum_{i=1}^{n}\Omega_i = 1$ and $\sum_{j=1}^{m}\Upsilon_j = 1$. So,

$$\begin{gathered} = \left\langle 1-\left(1-\mathcal{T}_{\check{c}_\delta}\right), \left(1-\mathcal{T}_{\check{c}_\delta}\right) - \left(1-\mathcal{T}_{\check{c}_\delta}-\mathcal{J}_{\check{c}_\delta}\right)\right\rangle \\ = \left(\mathcal{T}_{\check{c}_\delta}, \mathcal{J}_{\check{c}_\delta}\right) = \mathfrak{J}_{\check{c}_\delta}. \end{gathered}$$

□

3.2. Boundedness

Let $\mathfrak{J}_{\check{c}_{ij}}{}^{-} = \left\langle \binom{min}{j}\binom{min}{i}\left\{\mathcal{T}_{\check{c}_{ij}}\right\}, \binom{max}{j}\binom{max}{i}\left\{\mathcal{J}_{\check{c}_{ij}}\right\}\right\rangle$ and $\mathfrak{J}_{\check{c}_{ij}}{}^{+} = \left\langle \binom{max}{j}\binom{max}{i}\left\{\mathcal{T}_{\check{c}_{ij}}\right\}, \binom{min}{j}\binom{min}{i}\left\{\mathcal{J}_{\check{c}_{ij}}\right\}\right\rangle$. Then,

$$\mathfrak{J}_{\check{c}_{ij}}{}^{-} \leq IFHSIWA\left(\mathfrak{J}_{\check{c}_{11}}, \mathfrak{J}_{\check{c}_{12}}, \ldots, \mathfrak{J}_{\check{c}_{nm}}\right) \leq \mathfrak{J}_{\check{c}_{ij}}{}^{+}$$

Proof. We know $\mathfrak{J}_{\check{c}_{ij}} = \left(\mathcal{T}_{\check{c}_{ij}}, \mathcal{J}_{\check{c}_{ij}}\right)$ is an IFHSN. Then,

$$\begin{aligned}
&\binom{min}{j}\binom{min}{i}\left\{\mathcal{T}_{\check{c}_{ij}}\right\} \leq \mathcal{T}_{\check{c}_{ij}} \leq \binom{max}{j}\binom{max}{i}\left\{\mathcal{T}_{\check{c}_{ij}}\right\}\\
&\Rightarrow 1 - \binom{max}{j}\binom{max}{i}\left\{\mathcal{T}_{\check{c}_{ij}}\right\} \leq 1 - \mathcal{T}_{\check{c}_{ij}} \leq 1 - \binom{min}{j}\binom{min}{i}\left\{\mathcal{T}_{\check{c}_{ij}}\right\}\\
&\Leftrightarrow \left(1 - \binom{max}{j}\binom{max}{i}\left\{\mathcal{T}_{\check{c}_{ij}}\right\}\right)^{\Omega_i} \leq \left(1 - \mathcal{T}_{\check{c}_{ij}}\right)^{\Omega_i} \leq \left(1 - \binom{min}{j}\binom{min}{i}\left\{\mathcal{T}_{\check{c}_{ij}}\right\}\right)^{\Omega_i}\\
&\Leftrightarrow \left(1 - \binom{max}{j}\binom{max}{i}\left\{\mathcal{T}_{\check{c}_{ij}}\right\}\right)^{\sum_{i=1}^{n}\Omega_i} \leq \prod_{i=1}^{n}\left(1 - \mathcal{T}_{\check{c}_{ij}}\right)^{\Omega_i} \leq \left(1 - \binom{min}{j}\binom{min}{i}\left\{\mathcal{T}_{\check{c}_{ij}}\right\}\right)^{\sum_{i=1}^{n}\Omega_i}\\
&\Leftrightarrow \left(1 - \binom{max}{j}\binom{max}{i}\left\{\mathcal{T}_{\check{c}_{ij}}\right\}\right)^{\sum_{j=1}^{m}\gamma_j} \leq \prod_{j=1}^{m}\left(\prod_{i=1}^{n}\left(1 - \mathcal{T}_{\check{c}_{ij}}\right)^{\Omega_i}\right)^{\gamma_j} \leq \left(1 - \binom{min}{j}\binom{min}{i}\left\{\mathcal{T}_{\check{c}_{ij}}\right\}\right)^{\sum_{j=1}^{m}\gamma_j}
\end{aligned}$$

As $\sum_{i=1}^{n}\Omega_i = 1$ and $\sum_{j=1}^{m}\gamma_j = 1$. So,

$$\begin{aligned}
&\Leftrightarrow 1 - \binom{max}{j}\binom{max}{i}\left\{\mathcal{T}_{\check{\mu}_{ij}}\right\} \leq \prod_{j=1}^{m}\left(\prod_{i=1}^{n}\left(1 - \mathcal{T}_{\check{c}_{ij}}\right)^{\Omega_i}\right)^{\gamma_j} \leq 1 - \binom{min}{j}\binom{min}{i}\left\{\mathcal{T}_{\check{c}_{ij}}\right\}\\
&\Leftrightarrow \binom{min}{j}\binom{min}{i}\left\{\mathcal{T}_{\check{c}_{ij}}\right\} \leq 1 - \prod_{j=1}^{m}\left(\prod_{i=1}^{n}\left(1 - \mathcal{T}_{\check{c}_{ij}}\right)^{\Omega_i}\right)^{\gamma_j} \leq \binom{max}{j}\binom{max}{i}\left\{\mathcal{T}_{\check{c}_{ij}}\right\}\\
&\Leftrightarrow \binom{min}{j}\binom{min}{i}\left\{\mathcal{T}_{\check{c}_{ij}}\right\} \leq 1 - \prod_{j=1}^{m}\left(\prod_{i=1}^{n}\left(1 - \mathcal{T}_{\check{c}_{ij}}\right)^{\Omega_i}\right)^{\gamma_j} \leq \binom{max}{j}\binom{max}{i}\left\{\mathcal{T}_{\check{c}_{ij}}\right\}
\end{aligned} \tag{4}$$

Similarly,

$$\binom{min}{j}\binom{min}{i}\left\{\mathcal{J}_{\check{c}_{ij}}\right\} \leq \prod_{j=1}^{m}\left(\prod_{i=1}^{n}\left(1 - \mathcal{T}_{\check{c}_{ij}}\right)^{\Omega_i}\right)^{\gamma_j} - \prod_{j=1}^{m}\left(\prod_{i=1}^{n}\left(1 - \mathcal{T}_{\check{c}_{ij}} - \mathcal{J}_{\check{c}_{ij}}\right)^{\Omega_i}\right)^{\gamma_j} \leq \binom{max}{j}\binom{max}{i}\left\{\mathcal{J}_{\check{c}_{ij}}\right\} \tag{5}$$

Let $IFHSIWA\left(\mathfrak{J}_{\check{c}_{11}}, \mathfrak{J}_{\check{c}_{12}}, \ldots, \mathfrak{J}_{\check{c}_{nm}}\right) = \left\langle \mathcal{T}_{\check{c}_{\delta}}, \mathcal{J}_{\check{c}_{\delta}}\right\rangle = \mathfrak{J}_{\check{c}_{\delta}}$. Then, inequities (4) and (5) could be turned into the following form:
$\binom{min}{j}\binom{min}{i}\left\{\mathcal{T}_{\check{c}_{ij}}\right\} \leq \mathcal{T}_{\check{c}_{\delta}} \leq \binom{max}{j}\binom{max}{i}\left\{\mathcal{T}_{\check{c}_{ij}}\right\}$ and $\binom{min}{j}\binom{min}{i}\left\{\mathcal{J}_{\check{c}_{ij}}\right\} \leq \mathcal{J}_{\check{c}_{\delta}} \leq \binom{max}{j}\binom{max}{i}\left\{\mathcal{J}_{\check{c}_{ij}}\right\}$.
From Equation (1), we have
$S(\mathfrak{J}_{\check{c}_{\delta}}) = \{\mathcal{T}_{\check{c}_{\delta}}\} - \{\mathcal{J}_{\check{c}_{\delta}}\} \leq \binom{max}{j}\binom{max}{i}\left\{\mathcal{T}_{\check{c}_{ij}}\right\} - \binom{min}{j}\binom{min}{i}\left\{\mathcal{J}_{\check{c}_{ij}}\right\} = S\left(\mathfrak{J}_{\check{c}_{ij}}{}^{+}\right) \Rightarrow S(\mathfrak{J}_{\check{c}_{\delta}}) \leq S\left(\mathfrak{J}_{\check{c}_{ij}}{}^{+}\right)$
and $S(\mathfrak{J}_{\check{c}_{\delta}}) = \mathcal{T}_{\check{c}_{\delta}} - \mathcal{J}_{\check{c}_{\delta}} \geq \binom{min}{j}\binom{min}{i}\left\{\mathcal{T}_{\check{c}_{ij}}\right\} - \binom{max}{j}\binom{max}{i}\left\{\mathcal{J}_{\check{c}_{ij}}\right\} = S\left(\mathfrak{J}_{\check{c}_{ij}}{}^{-}\right) \Rightarrow S(\mathfrak{J}_{\check{c}_{\delta}}) \geq S\left(\mathfrak{J}_{\check{c}_{ij}}{}^{-}\right)$.
Now, there are two possibilities

If $S(\mathfrak{J}_{\check{c}_{\delta}}) < S\left(\mathfrak{J}_{\check{c}_{ij}}{}^{+}\right)$ and $S(\mathfrak{J}_{\check{c}_{\delta}}) > S\left(\mathfrak{J}_{\check{c}_{ij}}{}^{-}\right)$. Then,

$$\mathfrak{J}_{\check{c}_{ij}}{}^{-} < IFHSIWA\left(\mathfrak{J}_{\check{c}_{11}}, \mathfrak{J}_{\check{c}_{12}}, \ldots, \mathfrak{J}_{\check{c}_{nm}}\right) < \mathfrak{J}_{\check{c}_{ij}}{}^{+} \tag{6}$$

If $S(\mathfrak{J}_{\check{c}_{\delta}}) = S\left(\mathfrak{J}_{\check{c}_{ij}}{}^{+}\right)$, then

$\{\mathcal{T}_{\check{c}_{\delta}}\} - \{\mathcal{J}_{\check{c}_{\delta}}\} = \binom{max}{j}\binom{max}{i}\left\{\mathcal{T}_{\check{c}_{ij}}\right\} - \binom{min}{j}\binom{min}{i}\left\{\mathcal{J}_{\check{c}_{ij}}\right\}$. So, $\{\mathcal{T}_{\check{c}_{\delta}}\} = \binom{max}{j}\binom{max}{i}\left\{\mathcal{T}_{\check{c}_{ij}}\right\}$ and $\{\mathcal{J}_{\check{c}_{\delta}}\} = \binom{min}{j}\binom{min}{i}\left\{\mathcal{J}_{\check{c}_{ij}}\right\}$. Hence,

$$IFHSIWA\left(\mathfrak{J}_{\check{c}_{11}}, \mathfrak{J}_{\check{c}_{12}}, \ldots, \mathfrak{J}_{\check{c}_{nm}}\right) = \mathfrak{J}_{\check{c}_{ij}}{}^{+} \tag{7}$$

If $S(\mathfrak{J}_{\check{c}_{\delta}}) = S\left(\mathfrak{J}_{\check{c}_{ij}}{}^{-}\right)$, then

$\{\mathcal{T}_{\check{c}_\delta}\} - \{\mathcal{J}_{\check{c}_\delta}\} = \binom{min}{j}\binom{min}{i}\left\{\mathcal{T}_{\check{c}_{ij}}\right\} - \binom{max}{j}\binom{max}{i}\left\{\mathcal{J}_{\check{c}_{ij}}\right\}$. So, $\{\mathcal{T}_{\check{c}_\delta}\} = \binom{min}{j}\binom{min}{i}\left\{\mathcal{T}_{\check{c}_{ij}}\right\}$ and $\{\mathcal{J}_{\check{c}_\delta}\} = \binom{max}{j}\binom{max}{i}\left\{\mathcal{J}_{\check{c}_{ij}}\right\}$. Hence,

$$IFHSIWA(\mathfrak{J}_{\check{c}_{11}}, \mathfrak{J}_{\check{c}_{12}}, \ldots, \mathfrak{J}_{\check{c}_{nm}}) = \mathfrak{J}_{\check{c}_{ij}}{}^{-} \tag{8}$$

Therefore, using Equations (6)–(8), we have

$$\mathfrak{J}_{\check{c}_{ij}}{}^{-} \le IFHSIWA(\mathfrak{J}_{\check{c}_{11}}, \mathfrak{J}_{\check{c}_{12}}, \ldots, \mathfrak{J}_{\check{c}_{nm}}) \le \mathfrak{J}_{\check{c}_{ij}}{}^{+}.$$

□

3.3. Shift Invariance

If $\mathfrak{J}_{\check{c}_{ij}} = \left(\mathcal{T}_{\check{c}_{ij}}, \mathcal{J}_{\check{c}_{ij}}\right)$ is a collection of IFHSNs and $\mathfrak{J}_{\check{c}_\delta} = (\mathcal{T}_{\check{c}_\delta}, \mathcal{J}_{\check{c}_\delta})$ is an IFHSN, then

$$\begin{gathered}IFHSIWA(\mathfrak{J}_{c_{11}} \oplus \mathfrak{J}_{\check{c}_\delta}, \mathfrak{J}_{\check{c}_{12}} \oplus \mathfrak{J}_{\check{c}_\delta}, \ldots, \mathfrak{J}_{\check{c}_{nm}} \oplus \mathfrak{J}_{\check{c}_\delta}) = \\ IFHSIWA(\mathfrak{J}_{\check{c}_{11}}, \mathfrak{J}_{\check{c}_{12}}, \ldots, \mathfrak{J}_{\check{c}_{nm}}) \oplus \mathfrak{J}_{\check{c}_\delta}.\end{gathered}$$

Proof. We have

$$\mathfrak{J}_{c_{ij}} \oplus \mathfrak{J}_{\check{c}_\delta} = \left\langle 1 - \left(1 - \mathcal{T}_{\check{c}_{ij}}\right)(1 - \mathcal{T}_{\check{c}_\delta}), \left(1 - \mathcal{T}_{\check{c}_{ij}}\right)(1 - \mathcal{T}_{\check{c}_\delta}) - \left(1 - \mathcal{T}_{\check{c}_{ij}} - \mathcal{J}_{\check{c}_{ij}}\right)(1 - \mathcal{T}_{\check{c}_\delta} - \mathcal{J}_{\check{c}_\delta}) \right\rangle$$

So,

$$\begin{gathered}IFHSIWA\ (\mathfrak{J}_{c_{11}} \oplus \mathfrak{J}_{\check{c}_\delta},\ \mathfrak{J}_{\check{c}_{12}} \oplus \mathfrak{J}_{\check{c}_\delta}, \ldots,\ \mathfrak{J}_{\check{c}_{nm}} \oplus \mathfrak{J}_{\check{c}_\delta}) \\
= \left\langle \begin{array}{c} 1 - \prod_{j=1}^{m}\left(\prod_{i=1}^{n}\left(\left(1 - \mathcal{T}_{\check{c}_{ij}}\right)(1 - \mathcal{T}_{\check{c}_\delta})\right)^{\Omega_i}\right)^{\gamma_j}, \\ \prod_{j=1}^{m}\left(\prod_{i=1}^{n}\left(\left(1 - \mathcal{T}_{\check{c}_{ij}}\right)(1 - \mathcal{T}_{\check{c}_\delta})\right)^{\Omega_i}\right)^{\gamma_j} - \prod_{j=1}^{m}\left(\prod_{i=1}^{n}\left(\left(1 - \mathcal{T}_{\check{c}_{ij}} - \mathcal{J}_{\check{c}_{ij}}\right)(1 - \mathcal{T}_{\check{c}_\delta} - \mathcal{J}_{\check{c}_\delta})\right)^{\Omega_i}\right)^{\gamma_j} \end{array} \right\rangle \\
= \left\langle \begin{array}{c} \left(1 - \prod_{j=1}^{m}\left(\prod_{i=1}^{n}\left(1 - \mathcal{T}_{\check{c}_{ij}}\right)^{\Omega_i}\right)^{\gamma_j}\right)\left(1 - \left((1 - \mathcal{T}_{\check{c}_\delta})^{\Sigma_{i=1}^{n}\Omega_i}\right)^{\Sigma_{j=1}^{m}\gamma_j}\right), \\ \prod_{j=1}^{m}\left(\prod_{i=1}^{n}\left(1 - \mathcal{T}_{\check{c}_{ij}}\right)^{\Omega_i}\right)^{\gamma_j}\left((1 - \mathcal{T}_{\check{c}_\delta})^{\Sigma_{i=1}^{n}\Omega_i}\right)^{\Sigma_{j=1}^{m}\gamma_j} - \prod_{j=1}^{m}\left(\prod_{i=1}^{n}\left(1 - \mathcal{T}_{\check{c}_{ij}} - \mathcal{J}_{\check{c}_{ij}}\right)^{\Omega_i}\right)^{\gamma_j}\left((1 - \mathcal{T}_{\check{c}_\delta} - \mathcal{J}_{\check{c}_\delta})^{\Sigma_{i=1}^{n}\Omega_i}\right)^{\Sigma_{j=1}^{m}\gamma_j} \end{array} \right\rangle \\
= \left\langle \begin{array}{c} (1 - (1 - \mathcal{T}_{\check{c}_\delta}))\left(1 - \prod_{j=1}^{m}\left(\prod_{i=1}^{n}\left(1 - \mathcal{T}_{\check{c}_{ij}}\right)^{\Omega_i}\right)^{\gamma_j}\right), \\ (1 - \mathcal{T}_{\check{c}_\delta})\prod_{j=1}^{m}\left(\prod_{i=1}^{n}\left(1 - \mathcal{T}_{\check{c}_{ij}}\right)^{\Omega_i}\right)^{\gamma_j} - (1 - \mathcal{T}_{\check{c}_\delta} - \mathcal{J}_{\check{c}_\delta})\prod_{j=1}^{m}\left(\prod_{i=1}^{n}\left(1 - \mathcal{T}_{\check{c}_{ij}} - \mathcal{J}_{\check{c}_{ij}}\right)^{\Omega_i}\right)^{\gamma_j} \end{array} \right\rangle \\
= \left\langle \begin{array}{c} \left(1 - \prod_{j=1}^{m}\left(\prod_{i=1}^{n}\left(1 - \mathcal{T}_{\check{c}_{ij}}\right)^{\Omega_i}\right)^{\gamma_j}\right), \\ \prod_{j=1}^{m}\left(\prod_{i=1}^{n}\left(1 - \mathcal{T}_{\check{c}_{ij}}\right)^{\Omega_i}\right)^{\gamma_j} - \prod_{j=1}^{m}\left(\prod_{i=1}^{n}\left(1 - \mathcal{T}_{\check{c}_{ij}} - \mathcal{J}_{\check{c}_{ij}}\right)^{\Omega_i}\right)^{\gamma_j} \end{array} \right\rangle \oplus \left\langle \begin{array}{c} (1 - (1 - \mathcal{T}_{\check{c}_\delta})), \\ (1 - \mathcal{T}_{\check{c}_\delta}) - (1 - \mathcal{T}_{\check{c}_\delta} - \mathcal{J}_{\check{c}_\delta}) \end{array} \right\rangle \\
= IFHSIWA(\mathfrak{J}_{\check{c}_{11}}, \mathfrak{J}_{\check{c}_{12}}, \ldots, \mathfrak{J}_{\check{c}_{nm}}) \oplus \mathfrak{J}_{\check{c}_\delta}.\end{gathered}$$

□

3.4. Homogeneity

Let $\mathfrak{J}_{\check{c}_{ij}} = \left(\mathcal{T}_{\check{c}_{ij}}, \mathcal{J}_{\check{c}_{ij}}\right)$ be a family of IFHSNs, where $i = 1, 2, 3, \ldots, n; j = 1, 2, 3, \ldots, m$. Then,

$IFHSIWA(\alpha\mathfrak{J}_{\check{c}_{11}}, \alpha\mathfrak{J}_{\check{c}_{12}}, \ldots, \alpha\mathfrak{J}_{\check{c}_{nm}}) = \alpha IFHSIWA(\mathfrak{J}_{\check{c}_{11}}, \mathfrak{J}_{\check{c}_{12}}, \ldots, \mathfrak{J}_{\check{c}_{nm}})$, where $\alpha > 0$.

Proof. As $\mathfrak{J}_{\check{c}_{ij}} = \left(\mathcal{T}_{\check{c}_{ij}}, \mathcal{J}_{\check{c}_{ij}}\right)$ is a family of IFHSNs and $\alpha > 0$, then

$$\alpha\mathfrak{J}_{\check{c}_{ij}} = \left\langle 1 - \left(\left(1 - \mathcal{T}_{\check{c}_{ij}}\right)^{\alpha\Omega_i}\right)^{\gamma_j}, \left(\left(1 - \mathcal{T}_{\check{c}_{ij}}\right)^{\alpha\Omega_i}\right)^{\gamma_j} - \left(\left(1 - \mathcal{T}_{\check{c}_{ij}} - \mathcal{J}_{\check{c}_{ij}}\right)^{\alpha\Omega_i}\right)^{\gamma_j} \right\rangle$$

So,

$$IFHSIWA\left(\mathfrak{J}_{\check{c}_{11}}, \mathfrak{J}_{\check{c}_{12}}, \ldots, \mathfrak{J}_{\check{c}_{nm}}\right) = \left\langle 1 - \left(\left(1 - \mathcal{T}_{\check{c}_{ij}}\right)^{\alpha\Omega_i}\right)^{\gamma_j}, \left(\left(1 - \mathcal{T}_{\check{c}_{ij}}\right)^{\alpha\Omega_i}\right)^{\gamma_j} - \left(\left(1 - \mathcal{T}_{\check{c}_{ij}} - \mathcal{J}_{\check{c}_{ij}}\right)^{\alpha\Omega_i}\right)^{\gamma_j} \right\rangle$$
$$= \left\langle 1 - \left(\left(\left(1 - \mathcal{T}_{\check{c}_{ij}}\right)^{\Omega_i}\right)^{\gamma_j}\right)^{\alpha}, \left(\left(\left(1 - \mathcal{T}_{\check{c}_{ij}}\right)^{\Omega_i}\right)^{\gamma_j}\right)^{\alpha} - \left(\left(\left(1 - \mathcal{T}_{\check{c}_{ij}} - \mathcal{J}_{\check{c}_{ij}}\right)^{\Omega_i}\right)^{\gamma_j}\right)^{\alpha} \right\rangle$$
$$= \alpha IFHSIWA\left(\mathfrak{J}_{\check{c}_{11}}, \mathfrak{J}_{\check{c}_{12}}, \ldots, \mathfrak{J}_{\check{c}_{nm}}\right).$$

□

3.5. Monotonicity

Let $\mathfrak{J}_{\check{c}_{ij}} = \left(\mathcal{T}_{\check{c}_{ij}}, \mathcal{J}_{\check{c}_{ij}}\right)$ and $\mathfrak{J}^*_{\check{c}_{ij}} = \left(\mathcal{T}^*_{\check{c}_{ij}}, \mathcal{J}^*_{\check{c}_{ij}}\right)$ be the two distinct families of IFHSNs. Then,

$$IFHSIWA\left(\mathfrak{J}_{\check{c}_{11}}, \mathfrak{J}_{\check{c}_{12}}, \ldots, \mathfrak{J}_{\check{c}_{nm}}\right) \leq IFHSIWA\left(\mathfrak{J}^*_{\check{c}_{11}}, \mathfrak{J}^*_{\check{c}_{12}}, \ldots\ldots\ldots, \mathfrak{J}^*_{\check{c}_{nm}}\right), \text{ if } \mathfrak{J}_{\check{c}_{ij}} \leq \mathfrak{J}^*_{\check{c}_{ij}}\ \forall i, j.$$

Proof. Let $v = f(u) = \frac{1-u}{1+u}$, $u \in [0,1]$, and then $\frac{d}{d(u)}(f(u)) = \frac{-2}{(1+u)^2} < 0$, so $f(u)$ is a decreasing function on $[0,1]$. If $\mathfrak{J}_{\check{c}_{ij}} \leq \mathfrak{J}^*_{\check{c}_{ij}}$, then $f\left(\mathfrak{J}_{\check{c}_{ij}}\right) \leq f\left(\mathfrak{J}^*_{\check{c}_{ij}}\right)\ \forall i, j.$

$$\begin{aligned}
&\mathcal{T}_{\check{c}_{ij}} \leq \mathcal{T}^*_{\check{c}_{ij}} \\
&\Rightarrow 1 - \mathcal{T}^*_{\check{c}_{ij}} \leq 1 - \mathcal{T}_{\check{c}_{ij}} \\
&\Rightarrow 1 - \left(\mathcal{T}^*_{\check{c}_{ij}}\right)^q \leq 1 - \left(\mathcal{T}_{\check{c}_{ij}}\right)^q \\
&\Rightarrow 1 - \left(1 - \left(\mathcal{T}_{\check{c}_{ij}}\right)^q\right) \leq 1 - \left(1 - \left(\mathcal{T}^*_{\check{c}_{ij}}\right)^q\right)
\end{aligned}$$

as $\Omega_i > 0$, $\sum_{i=1}^{n} \Omega_i = 1$, and $\gamma_j > 0$, $\sum_{j=1}^{m} \gamma_j = 1$. So,

$$\begin{aligned}
&\Rightarrow 1 - \left(\left(1 - \mathcal{T}_{\check{c}_{ij}}\right)^{\sum_{i=1}^{n}\Omega_i}\right)^{\sum_{j=1}^{m}\gamma_j} \leq 1 - \left(\left(1 - \mathcal{T}^*_{\check{c}_{ij}}\right)^{\sum_{i=1}^{n}\Omega_i}\right)^{\sum_{j=1}^{m}\gamma_j} \\
&\Rightarrow 1 - \prod_{j=1}^{m}\left(\prod_{i=1}^{n}\left(1 - \mathcal{T}_{\check{c}_{ij}}\right)^{\Omega_i}\right)^{\gamma_j} \leq 1 - \prod_{j=1}^{m}\left(\prod_{i=1}^{n}\left(1 - \mathcal{T}^*_{\check{c}_{ij}}\right)^{\Omega_i}\right)^{\gamma_j}
\end{aligned}$$

Let $x = g(y) = \frac{2-y}{y}$, $y \in [0,1]$, and then $\frac{d}{dy}(g(y)) = -\frac{2}{y^2} < 0$. So, $g(y)$ is decreasing on $[0,1]$. If $\mathfrak{J}_{\check{c}_{ij}} \leq \mathfrak{J}^*_{\check{c}_{ij}}$, then $f\left(\mathfrak{J}_{\check{c}_{ij}}\right) \leq f\left(\mathfrak{J}^*_{\check{c}_{ij}}\right)\ \forall i, j.$

$$\begin{aligned}
&\mathcal{J}_{\check{c}_{ij}} \leq \mathcal{J}^*_{\check{c}_{ij}} \Rightarrow \mathcal{T}^*_{\check{c}_{ij}} - \mathcal{J}^*_{\check{c}_{ij}} \leq \mathcal{T}_{\check{c}_{ij}} - \mathcal{J}_{\check{c}_{ij}} \Rightarrow 1 - \mathcal{T}_{\check{c}_{ij}} - \mathcal{J}_{\check{c}_{ij}} \leq 1 - \mathcal{T}^*_{\check{c}_{ij}} - \mathcal{J}^*_{\check{c}_{ij}} \\
&\Rightarrow \left(1 - \mathcal{T}_{\check{c}_{ij}}\right) - \left(1 - \mathcal{T}_{\check{c}_{ij}} - \mathcal{J}_{\check{c}_{ij}}\right) \geq \left(1 - \mathcal{T}^*_{\check{c}_{ij}}\right) - \left(1 - \mathcal{T}^*_{\check{c}_{ij}} - \mathcal{J}^*_{\check{c}_{ij}}\right)
\end{aligned}$$

as $\Omega_i > 0$, $\sum_{i=1}^{n} \Omega_i = 1$, and $\gamma_j > 0$, $\sum_{j=1}^{m} \gamma_j = 1$. So,

$$\begin{aligned}
&\Rightarrow \left(\left(1 - \mathcal{T}_{\check{c}_{ij}}\right)^{\sum_{i=1}^{n}\Omega_i}\right)^{\sum_{j=1}^{m}\gamma_j} - \left(\left(1 - \mathcal{T}_{\check{c}_{ij}} - \mathcal{J}_{\check{c}_{ij}}\right)^{\sum_{i=1}^{n}\Omega_i}\right)^{\sum_{j=1}^{m}\gamma_j} \\
&\geq \left(\left(1 - \mathcal{T}^*_{\check{c}_{ij}}\right)^{\sum_{i=1}^{n}\Omega_i}\right)^{\sum_{j=1}^{m}\gamma_j} - \left(\left(1 - \mathcal{T}^*_{\check{c}_{ij}} - \mathcal{J}^*_{\check{c}_{ij}}\right)^{\sum_{i=1}^{n}\Omega_i}\right)^{\sum_{j=1}^{m}\gamma_j}
\end{aligned}$$

$$\Rightarrow \prod_{j=1}^{m}\left(\prod_{i=1}^{n}\left(1-\mathcal{T}_{\check{c}_{ij}}\right)^{\Omega_i}\right)^{\gamma_j} - \prod_{j=1}^{m}\left(\prod_{i=1}^{n}\left(1-\mathcal{T}_{\check{c}_{ij}}-\mathcal{J}_{\check{c}_{ij}}\right)^{\Omega_i}\right)^{\gamma_j}$$
$$\geq \prod_{j=1}^{m}\left(\prod_{i=1}^{n}\left(1-\mathcal{T}^*_{\check{c}_{ij}}\right)^{\Omega_i}\right)^{\gamma_j} - \prod_{j=1}^{m}\left(\prod_{i=1}^{n}\left(1-\mathcal{T}^*_{\check{c}_{ij}}-\mathcal{J}^*_{\check{c}_{ij}}\right)^{\Omega_i}\right)^{\gamma_j}$$

Hence,

$$IFHSIWA(\mathfrak{J}_{\check{c}_{11}},\mathfrak{J}_{\check{c}_{12}},\ldots,\mathfrak{J}_{\check{c}_{nm}}) \leq IFHSIWA\left(\mathfrak{J}^*_{\check{c}_{11}},\mathfrak{J}^*_{\check{c}_{12}},\ldots\ldots\ldots,\mathfrak{J}^*_{\check{c}_{nm}}\right).$$

□

Definition 8. *Let $\mathfrak{J}_{\check{c}_{ij}} = \left(\mathcal{T}_{\check{c}_{ij}}, \mathcal{J}_{\check{c}_{ij}}\right)$ be a family of IFHSNs. Then, the IFHSIOWA operator is defined as*

$$IFHSIOWA(\mathfrak{J}_{\check{c}_{11}},\mathfrak{J}_{\check{c}_{12}},\ldots,\mathfrak{J}_{\check{c}_{nm}})$$
$$= \gamma_1\Omega_1\mathfrak{J}_{\sigma(\check{c}_{\mathfrak{r}(1)\mathfrak{s}(1)})} \oplus \gamma_1\Omega_2\mathfrak{J}_{\sigma(\check{c}_{\mathfrak{r}(2)\mathfrak{s}(1)})} \oplus \ldots\ldots$$
$$\oplus\gamma_m\Omega_n\mathfrak{J}_{\sigma(\check{c}_{\mathfrak{r}(n)\mathfrak{s}(m)})} = \oplus_{j=1}^{m}\gamma_j\left(\oplus_{i=1}^{n}\Omega_i\mathfrak{J}_{\sigma(\check{c}_{\mathfrak{r}(i)\mathfrak{s}(j)})}\right)$$

$\sigma\left(\check{c}_{\mathfrak{r}(i)\mathfrak{s}(j)}\right)$ is the permutation, such as $\mathfrak{J}_{\sigma(\check{c}_{\mathfrak{r}(i)\mathfrak{s}(j)})} \leq \mathfrak{J}_{\sigma(\check{c}_{\mathfrak{r}(i-1)\mathfrak{s}(j)})}$ and $\mathfrak{J}_{\sigma(\check{c}_{\mathfrak{r}(i)\mathfrak{s}(j)})} \leq \mathfrak{J}_{\sigma(\check{c}_{\mathfrak{r}(i)\mathfrak{s}(j-1)})}$ $\forall\, i, j$. Ω_i and γ_j are the weights of experts and sub-attributes, such as $\Omega_i > 0$, $\sum_{i=1}^{n}\Omega_i = 1$, $\gamma_j > 0$, and $\sum_{j=1}^{m}\gamma_j = 1$.

Theorem 2. *Let $\mathfrak{J}_{\check{c}_{ij}} = \left(\mathcal{T}_{\check{c}_{ij}}, \mathcal{J}_{\check{c}_{ij}}\right)$ be a family of IFHSNs. Then, the aggregated value of the IFHSIOWA operator is also an IFHSN.*

$$\begin{gathered} IFHSIOWA\ (\mathfrak{J}_{\check{c}_{11}},\ \mathfrak{J}_{\check{c}_{12}},\ldots,\ \mathfrak{J}_{\check{c}_{nm}}) = \\ \left\langle \begin{gathered} 1-\prod_{j=1}^{m}\left(\prod_{i=1}^{n}\left(1-\mathcal{T}_{\sigma(\check{c}_{\mathfrak{r}(i)\mathfrak{s}(j)})}\right)^{\Omega_i}\right)^{\gamma_j}, \\ \prod_{j=1}^{m}\left(\prod_{i=1}^{n}\left(1-\mathcal{T}_{\sigma(\check{c}_{\mathfrak{r}(i)\mathfrak{s}(j)})}\right)^{\Omega_i}\right)^{\gamma_j} - \prod_{j=1}^{m}\left(\prod_{i=1}^{n}\left(1-\mathcal{T}_{\sigma(\check{c}_{\mathfrak{r}(i)\mathfrak{s}(j)})}-\mathcal{J}_{\sigma(\check{c}_{\mathfrak{r}(i)\mathfrak{s}(j)})}\right)^{\Omega_i}\right)^{\gamma_j} \end{gathered} \right\rangle \end{gathered} \quad (9)$$

where $\mathfrak{J}_{\sigma(\check{c}_{\mathfrak{r}(i)\mathfrak{s}(j)})}$ is the leading element of the ith row and the jth column in $(\mathfrak{J}_{\check{c}_{11}},\ \mathfrak{J}_{\check{c}_{12}},\ldots,\ \mathfrak{J}_{\check{c}_{nm}})$, such as $\mathfrak{J}_{\sigma(\check{c}_{\mathfrak{r}(i)\mathfrak{s}(j)})} \leq \mathfrak{J}_{\sigma(\check{c}_{\mathfrak{r}(i-1)\mathfrak{s}(j)})}$ and $\mathfrak{J}_{\sigma(\check{c}_{\mathfrak{r}(i)\mathfrak{s}(j)})} \leq \mathfrak{J}_{\sigma(\check{c}_{\mathfrak{r}(i)\mathfrak{s}(j-1)})}$ $\forall i, j$. Also, Ω_i and γ_j are the weights for experts and sub-attributes, such as $\Omega_i > 0$, $\sum_{i=1}^{n}\Omega_i = 1$, $\gamma_j > 0$, and $\sum_{j=1}^{m}\gamma_j = 1$.

Proof. Similar to Theorem 1.

Moreover, the IFHSIOWA operator presents notable properties similar to the IFHSIWA operator, such as idempotency, boundedness, homogeneity, monotonicity, and shift-invariance. The following sections contain a comprehensive explanation of these characteristics. □

Proposition 2. *Let $\mathfrak{J}_{\check{c}_{ij}} = \left(\mathcal{T}_{\check{c}_{ij}}, \mathcal{J}_{\check{c}_{ij}}\right)$ be a family of IFHSNs and Ω_i and γ_j be the weight vectors of experts and sub-attributes, respectively, such as $\Omega_i > 0, \sum_{i=1}^{n}\Omega_i = 1$ and $\gamma_j > 0, \sum_{j=1}^{m}\gamma_j = 1, \forall\ i,\ j$.*

3.6. Idempotency

If $\mathfrak{J}_{\check{c}_{ij}} = \mathfrak{J}_{\check{c}_{\delta}} = (\mathcal{T}_{\check{c}_{\delta}}, \mathcal{J}_{\check{c}_{\delta}})$, then

$$IFHSIOWA\ (\mathfrak{J}_{\check{c}_{11}},\ \mathfrak{J}_{\check{c}_{12}},\ldots,\ \mathfrak{J}_{\check{c}_{nm}}) = \mathfrak{J}_{\check{c}_{\delta}}$$

3.7. Boundedness

Let $\mathfrak{J}_{\check{c}_{ij}}^{-} = \left\langle \binom{min}{j}\binom{min}{i}\left\{\mathcal{T}_{\check{c}_{ij}}\right\}, \binom{max}{j}\binom{max}{i}\left\{\mathcal{J}_{\check{c}_{ij}}\right\}\right\rangle$ and $\mathfrak{J}_{\check{c}_{ij}}^{+} = \left\langle \binom{max}{j}\binom{max}{i}\left\{\mathcal{T}_{\check{c}_{ij}}\right\}, \binom{min}{j}\binom{min}{i}\left\{\mathcal{J}_{\check{c}_{ij}}\right\}\right\rangle$. Then,

$$\mathfrak{J}_{\check{c}_{ij}}^{-} \leq IFHSIOWA\left(\mathfrak{J}_{\check{c}_{11}}, \mathfrak{J}_{\check{c}_{12}}, \ldots, \mathfrak{J}_{\check{c}_{nm}}\right) \leq \mathfrak{J}_{\check{c}_{ij}}^{+}$$

3.8. Shift Invariance

Let $\mathfrak{J}_{\check{c}_{ij}} = \left(\mathcal{T}_{\check{c}_{ij}}, \mathcal{J}_{\check{c}_{ij}}\right)$ be a collection of IFHSNs and $\mathfrak{J}_{\check{c}_{\delta}} = \left(\mathcal{T}_{\check{c}_{\delta}}, \mathcal{J}_{\check{c}_{\delta}}\right)$ be an IFHSN. Then,

$$IFHSIOWA\left(\mathfrak{J}_{c_{11}} \oplus \mathfrak{J}_{\check{c}_{\delta}}, \mathfrak{J}_{\check{c}_{12}} \oplus \mathfrak{J}_{\check{c}_{\delta}}, \ldots, \mathfrak{J}_{\check{c}_{nm}} \oplus \mathfrak{J}_{\check{c}_{\delta}}\right) = IFHSIOWA\left(\mathfrak{J}_{\check{c}_{11}}, \mathfrak{J}_{\check{c}_{12}}, \ldots, \mathfrak{J}_{\check{c}_{nm}}\right) \oplus \mathfrak{J}_{\check{c}_{\delta}}$$

3.9. Homogeneity

Let $\mathfrak{J}_{\check{c}_{ij}} = \left(\mathcal{T}_{\check{c}_{ij}}, \mathcal{J}_{\check{c}_{ij}}\right)$ be a family of IFHSNs. Then,

$$IFHSIOWA\left(\alpha\mathfrak{J}_{\check{c}_{11}}, \alpha\mathfrak{J}_{\check{c}_{12}}, \ldots, \alpha\mathfrak{J}_{\check{c}_{nm}}\right) = \alpha IFHSIOWA\left(\mathfrak{J}_{\check{c}_{11}}, \mathfrak{J}_{\check{c}_{12}}, \ldots, \mathfrak{J}_{\check{c}_{nm}}\right)$$

where $\alpha > 0$.

3.10. Monotonicity

Let $\mathfrak{J}_{\check{c}_{ij}} = \left(\mathcal{T}_{\check{c}_{ij}}, \mathcal{J}_{\check{c}_{ij}}\right)$ and $\mathfrak{J}^{*}_{\check{c}_{ij}} = \left(\mathcal{T}^{*}_{\check{c}_{ij}}, \mathcal{J}^{*}_{\check{c}_{ij}}\right)$ be the two distinct families of IFHSNs. Then,
$IFHSIOWA\left(\mathfrak{J}_{\check{c}_{11}}, \mathfrak{J}_{\check{c}_{12}}, \ldots, \mathfrak{J}_{\check{c}_{nm}}\right) \leq IFHSIOWA\left(\mathfrak{J}^{*}_{\check{c}_{11}}, \mathfrak{J}^{*}_{\check{c}_{12}}, \ldots\ldots\ldots, \mathfrak{J}^{*}_{\check{c}_{nm}}\right)$, if $\mathfrak{J}_{\check{c}_{ij}} \leq \mathfrak{J}^{*}_{\check{c}_{ij}}\ \forall i, j$.

4. Interactive Geometric Aggregation Operators for Intuitionistic Fuzzy Hypersoft Sets

This section will present the IFHSIWG and IFHSIOWG operators specifically developed for IFHSNs and their essential characteristics.

Definition 9. *Let* $\mathfrak{J}_{\check{d}_{ij}} = \left(\mathcal{T}_{\check{d}_{ij}}, \mathcal{J}_{\check{d}_{ij}}\right)$ *be an IFHSN;* Ω_i *and* γ_j *represent the weights of experts and multi-sub-attributes along with stated conditions* $\Omega_i > 0$, $\sum_{i=1}^{n}\Omega_i = 1$, $\gamma_j > 0$, *and* $\sum_{j=1}^{m}\gamma_j = 1$. *Then, IFHSIWG,* $\Delta^n \to \Delta$, *is defined as follows:*

$$IFHSIWG\left(\mathfrak{J}_{\check{c}_{11}}, \mathfrak{J}_{\check{c}_{12}}, \ldots, \mathfrak{J}_{\check{c}_{nm}}\right) = \otimes_{j=1}^{m}\left(\otimes_{i=1}^{n}\left(\mathfrak{J}_{\check{c}_{nm}}\right)^{\Omega_i}\right)^{\gamma_j}$$

Theorem 3. *Let* $\mathfrak{J}_{\check{c}_{ij}} = \left(\mathcal{T}_{\check{c}_{ij}}, \mathcal{J}_{\check{c}_{ij}}\right)$ *be a family of IFHSNs. Then, the aggregated value of the IFHSIWG operator is also an IFHSN, and*

$$IFHSIWG\left(\mathfrak{J}_{\check{c}_{11}}, \mathfrak{J}_{\check{c}_{12}}, \ldots, \mathfrak{J}_{\check{c}_{nm}}\right) = \left\langle \begin{array}{c} \prod_{j=1}^{m}\left(\prod_{i=1}^{n}\left(\mathcal{T}_{\check{c}_{ij}}\right)^{\Omega_i}\right)^{\gamma_j}, \\ 1 - \prod_{j=1}^{m}\left(\prod_{i=1}^{n}\left(\mathcal{T}_{\check{c}_{ij}}\right)^{\Omega_i}\right)^{\gamma_j} - \prod_{j=1}^{m}\left(\prod_{i=1}^{n}\left(1 - \mathcal{T}_{\check{c}_{ij}} - \mathcal{J}_{\check{c}_{ij}}\right)^{\Omega_i}\right)^{\gamma_j} \end{array} \right\rangle \tag{10}$$

where Ω_i *and* γ_j *represent the experts' and sub-attribute weights within certain circumstances,* $\Omega_i > 0$, $\sum_{i=1}^{n}\Omega_i = 1$, $\gamma_j > 0$, *and* $\sum_{j=1}^{m}\gamma_j = 1$.

Proof. Similar to Theorem 1. □

Proposition 3. *Let $\mathfrak{J}_{\check{c}_{ij}} = \left(\mathcal{T}_{\check{c}_{ij}}, \mathcal{J}_{\check{c}_{ij}}\right)$ be a family of IFHSNs and Ω_i and γ_j be the weight vectors of experts and sub-attributes, respectively, such as $\Omega_i > 0, \sum_{i=1}^{n} \Omega_i = 1$ and $\gamma_j > 0, \sum_{j=1}^{m} \gamma_j = 1$.*

4.1. Idempotency

If $\mathfrak{J}_{\check{c}_{ij}} = \mathfrak{J}_{\check{c}_\delta} = \left(\mathcal{T}_{\check{c}_\delta}, \mathcal{J}_{\check{c}_\delta}\right)$, then

$$IFHSIWG\left(\mathfrak{J}_{\check{c}_{11}}, \mathfrak{J}_{\check{c}_{12}}, \ldots, \mathfrak{J}_{\check{c}_{nm}}\right) = \mathfrak{J}_{\check{c}_\delta}$$

4.2. Boundedness

Let $\mathfrak{J}_{\check{c}_{ij}}{}^{-} = \left\langle \left(\begin{smallmatrix} min \\ j \end{smallmatrix}\right)\left(\begin{smallmatrix} min \\ i \end{smallmatrix}\right)\left\{\mathcal{T}_{\check{c}_{ij}}\right\}, \left(\begin{smallmatrix} max \\ j \end{smallmatrix}\right)\left(\begin{smallmatrix} max \\ i \end{smallmatrix}\right)\left\{\mathcal{J}_{\check{c}_{ij}}\right\}\right\rangle$ and $\mathfrak{J}_{\check{c}_{ij}}{}^{+} = \left\langle \left(\begin{smallmatrix} max \\ j \end{smallmatrix}\right)\left(\begin{smallmatrix} max \\ i \end{smallmatrix}\right)\left\{\mathcal{T}_{\check{c}_{ij}}\right\}, \left(\begin{smallmatrix} min \\ j \end{smallmatrix}\right)\left(\begin{smallmatrix} min \\ i \end{smallmatrix}\right)\left\{\mathcal{J}_{\check{c}_{ij}}\right\}\right\rangle$. Then,

$$\mathfrak{J}_{\check{c}_{ij}}{}^{-} \leq IFHSIWG\left(\mathfrak{J}_{\check{c}_{11}}, \mathfrak{J}_{\check{c}_{12}}, \ldots, \mathfrak{J}_{\check{c}_{nm}}\right) \leq \mathfrak{J}_{\check{c}_{ij}}{}^{+}$$

4.3. Shift Invariance

Let $\mathfrak{J}_{\check{c}_{ij}} = \left(\mathcal{T}_{\check{c}_{ij}}, \mathcal{J}_{\check{c}_{ij}}\right)$ be a collection of IFHSNs and $\mathfrak{J}_{\check{c}_\delta} = \left(\mathcal{T}_{\check{c}_\delta}, \mathcal{J}_{\check{c}_\delta}\right)$ be an IFHSN. Then,

$$IFHSIWG\left(\mathfrak{J}_{c_{11}} \otimes \mathfrak{J}_{\check{c}_\delta}, \mathfrak{J}_{\check{c}_{12}} \otimes \mathfrak{J}_{\check{c}_\delta}, \ldots, \mathfrak{J}_{\check{c}_{nm}} \otimes \mathfrak{J}_{\check{c}_\delta}\right) = IFHSIWG\left(\mathfrak{J}_{\check{c}_{11}}, \mathfrak{J}_{\check{c}_{12}}, \ldots, \mathfrak{J}_{\check{c}_{nm}}\right) \otimes \mathfrak{J}_{\check{c}_\delta}$$

4.4. Homogeneity

Let $\mathfrak{J}_{\check{c}_{ij}} = \left(\mathcal{T}_{\check{c}_{ij}}, \mathcal{J}_{\check{c}_{ij}}\right)$ be a family of IFHSNs. Then,

$$IFHSIWG\left(\alpha\mathfrak{J}_{\check{c}_{11}}, \alpha\mathfrak{J}_{\check{c}_{12}}, \ldots, \alpha\mathfrak{J}_{\check{c}_{nm}}\right) = \alpha\, IFHSIWG\left(\mathfrak{J}_{\check{c}_{11}}, \mathfrak{J}_{\check{c}_{12}}, \ldots, \mathfrak{J}_{\check{c}_{nm}}\right)$$

where $\alpha > 0$.

4.5. Monotonicity

Let $\mathfrak{J}_{\check{c}_{ij}} = \left(\mathcal{T}_{\check{c}_{ij}}, \mathcal{J}_{\check{c}_{ij}}\right)$ and $\mathfrak{J}^*_{\check{c}_{ij}} = \left(\mathcal{T}^*_{\check{c}_{ij}}, \mathcal{J}^*_{\check{c}_{ij}}\right)$ be the two distinct families of IFHSNs. Then,

$$IFHSIWG\left(\mathfrak{J}_{\check{c}_{11}}, \mathfrak{J}_{\check{c}_{12}}, \ldots, \mathfrak{J}_{\check{c}_{nm}}\right) \leq IFHSIWG\left(\mathfrak{J}^*_{\check{c}_{11}}, \mathfrak{J}^*_{\check{c}_{12}}, \ldots\ldots\ldots, \mathfrak{J}^*_{\check{c}_{nm}}\right), \text{ if } \mathfrak{J}_{\check{c}_{ij}} \leq \mathfrak{J}^*_{\check{c}_{ij}}\ \forall i, j.$$

Definition 10. *Let $\mathfrak{J}_{\check{c}_{ij}} = \left(\mathcal{T}_{\check{c}_{ij}}, \mathcal{J}_{\check{c}_{ij}}\right)$ be a family of IFHSNs. Then, the IFHSIOWG operator is defined as*

$$IFHSIOWG\left(\mathfrak{J}_{\check{c}_{11}}, \mathfrak{J}_{\check{c}_{12}}, \ldots, \mathfrak{J}_{\check{c}_{nm}}\right) = \gamma_1\Omega_1\mathfrak{J}_{\sigma(\check{c}_{\mathfrak{r}(1)\mathfrak{s}(1)})} \bigotimes \gamma_1\Omega_2\mathfrak{J}_{\sigma(\check{c}_{\mathfrak{r}(2)\mathfrak{s}(1)})} \bigotimes \ldots\ldots \bigotimes \gamma_m\Omega_n\mathfrak{J}_{\sigma(\check{c}_{\mathfrak{r}(n)\mathfrak{s}(m)})} = \bigotimes_{j=1}^{m}\left(\left(\bigotimes_{i=1}^{n}\mathfrak{J}_{\sigma(\check{c}_{\mathfrak{r}(i)\mathfrak{s}(j)})}\right)^{\Omega_i}\right)^{\gamma_j}$$

$\sigma\left(\check{c}_{\mathfrak{r}(i)\mathfrak{s}(j)}\right)$ is the permutation, such as $\mathfrak{J}_{\sigma(\check{c}_{\mathfrak{r}(i)\mathfrak{s}(j)})} \leq \mathfrak{J}_{\sigma(\check{c}_{\mathfrak{r}(i-1)\mathfrak{s}(j)})}$ and $\mathfrak{J}_{\sigma(\check{c}_{\mathfrak{r}(i)\mathfrak{s}(j)})} \leq \mathfrak{J}_{\sigma(\check{c}_{\mathfrak{r}(i)\mathfrak{s}(j-1)})}\ \forall\, i, j$. Also, Ω_i and γ_j represent the weight vectors for experts and sub-attributes, such as $\Omega_i > 0$, $\sum_{i=1}^{n} \Omega_i = 1$, $\gamma_j > 0$, and $\sum_{j=1}^{m} \gamma_j = 1$.

Theorem 4. *Let $\mathfrak{J}_{\check{c}_{ij}} = \left(\mathcal{T}_{\check{c}_{ij}}, \mathcal{J}_{\check{c}_{ij}}\right)$ be a family of IFHSNs. Then, the aggregated value of the IFHSIOWG operator is also an IFHSN, and*

$$IFHSIOWG\left(\mathfrak{J}_{\check{c}_{11}}, \mathfrak{J}_{\check{c}_{12}}, \ldots, \mathfrak{J}_{\check{c}_{nm}}\right) = \left\langle \begin{array}{c} \prod_{j=1}^{m}\left(\prod_{i=1}^{n}\left(\mathcal{T}_{\sigma(\check{c}_{\mathfrak{r}(i)\mathfrak{s}(j)})}\right)^{\Omega_i}\right)^{\gamma_j}, \\ 1 - \prod_{j=1}^{m}\left(\prod_{i=1}^{n}\left(\mathcal{T}_{\sigma(\check{c}_{\mathfrak{r}(i)\mathfrak{s}(j)})}\right)^{\Omega_i}\right)^{\gamma_j} - \prod_{j=1}^{m}\left(\prod_{i=1}^{n}\left(1 - \mathcal{T}_{\sigma(\check{c}_{\mathfrak{r}(i)\mathfrak{s}(j)})} - \mathcal{J}_{\sigma(\check{c}_{\mathfrak{r}(i)\mathfrak{s}(j)})}\right)^{\Omega_i}\right)^{\gamma_j} \end{array} \right\rangle \quad (11)$$

where $\mathfrak{J}_{\sigma(\check{c}_{\mathfrak{r}(i)\mathfrak{s}(j)})}$ *is the largest element of the* i^{th} *row and the* j^{th} *column in* $(\mathfrak{J}_{\check{c}_{11}}, \mathfrak{J}_{\check{c}_{12}}, \ldots, \mathfrak{J}_{\check{c}_{nm}})$*, such as* $\mathfrak{J}_{\sigma(\check{c}_{\mathfrak{r}(i)\mathfrak{s}(j)})} \leq \mathfrak{J}_{\sigma(\check{c}_{\mathfrak{r}(i-1)\mathfrak{s}(j)})}$ *and* $\mathfrak{J}_{\sigma(\check{c}_{\mathfrak{r}(i)\mathfrak{s}(j)})} \leq \mathfrak{J}_{\sigma(\check{c}_{\mathfrak{r}(i)\mathfrak{s}(j-1)})}$ $\forall i, j$*. Also,* Ω_i *and* γ_j *are the weight vectors for experts and sub-attributes, such as* $\Omega_i > 0$*,* $\sum_{i=1}^{n} \Omega_i = 1$*,* $\gamma_j > 0$*, and* $\sum_{j=1}^{m} \gamma_j = 1$*.*

Proof. Similar to Theorem 1.

Moreover, the IFHSIOWG operator presents notable properties similar to the IFHSIWG operator, such as idempotency, boundedness, homogeneity, monotonicity, and shift-invariance. The following sections contain a comprehensive explanation of these characteristics. □

Proposition 4. *Let* $\mathfrak{J}_{\check{c}_{ij}} = \left(\mathcal{T}_{\check{c}_{ij}}, \mathcal{J}_{\check{c}_{ij}}\right)$ *be a family of IFHSNs and let* Ω_i*,* γ_j *be the weights of experts and sub-attributes, respectively, such as* $\Omega_i > 0, \sum_{i=1}^{n} \Omega_i = 1$ *and* $\gamma_j > 0, \sum_{j=1}^{m} \gamma_j = 1$*.*

4.6. Idempotency

If $\mathfrak{J}_{\check{c}_{ij}} = \mathfrak{J}_{\check{c}_{\delta}} = (\mathcal{T}_{\check{c}_{\delta}}, \mathcal{J}_{\check{c}_{\delta}})$, then

$$IFHSIOWG\,(\mathfrak{J}_{\check{c}_{11}}, \mathfrak{J}_{\check{c}_{12}}, \ldots, \mathfrak{J}_{\check{c}_{nm}}) = \mathfrak{J}_{\check{c}_{\delta}}$$

4.7. Boundedness

Let $\mathfrak{J}_{\check{c}_{ij}}{}^{-} = \left\langle \binom{min}{j}\binom{min}{i}\left\{\mathcal{T}_{\check{c}_{ij}}\right\}, \binom{max}{j}\binom{max}{i}\left\{\mathcal{J}_{\check{c}_{ij}}\right\}\right\rangle$ and $\mathfrak{J}_{\check{c}_{ij}}{}^{+} = \left\langle \binom{max}{j}\binom{max}{i}\left\{\mathcal{T}_{\check{c}_{ij}}\right\}, \binom{min}{j}\binom{min}{i}\left\{\mathcal{J}_{\check{c}_{ij}}\right\}\right\rangle$. Then,

$$\mathfrak{J}_{\check{c}_{ij}}{}^{-} \leq IFHSIOWG\,(\mathfrak{J}_{\check{c}_{11}}, \mathfrak{J}_{\check{c}_{12}}, \ldots, \mathfrak{J}_{\check{c}_{nm}}) \leq \mathfrak{J}_{\check{c}_{ij}}{}^{+}$$

4.8. Shift Invariance

Let $\mathfrak{J}_{\check{c}_{ij}} = \left(\mathcal{T}_{\check{c}_{ij}}, \mathcal{J}_{\check{c}_{ij}}\right)$ be a collection of IFHSNs and $\mathfrak{J}_{\check{c}_{\delta}} = (\mathcal{T}_{\check{c}_{\delta}}, \mathcal{J}_{\check{c}_{\delta}})$ be an IFHSN. Then,

$$IFHSIOWG\,(\mathfrak{J}_{c_{11}} \otimes \mathfrak{J}_{\check{c}_{\delta}}, \mathfrak{J}_{\check{c}_{12}} \otimes \mathfrak{J}_{\check{c}_{\delta}}, \ldots, \mathfrak{J}_{\check{c}_{nm}} \otimes \mathfrak{J}_{\check{c}_{\delta}}) = IFHSIOWG\,(\mathfrak{J}_{\check{c}_{11}}, \mathfrak{J}_{\check{c}_{12}}, \ldots, \mathfrak{J}_{\check{c}_{nm}}) \otimes \mathfrak{J}_{\check{c}_{\delta}}$$

4.9. Homogeneity

Let $\mathfrak{J}_{\check{c}_{ij}} = \left(\mathcal{T}_{\check{c}_{ij}}, \mathcal{J}_{\check{c}_{ij}}\right)$ be a family of IFHSNs. Then,

$$IFHSIOWG\,(\alpha\mathfrak{J}_{\check{c}_{11}}, \alpha\mathfrak{J}_{\check{c}_{12}}, \ldots, \alpha\mathfrak{J}_{\check{c}_{nm}}) = \alpha\; IFHSIOWG\,(\mathfrak{J}_{\check{c}_{11}}, \mathfrak{J}_{\check{c}_{12}}, \ldots, \mathfrak{J}_{\check{c}_{nm}})$$

where $\alpha > 0$.

4.10. Monotonicity

Let $\mathfrak{J}_{\check{c}_{ij}} = \left(\mathcal{T}_{\check{c}_{ij}}, \mathcal{J}_{\check{c}_{ij}}\right)$ and $\mathfrak{J}^{*}_{\check{c}_{ij}} = \left(\mathcal{T}^{*}_{\check{c}_{ij}}, \mathcal{J}^{*}_{\check{c}_{ij}}\right)$ be the two distinct families of IFHSNs. Then,

$IFHSIOWG(\mathfrak{J}_{\check{c}_{11}}, \mathfrak{J}_{\check{c}_{12}}, \ldots, \mathfrak{J}_{\check{c}_{nm}}) \leq IFHSIOWG\left(\mathfrak{J}^{*}_{\check{c}_{11}}, \mathfrak{J}^{*}_{\check{c}_{12}}, \ldots\ldots\ldots, \mathfrak{J}^{*}_{\check{c}_{nm}}\right)$, if $\mathfrak{J}_{\check{c}_{ij}} \leq \mathfrak{J}^{*}_{\check{c}_{ij}}$ $\forall i, j$.

5. EDAS Technique Based on Developed Interaction Aggregation Operators

Experts from different areas of expertise, each with different viewpoints and opinions, may have conflicting opinions on the most productive approaches to employ. Cooperative methods for decision making are beneficial in such circumstances. Experts frequently implement intuitionistic fuzzy hypersoft values to indicate complexity and unpredictability when stating opinions. This research describes the "intuitionistic fuzzy hypersoft interaction-based EDAS method" as an approach to deal with the obstacles faced in MAGDM problems.

The main objective of this technique is to build an effective basis for managing DM situations characterized by uncertainty and ambiguity. The impact of the EDAS technique is reliant upon two particular evaluations, the PDA and the NDA, which are effective at evaluating discrepancies between various alternatives. The EDAS approach reduces the need to systematically assess ideal and non-ideal solutions when applying MAGDM challenges. In this research, decision experts utilize IFHSNs to evaluate the combined values of the alternatives by assigning preference values. The following section describes the computational framework of the stated methodology.

Let $\mathcal{Q} = \{\mathcal{Q}^1, \mathcal{Q}^2, \mathcal{Q}^3, \ldots, \mathcal{Q}^s\}$ be a set of s alternatives and let $\mathcal{X} = \{\mathcal{X}_1, \mathcal{X}_2, \mathcal{X}_3, \ldots, \mathcal{X}_n\}$ be a group of n experts. The weights of experts are given as $\Omega = (\Omega_1, \Omega_1, \ldots, \Omega_n)^T$ and $\Omega_i > 0$, $\sum_{i=1}^{n} \Omega_i = 1$. Let $\mathfrak{L} = \{c_1, c_2, \ldots, c_m\}$ and $\mathfrak{L}' = \{(c_{1\rho} \times c_{2\rho} \times \ldots \times c_{m\rho})$ for all $\rho \in \{1, 2, \ldots, t\}\}$ represent the set of attributes with their corresponding multi-sub-attributes with weights $\gamma = \left(\gamma_{1\rho}, \gamma_{2\rho}, \gamma_{3\rho}, \ldots, \gamma_{m\rho}\right)^T$, such as $\gamma_\rho > 0$, $\sum_{\rho=1}^{t} \gamma_\rho = 1$, and they can be indicated as $\mathfrak{L}' = \{\check{c}_\partial : \partial \in \{1, 2, \ldots, m\}\}$. The group of experts $\{\mathcal{X}^i : i = 1, 2, \ldots, n\}$ assesses the alternatives $\left\{\mathcal{Q}^{(z)} : z = 1, 2, \ldots, s\right\}$ under the chosen sub-attributes $\{\check{c}_\partial : \partial = 1, 2, \ldots, k\}$ in the form of IFHSNs, such as $\left(\Xi^{(z)}_{\check{c}_{ij}}\right)_{n \times m} = \left(\mathcal{T}^{(z)}_{\check{c}_{ij}}, \mathcal{J}^{(z)}_{\check{c}_{ij}}\right)_{n \times m}$, where $0 \leq \mathcal{T}^{(z)}_{\check{c}_{ij}}, \mathcal{J}^{(z)}_{\check{c}_{ij}} \leq 1$ and $0 \leq \mathcal{T}^{(z)}_{\check{c}_{ij}} + \mathcal{J}^{(z)}_{\check{c}_{ij}} \leq 1 \,\forall\, i, j$.

Therefore, to put into practice such an effective approach, the implementation of the subsequent steps must be undertaken.

Step 1: Experts evaluate matrices for alternatives $\left\{\mathcal{Q}^{(z)} : z = 1, 2, \ldots, s\right\}$ in terms of IFHSNs under their considered sub-attributes.

$$\left[\Xi^{(z)}_{\check{c}_{ij}}\right]_{n \times m} = \begin{matrix} \mathcal{X}_1 \\ \mathcal{X}_2 \\ \vdots \\ \mathcal{X}_n \end{matrix} \begin{pmatrix} \left(\mathcal{T}^{(z)}_{\check{c}_{11}}, \mathcal{J}^{(z)}_{\check{c}_{11}}\right) & \left(\mathcal{T}^{(z)}_{\check{c}_{12}}, \mathcal{J}^{(z)}_{\check{c}_{12}}\right) & \cdots & \left(\mathcal{T}^{(z)}_{\check{c}_{1\partial}}, \mathcal{J}^{(z)}_{\check{c}_{1\partial}}\right) \\ \left(\mathcal{T}^{(z)}_{\check{c}_{21}}, \mathcal{J}^{(z)}_{\check{c}_{21}}\right) & \left(\mathcal{T}^{(z)}_{\check{c}_{22}}, \mathcal{J}^{(z)}_{\check{c}_{22}}\right) & \cdots & \left(\mathcal{T}^{(z)}_{\check{c}_{2\partial}}, \mathcal{J}^{(z)}_{\check{c}_{2\partial}}\right) \\ \vdots & \vdots & \vdots & \vdots \\ \left(\mathcal{T}^{(z)}_{\check{c}_{n1}}, \mathcal{J}^{(z)}_{\check{c}_{n1}}\right) & \left(\mathcal{T}^{(z)}_{\check{c}_{n2}}, \mathcal{J}^{(z)}_{\check{c}_{n2}}\right) & \cdots & \left(\mathcal{T}^{(z)}_{\check{c}_{n\partial}}, \mathcal{J}^{(z)}_{\check{c}_{n\partial}}\right) \end{pmatrix}$$

Step 2: Therefore, the matrix $\left[\Xi^{(z)}_{\check{c}_{ij}}\right]_{n \times \mathrm{m}}$ undergoes consideration by categorizing parameters into two different categories: those involving costs and those about benefits. Normalization is unnecessary when the parameters are identical. However, if the decisions have distinct kinds of attributes, it is necessary to use the normalization procedure to normalize them. The procedures of normalization are described in the following way:

$$\langle_{ij} = \begin{cases} \left(\Xi^{(z)}_{\check{c}_{ij}}\right)^c = \left(\mathcal{J}^{(z)}_{\check{c}_{ij}}, \mathcal{T}^{(z)}_{\check{c}_{ij}}\right); \text{ cost type parameter} \\ \Xi^{(z)}_{\check{c}_{ij}} = \left(\mathcal{T}^{(z)}_{\check{c}_{ij}}, \mathcal{J}^{(z)}_{\check{c}_{ij}}\right); \text{ benefit type parameter} \end{cases} \tag{12}$$

Step 3: Evaluate the ordered matrices for each alternative if using IFHSIOWA or IFHSIOWG operators. Otherwise, there is no need to find ordered matrices, and we can move forward to step 4 directly without finding ordered matrices.

Step 4: Equations (3), (9), (10), or (11) will determine aggregated decision matrices of the expert's assessment. This is carried out while considering the weight of all of the experts.

$$\left[\Xi^{(z)}_{\check{c}_{ij}}\right]_{n \times m} = \left(\mathcal{T}^{(z)}_{\check{c}_{ij}}, \mathcal{J}^{(z)}_{\check{c}_{ij}}\right)_{n \times m}, i = 1, 2, \ldots n, j = 1, 2, \ldots m$$

Step 5: Find the average solution matrix using the following equation.

$$AVS_h = [AVS_h]_{1\times m} = \frac{1}{n} \overset{n}{\underset{i=1}{\oplus}} \left(\Xi_{\check{c}_{ij}}^{(z)}\right) \tag{13}$$

Step 6: Evaluate the PDA using Equation (14) and the NDA using Equation (16). In some situations, it is not easy to justify which alternative has a higher appraisal value. So, in such scenarios, we cannot decide which alternative is most appropriate. We will use Equation (15) to find the PDA and Equation for this. We will use (17) to find the NDA. Moreover, D_b and D_c represent the beneficial and non-beneficial criteria, respectively.

$PDA = \left[\rho_{\check{c}_{ij}}\right]_{n\times m}$, $NDA = \left[\partial_{\check{c}_{ij}}\right]_{n\times m}$, where

$$\rho_{\check{c}_{ij}} = \begin{cases} \frac{max\left(0,\left(\Xi_{\check{c}_{ij}}-AVS_h\right)\right)}{S(AVS_h)} & for\ i = 1(1)n,\ j = 1(1)m\ and\ c_{mp} \in D_b \\ \frac{max\left(0,\left(AVS_h-\Xi_{\check{c}_{ij}}\right)\right)}{S(AVS_h)}, & for\ i = 1(1)n,\ j = 1(1)m\ and\ c_{mp} \in D_c \end{cases} \tag{14}$$

or

$$\rho_{\check{c}_{ij}} = \begin{cases} \frac{max\left(0,\left(\Xi_{\check{c}_{ij}}-AVS_h\right)\right)}{H(AVS_h)} & for\ i = 1(1)n,\ j = 1(1)m\ and\ c_{mp} \in D_b \\ \frac{max\left(0,\left(AVS_h-\Xi_{\check{c}_{ij}}\right)\right)}{H(AVS_h)}, & for\ i = 1(1)n,\ j = 1(1)m\ and\ c_{mp} \in D_c \end{cases} \tag{15}$$

and

$$\partial_{\check{c}_{ij}} = \begin{cases} \frac{max\left(0,\left(AVS_h-\Xi_{\check{c}_{ij}}\right)\right)}{S(AVS_h)} & for\ i = 1(1)n,\ j = 1(1)m\ and\ c_{mp} \in D_b \\ \frac{max\left(0,\left(\Xi_{\check{c}_{ij}}-AVS_h\right)\right)}{S(AVS_h)} & for\ i = 1(1)n,\ j = 1(1)m\ and\ c_{mp} \in D_c \end{cases} \tag{16}$$

or

$$\partial_{\check{c}_{ij}} = \begin{cases} \frac{max\left(0,\left(AVS_h-\Xi_{\check{c}_{ij}}\right)\right)}{H(AVS_h)} & for\ i = 1(1)n,\ j = 1(1)m\ and\ c_{mp} \in D_b \\ \frac{max\left(0,\left(\Xi_{\check{c}_{ij}}-AVS_h\right)\right)}{H(AVS_h)} & for\ i = 1(1)n,\ j = 1(1)m\ and\ c_{mp} \in D_c \end{cases} \tag{17}$$

Step 7: For both the IFHSIWA and IFHSIOWA operators, compute the positive weighted distance P_τ and the negative weighted distance Q_τ, which are given in Equations (18) and (19).

$$P_\tau = \overset{m}{\underset{j=1}{\oplus}} \gamma_j \left(\overset{n}{\underset{i=1}{\oplus}} \Omega_i \rho_{\check{c}_{ij}} \right) \tag{18}$$

$$Q_\tau = \overset{m}{\underset{j=1}{\oplus}} \gamma_j \left(\overset{n}{\underset{i=1}{\oplus}} \Omega_i \partial_{\check{c}_{ij}} \right) \tag{19}$$

Or, use IFHSIWG and IFHSIOWG operators given in Equations (20) and (21):

$$P_\tau = \bigotimes_{j=1}^{m} \gamma_j \left(\bigotimes_{i=1}^{n} \Omega_i \rho_{\check{c}_{ij}} \right) \tag{20}$$

$$Q_\tau = \bigotimes_{j=1}^{m} \gamma_j \left(\bigotimes_{i=1}^{n} \Omega_i \partial_{\check{c}_{ij}} \right) \tag{21}$$

where $\gamma_j > 0, \Omega_i > 0$, such as $\sum_{j=1}^{m} \gamma_j = 1$ and $\sum_{i=1}^{n} \Omega_i > 0$.

Step 8: Normalize the values of P_τ and Q_τ, $\tau = 1, 2, \ldots\ldots m$.

$$NP_\tau = \frac{P_\tau}{max(P_1, P_2, \ldots\ldots P_m)} \tag{22}$$

$$NQ_{\tau} = 1 - \frac{Q_{\tau}}{max(Q_1, Q_2, \ldots\ldots Q_m)} \tag{23}$$

where $max(P_1, P_2, \ldots\ldots P_m)$ and $max(Q_1, Q_2, \ldots\ldots Q_m)$ are the maximum distances.

Step 9: Derive the integrative appraisal score I

$$I = \frac{1}{2}(NP_{\tau} \oplus NQ_{\tau}) \tag{24}$$

where $I \in [0, 1]$.

Step 10: After determining their relative values, each alternative will be placed in descending order of importance. The option with the best potential score is perceived to be most efficient, while the other option with the lowest total rating is perceived as least productive. Figure 2 presents a visual representation of the planned method.

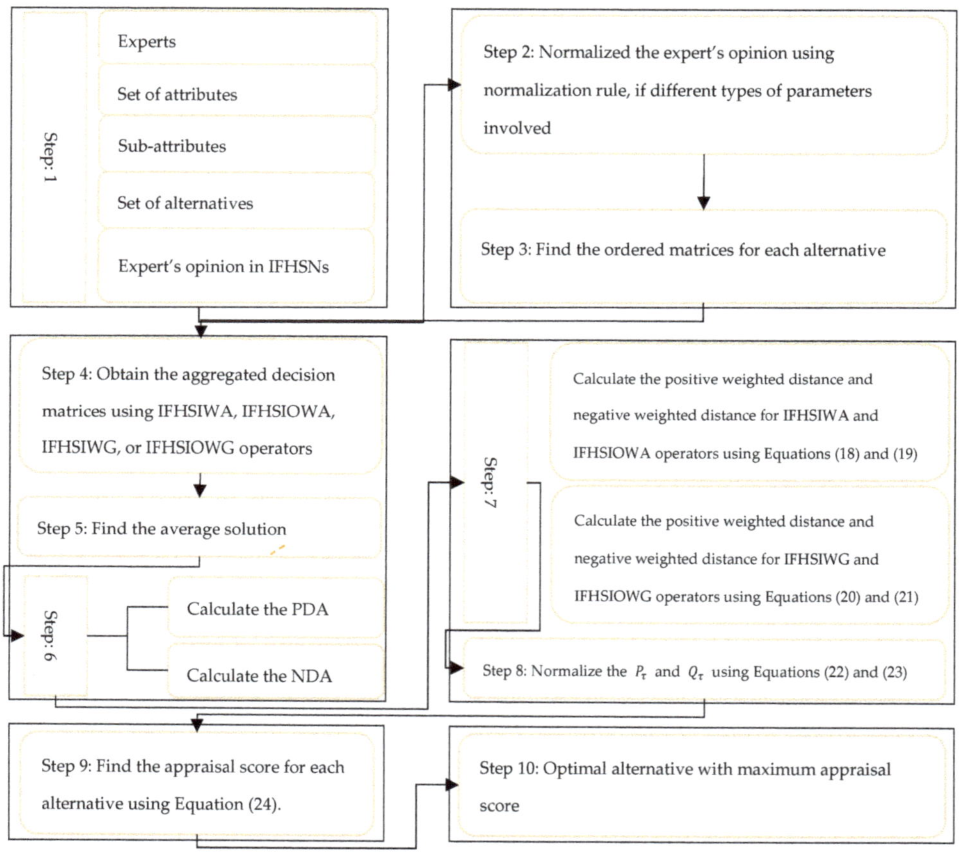

Figure 2. Flow chart of the proposed EDAS model.

6. Application of the Proposed EDAS Method in MSW Management

The following section explores the practical significance of this model in assessing the optimal thermochemical waste disposal techniques for the disposal of MSW.

6.1. Case Study of MSW Management

The removal and disposal of MSW constitute an extensive ecological problem. Various techniques, including thermochemical processes, are used to manage and alleviate the impact of waste. Thermochemical treatments utilize heat to influence the waste products' chemical structure. Several thermochemical techniques for disposal of MSW are outlined in the following order.

6.1.1. Incineration

One standard method for handling MSW is thermochemical incineration, which involves burning waste products at a high temperature with extra air present. Because inorganic substances are generally processed into ash through this procedure, organic trash ingredients oxidize into water and carbon dioxide. The eradication of potentially hazardous substances and the decrease in waste size are two advantages of the extreme temperatures obtained during combustion. Effective air quality management techniques, like scrubbers and electrostatic precipitation devices, must be developed to solve problems. After burning, hazardous ash may be left behind that needs to be managed and disposed of appropriately to prevent damage to surroundings. Also, how society perceives and absorbs a factor can lead to obstacles, as concerns regarding the purity of air and prospective safety risks can impact the authorization and execution of incineration sites.

6.1.2. Plasma Gasification

Plasma gasification is a complicated thermochemical mechanism that handles MSW. It uses elevated temperatures of ionized gases (plasma) to transform waste products into syngas, which are metallic substances and a glass-like remnant that is sterilized. In this method, waste is heated to high temperatures through an electromagnetic arc within a radiant flame, disintegrating complicated inorganic and organic substances. This method presents another benefit of reducing the quantity of trash and minimizing the harmful environmental effects attributed to conventional disposal methods. High initial expenditures, technological intricacy, and the requirement of complicated operation are some of the limitations that plasma gasification must overcome, considering its appealing characteristics. The project's sustainability is significantly impacted by financial viability because of the massive capital expenses required to build plasma gasification plants. This technology also requires operators and servicing experts with the necessary expertise.

6.1.3. Thermal Depolymerization

Thermal depolymerization is a novel technique employed for disposing of MSW, providing an environmentally friendly method for managing various types of garbage. Throughout this process, organic substances are exposed to severe pressures and temperatures in an oxygen-free environment, decomposing into less complex particles. The considerable energy expenditure needed to achieve and sustain the desired temperatures and pressures shows a financial obstacle, necessitating effective energy recovery techniques. The successful implementation of thermochemical thermal depolymerization in municipal solid waste management strategies requires striking a balance between the advantages of resource rehabilitation and the difficulties of consumption of energy and material fluctuations.

6.1.4. Pyrolysis

Gases, liquids, and char are byproducts of the thermochemical process known as pyrolysis used to dispose of MSW. This process occurs in an oxygen-free environment. The thermal decomposition process is carried out by treating waste products at high temperatures ranging from 300 °C to 800 °C. Biofuels, heat, and important pharmaceuticals are just a few possible applications for the byproducts. Strong administration is required to prevent reaction degradation and enhance energy extraction performance. Maintaining an integrated approach to technological developments and economic feasibility and managing possible adverse ecological effects are necessary for the effective execution of pyrolysis as a thermochemical technique for dealing with MSW.

6.1.5. Thermal Gasification

One viable method for handling waste is thermal gasification, demonstrating the potential for adequately handling MSW. Carbon monoxide and hydrogen, two of the main substances found in syngas, are usable to provide heat and power or to generate biofuels and chemicals. Thermal gasification's advantages are its ability to adapt to various municipal waste streams, its low negative environmental impact, and its great energy efficiency. Maintaining stability among scientific progress, financial stability, and environmental sustainability is crucial for effectively implementing thermal gasification as a thermochemical process for processing MSW. The flow chart of the thermochemical MSW disposal techniques and their benefits and limitations is presented in Figure 3.

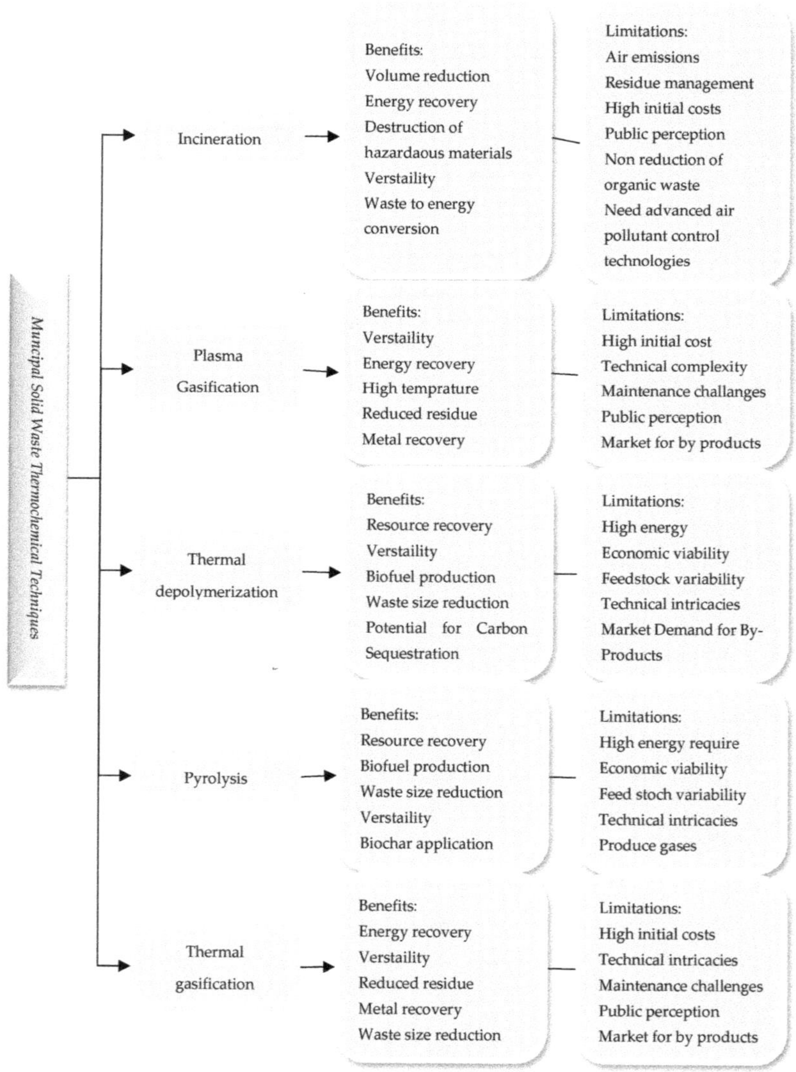

Figure 3. Benefits and limitations of the thermochemical MSW disposal techniques.

6.2. Criteria Description

A detailed investigation has been carried out to assess thermochemical disposal techniques in MSW management, considering a variety of attributes with their corresponding sub-attributes. Proper decision making includes carefully evaluating several parameters; the complicated framework of the evaluation procedure reflects the multi-faceted characteristic of MSW administration.

6.2.1. Energy Recovery (ℓ_1)

Energy recovery plays a significant role in analyzing the effectiveness of thermochemical techniques for handling MSW. The main objective is to efficiently transform the energy generated by solid waste into consumable kinds of power, including heat, power, or synthetic gasoline. Also, thermochemical processes' potential to reduce costs, cut emissions, and maintain ecological sustainability is essential for assessing how well they integrate power requirements with waste disposal. Economical and functional reliability, two of the most important sub-criteria in this assessment, are presented as follows.

Efficiency (c_{11}): The financial stability and environmental viability of waste-to-energy processes depend on the efficient use of the optimum proportion of energy and the smallest quantity of waste. Thermochemical method efficiency determines the feasibility of transforming waste material into useful products like heat, electricity, or synthetic fuels.

Integration Compatibility (c_{12}): The subsequent sub-criterion admits the significance of the kind of power generated and its compatibility with the prevailing energy systems. It realizes that distinct thermochemical methods may generate electricity, heat, or synthetic fuels.

6.2.2. Environmental Impact: (ℓ_2)

The environmental consequences and specifications for treating MSW using thermochemical techniques are necessary for the continued viability of waste disposal activities. It requires an accurate evaluation of potential ecological impacts and the execution of reliable procedures for dumping or recycling. Integrating measures that decrease the release of carbon dioxide into the atmosphere and adhere to existing environmental rules is essential for adopting an environmentally friendly approach to handling MSW using thermochemical techniques.

Control and reduction of emissions (c_{21}): The control and reduction of emissions are the top objectives when evaluating the environmental consequences of MSW waste disposal, particularly for thermochemical techniques. These techniques may culminate in the emission of pollutants, such as nitrogen oxides, sulfur dioxide, heavy metals, and dioxins.

6.2.3. Economic Viability (ℓ_3)

Financial viability factors are important in determining the practicality of disposing of MSW using thermochemical techniques. The competitiveness of thermochemical plants relies on effectively controlling expenses related to installing technology compared to the advantages obtained from energy recovery or the use of byproducts.

Revenue generation (c_{31}): Revenue-generating prospects, such as the sale of electricity or recovered materials, additionally enhance the fiscal viability of the plants. This demonstrates the necessity of ongoing technological innovation and optimization to reduce expenditures and improve the financial benefits of recycling waste into energy projects.

6.2.4. Resource Efficiency (ℓ_4)

The most critical consideration in assessing the treatment of MSW through thermochemical techniques is the sustainable efficiency of resources. This requires an integrated strategy to maximize energy, water, and auxiliary materials' consumption. Two crucial sub-criteria in this context are maintaining a balance in energy use and conserving water supply.

Energy Balance (c_{41})*:* An important component in thermochemical operations is to preserve a favorable balance of energy, where the energy generated by waste exceeds the energy required to execute thermochemical processes.

Water Conservation (c_{42})*:* An essential aspect of thermochemical operation is properly handling water by preventing water use while maximizing productivity. This occurs through the use of technological advances and procedures that minimize the use of water and endorse reuse and recycling.

6.2.5. Regulatory Compliance (ℓ_5)

Evaluating the implementation of thermochemical approaches for handling MSW requires a thorough review of compliance with regulators. It incorporates detailed environmental, medical care, and protection rules. Within this framework, two essential sub-criteria are emissions control and residue management.

Emissions Control (c_{51})*:* An essential component of regulatory compliance in thermochemical waste treatment is the effective administration and reduction of pollutants to properly locate and eliminate particulates.

Residue Management (c_{52})*:* Another critical component in thermochemical waste disposal is properly handling byproducts, such as ash or slag, and maintaining a commitment to waste management standards to prevent soil and water pollution.

6.3. Numerical Example

A numerical example presents the significance and usability of the established model under IFHSS, and the developed approach was utilized to select an effective thermochemical waste disposal technique.

The team of five experts $\mathcal{X} = \{\mathcal{X}_1, \mathcal{X}_2, \mathcal{X}_3, \mathcal{X}_4, \mathcal{X}_5\}$ had weights $(0.15, 0.25, 0.20, 0.22, 0.18)^T$ to evaluate the ranking of the thermochemical waste disposal techniques $\mathcal{Q} = \{\mathcal{Q}^1 = Incineration, \mathcal{Q}^2 = Plasma\ gasification, \mathcal{Q}^3 = Thermal\ depolymerization, \mathcal{Q}^4 = Pyrolysis, \mathcal{Q}^5 = Thermal\ gasification\}$. Professionals adopt the standards for the assortment of thermochemical waste disposal techniques, such as $\mathfrak{L} = \{\ell_1 = \text{Energy Recovery}, \ell_2 = \text{Environmental Impact}, \ell_3 = \text{Economic Viability}, \ell_4 = \text{Resource Efficiency}, \ell_5 = \text{Regulatory Compliance}\}$, which represents the attributes of the alternatives, and their corresponding sub-attribute is given as $\ell_1 = \text{Energy Recovery} = \{c_{11} = \text{efficiency}, c_{12} = \text{Integration compatibility}\}$, $\ell_2 = \text{Environmental Impact} = \{c_{21} = \text{control and reduction of emissions}\}$, $\ell_3 = \text{Economic Viability} = \{c_{31} = \text{Revenue generation}\}$, $\ell_4 = \text{Resource Efficiency} = \{c_{41} = \text{Energy balance}, c_{42} = \text{Water Conservation}\}$, and $\ell_5 = \text{Regulatory Compliance} = \{c_{51} = \text{Emissions Control}, c_{52} = \text{Residue Management}\}$. Let $\mathfrak{L}' = \ell_1 \times \ell_2 \times \ell_3 \times \ell_4 \times \ell_5$ be a set of sub-attributes $\mathfrak{L}' = \ell_1 \times \ell_2 \times \ell_3 \times \ell_4 \times \ell_5 = \{c_{11}, c_{12}\} \times \{c_{21}\} \times \{c_{31}\} \times \{c_{41}, c_{42}\} \times \{c_{51}, c_{52}\} = \left\{\begin{matrix} (c_{11}, c_{21}, c_{31}, c_{41}, c_{51}), (c_{11}, c_{21}, c_{31}, c_{41}, c_{52}), (c_{11}, c_{21}, c_{31}, c_{42}, c_{51}), (c_{11}, c_{21}, c_{31}, c_{42}, c_{52}), \\ (c_{12}, c_{21}, c_{31}, c_{41}, c_{51}), (c_{12}, c_{21}, c_{31}, c_{41}, c_{52}), (c_{12}, c_{21}, c_{31}, c_{42}, c_{51}), (c_{12}, c_{21}, c_{31}, c_{42}, c_{52}) \end{matrix}\right\}$ $\mathfrak{L}' = \{\check{c}_1, \check{c}_2, \check{c}_3, \check{c}_4, \check{c}_5, \check{c}_6, \check{c}_7, \check{c}_8\}$ is a set of all multi-sub-attributes with weights $(0.12, 0.18, 0.1, 0.15, 0.05, 0.22, 0.08, 0.1)^T$. Specialists have access to alternatives in the form of IFHSNs.

The stakeholders will prioritize among the five alternatives based on their preferences in IFHSNs after analyzing each based on these five criteria and their related sub-attributes. The procedure is shown in the following diagram (see Figure 4).

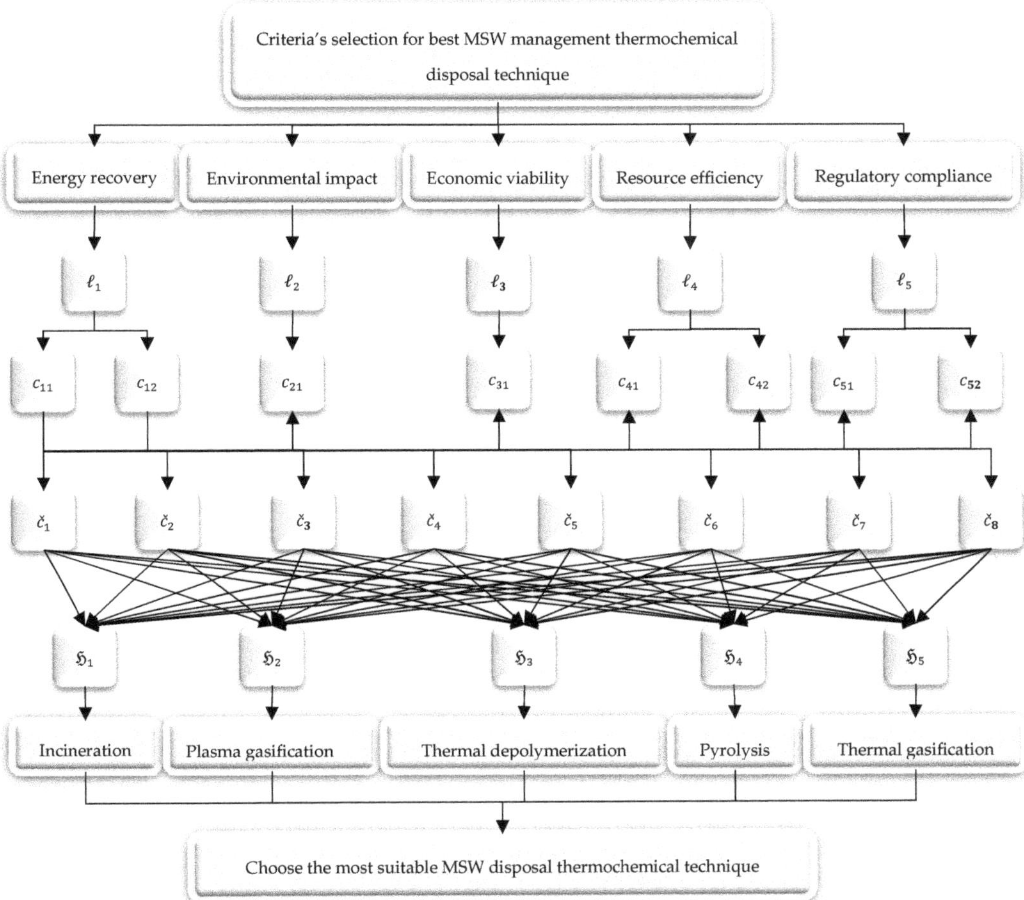

Figure 4. Selection of thermochemical technique for MSW disposal.

The technique presented here assists in discovering the most optimal solution using the provided criteria. The selection method focuses on preserving environmental stability and safety for both humans and plants. The main objective of the MSW thermochemical technique is to ensure sustainable economic and social development. However, this research identifies potential risks to people's safety, indicating its essential importance in promoting social prosperity. Through detailed study and meticulous evaluation of these variables, experts can recommend the most suitable and efficient thermochemical strategy for treating MSW. The experts' opinions for each choice in Table 1 are presented using IFHSNs and then contained in the EDAS approach. This strategy combines the interaction AOs and interaction ordered AOs presented in Sections 4 and 5, intending to determine the most economical thermochemical treatment process in MSW management.

Step 1. The expert clarifies that the intuitionistic fuzzy hypersoft decision matrices for all potential outcomes are shown in Table 1.

Table 1. Expert's opinion for each alternative.

$\mathcal{Q}^{(1)}$	$\check{c}_1$	$\check{c}_2$	$\check{c}_3$	$\check{c}_4$	$\check{c}_5$	$\check{c}_6$	$\check{c}_7$	$\check{c}_8$
$\mathcal{X}_1$	(0.3, 0.6)	(0.7, 0.3)	(0.6, 0.2)	(0.5, 0.4)	(0.2, 0.4)	(0.4, 0.6)	(0.1, 0.8)	(0.4, 0.3)
$\mathcal{X}_2$	(0.2, 0.5)	(0.4, 0.6)	(0.3, 0.4)	(0.5, 0.2)	(0.3, 0.6)	(0.2, 0.4)	(0.5, 0.5)	(0.4, 0.5)
$\mathcal{X}_3$	(0.7, 0.3)	(0.2, 0.5)	(0.1, 0.6)	(0.3, 0.4)	(0.4, 0.6)	(0.3, 0.4)	(0.4, 0.6)	(0.2, 0.5)
$\mathcal{X}_4$	(0.5, 0.4)	(0.2, 0.6)	(0.2, 0.4)	(0.4, 0.6)	(0.6, 0.3)	(0.2, 0.6)	(0.4, 0.5)	(0.6, 0.3)
$\mathcal{X}_5$	(0.6, 0.3)	(0.4, 0.6)	(0.2, 0.3)	(0.1, 0.7)	(0.4, 0.5)	(0.3, 0.6)	(0.2, 0.7)	(0.4, 0.6)
$\mathcal{Q}^{(2)}$								
$\mathcal{X}_1$	(0.5, 0.4)	(0.3, 0.5)	(0.5, 0.4)	(0.4, 0.3)	(0.3, 0.6)	(0.2, 0.4)	(0.5, 0.4)	(0.3, 0.5)
$\mathcal{X}_2$	(0.4, 0.5)	(0.3, 0.4)	(0.2, 0.5)	(0.5, 0.2)	(0.2, 0.4)	(0.3, 0.5)	(0.4, 0.5)	(0.5, 0.4)
$\mathcal{X}_3$	(0.3, 0.6)	(0.4, 0.5)	(0.3, 0.5)	(0.6, 0.2)	(0.4, 0.5)	(0.5, 0.4)	(0.3, 0.5)	(0.4, 0.5)
$\mathcal{X}_4$	(0.4, 0.5)	(0.3, 0.2)	(0.3, 0.5)	(0.5, 0.3)	(0.3, 0.5)	(0.2, 0.5)	(0.6, 0.2)	(0.2, 0.5)
$\mathcal{X}_5$	(0.2, 0.5)	(0.5, 0.2)	(0.4, 0.6)	(0.3, 0.3)	(0.2, 0.8)	(0.3, 0.6)	(0.5, 0.4)	(0.5, 0.5)
$\mathcal{Q}^{(3)}$								
$\mathcal{X}_1$	(0.4, 0.3)	(0.4, 0.5)	(0.2, 0.4)	(0.4, 0.3)	(0.1, 0.2)	(0.2, 0.4)	(0.3, 0.4)	(0.2, 0.4)
$\mathcal{X}_2$	(0.3, 0.5)	(0.2, 0.4)	(0.8, 0.1)	(0.5, 0.4)	(0.4, 0.3)	(0.4, 0.5)	(0.7, 0.2)	(0.6, 0.3)
$\mathcal{X}_3$	(0.6, 0.3)	(0.4, 0.5)	(0.6, 0.2)	(0.6, 0.4)	(0.1, 0.2)	(0.3, 0.4)	(0.5, 0.3)	(0.4, 0.5)
$\mathcal{X}_4$	(0.5, 0.4)	(0.1, 0.3)	(0.3, 0.5)	(0.5, 0.3)	(0.3, 0.5)	(0.8, 0.1)	(0.3, 0.5)	(0.2, 0.5)
$\mathcal{X}_5$	(0.4, 0.5)	(0.3, 0.5)	(0.4, 0.6)	(0.4, 0.5)	(0.2, 0.4)	(0.5, 0.4)	(0.4, 0.6)	(0.7, 0.2)
$\mathcal{Q}^{(4)}$								
$\mathcal{X}_1$	(0.3, 0.6)	(0.3, 0.4)	(0.4, 0.5)	(0.3, 0.7)	(0.5, 0.4)	(0.4, 0.6)	(0.1, 0.5)	(0.4, 0.5)
$\mathcal{X}_2$	(0.6, 0.2)	(0.5, 0.4)	(0.6, 0.2)	(0.5, 0.4)	(0.4, 0.5)	(0.5, 0.3)	(0.2, 0.7)	(0.3, 0.7)
$\mathcal{X}_3$	(0.3, 0.7)	(0.4, 0.5)	(0.4, 0.5)	(0.3, 0.4)	(0.2, 0.7)	(0.3, 0.4)	(0.8, 0.2)	(0.7, 0.2)
$\mathcal{X}_4$	(0.5, 0.4)	(0.6, 0.3)	(0.4, 0.3)	(0.4, 0.5)	(0.7, 0.2)	(0.2, 0.4)	(0.4, 0.6)	(0.3, 0.5)
$\mathcal{X}_5$	(0.4, 0.5)	(0.4, 0.6)	(0.2, 0.6)	(0.5, 0.3)	(0.4, 0.3)	(0.2, 0.7)	(0.5, 0.4)	(0.4, 0.5)
$\mathcal{Q}^{(5)}$								
$\mathcal{X}_1$	(0.4, 0.5)	(0.3, 0.6)	(0.2, 0.5)	(0.3, 0.6)	(0.5, 0.4)	(0.3, 0.5)	(0.1, 0.8)	(0.4, 0.6)
$\mathcal{X}_2$	(0.4, 0.3)	(0.5, 0.4)	(0.2, 0.8)	(0.2, 0.6)	(0.3, 0.5)	(0.5, 0.4)	(0.4, 0.5)	(0.2, 0.7)
$\mathcal{X}_3$	(0.3, 0.5)	(0.1, 0.7)	(0.4, 0.3)	(0.3, 0.5)	(0.2, 0.6)	(0.3, 0.4)	(0.6, 0.3)	(0.4, 0.6)
$\mathcal{X}_4$	(0.3, 0.6)	(0.3, 0.6)	(0.3, 0.5)	(0.4, 0.5)	(0.3, 0.6)	(0.3, 0.4)	(0.2, 0.5)	(0.4, 0.5)
$\mathcal{X}_5$	(0.4, 0.6)	(0.3, 0.5)	(0.2, 0.6)	(0.6, 0.3)	(0.4, 0.6)	(0.4, 0.5)	(0.3, 0.5)	(0.3, 0.6)

Step 2: Economic viability. Let ℓ_3 be the cost type parameter with sub-criteria c_{31}. As we know, $\{\check{c}_1, \check{c}_2, \check{c}_3, \check{c}_4, \check{c}_5, \check{c}_6, \check{c}_7, \check{c}_8\}$, the multi-sub-attributes and sub-criteria c_{31} are involved in every multi-sub attribute, which shows that each attribute contains this cost type sub-criteria. So, it is necessary to normalize the decision matrix using Equation (12). The normalized decision matrix is presented in Table 2.

Step 3: We will apply the IFHSIWA or IFHSIWG operators. Therefore, there is no need to find ordered matrices, and we can proceed to step 4 directly without finding ordered matrices.

Step 4: Equation (3) or (10) (IFHSIWA or IFHSIWG) operators will be used to determine the experts' assessments' aggregated decision matrices. Based on the defined parameters, the above operators verify all of the desired values for alternatives. The aggregated evaluation matrix implementing Equation (3) is laid out in Table 3.

Table 2. Normalized decision matrix.

$\mathcal{Q}^{(1)}$	$\check{c}_1$	$\check{c}_2$	$\check{c}_3$	$\check{c}_4$	$\check{c}_5$	$\check{c}_6$	$\check{c}_7$	$\check{c}_8$
$\mathcal{X}_1$	(0.6, 0.3)	(0.3, 0.7)	(0.2, 0.6)	(0.4, 0.5)	(0.4, 0.2)	(0.6, 0.4)	(0.8, 0.1)	(0.3, 0.4)
$\mathcal{X}_2$	(0.5, 0.2)	(0.6, 0.4)	(0.4, 0.3)	(0.2, 0.5)	(0.6, 0.3)	(0.4, 0.2)	(0.5, 0.5)	(0.5, 0.4)
$\mathcal{X}_3$	(0.3, 0.7)	(0.5, 0.2)	(0.6, 0.1)	(0.4, 0.3)	(0.6, 0.4)	(0.4, 0.3)	(0.6, 0.4)	(0.5, 0.2)
$\mathcal{X}_4$	(0.4, 0.5)	(0.6, 0.2)	(0.4, 0.2)	(0.6, 0.4)	(0.3, 0.6)	(0.6, 0.2)	(0.5, 0.4)	(0.3, 0.6)
$\mathcal{X}_5$	(0.3, 0.6)	(0.6, 0.4)	(0.3, 0.2)	(0.7, 0.1)	(0.5, 0.4)	(0.6, 0.3)	(0.7, 0.2)	(0.6, 0.4)
$\mathcal{Q}^{(2)}$								
$\mathcal{X}_1$	(0.4, 0.5)	(0.5, 0.3)	(0.4, 0.5)	(0.3, 0.4)	(0.6, 0.3)	(0.4, 0.2)	(0.4, 0.5)	(0.5, 0.3)
$\mathcal{X}_2$	(0.5, 0.4)	(0.4, 0.3)	(0.5, 0.2)	(0.2, 0.5)	(0.4, 0.2)	(0.5, 0.3)	(0.5, 0.4)	(0.4, 0.5)
$\mathcal{X}_3$	(0.6, 0.3)	(0.5, 0.4)	(0.5, 0.3)	(0.2, 0.6)	(0.5, 0.4)	(0.4, 0.5)	(0.5, 0.3)	(0.5, 0.4)
$\mathcal{X}_4$	(0.5, 0.4)	(0.2, 0.3)	(0.5, 0.3)	(0.3, 0.5)	(0.5, 0.3)	(0.5, 0.2)	(0.2, 0.6)	(0.5, 0.2)
$\mathcal{X}_5$	(0.5, 0.2)	(0.2, 0.5)	(0.6, 0.4)	(0.3, 0.3)	(0.8, 0.2)	(0.6, 0.3)	(0.4, 0.5)	(0.5, 0.5)
$\mathcal{Q}^{(3)}$								
$\mathcal{X}_1$	(0.3, 0.4)	(0.5, 0.4)	(0.4, 0.2)	(0.3, 0.4)	(0.2, 0.1)	(0.4, 0.2)	(0.4, 0.3)	(0.4, 0.2)
$\mathcal{X}_2$	(0.5, 0.3)	(0.4, 0.2)	(0.1, 0.8)	(0.4, 0.5)	(0.3, 0.4)	(0.5, 0.4)	(0.2, 0.7)	(0.3, 0.6)
$\mathcal{X}_3$	(0.3, 0.6)	(0.5, 0.4)	(0.2, 0.6)	(0.4, 0.6)	(0.2, 0.1)	(0.4, 0.3)	(0.3, 0.5)	(0.5, 0.4)
$\mathcal{X}_4$	(0.4, 0.5)	(0.3, 0.1)	(0.5, 0.3)	(0.3, 0.5)	(0.5, 0.3)	(0.1, 0.8)	(0.5, 0.3)	(0.5, 0.2)
$\mathcal{X}_5$	(0.5, 0.4)	(0.5, 0.3)	(0.6, 0.4)	(0.5, 0.4)	(0.4, 0.2)	(0.4, 0.5)	(0.6, 0.4)	(0.2, 0.7)
$\mathcal{Q}^{(4)}$								
$\mathcal{X}_1$	(0.6, 0.3)	(0.4, 0.3)	(0.5, 0.4)	(0.7, 0.3)	(0.4, 0.5)	(0.6, 0.4)	(0.5, 0.1)	(0.5, 0.4)
$\mathcal{X}_2$	(0.2, 0.6)	(0.4, 0.5)	(0.2, 0.6)	(0.4, 0.5)	(0.5, 0.4)	(0.3, 0.5)	(0.7, 0.2)	(0.7, 0.3)
$\mathcal{X}_3$	(0.7, 0.3)	(0.5, 0.4)	(0.5, 0.4)	(0.4, 0.3)	(0.7, 0.2)	(0.4, 0.3)	(0.2, 0.8)	(0.2, 0.7)
$\mathcal{X}_4$	(0.4, 0.5)	(0.3, 0.6)	(0.3, 0.4)	(0.5, 0.4)	(0.2, 0.7)	(0.4, 0.2)	(0.6, 0.4)	(0.5, 0.3)
$\mathcal{X}_5$	(0.5, 0.4)	(0.6, 0.4)	(0.6, 0.2)	(0.3, 0.5)	(0.3, 0.4)	(0.7, 0.2)	(0.4, 0.5)	(0.5, 0.4)
$\mathcal{Q}^{(5)}$								
$\mathcal{X}_1$	(0.5, 0.4)	(0.6, 0.3)	(0.5, 0.2)	(0.6, 0.3)	(0.4, 0.5)	(0.5, 0.3)	(0.8, 0.1)	(0.6, 0.4)
$\mathcal{X}_2$	(0.3, 0.4)	(0.4, 0.5)	(0.8, 0.2)	(0.6, 0.2)	(0.5, 0.3)	(0.4, 0.5)	(0.5, 0.4)	(0.7, 0.2)
$\mathcal{X}_3$	(0.5, 0.3)	(0.7, 0.1)	(0.3, 0.4)	(0.5, 0.3)	(0.6, 0.2)	(0.4, 0.3)	(0.3, 0.6)	(0.6, 0.4)
$\mathcal{X}_4$	(0.6, 0.3)	(0.6, 0.3)	(0.5, 0.3)	(0.5, 0.4)	(0.6, 0.3)	(0.4, 0.3)	(0.5, 0.2)	(0.5, 0.4)
$\mathcal{X}_5$	(0.6, 0.4)	(0.5, 0.3)	(0.6, 0.2)	(0.3, 0.6)	(0.6, 0.4)	(0.5, 0.4)	(0.5, 0.3)	(0.6, 0.3)

Table 3. Aggregated decision matrix using IFHSIWA operator.

	$\check{c}_1$	$\check{c}_2$	$\check{c}_3$	$\check{c}_4$	$\check{c}_5$	$\check{c}_6$	$\check{c}_7$	$\check{c}_8$
$\mathcal{X}_1$	(0.48461, 0.39082)	(0.47415, 0.52585)	(0.41760, 0.40398)	(0.48672, 0.51321)	(0.42573, 0.39258)	(0.50026, 0.49974)	(0.59887, 0.23213)	(0.47606, 0.52394)
$\mathcal{X}_2$	(0.41091, 0.39683)	(0.43540, 0.56459)	(0.45653, 0.54347)	(0.37421, 0.44999)	(0.45868, 0.34174)	(0.42819, 0.40128)	(0.50911, 0.49089)	(0.54369, 0.45626)
$\mathcal{X}_3$	(0.52458, 0.47542)	(0.54392, 0.32249)	(0.43563, 0.17177)	(0.37607, 0.62393)	(0.54396, 0.45604)	(0.40000, 0.37205)	(0.39065, 0.60935)	(0.46736, 0.53264)
$\mathcal{X}_4$	(0.46708, 0.43292)	(0.39828, 0.36429)	(0.44666, 0.31073)	(0.43741, 0.56259)	(0.43979, 0.42361)	(0.41499, 0.40705)	(0.46459, 0.53540)	(0.47412, 0.33493)
$\mathcal{X}_5$	(0.49482, 0.50518)	(0.48222, 0.51778)	(0.56497, 0.43503)	(0.42366, 0.39357)	(0.57339, 0.42661)	(0.57613, 0.32387)	(0.51747, 0.48253)	(0.48971, 0.51029)

Step 5: Find the average solution matrix using Equation (13).

$$[AVS_h]_{1\times 8} = \left[\begin{pmatrix} 0.47640, \\ 0.44023 \end{pmatrix} \begin{pmatrix} 0.46679, \\ 0.45900 \end{pmatrix} \begin{pmatrix} 0.46428, \\ 0.37299 \end{pmatrix} \begin{pmatrix} 0.41962, \\ 0.50866 \end{pmatrix} \begin{pmatrix} 0.48831, \\ 0.40812 \end{pmatrix} \begin{pmatrix} 0.46392, \\ 0.40079 \end{pmatrix} \begin{pmatrix} 0.49614, \\ 0.47006 \end{pmatrix} \begin{pmatrix} 0.49019, \\ 0.47161 \end{pmatrix} \right]$$

Step 6: Evaluate the PDA using Equation (15) and the NDA using Equation (17). We cannot use Equations (14) or (16) to compute the PDA or the NDA, respectively. The PDA and NDA matrices are given in Tables 4 and 5.

Table 4. PDA matrix.

	$\check{c}_1$	$\check{c}_2$	$\check{c}_3$	$\check{c}_4$	$\check{c}_5$	$\check{c}_6$	$\check{c}_7$	$\check{c}_8$
$\mathcal{X}_1$	(0, 0.05391)	(0, 0)	(0.05575, 0)	(0, 0)	(0.06981, 0.01733)	(0, 0)	(0, 0.24626)	(0.01447, 0)
$\mathcal{X}_2$	(0.07144, 0.04735)	(0.03391, 0)	(0.00926, 0)	(0.04892, 0.06320)	(0.03305, 0.07405)	(0.04131, 0)	(0, 0)	(0, 0.01596)
$\mathcal{X}_3$	(0, 0)	(0, 0.14745)	(0.03421, 0.24033)	(0.04691, 0)	(0, 0)	(0.07392, 0.033246)	(0.10918, 0)	(0.02376, 0)
$\mathcal{X}_4$	(0.01017, 0.00798)	(0.07401, 0.10230)	(0.02105, 0.07437)	(0, 0)	(0.05413, 0)	(0.05658, 0)	(0.03264, 0)	(0.01671, 0.14211)
$\mathcal{X}_5$	(0, 0)	(0, 0)	(0, 0)	(0, 0.12398)	(0, 0)	(0, 0.08896)	(0, 0)	(0.00049, 0)

Table 5. NDA matrix.

	$\check{c}_1$	$\check{c}_2$	$\check{c}_3$	$\check{c}_4$	$\check{c}_5$	$\check{c}_6$	$\check{c}_7$	$\check{c}_8$
$\mathcal{X}_1$	(0.00896, 0)	(0.00794, 0.07221)	(0, 0.03700)	(0.07229, 0.00490)	(0, 0)	(0.04203, 0.11442)	(0.10633, 0)	(0, 0.05441)
$\mathcal{X}_2$	(0, 0)	(0, 0.11406)	(0, 0.20361)	(0, 0)	(0, 0)	(0, 0.00056)	(0.01342, 0.02156)	(0.05563, 0)
$\mathcal{X}_3$	(0.05256, 0.03839)	(0.08331, 0)	(0, 0)	(0, 0.12418)	(0.06208, 0.053347)	(0, 0)	(0, 0.14416)	(0, 0.06345)
$\mathcal{X}_4$	(0, 0)	(0, 0)	(0, 0)	(0.01917, 0.05810)	(0, 0.01728)	(0, 0.00723)	(0, 0.06763)	(0, 0)
$\mathcal{X}_5$	(0.02009, 0.07085)	(0.01666, 0.06349)	(0.12027, 0.07409)	(0.00436, 0)	(0.09492, 0.02063)	(0.12977, 0)	(0.02208, 0.01291)	(0, 0.04021)

Step 7: Compute the positive weighted distance P_τ using Equation (18) and the negative weighted distance Q_τ using Equation (19).

$P_1 = (1.13162, 1.14138)$, $P_2 = (1.10292, 1.30423)$, $P_3 = (1.27992, 1.15422)$, $P_4 = (1.40578, 1.25593)$, $P_5 = (0.87198, 1.05101)$, $Q_1 = (1.17884, 1.18842)$, $Q_2 = (1.09874, 1.25203)$, $Q_3 = (1.17206, 1.31196)$, $Q_4 = (0.87766, 1.23531)$, and $Q_5 = (1.31715, 1.30362)$.

Step 8: Normalize the positive weighted distance P_τ using Equation (22) and the negative weighted distance Q_τ using Equation (23), where, $\tau = 1, 2, \ldots\ldots m$ are given as $NP_1 = (0.80498,\ 0.87514)$, $NP_2 = (0.99797,\ 1)$, $NP_3 = (0.91047,\ 0.88498)$, $NP_4 = (1,\ 0.96297)$, $NP_5 = (0.62028,\ 0.80585)$, $NQ_1 = (0.10501,\ 0.09417)$, $NQ_2 = (0.16582,\ 0.04568)$, $NQ_3 = (0.11016,\ 0)$, $NQ_4 = (0.33367,\ 0.05842)$, and $NQ_5 = (0,\ 0.00635)$.

Step 9: Calculate the appraisal value using Equation (24), such as $I_1 = 0.45393$, $I_2 = 0.52199$, $I_3 = 0.46017$, $I_4 = 0.52813$, and $I_5 = 0.31270$.

Step 10: Each alternative is rated in descending order of appraisal value. The feasible choice with the highest and lowest appraisal scores appears to be the poorest, $\mathcal{Q}_4 > \mathcal{Q}_2 > \mathcal{Q}_3 > \mathcal{Q}_1 > \mathcal{Q}_5$.

The suggested method presents substantial improvements over the existing EDAS strategy. Employing real numbers, the prevailing EDAS methods do not express the analyzed information's multi-sub-parametric MD and NMD values. Meanwhile, the proposed model enables experts to present their opinions more easily.

7. Comparative Analysis and Theoretical Implications

This section will perform a comprehensive evaluation by comparing and analyzing prior studies and investigating the implications of the proposed strategy.

7.1. Comparative Analysis

This subsection will analyze our proposed framework from three different viewpoints. We investigate our proposed EDAS technique against existing EDAS techniques in different fuzzy environments. Additionally, we will analyze the newly introduced EDAS model with relevant AOs within the IFHSS context. Finally, we examine our proposed approach with the TOPSIS approach in the IFHSS structure.

7.1.1. Comparison with Existing EDAS Approaches

It is observed that prevalent techniques, such as the fuzzy EDAS [65], exclusively compensate for the MD of alternates. In comparison, particular mathematical models, such as the IFS EDAS [66], incorporate the advantage of the MD and NMD while analyzing alternatives. Liu et al. [69] and Paul et al. [29] proposed the EDAS strategy for PFS and cubic PFS. Zhang et al. [71] presented the EDAS technique for picture fuzzy sets. Darko and Liang [72] enlarged the EDAS method using the q-rung orthopair fuzzy context. However, such techniques lack barriers concerning how to parameterize alternatives. A conceptual assessment of our novel EDAS approach to existing EDAS strategies for different fuzzy set configurations is discussed in Table 6.

Table 6. Comparative analysis with existing EDAS models in different structures. n/a: not applicable.

Structure	Appraisal Score					Ranking
Fuzzy EDAS [65]			n/a			n/a
IFS EDAS [66]			n/a			n/a
PFS EDAS [69]			n/a			n/a
Cubic PFS EDAS [29]			n/a			n/a
Picture fuzzy EDAS [71]			n/a			n/a
q-ROFS EDAS [72]			n/a			n/a
Proposed EDAS	0.45393	0.52199	0.46017	0.52813	0.31270	$\mathcal{Q}_4 > \mathcal{Q}_2 > \mathcal{Q}_3 > \mathcal{Q}_1 > \mathcal{Q}_5$

The advanced EDAS technique effectively solves the essential difficulties raised by conventional EDAS strategies, particularly for maintaining the parameterized nature of alternates. Also, it efficiently addresses data related to sub-parameter forms and multi-sub-attributes of alternatives, which frequently get overlooked or not dealt with in other EDAS techniques. Given the importance of such variables in decision making, the proposed EDAS model addresses issues precisely. The feasibility of the developed EDAS model is shown by its ability to determine the most productive thermochemical method for disposing of MSW. Table 6 demonstrates that the proposed technique prevails over prior techniques in selecting the most optimal choice from the relevant prospects.

7.1.2. Comparison with Different Aggregation Operators

This method is extended within the framework of the intuitionistic fuzzy hypersoft set. So, this strategy is comparable to the multi-criteria decision-making approach developed by Zulqarnain et al. [45], which is based on the intuitionistic fuzzy hypersoft weighted average and weighted geometric AOs. The intuitionistic fuzzy hypersoft number is an extension of the intuitionistic fuzzy number. A comparison of our proposed approach with some prevalent operators under the same structure is given in Table 7, where score values and ranking orders are provided.

Table 7. Comparison with different operators.

Operators	Alternatives Score Values/Appraisal Score					Ranking
IFHSWA [45]	0.52647	0.54768	0.53082	0.56397	0.51837	$\mathcal{Q}_4 > \mathcal{Q}_2 > \mathcal{Q}_3 > \mathcal{Q}_1 > \mathcal{Q}_5$
IFHSWG [45]	0.46926	0.47835	0.46052	0.49573	0.44673	$\mathcal{Q}_4 > \mathcal{Q}_2 > \mathcal{Q}_1 > \mathcal{Q}_3 > \mathcal{Q}_5$
Proposed method	0.45393	0.52199	0.46017	0.52813	0.31270	$\mathcal{Q}_4 > \mathcal{Q}_2 > \mathcal{Q}_3 > \mathcal{Q}_1 > \mathcal{Q}_5$

Table 7 demonstrates that pyrolysis is the optimum alternative for the thermochemical disposal approach to MSW in the decision-making procedures given in [45]. Figure 5 graphically represents the proposed EDAS method with different operators.

7.1.3. Comparison with the TOPSIS Method

However, we explore the resemblances and disparities between the proposed strategy and the TOPSIS method [44]. $\mathcal{X}_4$ was determined to be the most suitable alternative for the thermochemical technique of MSW disposal in the TOPSIS method [44]. The difference appears in the order of alternatives as an outcome of the procedure used to identify positive and negative ideal solutions, which is subsequently examined by considering the relative distances from these possible solutions. On the other hand, our proposed method depends on the principle of average solutions instead of implementing the optimal strategies, which includes the mathematical computation of comparable distances. The comparison of the results of Table 8 demonstrates that the developed EDAS approach, which entails the implementation of IFHSNs, delivers better precision in choosing the most suitable alternative for thermochemical MSW disposal processes. The distinction pertains to integrating more detailed data compared to IFS and IFSS and employing a unique technique to determine average solutions and calculate positive and negative distance measures.

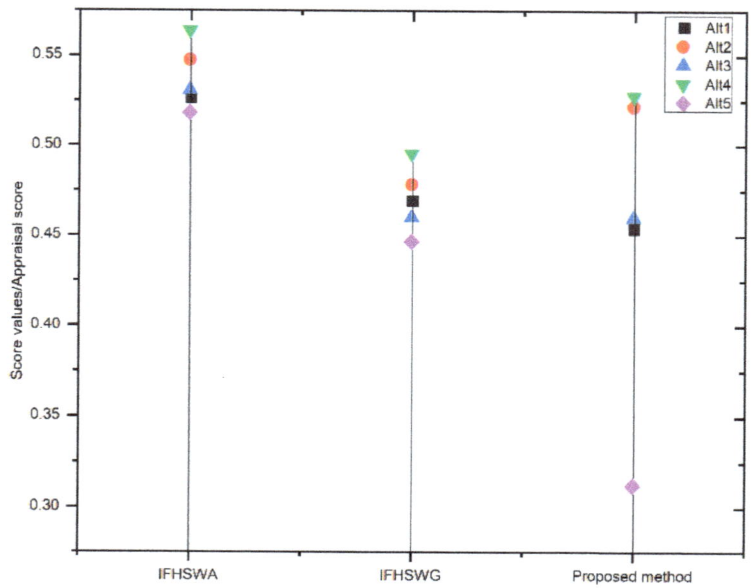

Figure 5. Comparison with different operators.

Table 8. Comparison with the TOPSIS method.

Method	Appraisal Score/Closeness Coefficient					Ranking
TOPSIS [44]	0.62497	0.56289	0.67835	0.71509	0.49385	$\mathcal{Q}_4 > \mathcal{Q}_3 > \mathcal{Q}_1 > \mathcal{Q}_2 > \mathcal{Q}_5$
Proposed EDAS	0.45393	0.52199	0.46017	0.52813	0.31270	$\mathcal{Q}_4 > \mathcal{Q}_2 > \mathcal{Q}_3 > \mathcal{Q}_1 > \mathcal{Q}_5$

The graphical comparison of the proposed approach and the TOPSIS [44] method is demonstrated in Figure 6.

7.2. Discussion

After a comprehensive review of various research studies, it is determined that the EDAS approach used in this research is comparable to prior methods. However, the most significant advantage is integrating more details about the sub-parameters of alternatives, which attempts to address the fundamental confusion in empirical data. This approach delivers reliable and credible insights into objects, illuminating unclear and inaccurate aspects of the decision-making process. The stated EDAS approach provides greater effectiveness over existing methodologies and associated metrics by analyzing the level of perceptions and consistency across interpretations. Furthermore, it properly tackles issues related to parameterizing decisions and managing multiple sub-attributes, problems other EDAS techniques have not yet solved.

The assessments of the comparative study indicate a consensus between the ideal alternative identified by the proposed approach and the existing systems, validating the reliability and effectiveness of the proposed EDAS model. Therefore, the suggested EDAS strategy is a significant asset for decision makers, presenting more accurate and reliable findings than previous methodologies when managing instabilities and multiple sub-attributes.

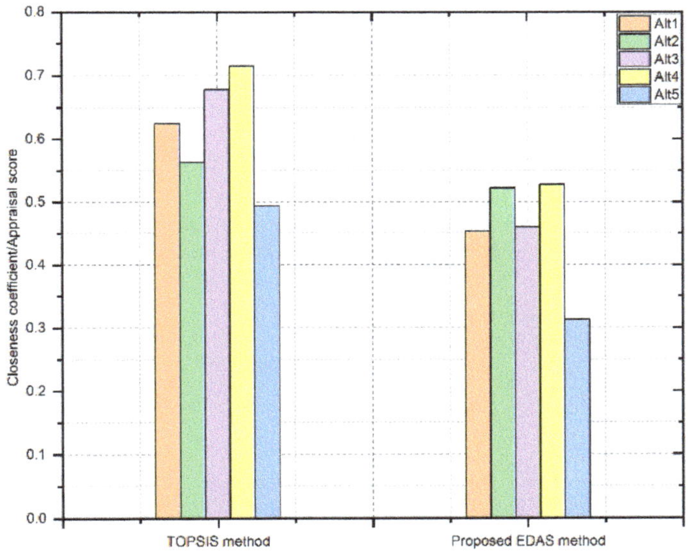

Figure 6. Comparison with the TOPSIS method.

7.3. Research Implications

The most important purpose of this paper is to present an adequate thermochemical waste disposal method for municipalities in Pakistan. Five thermochemical alternatives were explored for assessment. The conventional EDAS approach is frequently implemented as a multi-criteria decision-making strategy, and a substantial quantity of literature on the technique is available. This research uses a generalized distance function to extend the EDAS technique in the IFHSS structure. Several implications of this planned improvement are presented as follows.

7.3.1. Methodological Implications

The diagram shown in Figure 2, along with the details described in Section 5, demonstrate the sequential steps of the proposed approach. This approach uses the interaction aggregation and interaction-ordered AOs presented in Sections 3 and 4 to organize the five experts' decision matrices. The consequent fused decision matrix is applied to determine the average solution. This approach mainly uses the EDAS technique in an IFHSS framework. IFHSS–EDAS is characterized by its insistence on the average solution, which varies from the positive and negative ideal solutions determined in the TOPSIS approach [44]. This property imparts more benefit when comparing FS, IFS, FSS, and IFSS, stimulating the practicality of decision-making processes. The method's applicability is demonstrated by its ability to be used by all municipal governments in Pakistan. The technique proposed here indicates the prospect of future study efforts, adding to the developing field of decision-making strategies.

7.3.2. Theoretical Implications

This study concentrates on an MAGDM problem regarding the most effective decisions about MSW disposal methods for municipalities in Pakistan. The theoretical applications of this hybrid approach will be stated in the following way.

Integrating IFHSS in the EDAS approach helps with the modeling of integral unpredictability and inaccuracy in the decision-making process. IFHSS effectively conveys complicated fuzzy facts by combining HSS and IFS knowledge. It enables amplification of the unpredictability of the data to calculate criteria weights.

The merging of the dynamically optimized methodology with IFHSS–EDAS improves its theoretical potential by effectively resolving the convoluted interactions between the parameters. When determining the best way to dump MSW, several factors require being taken into perspective, such as its impact on the surroundings, its affordability, its energy effectiveness, and its ability.

A detailed assessment of MSW disposal methods can be executed using the IFHSS–EDAS technique and irregular optimization. By implementing compromise research, stakeholders can review the functionality of various techniques over distinct parameters while simultaneously focusing on several criteria for assessment.

The execution of the IFHSS–EDAS technique complemented by dynamic stabilization indicates theoretical development regarding decision-making techniques for MSW. The theoretical repercussions include the ability to boost and enhance the approach, presenting prospects for future research.

7.3.3. Managerial Implications

This research investigates five thermochemical techniques for treating MSW: incineration, plasma arc gasification, thermal depolymerization, pyrolysis, and thermal gasification. However, according to the IFHSS–EDAS approach, pyrolysis ($\mathcal{X}_4$) is the most effective technique for managing MSW based on the specified factors and sub-factors in Section 6.2.

Pyrolysis is when organic materials are heated without oxygen, causing them to disintegrate into smaller particles, including gases and liquid fuels. The process occurs inside of a pyrolysis reactor, a specially built container that operates at high temperatures (400–800 °C) and pressure. Pyrolysis is a flexible waste-to-energy approach capable of dealing with a wide range of resources, such as items that are difficult to recycle or dispose of, like rubber and agricultural residue.

Several managerial implications relate to the pyrolysis technique and need meticulous consideration. Here are a few of those features: it minimizes the detrimental impact of waste treatment on the surroundings by limiting the quantity of greenhouse gases released into the atmosphere and reducing the amount of fossil fuels used.

Using pyrolysis as an MSW management approach requires significant investments in technology and appliances. Therefore, it is important to precisely assess operational problems, such as the availability of adequate resources and the financial implications relating to gear and administration.

It may be essential to devote quite a bit of capital initially to implement pyrolysis as an MSW disposal method. However, it presents the prospect of financial rewards in the future. Alternatives include eliminating energy expenditures, minimizing trash disposal, and earning money from resale of waste products.

Before implementing pyrolysis, authorization from the regulatory organization with authority over waste and environmental handling techniques must be secured. This authorization is vital to prevent potential repercussions or legal problems.

Local communities and regulatory bodies are among several organizations that must be engaged in properly implementing pyrolysis. Although thoroughly describing precautions and protecting the environment is necessary, informing these stakeholders of pyrolysis's conceivable benefits and risks is needed.

Workers and other stakeholders need educational and training initiatives to make the pyrolysis manufacturing process. Training personnel on the correct use of pyrolysis equipment and providing information about the method's benefits and downsides are all part of this process.

Consequently, pyrolysis has been chosen for MSW management in part because of the EDAS approach. In light of the preceding information, pyrolysis can be implemented by Pakistani municipalities as an efficient and secure means of disposal of MSW, with an emphasis on environmental preservation and respect for the requirements and worries of all parties concerned.

Although our strategy enables mobility and resilience, it confronts limitations in complicated and unpredictable scenarios, particularly for computing expenses and manageability as the variety of feasible inputs grows.

8. Conclusions

The intuitionistic fuzzy hypersoft structure describes a novel technique in multi-attribute group decision-making processes, greatly enlarging the potential to integrate fuzzy data in a precise, logical paradigm. The proper handling of MSW persists as an important issue in Pakistan and other emerging economies due to poor policies and practices, resulting in adverse environmental effects. Prioritizing environmental concerns, cost-effectiveness, and extracting alternative energy sources from organic waste are essential to proper MSW disposal techniques. This research concentrates on five thermochemical MSW methods, integrating five criteria and their corresponding sub-criteria for classification.

We present interaction operational laws for IFHSNs and develop interactive aggregation and ordered aggregation operators, demonstrating their fundamental characteristics. The novel EDAS approach is presented to deal with the obstacles of MAGDM in the IFHSS structure using these interaction operators. The IFHSS–EDAS algorithm is implemented to determine the hierarchy of five alternatives, consequently determining pyrolysis as the best and most economical thermochemical MSW disposal approach in Pakistan. The comparative research and arguments demonstrate the viability and practicality of the presented approach, promoting its appropriateness to be implemented in all municipalities, taking into account financial and environmental considerations.

Moreover, the proposed research extended its practicality to the selection of sustainable resources, the selection of portfolios, the utilization of carbon capture, and the selection of storage sites. This study indicates the prospect of future research attempts because it implemented the novel idea of EDAS in the IFHSS environment and various optimization strategies. Further research will explore the distinct characteristics of Einstein-ordered and Einstein hybrid aggregation operators. Also, the technique stated in this article will be extended to include Pythagorean fuzzy hypersoft sets and q-rung orthopair fuzzy hypersoft sets. Moreover, studies will be performed to integrate IFHSNs with other MAGDM methods, evaluating their implications in medical diagnostics, material selection, pattern identification, and biomedical waste management. The prospective development of multiple topological, algebraic, and ordered structures in the IFHSS context will boost decision-making approaches in these areas.

Author Contributions: R.M.Z.: methodology, conceptualization, design, analysis, writing—original draft, writing—review and editing. H.W.: methodology, supervision, writing—original draft, writing—review and editing. I.S.: validation, design, analysis, writing—original draft. R.A.: funding, formal analysis, writing—original draft. H.N.: software, analysis, formal analysis, validation, writing—original draft. S.A.V.: writing—original draft, methodology, validation, formal analysis. M.I.A.: conceptualization, design, analysis, writing—original draft. All authors have read and agreed to the published version of the manuscript.

Funding: This research was funded by King Khalid University grant number RGP2/218/45.

Institutional Review Board Statement: Not applicable.

Informed Consent Statement: Not applicable.

Data Availability Statement: The datasets generated during and/or analyzed during the current study are available within the article.

Acknowledgments: The authors extend their appreciation to the Deanship of Research and Graduate Studies at King Khalid University for funding this work through a Large Group Project under grant number RGP2/218/45.

Conflicts of Interest: The authors declare that they do not have any conflicts of interest.

References

1. Ngoc, U.N.; Schnitzer, H. Sustainable solutions for solid waste management in Southeast Asian countries. *Waste Manag.* **2009**, *2* 1982–1995. [CrossRef]
2. Gautam, M.; Agrawal, M. Greenhouse gas emissions from municipal solid waste management: A review of global scenario. I *Carbon Footprint Case Studies: Municipal Solid Waste Management, Sustainable Road Transport and Carbon Sequestration*; Springe Berlin/Heidelberg, Germany, 2021; pp. 123–160.
3. Cetrulo, T.B.; Marques, R.C.; Cetrulo, N.M.; Pinto, F.S.; Moreira, R.M.; Mendizábal-Cortés, A.D.; Malheiros, T.F. Effectiveness solid waste policies in developing countries: A case study in Brazil. *J. Clean. Prod.* **2018**, *205*, 179–187. [CrossRef]
4. Aleluia, J.; Ferrão, P. Characterization of urban waste management practices in developing Asian countries: A new analytic framework based on waste characteristics and urban dimension. *Waste Manag.* **2016**, *58*, 415–429. [CrossRef] [PubMed]
5. Ma, Q.; Zhang, Y.; Hu, F.; Zhou, H. Can the energy conservation and emission reduction demonstration city policy enhance urba domestic waste control? Evidence from 283 cities in China. *Cities* **2024**, *154*, 105323. [CrossRef]
6. Marshall, R.E.; Farahbakhsh, K. Systems approaches to integrated solid waste management in developing countries. *Waste Mana* **2013**, *33*, 988–1003. [CrossRef] [PubMed]
7. Wang, J.; Liu, X.; Zhang, Z.; Liu, Y. Synergistic utilization, critical mechanisms, and environmental suitability of bauxite residu (red mud) based multi-solid wastes cementitious materials and special concrete. *J. Environ. Manag.* **2024**, *361*, 121255. [CrossRe [PubMed]
8. Mir, I.S.; Cheema, P.P.S.; Singh, S.P. Implementation analysis of solid waste management in Ludhiana city of Punjab. *Enviro Chall.* **2021**, *2*, 100023. [CrossRef]
9. Wu, S.; Cao, J.; Shao, Q. How to select remanufacturing mode: End-of-life or used product? In *Environment, Development an Sustainability*; Springer: Berlin/Heidelberg, Germany, 2024; pp. 1–21.
10. Sohoo, I.; Ritzkowski, M.; Sohu, Z.A.; Cinar, S.Ö.; Chong, Z.K.; Kuchta, K. Estimation of methane production and electrical energ generation from municipal solid waste disposal sites in Pakistan. *Energies* **2021**, *14*, 2444. [CrossRef]
11. Pan, X.R.; Shang-Guan, P.K.; Li, S.H.; Zhang, C.H.; Lou, J.M.; Guo, L.; Liu, L.; Lu, Y. The influence of carbon dioxide o fermentation products, microbial community, and functional gene in food waste fermentation with uncontrol pH. *Environ. Re* **2024**, *267*, 120645. [CrossRef] [PubMed]
12. Amin, N.; Aslam, M.; Yasin, M.; Hossain, S.; Shahid, M.K.; Inayat, A.; Samir, A.; Ahmad, R.; Murshed, M.N.; Khurram, M.S.; et a Municipal solid waste treatment for bioenergy and resource production: Potential technologies, techno-economic-environment aspects and implications of membrane-based recovery. *Chemosphere* **2023**, *323*, 138196. [CrossRef] [PubMed]
13. Sohoo, I.; Ritzkowski, M.; Guo, J.; Sohoo, K.; Kuchta, K. Municipal solid waste management through sustainable landfilling: I view of the situation in Karachi, Pakistan. *Int. J. Environ. Res. Public Health* **2022**, *19*, 773. [CrossRef] [PubMed]
14. Aslam, S.; Ali, F.; Naseer, A.; Sheikh, Z. Application of material flow analysis for the assessment of current municipal solid wast management in Karachi, Pakistan. *Waste Manag. Res.* **2022**, *40*, 185–194. [CrossRef]
15. Iqbal, A.; Yasar, A.; Tabinda, A.B.; Haider, R.; Sultan, I.A.; Kedwii, A.A.; Chaudhary, M.M.; Sheikh, M.M.; Nizami, A.S. Waste a Resource for Pakistan: An Innovative Business Model of Regenerative Circular Economy to Integrate Municipal Solid Wast Management Sector. *Sustainability* **2023**, *15*, 6281. [CrossRef]
16. Sohoo, I.; Ritzkowski, M.; Heerenklage, J.; Kuchta, K. Biochemical methane potential assessment of municipal solid wast generated in Asian cities: A case study of Karachi, Pakistan. *Renew. Sustain. Energy Rev.* **2021**, *135*, 110175. [CrossRef]
17. Han, Y.; Zhang, Y.; Yang, Z.; Zhang, Q.; He, X.; Song, Y.; Tian, L.; Wu, H. Improving aerobic digestion of food waste by adding personalized microbial inoculum. *Curr. Microbiol.* **2024**, *81*, 277. [CrossRef] [PubMed]
18. Liu, R.; Jiang, S.; Ou, J.; Kouadio, K.L.; Xiong, B. Multifaceted anomaly detection framework for leachate monitoring in landfill *J. Environ. Manag.* **2024**, *368*, 122130. [CrossRef] [PubMed]
19. Al-Salem, S.M.; Antelava, A.; Constantinou, A.; Manos, G.; Dutta, A. A review on thermal and catalytic pyrolysis of plastic soli waste (PSW). *J. Environ. Manag.* **2017**, *197*, 177–198. [CrossRef] [PubMed]
20. Wang, N.; Zhang, Z.; Zhang, Y.; Xu, X.; Guan, Q. Fe-Mn oxide activating persufate for the in-situ chemical remediation of organi contaminated groundwater. *Sep. Purif. Technol.* **2025**, *355*, 129566. [CrossRef]

Meng, L.; Liang, L.; Shi, Y.; Yin, H.; Li, L.; Xiao, J.; Huang, N.; Zhao, A.; Xia, Y.; Hou, J. Biofilms in plastisphere from freshwater wetlands: Biofilm formation, bacterial community assembly, and biogeochemical cycles. *J. Hazard. Mater.* **2024**, *476*, 134930. [CrossRef]

Arıkan, E.; Şimşit-Kalender, Z.T.; Vayvay, Ö. Solid waste disposal methodology selection using multi-criteria decision making methods and an application in Turkey. *J. Clean. Prod.* **2017**, *142*, 403–412. [CrossRef]

Zhuang, Q.; Li, X.; Lian, X.; Hu, H.; Wang, N.; Wu, J.; Miao, K.; Feng, G.; Luo, X. Catalysis Enhancement of Co_3O_4 through the Epitaxial Growth of Inert ZnO in Peroxymonosulfate Activation: The Catalytic Mechanism of Surface Hydroxyls in Singlet Oxygen Generation. *Cryst. Growth Des.* **2024**, *25*, 319–329. [CrossRef]

Mir, M.A.; Ghazvinei, P.T.; Sulaiman, N.M.N.; Basri, N.E.A.; Saheri, S.; Mahmood, N.Z.; Jahan, A.; Begum, R.A.; Aghamohammadi, N.J.J.O.E.M. Application of TOPSIS and VIKOR improved versions in a multi criteria decision analysis to develop an optimized municipal solid waste management model. *J. Environ. Manag.* **2016**, *166*, 109–115.

Wang, Z.; Ren, J.; Goodsite, M.E.; Xu, G. Waste-to-energy, municipal solid waste treatment, and best available technology: Comprehensive evaluation by an interval-valued fuzzy multi-criteria decision making method. *J. Clean. Prod.* **2018**, *172*, 887–899. [CrossRef]

Coban, A.; Ertis, I.F.; Cavdaroglu, N.A. Municipal solid waste management via multi-criteria decision making methods: A case study in Istanbul, Turkey. *J. Clean. Prod.* **2018**, *180*, 159–167. [CrossRef]

Afrane, S.; Ampah, J.D.; Agyekum, E.B.; Amoh, P.O.; Yusuf, A.A.; Fattah, I.M.R.; Agbozo, E.; Elgamli, E.; Shouran, M.; Mao, G.; et al. Integrated AHP-TOPSIS under a fuzzy environment for the selection of waste-to-energy technologies in Ghana: A performance analysis and socio-enviro-economic feasibility study. *Int. J. Environ. Res. Public Health* **2022**, *19*, 8428. [CrossRef] [PubMed]

Akram, M.; Ramzan, N.; Deveci, M. Linguistic Pythagorean fuzzy CRITIC-EDAS method for multiple-attribute group decision analysis. *Eng. Appl. Artif. Intell.* **2023**, *119*, 105777. [CrossRef]

Paul, T.K.; Jana, C.; Pal, M. Multi-criteria group decision-making method in disposal of municipal solid waste based on cubic Pythagorean fuzzy EDAS approach with incomplete weight information. *Appl. Soft Comput.* **2023**, *144*, 110515. [CrossRef]

Roy, J.; Adhikary, K.; Kar, S. Credibilistic TOPSIS model for evaluation and selection of municipal solid waste disposal methods. In *Advances in Waste Management: Select Proceedings of Recycle 2016*; Springer: Singapore, 2019; pp. 243–261.

Rahimi, S.; Hafezalkotob, A.; Monavari, S.M.; Hafezalkotob, A.; Rahimi, R. Sustainable landfill site selection for municipal solid waste based on a hybrid decision-making approach: Fuzzy group BWM-MULTIMOORA-GIS. *J. Clean. Prod.* **2020**, *248*, 119186. [CrossRef]

Rani, P.; Mishra, A.R.; Krishankumar, R.; Ravichandran, K.S.; Gandomi, A.H. A new Pythagorean fuzzy based decision framework for assessing healthcare waste treatment. *IEEE Trans. Eng. Manag.* **2020**, *69*, 2915–2929. [CrossRef]

Garg, H.; Rani, D. An efficient intuitionistic fuzzy MULTIMOORA approach based on novel aggregation operators for the assessment of solid waste management techniques. *Appl. Intell.* **2022**, *52*, 4330–4363. [CrossRef]

Al-Barakati, A.; Rani, P. Assessment of healthcare waste treatment methods using an interval-valued intuitionistic fuzzy double normalization-based multiple aggregation approach. *Environ. Dev. Sustain.* **2023**, *26*, 1–28. [CrossRef]

Zadeh, L.A. Fuzzy sets. *Inf. Control* **1965**, *8*, 338–353. [CrossRef]

Atanassov, K.T. Intuitionistic Fuzzy Sets. *Fuzzy Sets Syst.* **1986**, *20*, 87–96. [CrossRef]

Molodtsov, D. Soft set theory—First results. *Comput. Math. Appl.* **1999**, *37*, 19–31. [CrossRef]

Maji, P.K.; Biswas, R.; Roy, A.R. Fuzzy soft sets. *J. Fuzzy Math.* **2001**, *9*, 589–602.

Maji, P.K.; Biswas, R.; Roy, A.R. Intuitionistic fuzzy soft sets. *J. Fuzzy Math.* **2001**, *9*, 677–692.

Smarandache, F. Extension of soft set to hypersoft set, and then to plithogenic hypersoft set. *Neutrosophic Sets Syst.* **2018**, *22*, 168–170.

Saeed, M.; Ahsan, M.; Abdeljawad, T. A development of complex multi-fuzzy hypersoft set with application in MCDM based on entropy and similarity measure. *IEEE Access* **2021**, *9*, 60026–60042. [CrossRef]

Rahman, A.U.; Saeed, M.; Smarandache, F. A theoretical and analytical approach to the conceptual framework of convexity cum concavity on fuzzy hypersoft sets with some generalized properties. *Soft Comput.* **2022**, *26*, 4123–4139. [CrossRef]

Yolcu, A.; Smarandache, F.; Öztürk, T.Y. Intuitionistic fuzzy hypersoft sets. *Commun. Fac. Sci. Univ. Ank. Ser. A1 Math. Stat.* **2021**, *70*, 443–455. [CrossRef]

Zulqarnain, R.M.; Xin, X.L.; Saeed, M. Extension of TOPSIS method under intuitionistic fuzzy hypersoft environment based on correlation coefficient and aggregation operators to solve decision making problem. *AIMS Math.* **2020**, *6*, 2732–2755. [CrossRef]

Zulqarnain, R.M.; Siddique, I.; Ali, R.; Pamucar, D.; Marinkovic, D.; Bozanic, D. Robust aggregation operators for intuitionistic fuzzy hypersoft set with their application to solve MCDM problem. *Entropy* **2021**, *23*, 688. [CrossRef]

46. He, Y.; Chen, H.; Zhou, L.; Liu, J.; Tao, Z. Intuitionistic fuzzy geometric interaction averaging operators and their application multi-criteria decision making. *Inf. Sci.* **2014**, *259*, 142–159. [CrossRef]
47. He, Y.; Chen, H.; Zhou, L.; Han, B.; Zhao, Q.; Liu, J. Generalized intuitionistic fuzzy geometric interaction operators and th application to decision making. *Expert Syst. Appl.* **2014**, *41*, 2484–2495. [CrossRef]
48. Garg, H. Some series of intuitionistic fuzzy interactive averaging aggregation operators. *SpringerPlus* **2016**, *5*, 999. [CrossRef]
49. Garg, H. Generalized interaction aggregation operators in intuitionistic fuzzy multiplicative preference environment and th application to multicriteria decision-making. *Appl. Intell.* **2018**, *48*, 2120–2136. [CrossRef]
50. Garg, H. Generalized intuitionistic fuzzy multiplicative interactive geometric operators and their application to multiple criter decision making. *Int. J. Mach. Learn. Cybern.* **2016**, *7*, 1075–1092. [CrossRef]
51. Liu, P.; Wang, Y. Intuitionistic fuzzy interaction Hamy mean operators and their application to multi-attribute group decisi making. *Group Decis. Negot.* **2019**, *28*, 197–232. [CrossRef]
52. Garg, H. Generalized intuitionistic fuzzy interactive geometric interaction operators using Einstein t-norm and t-conorm ar their application to decision making. *Comput. Ind. Eng.* **2016**, *101*, 53–69. [CrossRef]
53. Liu, P. Multiple attribute decision-making methods based on normal intuitionistic fuzzy interaction aggregation operato *Symmetry* **2017**, *9*, 261. [CrossRef]
54. Garg, H.; Arora, R. Novel scaled prioritized intuitionistic fuzzy soft interaction averaging aggregation operators and the application to multi criteria decision making. *Eng. Appl. Artif. Intell.* **2018**, *71*, 100–112. [CrossRef]
55. Wei, G. Pythagorean fuzzy interaction aggregation operators and their application to multiple attribute decision making. *J. Int Fuzzy Syst.* **2017**, *33*, 2119–2132. [CrossRef]
56. Gao, H.; Lu, M.; Wei, G.; Wei, Y. Some novel Pythagorean fuzzy interaction aggregation operators in multiple attribute decisi making. *Fundam. Informaticae* **2018**, *159*, 385–428. [CrossRef]
57. Wang, L.; Garg, H.; Li, N. Pythagorean fuzzy interactive Hamacher power aggregation operators for assessment of expre service quality with entropy weight. *Soft Comput.* **2021**, *25*, 973–993. [CrossRef]
58. Zulqarnain, R.M.; Xin, X.L.; Garg, H.; Ali, R. Interaction aggregation operators to solve multi criteria decision making proble under pythagorean fuzzy soft environment. *J. Intell. Fuzzy Syst.* **2021**, *41*, 1151–1171. [CrossRef]
59. Riaz, M.; Garg, H.; Farid HM, A.; Aslam, M. Novel q-rung orthopair fuzzy interaction aggregation operators and their applicati to low-carbon green supply chain management. *J. Intell. Fuzzy Syst.* **2021**, *41*, 4109–4126. [CrossRef]
60. Farid HM, A.; Riaz, M. Some generalized q-rung orthopair fuzzy Einstein interactive geometric aggregation operators wi improved operational laws. *Int. J. Intell. Syst.* **2021**, *36*, 7239–7273. [CrossRef]
61. Zulqarnain, R.M.; Siddique, I.; Eldin, S.M.; Gurmani, S.H. Extension of interaction aggregation operators for the analysis cryptocurrency market under q-rung orthopair fuzzy hypersoft set. *IEEE Access* **2022**, *10*, 126627–126650. [CrossRef]
62. Zulqarnain, R.M.; Siddique, I.; Iampan, A.; Awrejcewicz, J.; Bednarek, M.; Ali, R.; Asif, M. Novel multicriteria decision maki approach for interactive aggregation operators of q-rung orthopair fuzzy soft set. *IEEE Access* **2022**, *10*, 59640–59660. [CrossRe
63. Keshavarz Ghorabaee, M.; Zavadskas, E.K.; Olfat, L.; Turskis, Z. Multi-criteria inventory classification using a new method evaluation based on distance from average solution (EDAS). *Informatica* **2015**, *26*, 435–451. [CrossRef]
64. Ghorabaee, M.K.; Zavadskas, E.K.; Amiri, M.; Turskis, Z. Extended EDAS method for fuzzy multi-criteria decision-making: A application to supplier selection. *Int. J. Comput. Commun. Control* **2016**, *11*, 358–371. [CrossRef]
65. Srivastava, P.; Mustafa, A.; Khanduja, D.; Chowdhary, S.K.; Kumar, N.; Shukla, R.K. Prioritizing autonomous maintenance syste attributes using fuzzy EDAS approach. *Procedia Comput. Sci.* **2020**, *167*, 1941–1949. [CrossRef]
66. Kahraman, C.; Keshavarz Ghorabaee, M.; Zavadskas, E.K.; Cevik Onar, S.; Yazdani, M.; Oztaysi, B. Intuitionistic fuzzy EDA method: An application to solid waste disposal site selection. *J. Environ. Eng. Landsc. Manag.* **2017**, *25*, 1–12. [CrossRef]
67. Mishra, A.R.; Mardani, A.; Rani, P.; Zavadskas, E.K. A novel EDAS approach on intuitionistic fuzzy set for assessment health-care waste disposal technology using new parametric divergence measures. *J. Clean. Prod.* **2020**, *272*, 122807. [CrossRef
68. Schitea, D.; Deveci, M.; Iordache, M.; Bilgili, K.; Akyurt, I.Z.; Iordache, I. Hydrogen mobility roll-up site selection usin intuitionistic fuzzy sets based WASPAS, COPRAS and EDAS. *Int. J. Hydrogen Energy* **2019**, *44*, 8585–8600. [CrossRef]
69. Liu, C.; Rani, P.; Pachori, K. Sustainable circular supplier selection and evaluation in the manufacturing sector using Pythagorea fuzzy EDAS approach. *J. Enterp. Inf. Manag.* **2022**, *35*, 1040–1066. [CrossRef]
70. Li, P.; Liu, J.; Wei, C.; Liu, J. A new EDAS method based on prospect theory for Pythagorean fuzzy set and its application i selecting investment projects for highway. *Kybernetes* **2022**, *51*, 2636–2651. [CrossRef]
71. Zhang, S.; Wei, G.; Gao, H.; Wei, C.; Wei, Y. EDAS method for multiple criteria group decision making with picture fuzz information and its application to green suppliers selections. *Technol. Econ. Dev. Econ.* **2019**, *25*, 1123–1138. [CrossRef]

. Darko, A.P.; Liang, D. Some q-rung orthopair fuzzy Hamacher aggregation operators and their application to multiple attribute group decision making with modified EDAS method. *Eng. Appl. Artif. Intell.* **2020**, *87*, 103259. [CrossRef]

. Mishra, A.R.; Rani, P.; Pandey, K. Fermatean fuzzy CRITIC-EDAS approach for the selection of sustainable third-party reverse logistics providers using improved generalized score function. *J. Ambient Intell. Humaniz. Comput.* **2022**, *13*, 295–311. [CrossRef]

sustainability

MDPI

Article

Presence of Humic Acids in Landfill Leachate and Treatment by Flocculation at Low pH to Reduce High Pollution of This Liquid

Carlos Costa *, M. Laura Pinedo and Brayan D. Riascos

Chemical Engineering Department, Faculty of Chemical Sciences, University of Salamanca, Plaza de la Merced s/n, 37008 Salamanca, Spain; lpinedomaraz@gmail.com (M.L.P.); riascosartea@wisc.edu (B.D.R.)

* Correspondence: ccosta@usal.es; Tel.: +34-923-294479; Fax: +34-923-294574

Abstract: Humic substances are abundant in landfill leachate, especially humic acids, which are insoluble at low pH in aqueous solutions. Focusing on the chemical properties of humic acids, we describe in this work a new method for a sustainable treatment of landfill leachate originated from solid wastes, which consists of the reduction of organic load (COD, chemical oxygen demand) and colour and is based in the gradual decrease in pH to the value in which HAs are insoluble in water solution. Zeta potential values mark the chemical stage of humic acids because of ionisation–protonation of the phenolic and carboxylic groups, and this parameter is monitored during flocculation, changing from −16.8 mV at pH 7.7 to 0.0068 mV at pH 2.0, when HAs precipitate. The final result is the reduction in the organic matter content (COD) and colour in the leachate, 86.1% and 84.7%, respectively. Solids produced by precipitation during the acidification treatment have been characterized by elemental chemical analysis and Fourier transform infrared spectrometry, concluding a high similarity in chemical composition with commercial and natural humic acids. Protonated humic acids at low pH can interact with other molecules by hydrogen bonds and form bigger molecular structures much more unstable in suspension, which conduct to precipitation. The mean diameter of the humic acids aggregates was measured, detecting the formation of big molecular structures at low pH. This process is analysed and compared economically with other processes proposed for landfill leachate treatment, resulting in a promising technique for the management of this residue.

Keywords: waste management; landfill leachate treatment; flocculation; humic acids; pH

1. Introduction

Landfill leachate is a liquid produced by concentration and storage of the highly polluted liquid originated from solid wastes deposited in the landfill and is considered a complex liquid with high content of organic matter, ammonia, sulphate, chloride, and metals [1,2]. This liquid can be treated by biological treatment to reduce organic matter and toxicity [3,4] or by other treatment processes like flocculation, chemical oxidation, or reverse osmosis [5–7]. The nature and concentration of the organic matter in landfill leachate are the main factors affecting the successful treatment of this liquid.

Concentration of organic matter can be very different in landfill leachates, and normally, when the COD value is low, landfill leachate can be processed by biological treatment [1,2,8,9]. In high organic loaded leachates, the reduction of COD is normally not well performed [10–12]. Organic matter present in these high organic loaded leachates has the origin mostly in composting of organics from solid wastes (composting leachate) [13–15].

Biological treatment combined with flocculation or oxidation is accepted currently as the most effective treatment of landfill leachates [16–19]. Flocculation is economically sustainable, compared to other processes like ozonation or Fenton oxidation [20–22]. If biological treatment is planned to be applied, the biodegradability of the organic compounds present in the leachate has to be considered [23].

In landfill leachates, the presence of humic substances (HSs) has been described in the literature [24–26]. Especially in composting leachate, the odour which is proximate to wood soil and the dark yellow and black colours of most landfill leachates are produced by HSs [27–29]. HSs are known to be recalcitrant compounds [30,31] and conduct to low reduction of COD in the biological treatment. They have been cited to be abundant in landfill leachates, reaching sometimes 60% of the total organic matter [30,32].

HSs are divided into two fractions: humic and fulvic acids (HAs and FAs), with different solubility pertaining to pH [33]. HAs are insoluble at low pH and FAs are soluble at low, medium, and high pH [24,32,34,35]. The insolubility of HAs when the pH is low is due to the protonation under acidic conditions of the functional groups which are present in these substances: carboxylic acids and phenolic alcohols. The difference in the pH-dependent forms of the carboxylic acids and phenols produces a Zeta potential variation due to ionisation–protonation [36]. In addition, protonation of the functional groups of the HAs at low pH can facilitate the formation of H-bonds among different molecules and induce flocculation of the intermolecular structures formed. At a higher pH (slightly acid or neutral), carboxylic acids and phenolic alcohols are ionized, and the Zeta potential increases because of the formation of positive ions in the HA surface. This increase in the Zeta potential absolute value as a consequence of ionisation of the functional groups of HAs can maintain in suspension HAs because of electrostatic repulsion between different molecules. In the literature, ionisation of carboxylic acids at a pH over 3.5–4.0 and phenolic groups over 5.5–6.0 is reported [37]. In accordance with these considerations, flocculation of HAs from landfill leachate has to be performed at low pH, because at a slightly acidic, neutral, or high pH, HAs will be maintained in suspension [38].

In this article, we prove an important presence of HAs in landfill leachate, which proceed from composting leachate, a frequent situation in high-COD leachates. These recalcitrant organics are flocculated at low pH in accordance with their chemical properties and stability of aqueous suspensions. Flocculation at low pH is the main treatment for the reduction in recalcitrant compounds (HAs), and the explanation of this novel procedure is focused on the chemical properties of HAs at different pH values. Flocculation at low pH significantly reduces the COD value of landfill leachates, especially with high organic load, and is considered an effective and sustainable treatment of this polluted liquid. The presence of HAs in landfill leachate was proved via elemental chemical analysis (CHNS) and infrared spectrometry (FT-IR), and the evolution of the molecular size of the HAs and aggregates are monitored by the analysis of the mean diameter during the flocculation at low pH.

2. Materials and Methods

2.1. Characterisation of Landfill Leachate

Landfill leachate was characterized in a laboratory analysing the samples collected in the leachate deposit (March–June) of the Waste Treatment Centre (WTC) of Salamanca, Spain. In the characterisation of this highly polluted liquid, pH, organic load (chemical oxygen demand, COD), colour, solids (total solids, TS and volatile solids, VS), nitrogen compounds (organic nitrogen, ammonia, nitrates, and nitrites), phosphate, sulphate, chloride, and metals were measured by the standard methods described in the literature [39].

The pH was measured using a pHmeter with precision 0.1 pH units (CRISON MicropH 2000, Crison, Barcelona, Spain).

COD and colour were chosen for monitoring the effectivity in the treatment of landfill leachate by flocculation at low pH. COD was measured in triplicate by the colorimetric closed reflux method (5220D, [39]), using $HgSO_4$ to avoid the interference of Cl^- as described in the standard method (precipitation of $HgCl_2$). Because of the high concentration of organics in leachate (high COD value), samples were diluted 8 times, and a pattern calibration line was used for COD values in the range 2000–20,000 mg/L (potassium hydrogen phthalate). Because of the presence of HSs in the landfill leachate which can interfere in colour analysis, the spectrophotometric method was selected for measuring colour by detecting the maximum absorbance wavelength (chromatic method, method 2120C, [39]).

Solids were analysed by the standard methods described in the literature for total and volatile solids, based on thermal treatment at 105 °C and 550 °C (TS: 2540B, VS: 2540E [39]).

Organic nitrogen (N_{org}) and ammonia (NH_4^+) were analysed with the standard method (Kjeldahl, 4500-NH_3 B and E, [39]) after filtration by Millipore filters (0.45 µm). N_{org} analysis was performed performing a previous digestion in H_2SO_4 before Kjeldahl distillation (4500-N_{org} B, [39]).

Samples were also filtered using Millipore filters (0.45 µm) for NO_3^-, NO_2^-, SO_4^{2-}, PO_4^{3-}, and Cl^- analysis and measured via ionic chromatography (ICS 2000 ISO 10304-1 [40], Dionex, Thermo Fisher Scientific, Madrid, Spain). ICS 2000 measurement conditions for 10 µL injection volume were as follows: cell temperature: 35 °C, column temperature: 30 °C, eluent flow rate: 1.00 mL/min (35 mM KOH), and applied current: 50 mA. The column used was 4 × 250 mm IonPac AS19, and the guard was 4 × 50 mm IonPac AG19 (Dionex, Thermo Fisher Scientific, Madrid, Spain). The software Chromeleon 6.5 SP2 of Dionex was employed for calibration and measurement of all these anions together in a unique chromatogram.

Metals present in leachate were analysed via ICP-MS (Agilent 7800, Agilent Technologies, Madrid, Spain, Chemical Analysis Service of the University of Salamanca). Because of the high organic content in landfill leachate, the samples were treated with concentrated nitric acid in a closed Teflon glass (50 mL final sample volume). The ICP-MS spectrometer was calibrated with two multi-element patterns prepared from certificated Panreac standard solutions (1000 mg/L). Measurement flux conditions of the ICP-MS were as follows: plasma gas: 15 L/min, auxiliary gas: 0.9 L/min, nebulisation gas: 0.99 L/min, and cell gas (Helio): 4.5 mL/min.

2.2. Analysis of Humic Acids

HAs analysis in the literature is based on a suggested method with two variants, depending in concentration (5510, [39]). In natural waters with a low HSs content, the method proposed is a concentration of HSs by using ionic exchange columns and the measurement of dissolved organic carbon (DOC) [41]. In liquids with high HSs content, HAs have to be precipitated at acid pH, attending the chemical properties of HAs by the method described by Christensen et al., 1998 [30,32,34]. Following the method of Christensen et al., 200 mL leachate samples were acidified with HCl concentrated to pH = 1.0 and placed for 24 h. The precipitate after this time was purified twice during 48 h in total with 100 mL acid solution (0.5% HF v/v and 0.5% HCl v/v). The resulting precipitate, filtered by Millipore filters (0.45 µm), was purified 3 times with HCl acid solution (pH 2.0), separated in small portions, and placed for a further analysis in an incubator (35 °C, 5 days).

HAs obtained by the procedure described before were characterized via CHNS analysis using a LECO CHNS-932 analyzer (Leco Instruments, Madrid, Spain, Elemental Microanalysis Service of the UCM, Madrid University) and infrared spectrometry (FT-IR,

Perkin Elmer SpectrumTwo, Perkin Elmer-Pharmatech, Madrid, Spain). Infrared spectra were obtained from samples prepared in pellets of the HAs from leachate and commercial AHA (Aldrich humic acid) in KBr, measuring the absorbance in the range of wavenumber 4000–400 cm^{-1}.

2.3. Reduction in pH

The reduction in pH in landfill leachate was performed using H_2SO_4 9 M for acidification. Samples were collected during pH reduction separated by 0.5 units of pH and COD, colour, and Z potential were quantified in triplicate to monitor leachate treatment related to the chemical properties of the HAs.

Zeta potential and mean diameter were analysed in samples using the same equipment (Malvern Zetasizer Nano ZS, Malvern Panalytical, Massy, France).

3. Results and Discussion

3.1. Composition of Landfill Leachate

Composition of landfill leachate is shown in Table 1 where most important parameters are recorded (pH, organic load, solids, colour, and nutrients). COD value is very high: 42,000 ± 2516 mg/L, in view of other data from the literature [1,2,4] but is coincident with values for leachates with a high content of organic matter [24,42]. The high value of the organic matter concentration is due to the characteristics of this liquid, produced from the solid wastes deposited in the landfill and the liquid originated from them, and mixed with the fluid residue proceed from composting of the organic fraction of these solids (composting leachate) [43]. As a consequence of aerobic oxidation of organic matter, composting process produces the formation of HSs, conducing to the formation of alcohols (phenols) and especially carboxylic acids [44].

Table 1. Characterisation of landfill leachate described by the most important parameter values: pH, organic load (COD), solids (total and volatile solids, TS and VS), colour, nutrients (nitrogen and phosphorous compounds), sulphate and chloride. In nutrient analysis (N-compounds, SO_4^{2-}, PO_4^{3-} and Cl^-) five independent samples of landfill leachate were obtained in spring (March–June) from the WTC. Samples of pH, COD, TS, VS, and colour were analysed in triplicate from one sample from the WTC.

Parameter	Mean ± SD
pH	7.7 ± 0.1
COD	42,000 ± 2516
TS	16,410 ± 486
VS	10,076 ± 620
Colour	0.537 ± 0.000
N_{org}	3795 ± 314
NH_4^+	2484 ± 340
NO_3^-	417 ± 115
NO_2^-	11 ± 1
SO_4^{2-}	223 ± 64
PO_4^{3-}	39 ± 18
Cl^-	11,710 ± 1978

Concentrations in mg/L, except pH (pH units) and colour (ABS_{474nm}).

The value of solids is high in leachate (total solids, TS: 16,410 ± 486 mg/L), with high proportion of organic solids: more than 60% of TS are vs. (VS = 10,076 ± 620 mg/L, Table 1).

Colour in landfill leachate has been analysed frequently in the literature by the method of the Pt-Co scale [45,46], which compare the tonality of the liquid with prepared standard solutions of K_2PtCl_6 with $CoCl_3$. This method does not reflect the real colour of this liquid because the maximum wavelength absorbed is not measured.

The analysis of colour in the leachate was performed by the chromatic method described in Section 2.1, which obtains the predominant wavelength absorbed by this liquid, shown in Figure 1. Although colour in the leachate can change after flocculation at low pH, by this method, the original colour of the liquid was measured. This predominant wavelength was 474 nm, which is assigned in the chromaticity diagram to dark blue tonalities. The quantitative value of colour obtained is high in Table 1 (ABS = 0.537), and the external appearance of this liquid is dense with a dark brown colour and a blue-yellowish iridescence.

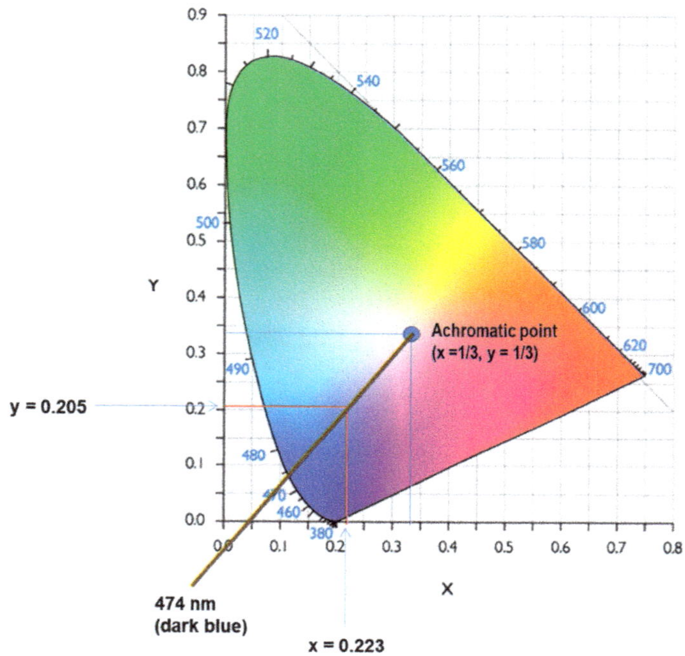

Figure 1. Chromaticity diagram for landfill leachate, where x = 0.223 and y = 0.205 are the trichromatic coefficients obtained by spectrophotometric method for colour analysis (2120C, [39]). Dominant wavelength, obtained by the union of the achromatic point and the trichromatic coefficients point, is 474 nm (dark blue colour).

The concentrations of nutrients reported in Table 1 show high content of Cl^-, NH_4^+, NO_3^-, and SO_4^{2-}, the most abundant negative ions in this liquid. Cl^- is extremely abundant, reaching a concentration close to 12 g/L; also, ammonia is highly abundant (2.5 g/L), and PO_4^{3-} and NO_2^- are present in low concentration [47]. These concentrations for nutrients are consistent with the values obtained by other authors [24,42,48,49].

Metal ions content in landfill leachate shows high concentrations of alkali metals Na^+ and K^+ (>2000 mg/L) and abundance of alkaline earth metals Ca^{2+} and Mg^{2+} (150–600 mg/L, Figure 2). P(V), Al^{3+}, and Fe^{3+} appear in this liquid with visible concentration (10–40 mg/L). Zn^{2+}, Mn^{2+}, Cr^{3+}, and Cu^{2+} contribute in very low concentrations (1.2–2.1 mg/L) and other cations (Ni^{2+}, As^{5+} and Pb^{2+}) below 1 mg/L. The content of metal ions in this landfill leachate can be considered normal in accordance with data from literature [42,49,50].

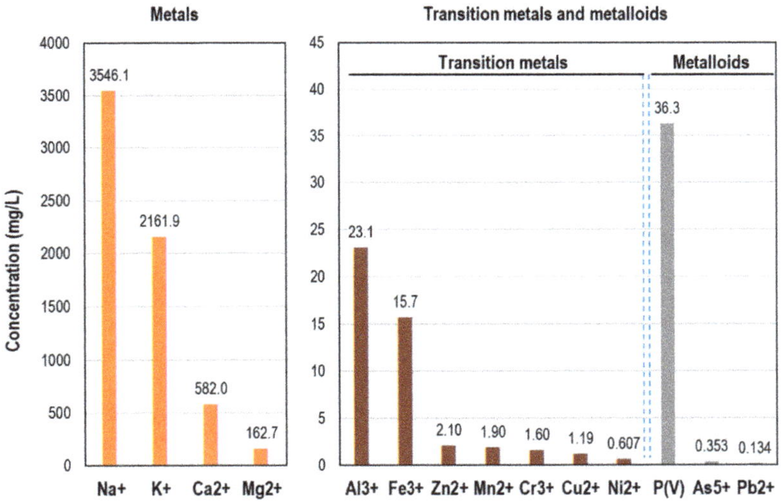

Figure 2. Concentration of metal ions in landfill leachate (metals, transition metals, and metalloids), obtained in one sample from the WTC (average relative SD 2%). Metals and metalloids are shown in the horizontal axe and concentration in leachate (mg/L) in the vertical axe, with the numerical value of concentration over the corresponding bar. Cd^{2+} and Hg^{2+} were detected in trace concentrations (<3 μg/L and <0.1 μg/L, respectively).

3.2. Flocculation of Landfill Leachate at Low pH

Flocculation of landfill leachate at low pH was performed by using H_2SO_4 9 M for a gradual acidification. This liquid flocculates at pH 5.0 (Figure 3a), and COD is reduced from an initial value of 42,000 ± 2516 mg/L to a final value of 5850 ± 923 mg/L at pH 2.0 (86.1% reduction, numerical values and analysis reproducibility in Supplementary Materials, Table S1).

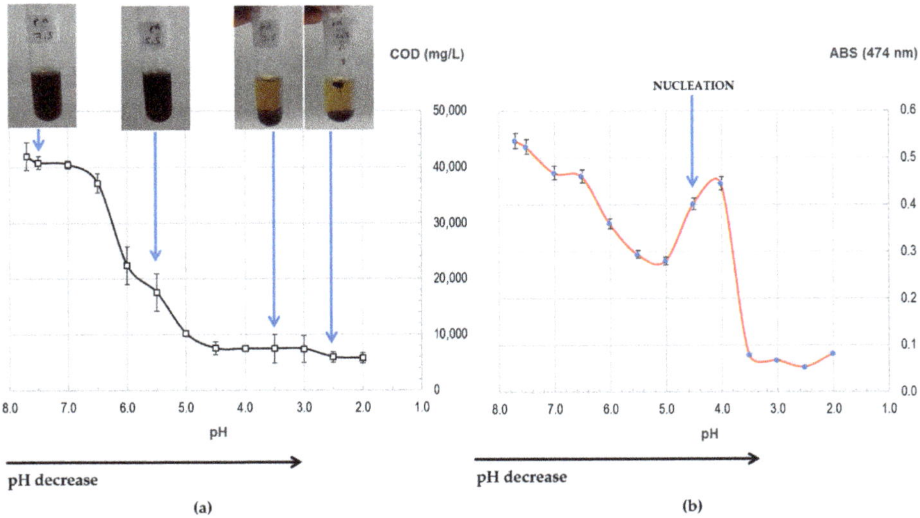

Figure 3. Results of flocculation of landfill leachate at low pH in COD (**a**) and colour reduction (**b**). Error bars represent SD of three independent measurements. Nucleation occurs in colour reduction graph (**b**) at pH 4.5. Numerical values of the graphs in Supplementary Materials, Table S1.

Colour reduction is observed during flocculation at low pH in Figure 3b from initial ABS_{474nm} = 0.537 ± 0.000 (pH 7.7) to 0.082 ± 0.000 (pH 2.0, 84.7% reduction, numerical values and analysis reproducibility in Supplementary Materials, Table S1). ABS increases at pH 4.5 and more visible at pH 4.0, and this is explained by nucleation, the beginning of flocculation, when the incipient formed flocs increase ABS and precipitation starts. This flocculation experiment was repeated three times, and the behaviour observed was the same, producing clearly flocculation and precipitation with formation of an important quantity of foam.

During flocculation at low pH, Z potential value was analysed (Figure 4). The initial value of Z potential in landfill leachate is −16.8 ± 1.4 mV at pH 7.7 and when the pH value decreases below 5.0, Z potential diminish quickly to zero (Figure 4 and numerical values and analysis reproducibility in Supplementary Materials, Table S1). The decay in absolute value in Z potential is highly evident at pH below 3.5. This initial value of Z potential in leachate (−16.8 ± 1.4 mV), in the range −15 to −25 mV, is in accordance with the value obtained recently by other authors [46,51]. Decay of Z potential to zero is the clear evidence of protonation of all the functional groups of HAs in landfill leachate, because is the demonstration of the neutral surface of the molecules without ionic charges.

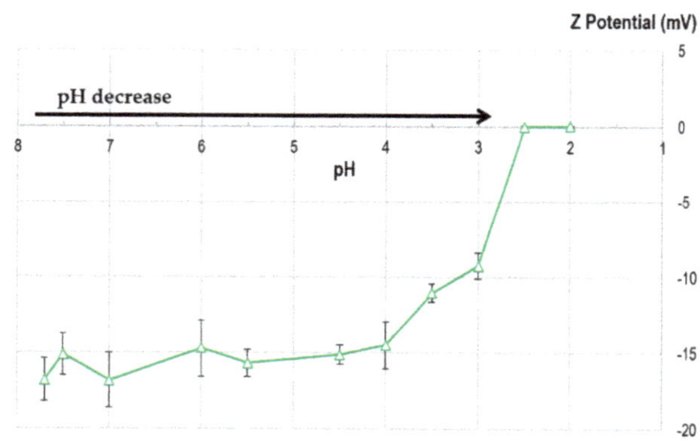

Figure 4. Results of the analysis of Z potential during the flocculation of landfill leachate at low pH, from the initial value of pH 7.7 in the leachate, to the final value of pH 2.0, in the graph from the left to the right. Horizontal axis represents pH, and vertical axis on the right refers to Z potential value in mV. Error bars represent SD of three independent measurements. Numerical values of Z potential in Supplementary Materials, Table S1.

Mean diameter was measured during flocculation at low pH to follow particle size evolution in landfill leachate. The initial distribution of particles in this liquid is not very diverse attending mean ± SD values in leachate in Figure 5 (83.03 ± 53.52 nm), because the standard deviation is lower than expected. During flocculation at low pH procedure, particle size increases at pH = 5.0 (from 83.03 ± 53.52 nm to 162.70 ± 62.35 nm) and big particles are formed at pH 3.0 and 2.0 in Figure 5 (921.5 ± 411.1 nm, 6.6% and 1780.0 ± 608.5 nm, 5.0%, respectively). The explanation of the formation of these big aggregates is assumed to be the intermolecular interaction of different molecules of HAs by formation of H-bonds. Mean diameter was measured in the precipitate, resulting in a solid composed by much bigger particles: 5590 ± 579.8 nm (100%). This value in the size of the particles (5.6 μm) is coincident with the observed value in the literature for the aggregates of natural HAs, several micrometres [52].

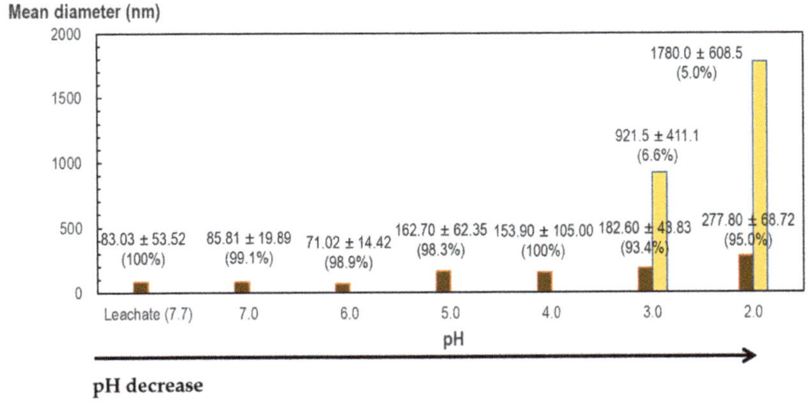

Figure 5. Particle size (mean diameter, nm) during flocculation at low pH of landfill leachate, mean ± SD (triplicate). Formation of big aggregates by intermolecular H-bonds is confirmed at pH 3.0 and 2.0 (yellow bars).

Flocculation of landfill leachate with the decrease in pH is explained because of the presence of HAs in this residue, with an important fraction of the liquid produced in composting of organic matter from solid wastes. The typical molecular structure of a HA is a long aromatic chain with some ramifications in which the functional groups are located. These functional groups are carboxylic acids normally situated at the end of the aromatic chain, phenolic groups distributed in the molecule, and also few sugar and peptides located in the central part of the molecule (Figure 6, HA).

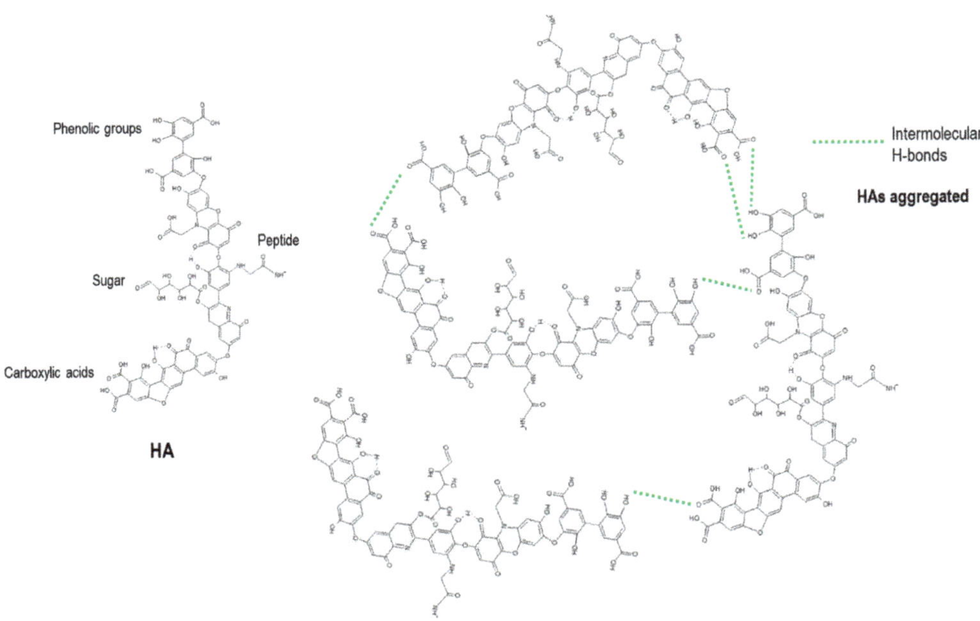

Figure 6. Proposal of the aggregation of humic acid molecules (HA) from landfill leachate protonated at low pH by intermolecular interaction and the formation of H-bonds (interrupted green lines).

In accordance with the chemical properties of HAs, in the gradual decrease in pH, phenolic alcohols are firstly protonated at pH below 5.5 and carboxylic acids lately at pH below 3.5 [37]. When the pH reaches 2.0 and the functional groups of the HAs are protonated, the abundance of protons in the surface of the HA molecules favours the aggregation by interaction of intermolecular H-bonds [37,53]. In Figure 6, this aggregation has been visualized, showing the interaction of four molecules of HAs by intermolecular H-bonds for the formation of a round structure with much higher molecular weight. This structure formed by the union of HAs, with the functional groups protonated, will be maintained hardly in suspension and will precipitate.

While the precipitation process of HAs is producing, other organic molecules are dragged, and the reduction in the global COD value can be highly increased (86.1%). This high reduction in the COD value is much more than the expected in accordance with the percentage of HAs in the total organic matter [30].

The visual appearance of landfill leachate changes significantly after flocculation at low pH. Clarification is produced from a dark brown, the initial colour observed in leachate, to a light yellowish-brown colour of a cleaner liquid (Figure 3a). During precipitation, the formation of foam is observed and is attributed to the formation of HA aggregates.

3.3. Humic Acids Characterisation

Samples collected from the precipitate and washed in HF-HCl acid medium (Section 2.2) were analysed after flocculation at low pH, in which HCl was used to avoid the interference of H_2SO_4 in CHNS chemical analysis. These samples were analysed via the CHNS analysis and FT-IR.

Figure 7 records CHNS elemental chemical analysis, where two independent samples of the precipitate (HA1 and HA2) were measured to assure reproducibility of the results obtained and HA from Aldrich (AHA), for comparison, was analysed and also compared with data of commercial (H Aldrich) and natural (HA aquatic and HA terrestrial) humic acids from literature. The results were normalized to 100% of total percentage of organic components, and oxygen (O) was obtained by the difference to 100% of the sum of the other four elements in percentage (C, H, N, and S). The ratios of hydrogen, oxygen, and nitrogen to carbon (H/C, O/C, and N/C) were obtained from the results of element analysis values. The carbon percentage in HAs from landfill leachate is in between the percentage of aquatic and terrestrial HAs and AHA (commercial) and close to those values and the value of H Aldrich from the literature [24]. The percentage of nitrogen in HAs is higher than the other references (natural and commercial HAs), due to amines and amides present in HA molecules and the important concentration of N-ammonia in the samples of landfill leachate (Table 1), which are coincident with the literature [24,54]. Oxygen, present in the main functional groups of the HAs (carboxylic acids and phenols), is very similar in percentage in HAs from landfill leachate and natural HAs, aquatic and terrestrial, especially aquatic. The sulphur percentage reference value has not been found in data from the literature, and the result measured from commercial HA (AHA) is below detection limit in Figure 7 (numerical values in Supplementary Materials, Table S2). The presence of S in the samples of HAs from landfill leachate ($1.60 \pm 0.35\%$, Supplementary Materials, Table S2) can be attributed to sulphate, which is abundant in leachate (Table 1).

With regard to element ratios in Figure 7, H/C ratio value close to 1.0 is indicative of a chemical structure with abundance of aromatic rings (AHA and natural HAs). When the value of H/C is over 1.0 (HA1 and HA2, landfill leachate), this is indicative of the abundance of aliphatic groups. O/C ratios in HAs from landfill leachate are higher related to natural humic acids (abundance of carboxylic acids and phenolic groups) but coincident with H Aldrich value from the literature and lower compared to analysed AHA value. N/C ratios are higher in HAs from leachate compared to the other values from natural and commercial humic acids, as commented before (ammonia in leachate, Table 1).

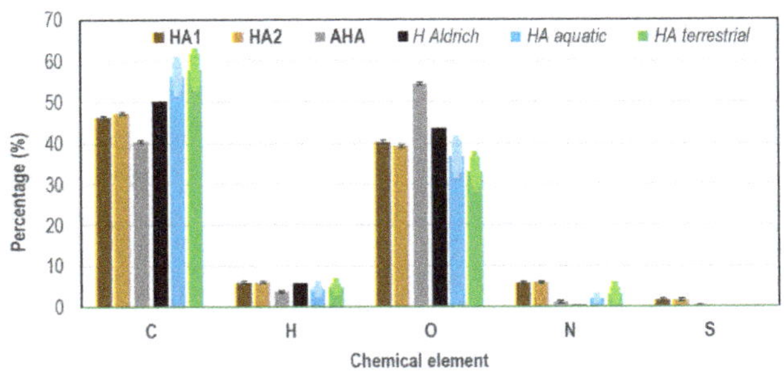

Ratio	HA1	HA2	AHA	H Aldrich	HA aquatic	HA terrestrial
H/C	1.53	1.52	1.08	1.35	0.74–1.15	0.56–1.29
O/C	0.65	0.62	1.01	0.65	0.42–0.58	0.36–0.51
N/C	0.11	0.11	0.02	0.01	0.029–0.034	0.022–0.076

Figure 7. Elemental chemical analysis (carbon, hydrogen, nitrogen, and sulphur: CHNS) of the precipitate of the landfill leachate after treatment at low pH (humic acids). The vertical axis represents percentage of the element included in the horizontal axis. HA1 and HA2 are two independent samples collected from the leachate precipitate. Aldrich humic acid (AHA) and humic acids from the literature are included for comparison (H Aldrich, HA aquatic, and HA terrestrial, [24]). Analysis of HAs (HA1 and HA2) and AHA were performed in triplicate and error bars record SD. For the external data (*HA aquatic* and *terrestrial*, from the literature), error bars represent the range in which the value is recorded.

The result of FT-IR spectrometric characterisation of HAs in the precipitate of landfill leachate after pH reduction treatment is shown in Figure 8, where FT-IR spectra of HAs (red line) and Aldrich humic acid (black line) are recorded.

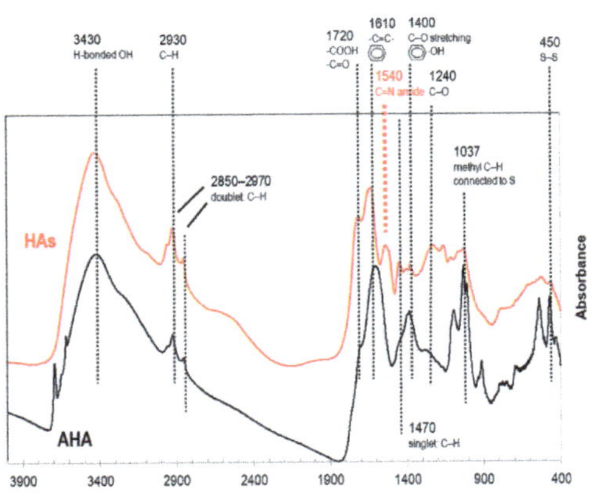

Figure 8. FT-IR spectra of humic acids proceed from the precipitate of landfill leachate (HAs, red line) and Aldrich humic acid (AHA, black line). Characteristic signals of humic acids are highlighted, red dashed line only for HAs in landfill leachate.

The most representative signals of HAs are assigned and explained in detail in accordance with the literature data [24,30,55]. There is a wide band (3400–3000 cm^{-1}) with the maximum at 3430 cm^{-1}, which is assigned to H-bonds formed between different molecules (aggregation of HAs) [30], and also, this signal is indicative of phenol, hydroxyl, and carboxyl functional groups (O-H stretching). There are two other significative signals appearing in HAs and AHA spectra and assigned to aliphatic groups (C-H) of organic structures: the singlet at 1470 cm^{-1} and especially the doublet at 2970–2850 cm^{-1}. In the FT-IR spectra of HAs in leachate and AHA, the characteristic peaks of C=O bonds

(1720 cm^{-1}) in ketonic and protonated carboxylic acid groups are visible, and the signals of stretching of double carbon bonds (C=C) are also highly visible (1610 cm^{-1}), typical of aromatic rings. There is a signal at 1540 cm^{-1} only visible in HAs and not AHA attributed to amide groups (C=N stretching), which is not visible in AHA due to a lower content of N (Figure 7). The peak at 1400 cm^{-1} is assigned to phenolic OH (C-O stretching and deformation of O-H). The slight signal located at 1240 cm^{-1} is attributed to the C-O bond (stretching). The peak at 1037 cm^{-1}, present specially in AHA and also visible in HAs, is assigned to rocking vibrations of aliphatic groups (C-H) connected to S. The S content observed in CHNS analysis in AHA was below the detection limit and was detected in HAs (Figure 7). Finally, the peak at 450 cm^{-1} has been assigned to S-S bonds [30], slightly visible in HAs from leachate and much more visible in AHA, with much lower S content.

3.4. Application, Scalability, and Sustainability

Flocculation at low pH of landfill leachate is a sustainable process for the treatment of this highly polluted liquid, with content in composting leachate rich in humic acids, which actually is processed by ozonation or Fenton oxidation, as described in the introduction [56,57] or these processes combined with coagulation [58,59]. Reverse osmosis has also recently been described to reduce pollution and toxicity of this liquid [60] and the use of natural resources, like extracts of plants, were also applied as tentative processes, for example, in coagulation of high charged compost leachate (63,500 mg/L COD), reaching 39.8% in COD removal [61]. Most effective treatments for landfill leachate are currently proposed as combined or hybrid processes which enclose various operations: flocculation–ultrafiltration with 400 mg/L $FeCl_3$ and ultrafiltration membranes for industrial landfill leachate [62], Fenton oxidation–adsorption using biochar as adsorbent, with 82.6% of COD removal for an initial value of 3155 mg/L [63], advance electrolysis after nanofiltration (1281 mg/L COD, with removal of 87%) [64], and, more recently, the combination of physicochemical and biological treatments [65–67]. These treatments are less sustainable, produce more residues and with higher economical costs, and normally have been proven for leachates with lower organic charge and content in HAs.

In the process described in this article, the concentration of H_2SO_4 required for pH 2.0 is 0.01 M, and assuming a price of 10 euros/L of this chemical compound (36 M, 0.28 euros/mol), the price for the treatment of the landfill leachate is 2.8 euros/m^3. This cost is much lower compared to the processes proposed for COD reduction in landfill leachate, for example, electrocoagulation–oxidation [65] with 19 euros/m^3 in energy (32 kWh/kg COD) for a liquid with COD = 6 197.5 $\pm$ 539.9 mg/L or advance electrolysis with 7.2 euros/m^3 (91.9 kWh/kg COD for 1281 mg/L COD) [64]. In the coagulation–ozonation process [59], using polymeric ferric sulphate (250 mg/L) at pH 4.0 and ozone, the reduction in COD obtained was 88.3%, and the cost of ozone generation could be estimated from data for wastewater treatment in 100 kWh/(kg ozone) for 20 mg/L of ozone concentration, as described in the article: 0.2 euros/m^3 (COD = 3855 mg/L). The total cost of adding the flocculant (20 euros/kg, 5 euros/m^3) would be 5.2 euros/m^3.

Comparing flocculation at low pH with other flocculation processes adding chemicals like 700 mg/L alum to reach 54% of COD removal for a very low concentrated landfill leachate (1 020 mg/L COD [68]), the reduction in pH by the addition of H_2SO_4 0.01 M (980 mg/L) with 86.1% of COD removal for a high concentrated landfill leachate proposed in this work appears much more competitive and does not increase the metal concentration in the liquid favouring downstream processes.

During flocculation at low pH, the estimated volume of HAs produced is about 1/30 the volume of leachate treated (30 L for 1 m^3 of landfill leachate), and this precipitate is proposed to be applied in the amendment of agricultural lands, like is actually conducted with solid sludges from wastewater treatment plants, although HAs are considered valuable products and could be used in a circular/sustainable process for the removal of heavy metals in wastewater [34,69].

Landfill leachate after flocculation at low pH could be biologically treated in a combined binary process flocculation–biological treatment for a further COD reduction [18], and in this case, pH has to be readjusted to a neutral value. No toxic/corrosive effects are expected because of sulfuric acid addition in low concentration (pH 2.0, H_2SO_4 0.01 M), only the increase in sulphate concentration in the liquid.

4. Conclusions

In this article, we present a new method for the treatment of highly polluted landfill leachate for a sustainable use and treatment of this residue, consisting in the flocculation at low pH. Highly polluted landfill leachate has a high proportion of composting leachate (composting of solid wastes), which means the abundance of humic acids in this liquid. In view of the chemical properties in aqueous solutions of these HAs, which are protonated at low pH and are ionized at medium–high pH, landfill leachate is effectively flocculated at pH 2.0.

Under acidic conditions, phenolic alcohols present in HAs are protonated at pH levels below 5.5 and carboxylic acids at pH levels below 3.5. The protonation of the functional groups of the HA molecules reduces the high Zeta potential absolute value to zero and conducts to the formation of intermolecular H-bonds, resulting in big aggregates which are unstable in solution and precipitate. During the precipitation process of HAs, other organic molecules are dragged, and COD and colour are reduced in high extent (84–86%).

The mean diameter of the molecules has been measured during flocculation at low pH, and the formation of big aggregates has been detected at pH 3.0 and 2.0 and also in the precipitate. This precipitate was analysed via the CHNS chemical analysis and infrared spectrometry (FT-IR), resulting very closed in chemical composition to commercial and natural HAs and showing in the FT-IR spectrum significative signals coincident with commercial HA, especially with carboxylic acids and phenolic alcohols, which are characteristic of the chemical structure of a humic acid. The formation of intermolecular H-bonds between HAs also has been detected by FT-IR spectrometry.

The economic analysis of flocculation at low pH applied to composting landfill leachate seems to describe a promising and sustainable technique compared with the actual methods used (flocculation, adsorption, oxidation, electrolysis, reverse osmosis) for COD and colour/turbidity removal.

Supplementary Materials: The following supporting information can be downloaded at https://www.mdpi.com/article/10.3390/su17020481/s1, Table S1: Numerical values for COD, colour (474 nm) and redox potential obtained in triplicate during flocculation of landfill leachate at low pH; Table S2: Numerical values of elemental analysis of HAs from landfill leachate (HA1 and HA2) and Aldrich HA (AHA) and data from literature. Analysed data (HA1, HA2 and AHA) are expressed as mean ± SD of three independent measurements.

Author Contributions: C.C.: conceptualisation, methodology, validation, formal analysis, data curation, supervision, funding acquisition, writing—original draft preparation, writing—review and editing. M.L.P. and B.D.R.: methodology, software, formal analysis, investigation and data curation. All authors have read and agreed to the published version of the manuscript.

Funding: The funding sources of this research were "Junta de Castilla y León (ADE)" (project 04/09/VA/0010), "Laboratorio Castilla y León, Valladolid Labaqua" and University of Salamanca, Strategic Plan of Research and Knowledge Transfer, C1 Research Projects 2021, project 18KA5N/463AC01.

Institutional Review Board Statement: Not applicable.

Informed Consent Statement: Not applicable.

Data Availability Statement: All data used are reported in the manuscript and Supplementary Material. Additional data, if necessary, will be purchased under request.

Acknowledgments: We acknowledge specially the staff of the WTC of Salamanca: Roberto Hernández and Javier Vázquez, for their valuable help in facilitate landfill leachate sample collection and solve technical questions. Milena A. Vega, from the University of Salamanca, is also gratefully acknowledged for technical support and analysis performance of FT-IR spectrometry, Z potential and mean diameter. Elemental Microanalysis Service of Madrid University (UCM) is acknowledged for assistance and performance of CHNS elemental chemical analysis, and the Chemical Analysis Service of the University of Salamanca (Emilio Romero) is also acknowledged for metal analysis.

Conflicts of Interest: The authors declare no conflicts of interest.

References

1. Renou, S.; Givaudan, J.G.; Poulain, S.; Dirassouyan, F.; Moulin, P. Landfill leachate treatment: Review and opportunity. *J. Hazar Mater.* **2008**, *150*, 468–493. [CrossRef] [PubMed]
2. Wiszniowski, J.; Robert, D.; Surmacz-Gorska, J.; Miksch, K.; Weber, J.V. Landfill leachate treatment methods: A review. *Enviro Chem. Let.* **2006**, *4*, 51–61. [CrossRef]
3. Zhang, S.; Peng, Y.; Wang, S.; Zheng, S.; Gou, J. Organic matter and concentrated nitrogen removal by shortcut nitrification an denitrification from municipal landfill leachate. *J. Environ. Sci.* **2007**, *19*, 647–651. [CrossRef]
4. Kheradmand, S.; Karimi-Jashni, A.; Sartaj, M. Treatment of municipal landfill leachate using a combined anaerobic digester an activated sludge system. *Waste Manag.* **2010**, *30*, 1025–1031. [CrossRef] [PubMed]
5. Li, H.; Zhou, S.; Sun, Y.; Feng, P.; Li, J. Advanced treatment of landfill leachate by a new combination process in a full-scale pla *J. Hazard. Mater.* **2009**, *172*, 408–415. [CrossRef] [PubMed]
6. Wu, Y.; Zhou, S.; Ye, X.; Chen, D.; Zheng, K.; Qin, F. Transformation of pollutants in landfill leachate treated by a combine sequence batch reactor, coagulation, Fenton oxidation and biological aerated filter technology. *Proc. Saf. Environ. Protec.* **2011**, *8* 112–120. [CrossRef]
7. Mahmud, K.; Hossain, M.D.; Shams, S. Different treatment strategies for highly polluted landfill leachate in developing countrie *Waste Manag.* **2012**, *32*, 2096–2105. [CrossRef] [PubMed]
8. Ding, A.; Zhang, Z.; Fu, J.; Cheng, L. Biological control of leachate from municipal landfills. *Chemosphere* **2001**, *44*, 1–8. [CrossRef] [PubMe
9. Chaturvedi, H.; Kaushal, P. Comparative study of different biological processes for non-segregated municipal solid waste (MSV leachate treatment. *Environ. Technol. Innov.* **2018**, *9*, 134–139. [CrossRef]
10. Pi, K.W.; Li, Z.; Wan, D.J.; Gao, L.X. Pretreatment of municipal landfill leachate by a combined process. *Proc. Saf. Environ. Prote* **2009**, *87*, 191–196. [CrossRef]
11. Cotman, M.; Gotvajn, A.Z. Comparison of different physico-chemical methods for the removal of toxicants from landfill leachat *J. Hazard. Mater.* **2010**, *178*, 298–305. [CrossRef] [PubMed]
12. Martins, C.L.; Fernandes, H.; Costa, R.H.R. Landfill leachate treatment as measured by nitrogen transformations in stabilizatio ponds. *Bioresour. Technol.* **2013**, *147*, 562–568. [CrossRef] [PubMed]
13. Tamrat, M.; Costa, C.; Márquez, M.C. Biological treatment of leachate from solid wastes: Kinetic study and simulation. *Biochen Eng. J.* **2012**, *66*, 46–51. [CrossRef]
14. Costa, C.; Domínguez, J.; Autrán, B.; Márquez, M.C. Dynamic Modeling of biological treatment of leachates from solid waste *Environ. Model. Assess.* **2018**, *23*, 165–173. [CrossRef]
15. Domínguez, J.; Costa, C.; Autrán, B.; Márquez, M.C. Characterization and biological stabilization of leachates from solid waste in north-centre Spain for agricultural application. *Waste Biomass Valorization* **2019**, *10*, 167–178. [CrossRef]
16. Marañón, E.; Castrillón, L.; Fernández-Nava, Y.; Fernández-Méndez, A.; Fernández-Sánchez, A. Coagulation-flocculation as a pretreatmer Process at a landfill leachate nitrification-denitrification plant. *J. Hazard. Mater.* **2008**, *156*, 538–544. [CrossRef] [PubMed]
17. Salem, Z.; Hamouri, K.; Djemaa, R.; Allia, K. Evaluation of landfill leachate pollution and treatment. *Desalination* **2008**, *22* 108–114. [CrossRef]
18. Castrillón, L.; Fernández-Nava, Y.; Ulmanu, M.; Anger, I.; Marañón, E. Physico-chemical and biological treatment of MSW landfi leachate. *Waste Manag.* **2010**, *30*, 228–235. [CrossRef] [PubMed]
19. Chemlal, R.; Azzouz, L.; Kernani, R.; Abdi, N.; Lounici, H.; Grib, H.; Mameri, N.; Drouiche, N. Combination of advance oxidation and biological processes for the leachate treatment. *Ecol. Eng.* **2014**, *73*, 281–289. [CrossRef]

Marañón, E.; Castrillón, L.; Fernández-Nava, Y.; Fernández-Méndez, A.; Fernández-Sánchez, A. Colour, turbidity and COD removal from old landfill leachate by coagulation-flocculation treatment. *Waste Manag. Res.* **2010**, *28*, 731–737. [CrossRef] [PubMed]

De Torres-Socías, E.; Prieto-Rodríguez, L.; Zapata, A.; Fernández-Calderero, I.; Oller, I.; Malato, S. Detailed treatment line for a specific landfill leachate remediation. Brief economic assessment. *Chem. Eng. J.* **2015**, *261*, 60–66. [CrossRef]

Oloibiri, V.; Chys, M.; De Wandel, S.; Demeestere, K.; Van Hulle, S.W.H. Removal of organic matter and ammonium from landfill leachate through different scenarios: Operational cost evaluation in a full-scale case study of Flemish landfill. *J. Environ. Manag.* **2017**, *203*, 774–781. [CrossRef] [PubMed]

Jiang, Z.; Yang, H.; Sun, L.; Shi, S. Integrated assessment for aerobic biodegradability of organic substances. *Chemosphere* **2002**, *48*, 133–138. [CrossRef] [PubMed]

Kang, K.-H.; Shin, H.S.; Park, H. Characterization of humic substances present in landfill leachates with different landfill ages and its implications. *Water Res.* **2002**, *36*, 4023–4032. [CrossRef] [PubMed]

Sír, M.; Podhola, M.; Patocka, T.; Honzajkova, Z.; Kocurek, P.; Kubal, M.; Kuras, M. The effect of humic acids on the reverse osmosis treatment of hazardous landfill leachate. *J. Hazard. Mater.* **2012**, *207–208*, 86–90. [CrossRef] [PubMed]

Torretta, V.; Ferronato, N.; Katsoyiannis, I.A.; Tolkou, A.K.; Airoldi, M. Novel and conventional technologies for landfill leachate treatment: A review. *Sustainability* **2017**, *9*, 9. [CrossRef]

Zouboulis, A.I.; Chai, X.-L.; Katsoyiannis, I.A. The application of bioflocculant for the removal of humic acids from stabilized landfill leachates. *J. Environ. Manag.* **2004**, *70*, 35–41. [CrossRef]

Tong, H.; Yin, K.; Ge, L.; Giannis, A.; Chuan, V.W.L.; Wang, J.-Y. Monitoring transitory profiles of leachate humic substances in landfill aeration reactors in mesophilic and thermophilic conditions. *J. Hazard. Mater.* **2015**, *287*, 342–348. [CrossRef]

Oulego, P.; Collado, S.; Laca, A.; Díaz, M. Impact of leachate composition on the advanced oxidation treatment. *Water Res.* **2016**, *88*, 389–402. [CrossRef] [PubMed]

Qi, G.; Yue, D.; Nie, Y. Characterization of humic substances in bio-treated municipal solid waste landfill leachate. *Front. Environ. Sci. Eng.* **2012**, *6*, 711–716. [CrossRef]

Zhang, D.; Vahala, R.; Wang, Y.; Smets, B.F. Microbes in biological processes for municipal landfill leachate treatment: Community, function and interaction. *Int. Biodeterior. Biodegrad.* **2016**, *113*, 88–96. [CrossRef]

Christensen, J.B.; Jensen, D.L.; Grøn, C.; Filip, Z.; Christensen, T.H. Characterization of the dissolved organic carbon in landfill leachate polluted groundwater. *Water Res.* **1998**, *32*, 125–135. [CrossRef]

Graber, E.R.; Rudich, Y. Atmospheric HULIS: How humic-like are they? A comprehensive and critical review. *Atmos. Chem. Phys.* **2006**, *6*, 729–753. [CrossRef]

Gomes de Melo, B.A.; Lopes Motta, F.; Andrade Santana, M.H. Humic acids: Structural properties and multiple functionalities for novel technological developments. *Mater. Sci. Eng. C* **2016**, *62*, 967–974. [CrossRef] [PubMed]

Dia, O.; Drogui, P.; Buelna, G.; Dubé, R. Hybrid process, electrocoagulation-biofiltration for landfill leachate treatment. *Waste Manag.* **2018**, *75*, 391–399. [CrossRef] [PubMed]

Motta, F.L.; Melo, B.A.G.; Santana, M.H.A. Deprotonation and protonation of humic acids as a strategy for the technological development of pH-responsive nanoparticles with fungicidal potential. *New Biotechnol.* **2016**, *33*, 773–780. [CrossRef] [PubMed]

Álvarez-Puebla, R.A.; Garrido, J.J. Effect of pH on the aggregation of a gray humic acid in colloidal and solid states. *Chemosphere* **2005**, *57*, 659–667. [CrossRef]

Souza, F.D.; Roca Bragança, S. Humic Acid as Dispersant of an Alumina Suspension and its Rheological Behaviour. *Mater. Res.* **2018**, *21*, e20170759. [CrossRef]

APHA; AWWA; WEF. *Standard Methods for the Examination of Water and Wastewater*, 23rd ed.; American Public Health Association: Washington DC, USA, 2017.

UNE-EN ISO 10304-1:2009; Determination of Dissolved Anions by Liquid Chromatography of Ions—Part 1: Determination of Bromide, Chloride, Fluoride, Nitrate, Nitrite, Phosphate and Sulphate. AENOR: Madrid, Spain, 2009.

Abbt-Braun, G.; Lankes, U.; Frimmel, F.H. Structural characterization of aquatic humic substances–The need for a multiple method approach. *Aquat. Sci.* **2004**, *66*, 151–170. [CrossRef]

Ziyang, L.; Bin, D.; Xiaoli, C.; Yu, S.; Youcai, Z.; Nanwen, Z. Characterization of refuse Landfill leachates of three different stages in landfill stabilization process. *J. Environ. Sci.* **2009**, *21*, 1309–1314. [CrossRef]

Vaverková, M.D.; Elbl, J.; Koda, E.; Adamcová, D.; Bilgin, A.; Lukas, V.; Podlasek, A.; Kintl, A.; Wdowska, M.; Brtnický, M. Chemical Composition and Hazardous Effects of Leachate from the Active Municipal Solid Waste Landfill Surrounded by Farmlands. *Sustainability* **2020**, *12*, 4531. [CrossRef]

Guo, X.; Liu, H.; Wu, S. Humic substances developed during organic waste composting: Formation mechanisms, structural properties, and agronomic functions. *Sci. Total Environ.* **2019**, *662*, 501–510. [CrossRef] [PubMed]

Tejera, J.; Miranda, R.; Hermosilla, D.; Urra, I.; Negro, C.; Blanco, A. Treatment of a Mature Landfill Leachate: Comparison between Homogeneous and Heterogeneous Photo-Fenton with Different Pretreatments. *Water* **2019**, *11*, 1849. [CrossRef]

46. Aziz, H.A.; Rahim, N.A.; Ramli, S.F.; Alazaiza, M.Y.D.; Omar, F.M.; Hung, Y.-T. Potential Use of *Dimocarpus longan* Seeds as Flocculant in Landfill Leachate Treatment. *Water* **2018**, *10*, 1672. [CrossRef]
47. Matovelle, C.; Quinteros, M.; Heras, D. Machine Learning Techniques in Dosing Coagulants and Biopolymers for Treati Leachate Generated in Landfills. *Water* **2023**, *15*, 4200. [CrossRef]
48. Kamal, A.; Makhatova, A.; Yergali, B.; Baidullayeva, A.; Satayeva, A.; Kim, J.; Inglezakis, V.J.; Poulopoulos, S.G.; Arkhangelsky, Biological Treatment, Advanced Oxidation and Membrane Separation for Landfill Leachate Treatment: A Review. *Sustainabil* **2022**, *14*, 14427. [CrossRef]
49. Remmas, N.; Manfe, N.; Zerva, I.; Melidis, P.; Raga, R.; Ntougias, S. A Critical Review on the Microbial Ecology of Landf Leachate Treatment Systems. *Sustainability* **2023**, *15*, 949. [CrossRef]
50. Wdowczyk, A.; Szymańska-Pulikowska, A. Differences in the Composition of Leachate from Active and Non-Operation Municipal Waste Landfills in Poland. *Water* **2020**, *12*, 3129. [CrossRef]
51. Ramli, S.F.; Aziz, H.A.; Omar, F.M.; Yusoff, M.S.; Halim, H.; Kamaruddin, M.A.; Ariffin, K.S.; Hung, Y.-T. Reduction of COD a Highly Coloured Mature Landfill Leachate by Tin Tetrachloride with Rubber Seed and Polyacrylamide. *Water* **2021**, *13*, 30 [CrossRef]
52. Baalousha, M.; Motelica-Heino, M.; Le Coustumer, P. Conformation and size of humic substances: Effects of major cati concentration and type, pH, salinity, and residence time. *Colloids Surf. A Physicochem. Eng. Asp.* **2006**, *272*, 48–55. [CrossRef]
53. Pinedo, L.M.; Riascos, B.D.; Quintero, X.E.; Costa, C. Mechanism of pH sensitive flocculation for organic load and colour reducti in landfill leachate. *Waste Manag.* **2022**, *144*, 163–172. [CrossRef]
54. De Castro, A.L.; dos Santos, R.F.; de Souza, J.C.; Izário, H.J. Characterization of controlled landfill leachate from the city Guaratinguetá-SP, Brazil. *Ambiente Agua-Interdiscip. J. Appl. Sci.* **2018**, *13*, 1. [CrossRef]
55. Tanaka, T. Functional groups and reactivity of size-fractionated Aldrich humic acid. *Thermochim. Acta* **2012**, *532*, 60–64. [CrossRe
56. Weiming, C.; Qibin, L. Elimination of UV-quenching substances from MBR- and SAARB-treated mature landfill leachates in ozonation process: A comparative study. *Chemosphere* **2020**, *242*, 125256. [CrossRef]
57. Pisharody, L.; Gopinath, A.; Malhotra, M.; Nidheesh, P.V.; Kumar, M.S. Occurrence of organic micropollutants in municip landfill leachate and its effective treatment by advanced oxidation processes. *Chemosphere* **2022**, *287*, 132216. [CrossRef]
58. Liu, Y.; Chen, Y.; Da, Y.; Xie, F.; Wang, J. Advanced treatment of landfill leachate using integrated coagulation/photo-Fent process through in-situ generated nascent Al^{3+} and H_2O_2 by Cl, N co-doped aluminum-graphite composite. *Appl. Catal. Environ.* **2022**, *304*, 121003. [CrossRef]
59. Chen, W.; Gu, Z.; Wen, P.; Li, Q. Degradation of refractory organic contaminants in membrane concentrates from landfill leacha by a combined coagulation-ozonation process. *Chemosphere* **2019**, *217*, 411–422. [CrossRef] [PubMed]
60. Wang, H.; Xiao, W.; Zhang, C.; Sun, Y.; Wang, N.Y.; Gong, Z.; Zhan, M.; Fu, Y.; Liu, K. Effective removal of refractory organ contaminants from reverse osmosis concentrated leachate using PFS-nZVI/PMS/O_3 process. *Waste Manag.* **2021**, *128*, 55–6 [CrossRef] [PubMed]
61. Tawakkoly, B.; Alizadehdakhel, A.; Dorosti, F. Evaluation of COD and turbidity removal from compost leachate wastewat using *Salvia hispanica* as a natural coagulant. *Ind. Crops Prod.* **2019**, *137*, 323–331. [CrossRef]
62. Nazia, S.; Sahu, N.; Jegatheesan, V.; Bhargava, S.K.; Sridhar, S. Integration of ultrafiltration membrane process with chemic coagulation for proficient treatment of old industrial landfill leachate. *Chem. Eng. J.* **2021**, *412*, 128598. [CrossRef]
63. Kwarciak-Kozłowska, A.; Fijałkowski, K.L. Efficiency assessment of municipal landfill leachate treatment during advance oxidation process (AOP) with biochar adsorption (BC). *J. Environ. Manag.* **2021**, *287*, 112309. [CrossRef] [PubMed]
64. Xue, W.-J.; Cui, Y.-H.; Liu, Z.-Q.; Yang, S.-Q.; Li, J.-Y.; Guo, X.-L. Treatment of landfill leachate nanofiltration concentrate aft ultrafiltration by electrochemically assisted heat activation of peroxydisulfate. *Sep. Purif. Technol.* **2020**, *231*, 115928. [CrossRef
65. He, H.; Zhang, C.; Yang, X.; Huang, B.; Zhe, J.; Lai, C.; Liao, Z.; Pan, X. The efficient treatment of mature landfill leachate using tow bipolar electrode flocculation-oxidation combined with electrochemical biofilm reactors. *Water Res.* **2020**, *230*, 119544. [CrossRef]
66. Gao, B.; Yang, X.; Dasi, E.A.; Lam, T.; Arias, M.E.; Ergas, S.J. Enhanced landfill leachate treatment in sequencing batch biofil reactors (SBBRs) amended with zeolite and biochar. *J. Chem. Technol. Biotechnol.* **2022**, *97*, 759–770. [CrossRef]
67. Long, C.; Zhu, B.; Liu, W.; Li, Q. Coagulation combined with electro-Fe^0/H_2O_2 reaction for effective treatment of landfill leachat effluent of membrane bioreactor. *Water* **2023**, *15*, 1158. [CrossRef]
68. Cheng, S.Y.; Show, P.; Juan, J.C.; Chang, J.; Lau, B.F.; Lai, S.H.; Ng, E.P.; Yian, H.C.; Ling, T.C. Landfill leachate wastewate treatment to facilitate resource recovery by a coagulation-flocculation process via hydrogen bond. *Chemosphere* **2021**, *262*, 12782 [CrossRef] [PubMed]
69. Christl, I. Magnesium binding by terrestrial humic acids. *Environ. Chem.* **2018**, *15*, 317–324. [CrossRef]

sustainability

rticle

'racking of Antibiotic Resistance Genes in Sludge from 'reatment Plants for Organic Fraction of Municipal Solid Waste

lena Franchitti *, Antonino Mafrica and Deborah Traversi

Department of Public Health and Pediatrics, University of Turin, Piazza Polonia 92, 10126 Turin, Italy; antonino.mafrica@edu.unito.it (A.M.); deborah.traversi@unito.it (D.T.)
* Correspondence: elena.franchitti@unito.it

Abstract: Bacteria quickly acquire the ability to survive or grow in the presence of an antibacterial agent that should be able to inhibit or kill them, leading to increased mortality caused by infective diseases. The digestate from the anaerobic digestion (AD) of the organic fraction of municipal solid waste (OFMSW) is spread on soil, but the knowledge on the presence and persistence of the antibiotic resistance genes (ARGs) is limited. Thus, this study aims to evaluate the presence of seven ARGs (*bla*$_{TEM}$, *bla*$_{OXA}$, *ermB*, *qnrB*, *sulI*, *sulII*, *tetA*, and *tetW*) in the OFMSW and their persistence after the AD, using the innovative droplet digital Polymerase Chain Reaction (ddPCR), not yet used on this matrix. A total of 31 samples were collected from the influent and effluent of the AD in two plants located in Northern Italy. ARG concentration ranged between 4 and 9 Log gene copies/kg of sludge. A limited reduction in the concentration of antibiotic resistance targets given by AD was observed in the study (<1 Log). The persistence of ARGs after mesophilic digestion of the OFMSW suggests a risk of horizontal transmission when directly spread on soil. Further evaluations are needed for safe and sustainable reuse of such sludges.

Keywords: antimicrobial resistance; antibiotic resistance genes; ddPCR; anaerobic digestion; organic fraction of municipal solid waste; waste management; circular economy

1. Introduction

Nowadays, antibiotic resistance is an increasingly alarming problem for human health. Antibiotic resistance consists of the adaptation of certain microorganisms that become able to survive or grow in the presence of a specific concentration of an antibacterial agent, usually sufficient to inhibit microorganisms of the same species. This phenomenon is a consequence of the misuse of antibiotics not only in human health care, but also in veterinary and agricultural practices, also producing an environmental accumulation [1,2]. Infections of resistant microorganisms cannot be treated with standard therapies (as they are ineffective). As a consequence, the following may occur: prolongation of disease, occurrence of complications, increased risk of death, and increased risk of epidemic outbreaks. In addition, health care costs also increase [3]. Antibiotic resistance also develops because some bacterial genes can shift between chromosomal and extra-chromosomal DNA, even between bacteria of different species and genera. Gene transfer can occur through several mechanisms: transformation, transduction, and conjugation [4].

Antibiotics can persist for a long time in the environment. They are excreted from the human body, the most important reservoir, with resistant bacteria or with biological samples such as food waste and fomites [2]. Recently, antibiotic compounds have been included in the watch list of the emerging compounds for environmental pollution mitigation [5]. The EU commission highlighted the necessity of better addressing the role of the environment in the "One Health Action Plan against AMR", with a focus on environmental monitoring [6]. Municipal solid waste dumps can also be considered as reservoirs of antibiotics and a vehicle for the spread of antibiotic resistance genes as they contain a great amount of food waste from both animal and vegetal origins [1]. The amount of the organic fraction of the municipal solid waste (OFMSW) has increased over time due to the improvement of waste sorting, reaching an annual production of 71 million of tons in the EU, with 35% of such organic materials treated in plants by anerobic digestion (AD). Anaerobic digestion is a green biotechnology carried out in extreme conditions, such as absence of oxygen and high temperatures, in which a microbial community decomposes the organic matter present in the sludge to produce biogas. The biogas is mainly composed of methane (CH_4) and carbon dioxide (CO_2), commonly used to produce heat and electricity. New or updated treatment plants with higher capacities were installed during recent decades all over the world, and their number could be above 11,600 plants by 2035, two times the number of plants already present [7]. However, the occurrence and the removal efficiency of AMR in OFMSW is not well known. Various articles showed an effective reduction of the resistant bacteria and a reduction of some AMR genes (ARGs) and an increment of others [8]. Despite these findings, an exhaustive risk assessment is missing. Antibiotics have also been included in various research studies to increase the biogas yield of anaerobic digestion through bacterial selection [9,10], but the use of such compounds must be carefully evaluated when not aimed at sanitary purposes. Therefore, proper organic waste management is crucial to avoid environmental spread of resistant microbes but also of AMR genetic elements that can potentially be transferred to other microbes. However, the health risk related to the presence of ARGs in the effluent of the OFMSW treatment is not yet fully known and, generally, culture methods are used to gather information on the potential pathogenicity and infectivity. Non-cultural methods based on the detection of AMR genes are now also applied to the sludge coming from the AD process. Among these, generally, real-time quantitative Polymerase Chain Reaction (qPCR) has been used, although its application is hindered by technical disadvantages, such as the presence of inhibitors, including metals and humic acids, of the reaction [11]. New quantification methods based on PCR, which are rarely used on this kind of matrices, can help to overcome some limits. The technical development of biomolecular methods is very fast and laboratory evidence has also been produced starting from digestate samples [12].

This study aims to screen the occurrence and persistence, after AD, of the most prevalent ARGs in organic waste sludges, in full-scale settings, using innovative biomolecular methods.

2. Materials and Methods

2.1. OFMSW Plants and Samplings

Two of the main OFMSW treatment plants of Northern Italy were involved in this study. The OFMSW-A receives 60,000 tons/year, OFMSW-F 45,000 tons/year. In such technologically advanced plants, an AD phase is included. The sampling activities covered a 6-month period from November 2021 to April 2022, in which both implants provided their samples consecutively each week. From the OFMSW plants, 16 samples from the influent of the digester and 16 digestate samples were collected by the plants' workers. Chemical and physical data on the samples, including T (temperature), pH, percentage of Total Suspended Solids (%TSS), and percentage of Volatile Suspended Solids (%VSS), were provided by the involved plant with standard internally validated methods and are reported in Tables 1 and S1.

Table 1. OFMSW plant. Physical and chemical characteristics of the sludge and the concentration of DNA extracted from the samples.

	OFMSW-A		OFMSW-F	
	Medio	Dev. St.	Medio	Dev. St.
pH	6.83	1.55	6.93	1.21
Temperature	42.34	3.78	40.90	0.24
%TSS	5.86	2.98	4.61	1.18
%VSS	4.54	2.98	3.20	1.19
DNA ng/μL	55.97	47.54	66.61	43.51

Once collected, samples were delivered to the Environmental Hygiene Section of the Department of Public Health and Pediatrics of the University of Turin in one-liter containers.

2.2. DNA Extraction and ARGs Quantification

The extraction of nucleic acids from the samples took place within 24 h of sampling. Nucleic acid extraction from samples was conducted in duplicate using a DNeasy Powersoil Pro Kit (QIAGEN, Venlo, The Netherlands) following the manufacturer's instructions. The chosen extraction methodology was based on column filtration. The samples' cells were lysed through chemical and mechanical homogenization, the lysis buffer was added to a mixed zirconium bead tube containing the 250 µL of sample and vortexed. The supernatant was then subjected to inhibitor removal for clean-up. Then, purified lysate was mixed with an equal volume of DNA binding solution and passed through a silica spin filter membrane. That membrane was washed with a two-step washing regime. The DNA bound to the silica was then eluted using a 10 mM Tris elution buffer. DNA was eluted in 100 µL of elution solution and quantified using a TECAN Infinite M200 Pro spectrophotometer, a NanoQuant Plate (Tecan Group Ltd., Männedorf, Swizerland), and i-Control™ software (version 1.11.10). Extracts were stored at −80 °C for subsequent biomolecular analysis. The heterogeneity of the matrix was high so the extraction was not always optimal for biomolecular analysis, even though the sample was homogenized. To select extracts on which to conduct further studies, excluding very poor extractions, a cut-off value for DNA of 5 ng μL^{-1} was set [13], finally obtaining 22 samples.

In this paper, a target-specific analysis was chosen to select the genes of interest, and two criteria were followed: (i) it should be one of the ARGs that induce a resistance ability against the most prescribed groups of antibiotics in Italy according to the European Centre for Disease Control (ECDC) [14] (Table 2), and (ii) the literature should include a solid PCR protocol with primers published in more papers. Consequently, seven ARGs were selected: bla_{OXA}, bla_{TEM}, *ermB*, *qnrB*, *sulI*, *sulII*, *tetA*, and *tetW*. *BlaOXA*, which encodes for carbapenemases, results in resistance to carbapenems and penicillins, both subclasses of β-lactam antibiotics. The carbapenemases, which are members of the β-lactamases, destroy the β-lactam ring. *BlaTEM*, which gives resistance to ampicillin and penicillin G through β-lactamase production, acts similarly to *BlaOXA*. *ErmB* and *qnrB* confer resistance toward macrolides and fluoroquinolones through a ribosomal protection mechanism. *SulI* and *sulII* both carry resistance to sulfonamides through a process of enzymatic inactivation that alters the chemical structure of the antibiotic molecules. Finally, *tetA* and *tetW* both bring resistance toward tetracyclines, but by two different mechanisms: *tetA* through increased expression of efflux pumps that expel the antibiotic from the bacterial cell, while *tetW* through a ribosomal protection mechanism. According to the most recent data from the European Surveillance of Antimicrobial Consumption Network, the most consumed antimicrobials in Italy in 2023 were penicillins and β-lactam antibiotics (399.03 tonnes per year), followed by macrolides, lincosamides, and streptogramins (40.49 tonnes per year), and, finally, other β-lactam antibiotics (58.20 tonnes per year) and quinolones (30.04 tonnes per year), sulfonamides (50.05 tonnes per year), and tetracyclines (3.31 tonnes per year) [14]. The gene encoding the *16S rRNA* V3-V4 region was added to evaluate the bacterial presence. Primers were selected by comparing more than three papers and after Primer BLAST alignment. After a screening phase of the positivity with traditional PCR, conducted by amplifying the ARGs with a T100 thermal cycler (Bio-Rad, Hercules, CA, USA) and evaluated by the presence of bands in gel electrophoresis, antimicrobial resistance was measured using the biomolecular method of the droplet digital PCR (ddPCR, Bio-Rad, Hercules, CA, USA) on the DNA extracts.

Table 2. List of antibiotic resistance genes. The primers and the thermal protocols are listed.

Antibiotic Resistance Gene	Primers	Annealing Temperature	Amplicon Size (bp)	Reference
16S rRNA	F:5′-ACTCCTACGGGAGGCAGCAG-3′ R:5′-ATTACCGCGGCTGCTGG-3′	61 °C	2736	[15]
blaOXA	F: 3′-TATCTACAGCAGCGCCAGTG-5′ R: 3′-CGCATCAAATGCCATAAGTG-5′	61 °C	199	[16]
bla_{TEM}	F: 3′-CATTTTCGTGTCGCCCTTAT-5′ R: 3′-GGGCGAAAACTCTCAAGGAT-5′	61 °C	167	[16,17]
ermB	F: 3′-AAAACTTACCCGCCATACCA-5′ R: 3′-TTTGGCGTGTTTCATTGCTT-5′	61 °C	139	[18]
qnrB	F:3′-GCGACGTTCAGTGGTTCAG-5′ R: 3′-TGTCCAACTTAACGCCTTGTAA-5′	61 °C	148	[19]
sulI	F: 3′-CGGCGTGGGCTACCTGAACG-5′ R: 3′-GCCGATCGCGTGAAGTTCCG-5′	60 °C	433	[20]
sulII	F: 3′-GCGCTCAAGGCAGATGGCATT-5′ R: 3′-GCGTTTGATACCGGCACCCGT-5′	61 °C	293	[20]
tetA	F: 3′-GCGCGATCTGGTTCACTCG-5′ R: 3′-AGTCGACAGYRGCGCCGGC-5′	62.5 °C	164	[21]
tetW	F: 3′-GAGAGCCTGCTATATGCCAGC-5′ R: 3′-GGGCGTATCCACAATGTTAAC-5′	61 °C	168	[22]

The thermal protocol used was 95 °C for 5 min, 94 °C for 1 min, annealing temperature for 1 min, and 72 °C for 3 min, for 35 cycles, followed by a post-cycling step at 4 °C for 5 min. The detailed annealing temperatures are listed in Table 2. The LOQ for each target is included in Table 3.

Table 3. LOQ calculated for each target. The percentage of the samples in which the level was above the LOQ before and after the AD for total bacteria (*16S rRNA*), for each target, and samples positive for at least 6 and for all the included targets. The difference between influent and effluent, as well as its statistical significance, is shown in the last two columns.

		% Samples > LOQ			
	LOQ (Log Gene Copies/mg Sludge)	Influent	Effluent	Δ(in-ef)	χ^2 *p*-Value
16S rRNA	6.60	100%	100%	0%	-
bla_{OXA}	4.54	63.6%	90.9%	−27%	0.127
bla_{TEM}	5.74	100%	100%	0%	-
ermB	5.28	100%	100%	0%	-
qnrB	3.92	100%	90.9%	9.1%	0.306
sulI	6.12	100%	100%	0%	-
sulII	4.74	100%	81.8%	18.2%	0.138
tetA	5.45	100%	100%	0%	-
tetW	5.43	72.7%	81.8%	−9.07%	0.611
Positivity to at least 6 ARGs		100%	90.9%	−9.1%	0.306
Positivity to all the ARGs		45.5%	63.6%	−18.1%	0.392

2.3. Statistical Analyses

Statistical analysis was performed using IBM SPSS Statistics (version 29.0). ARG concentrations were expressed as Log of gene copies per kg of sludge and as ratios of the Log of AMR gene copies on the 16S rRNA copies. The data, adjusted considering the bacterial presence, were then used in all statistical analyses. A descriptive analysis of all the variables was performed. We reported categorical variables as absolute numbers and percentages and as mean and standard deviations for continuous variables. Statistical analyses for categorical variables were conducted using the χ^2. Differences between gene concentrations and chemical and physical parameters were analyzed using nonparametric tests (Mann–Whitney U test and Kruskal–Wallis) for independent samples by ranks, comparing the plant type and the type of sludge. Statistical significance was evaluated with an ANOVA test. The Spearman rank-order correlation coefficient was determined. A p-value of $p < 0.05$ was considered significant, a value <0.01 was considered highly significant.

3. Results and Discussion

The OFMSW plants involved in the study provided the physical–chemical analyses, including pH (6.1 ± 1.4), temperature (41.9 ± 3.2 °C), TSS (5.5 ± 2.6%), and VSS (4.1 ± 2.6%). Overall, no significant differences were observed between the two plants. However, the main difference was the temperature of the anaerobic digestion, whereby the OFMSW-A plant produces a digestion oriented towards thermophilicity. As can be observed in Table 1, the temperature was, on average, 2 degrees higher but variable (Mann–Whitney U test, $p > 0.05$).

Higher concentration of solids (>8%) was inversely proportional to the hydrolysis velocity. This presents benefits such as lower water and working volume requirements, however, it can only be achieved under constant thermophilic conditions. High solids anaerobic digestion can also produce a higher amount of biogas, but often fails due to the rapid buildup of volatile fatty acids (VFAs) and the associated drop in pH, producing a severe acidosis (pH < 4) of the digester and the necessity to stop the plant working to remove the materials from the digester. As a consequence, the OFMSW plants generally choose a lower but more easily manageable organic load for the digester.

In the Table 3, the percentage of positive samples for each included gene (>LOQ) is shown, including both in influents and effluents from the AD. Moreover, the percentage of samples positive for more than one ARG are included. Several studies indicate that multidrug resistance is ubiquitous in environmental matrices [23], with a relatively high abundance, increasing the risk of multi antimicrobial resistance transmission.

The average concentration observed in the organic fraction prior to the anaerobic digestion was estimated at 9.45 ± 0.78 Log *ermB* copies/kg of sludge, 8.01 ± 0.49 for *sulII*, 7.66 ± 0.53 for *sulI*, 7.62 ± 0.57 for *tetA*, 7.23 ± 0.37 for bla_{TEM}, 7.11 ± 0.53 for *qnrB*, 6.03 ± 4.01 for *tetW*, and 4.11 ± 3.30 for bla_{OXA}. Significant differences were observed for *ermB*, *sulI*, *sulII*, *tetA*, *qnrB*, and bla_{TEM} (Mann–Whitney U test, $p = 0.04$, $p = 0.01$, $p = 0.007$, $p = 0.004$, $p < 0.001$, and $p < 0.001$, respectively), but the reduction caused by the AD observed was limited (<1 Log unit) (Figure 1). On the other hand, bla_{OXA} and *tetW* seemed not to be affected by the process (Figure 1). In line with the literature, at higher temperatures (40–50 °C) *tetW*—which encodes for an efflux pump—increased [24–26].

In order to evaluate the viability of the bacteria present in the sludge, *16S rRNA* was also measured at a concentration of 10.41 ± 0.56. The AD seemed to have no impact on the concentration of the total bacteria, as in the digestate, the concentration of *16S rRNA* was 10.57 ± 0.22. The *16S rRNA* gene was also included as internal control of the analytic process, the concentrations of the ARGs were normalized to the internal control dividing the Log ARGs copies/kg of sludge by Log *16S rRNA* copies/kg of sludge. Statistical analyses on normalized data (with respect to *16s rRNA*) produced quite similar results to those for sludge mass unit (Table 4).

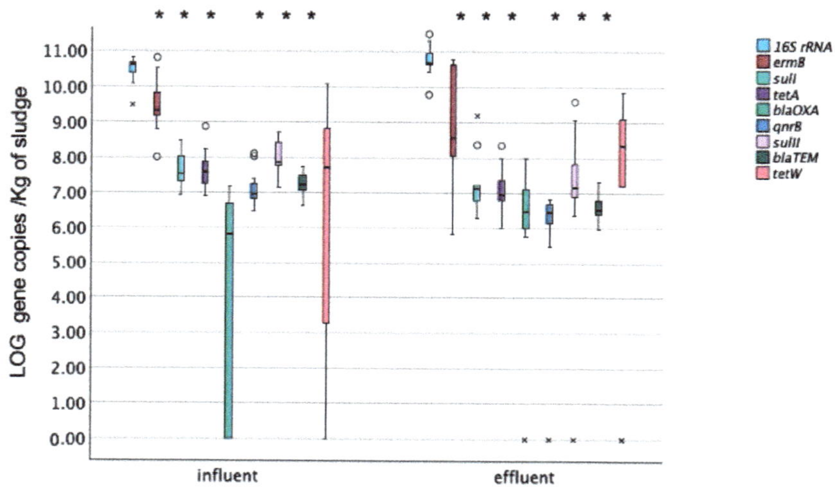

Figure 1. Effect of the AD on the concentrations of the total bacteria and ARGs. Boxplot of the total bacteria and ARG concentrations. Circles show the outliers and x shows the extreme values. Significant differences (Mann–Whitney U test, $p < 0.05$) between the influent and the digestate are marked with *.

Table 4. Effect of the AD on the ARG reduction. The concentration is reported as the ratio between the Log ARG copies/kg of sludge and the Log *16S rRNA* copies/kg of sludge. The effect of the AD is quantified as the difference between the influent and the effluent of the digester. Significant differences (*) were evaluated using the Mann–Whitney U test, $p < 0.05$.

	Influent		Effluent			
	Medium	Dev. St.	Medium	Dev. St.	Δ(in-ef)	*p*-Value
bla_{OXA}	0.40	0.32	0.57	0.21	−0.17	0.27
bla_{TEM}	0.69	0.06	0.62	0.04	0.07	<0.001 *
ermB	0.91	0.08	0.81	0.13	0.10	0.04 *
qnrB	0.68	0.06	0.54	0.19	0.14	<0.001 *
sulI	0.73	0.06	0.67	0.07	0.06	0.01 *
sulII	0.77	0.06	0.58	0.30	0.19	0.007 *
tetA	0.73	0.61	0.66	0.05	0.07	0.004 *
tetW	0.58	0.40	0.63	0.32	−0.05	0.898

As already mentioned, the main difference between the plants is the temperature in the digester. Analyzing the data divided by plant, the significant differences between influent and effluent were observed just for the OFMSW-A: *ermB, sulI, tetA, qnrB, sulII,* and *blaTEM* (Mann–Whitney U test, $p = 0.07$, $p = 0.002$, $p = 0.001$, $p < 0.001$, $p < 0.001$, and $p = 0.002$, respectively) (Figure 2). Thermophilic conditions have higher hygienic potential, reducing the presence of pathogens and ARGs [27,28]. According to the literature, temperature is a key factor in the reduction of potential biological health threats. Pathogenic spore-forming bacteria, such as Clostridium perfringens or Bacillus cereus, seem less affected by the extreme conditions of the digester and they could be found at high concentrations also in the digestate [29,30]. Flores-Orozco et al. suggested that the temperature induces a selection of the ARGs based on their mechanisms of action; indeed, genes encoding for antibiotic inactivators are reduced in mesophilic conditions. The thermophilic conditions reduce the ARGs concentration but the level of total ARGs remaining are similar [29].

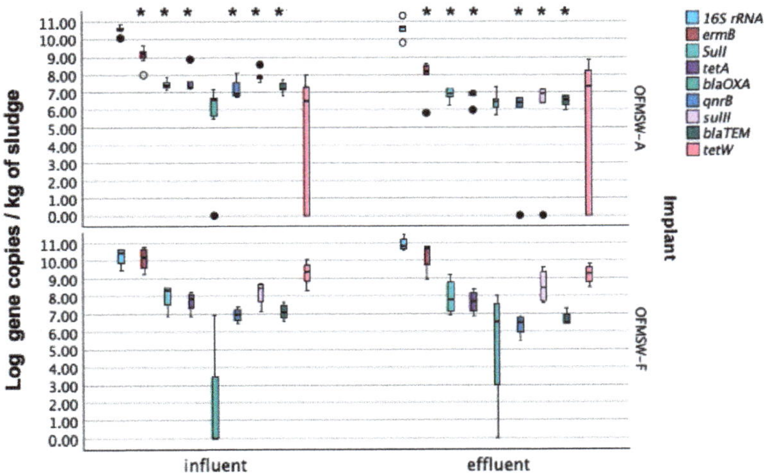

Figure 2. Temperature is a key factor in ARG concentration. Boxplot representing ARG concentration expressed in Log gene copies per kg of sludge. Significant differences between influent and digestate (Mann–Whitney U test, $p < 0.05$) are marked with *. The white circles represent the outliers, the black ones the extremes.

The %VSS is a driven parameter for the amount of biological molecules in the sludge. According to the literature, the solid component of the wastewater, due to its physical–chemical composition, shows a higher affinity for the biological compound [31,32]. In this study, a positive correlation among ARGs like *ermB*, *tetA*, *qnrB*, *sulI*, *sulII*, *blaTEM*, and %VSS is reported (Spearman's Rho, *ermB*: 0.582, $p = 0.007$, *tetA*: 0.731, $p < 0.001$, *qnrB*: 0.743, $p < 0.001$, *sulI*: 0.622, $p = 0.003$, *sulII*: 0.699, $p < 0.001$, *blaTEM*: 0.753, $p < 0.001$).

The persistence of the ARGs in the digestate implies a potential risk of horizontal transmission to the bacteria present in the soil. Although the decision making has admitted the relevance of the pollution linked to the presence of pharmaceuticals in the environmental matrix, only few references are present for water and referring to the watch list compounds, in which three antibiotics are listed [6]. A lack of regulation for limiting antibiotics and the presence of ARGs in the digestate destined for agricultural use has been reported in the European Union and other developed countries. According to the latest regulation on organic fertilizers [33], even though references to biological targets are present (for example *Salmonella* spp.), no specific statements are reported for antibiotics or ARGs.

The OFMSW is mainly composed of fruit and vegetable leftovers, as well as animal waste from meat and fish products. The main antibiotics used on crops are streptomycin and oxytetracycline [34]. According to the European Medicine Agency, in 2022 in the food-producing animals, the most administered antibiotics were penicillin and tetracyclines [35]. Due to the broad-spectrum properties of the above-mentioned antibiotics and their use in the agricultural setting, it is worth noticing that in this paper, ARGs against the most-used antibiotics in agriculture were the three most present antibiotics (5.74, 5.45, 5.43 Log gene copies/Kg of sludge for *blaTEM*, *tetA* and *tetW* respectively) (Table 3).

Biochar, eggshell, and biochar-based composites as other porous materials are able to reduce ARG transfer, decreasing mobility of heavy metals and antibiotics through adsorption and lowering the selection pressure. Moreover, bacteriophage therapy improvement and diffusion can also be an alternative to inhibit the proliferation and diffusion of ARGs [36]. However, more effort is needed to investigate the AMR removal mechanisms, and the interactions between both environmental matrices, to inhibit the proliferation and diffusion of ARGs.

According to the literature, the use of organic fertilizers (compost or digestate) favors vegetable growth [37]. Spreading contaminated digestate, on agricultural soil, due to the limited effect of the AD, could increase the pollution in soil, water, and food destined for human consumption. The soil microbial community, especially the rhizosphere environment, can mitigate the dissemination of ARGs [37,38]; on the other hand, the bacteria present in the soil can acquire the ARGs from the digestate, developing multi-drug resistance [38]. The limited reduction of the ARGs in the digestate can be partially overcome by the subsequent destination of the effluent. Frequently, the final product has two potential destinations: (i) dehydration for storing and spreading on the soil or (ii) dehydration followed by the aerobic composting plant for further treatment. The OFMSW plants tested in the experiment send the dehydrated digestate to the aerobic composting plants. This additional step, due to the prolonged high temperature reached in the process, inactivates more ARGs and pathogens, boosting its reuse in agriculture [39,40].

Both biomolecular and culture methods have been previously generally employed in AMR detection. However, the cultural methods require a lot of time, and they are able to detect only a fraction of the potential antibiotic resistance risk. In this study, biomolecular analyses have been conducted using the ddPCR to screen for the presence of AMR genes. This method accounts for several advantages, including higher sensitivity and sensibility, reduced effect of the chemical inhibitors presents in the environmental matrix, and high throughput without requiring standard curves. This last point is crucial for ARG research. Moreover, the use of the ddPCR allows a quantitative assessment of the occurrence and potential horizontal transfer of ARGs.

4. Conclusions

The detection and quantification of the antibiotics and the characterization of their behavior in the environment are limited and mainly refer to water. On the other hand, ARGs can cross environmental and species boundaries. An effective one health approach is crucial for health management in landfill systems. The one health pillars are the monitoring of human, veterinary, and environmental dissemination of ARGs. Surveillance of the first two is already ongoing, but few matrices are tested for the third one, with wastewater being the most well-known. This paper focused on the sludge as an environmental matrix with a high potential biological risk. Although the misuse of antibiotics in human and animal health is widely recognized in the literature that, it is only in recent documents from European institutions, such as WHO Europe, the Council of the European Union (EU), the European Commission, and the European Environment Agency (EEA), that the agricultural dimension of the problem has been addressed. In the "A European One Health Action Plan against Antimicrobial Resistance (COM/2017/0339)", references to the agricultural sector were limited to livestock and aquaculture [6]. However, the "Council Recommendation on stepping up EU actions to combat antimicrobial resistance in a One Health approach (2023/C 220/01)" indicated a notable advancement in 2023 by recognizing the importance of monitoring agricultural residues to better understand the role of microbial contaminants in the environment [41]. While the recommendation emphasizes the prudent management of manure and sewage sludge, it does not provide detailed guidelines and instead highlights the need for further data to fully assess the issue. Similar considerations are also reflected in national plans, such as the Italian National Action Plan on Antimicrobial Resistance (PNCAR) 2022–2025, which highlights the importance of fostering cooperation between the veterinary and agricultural sectors to develop effective strategies for reducing antimicrobial consumption [42]. Further general indications are provided in the "Roadmap on Antimicrobial Resistance for the WHO European Region (2023–2030)", which focuses on the risks of antifungal use in agriculture, particularly those similar to compounds critical for human health. The roadmap also identifies agricultural workers, including those in food animal production and meat-processing sectors, as a high-risk group for AMR exposure [43]. The availability of validated monitoring data could support future regulatory decisions.

The ddPCR presents several advantages, including the possibility to not include standards for the quantification of the target of interest, but also the lower sensitivity to inhibitors present in complex matrices like sludge. These biomolecular methods produce reliable datasets for further analyses.

From a sustainability perspective, this study provides new knowledge on the presence and persistence of the ARGs in the digestate for safe reuse of the organic wastes. The extreme conditions of the anaerobic process led to a significant reduction of the ARGs, observable only in thermophilic conditions. The reuse of the digestate in agriculture shows promising advantages as a soil amendment or fertilizer [44]. This practice mitigates the impact of the OFMSW on the environment following a circular economy approach. In this study, when the AD was conducted under thermophilic conditions, it seemed to be responsible for ARG spread, even if a significant reduction was observed. Further studies are required for an exhaustive risk assessment for human health linked to the residual presence of the antibiotics and AMR microbials in the digestate, as well as the potential horizontal transmission of the ARGs from OFMSW to the environment.

Supplementary Materials: The following supporting information can be downloaded at: https://www.mdpi.com/article/10.3390/su162410922/s1, Table S1: Dataset for data availability.

Author Contributions: Conceptualization, D.T.; methodology, E.F. and D.T.; software, E.F.; validation, D.T.; formal analysis, A.M.; resources, D.T.; data curation, A.M. and E.F.; writing—original draft preparation, A.M. and E.F.; writing—review and editing, D.T.; visualization, A.M. and E.F.; supervision, D.T.; project administration, D.T.; funding acquisition, D.T. All authors have read and agreed to the published version of the manuscript.

Funding: This research was funded by the University of Turin in the context of the finalized local research program.

Institutional Review Board Statement: Not applicable.

Informed Consent Statement: Not applicable.

Data Availability Statement: The original contributions presented in this study are included in the Supplementary Material. Further inquiries can be directed to the corresponding author(s).

Acknowledgments: The authors would like to thank the plants involved and their workers for collaboration on this project.

Conflicts of Interest: The authors declare no conflicts of interest.

eferences

Haenni, M.; Dagot, C.; Chesneau, O.; Bibbal, D.; Labanowski, J.; Vialette, M.; Bouchard, D.; Martin-Laurent, F.; Calsat, L.; Nazaret, S.; et al. Environmental Contamination in a High-Income Country (France) by Antibiotics, Antibiotic-Resistant Bacteria, and Antibiotic Resistance Genes: Status and Possible Causes. *Environ. Int.* **2022**, *159*, 107047. [CrossRef]

Miłobedzka, A.; Ferreira, C.; Vaz-Moreira, I.; Calderón-Franco, D.; Gorecki, A.; Purkrtova, S.; Bartacek, J.; Dziewit, L.; Singleton, C.M.; Nielsen, P.H.; et al. Monitoring Antibiotic Resistance Genes in Wastewater Environments: The Challenges of Filling a Gap in the One-Health Cycle. *J. Hazard. Mater.* **2022**, *424*, 127407. [CrossRef]

Ministero della Salute Antibiotico-Resistenza nel Settore Umano. Available online: https://www.salute.gov.it/portale/antibioticoresistenza/dettaglioContenutiAntibioticoResistenza.jsp?lingua=italiano&id=5282&area=antibiotico-resistenza&menu=vuoto (accessed on 19 September 2024).

Arnold, B.J.; Huang, I.T.; Hanage, W.P. Horizontal Gene Transfer and Adaptive Evolution in Bacteria. *Nat. Rev. Microbiol.* **2022**, *20*, 206–218. [CrossRef] [PubMed]

Gomez Cortes, L.; Marinov, D.; Sanseverino, I.; Navarro Cuenca, A.; Niegowska, M.; Porcel Rodriguez, E.; Stefanelli, F.; Lettieri, T. *Selection of Substances for the 4th Watch List under the Water Framework Directive*; Publications Office of the European Union: Luxembourg, 2022.

European Committee. *A European One Health Action Plan against Antimicrobial Resistance (AMR)*; European Committee: Brussels, Belgium, 2017.

Gilbert, J.; Siebert, S. *Ecn Data Report 2022 Compost and Digestate for a Circular Bioeconomy*; European Compost Network ECN e.V: Bochum, Germany, 2023.

Tong, J.; Lu, X.; Zhang, J.; Angelidaki, I.; Wei, Y. Factors Influencing the Fate of Antibiotic Resistance Genes during Thermochemical Pretreatment and Anaerobic Digestion of Pharmaceutical Waste Sludge. *Environ. Pollut.* **2018**, *243*, 1403–1413. [CrossRef] [PubMed]

9. Wu, Q.; Zou, D.; Zheng, X.; Liu, F.; Li, L.; Xiao, Z. Effects of Antibiotics on Anaerobic Digestion of Sewage Sludge: Performance Anaerobic Digestion and Structure of the Microbial Community. *Sci. Total Environ.* **2022**, *845*, 157384. [CrossRef]
10. Xiao, L.; Wang, Y.; Lichtfouse, E.; Li, Z.; Kumar, P.S.; Liu, J.; Feng, D.; Yang, Q.; Liu, F. Effect of Antibiotics on the Microb Efficiency of Anaerobic Digestion of Wastewater: A Review. *Front. Microbiol.* **2021**, *11*, 611613. [CrossRef]
11. Li, S.; Ondon, B.S.; Ho, S.H.; Jiang, J.; Li, F. Antibiotic Resistant Bacteria and Genes in Wastewater Treatment Plants: Fro Occurrence to Treatment Strategies. *Sci. Total Environ.* **2022**, *838*, 156544. [CrossRef] [PubMed]
12. Wybraniec, C.; Cournoyer, B.; Moussard, C.; Beaupère, M.; Lusurier, L.; Leriche, F.; Fayolle, K.; Sertillanges, N.; Haudin, C. Houot, S.; et al. Occurrence of 40 Sanitary Indicators in French Digestates Derived from Different Anaerobic Digestion Process and Raw Organic Wastes from Agricultural and Urban Origin. *Front. Microbiol.* **2024**, *15*, 1346715. [CrossRef] [PubMed]
13. Botes, M.; De Kwaadsteniet, M.; Cloete, T.E. Application of Quantitative PCR for the Detection of Microorganisms in Water. *An Bioanal. Chem.* **2013**, *405*, 91–108. [CrossRef] [PubMed]
14. European Centre for Disease Prevention and Control ECDC AMC2 Dashboard. Available online: https://qap.ecdc.europa.e public/extensions/AMC2_Dashboard/AMC2_Dashboard.html#eu-consumption-tab (accessed on 10 April 2024).
15. Harbison, J.E.; Roth-Schulze, A.J.; Giles, L.C.; Tran, C.D.; Ngui, K.M.; Penno, M.A.; Thomson, R.L.; Wentworth, J.M.; Colma P.G.; Craig, M.E.; et al. Gut Microbiome Dysbiosis and Increased Intestinal Permeability in Children with Islet Autoimmuni and Type 1 Diabetes: A Prospective Cohort Study. *Pediatr. Diabetes* **2019**, *20*, 574–583. [CrossRef] [PubMed]
16. Wang, Y.; Wang, C.; Song, L. Distribution of Antibiotic Resistance Genes and Bacteria from Six Atmospheric Environmen Exposure Risk to Human. *Sci. Total Environ.* **2019**, *694*, 133750. [CrossRef]
17. Markowicz, A.; Bondarczuk, K.; Wiekiera, A.; Sułowicz, S. Is Sewage Sludge a Valuable Fertilizer? A Soil Microbiome ar Resistome Study under Field Conditions. *J. Soils Sediments* **2021**, *21*, 2882–2895. [CrossRef]
18. Knapp, C.W.; Dolfing, J.; Ehlert, P.A.I.; Graham, D.W. Evidence of Increasing Antibiotic Resistance Gene Abundances in Archive Soils since 1940. *Environ. Sci. Technol.* **2010**, *44*, 580–587. [CrossRef]
19. Chen, Z.; Li, Y.; Ye, C.; He, X.; Zhang, S. Fate of Antibiotics and Antibiotic Resistance Genes during Aerobic Co-Composting Food Waste with Sewage Sludge. *Sci. Total Environ.* **2021**, *784*, 146950. [CrossRef]
20. Kerrn, M.B.; Klemmensen, T.; Frimodt-Möller, N.; Espersen, F. Susceptibility of Danish Escherichia Coli Strains Isolated fro Urinary Tract Infections and Bacteraemia, and Distribution of Sul Genes Conferring Sulphonamide Resistance. *J. Antimicro Chemother.* **2002**, *50*, 513–516. [CrossRef]
21. Aminov, R.I.; Chee-Sanford, J.C.; Garrigues, N.; Teferedegne, B.; Krapac, I.J.; White, B.A.; Mackie, R.I. Development, Validatio and Application of PCR Primers for Detection of Tetracycline Efflux Genes of Gram-Negative Bacteria. *Appl. Environ. Microbi* **2002**, *68*, 1786–1793. [CrossRef]
22. Aminov, R.I.; Garrigues-Jeanjean, N.; Mackie, R.I. Molecular Ecology of Tetracycline Resistance: Development and Validation Primers for Detection of Tetracycline Resistance Genes Encoding Ribosomal Protection Proteins. *Appl. Environ. Microbiol.* **200** *67*, 22–32. [CrossRef] [PubMed]
23. Song, L.; Yang, S.; Gong, Z.; Wang, J.; Shi, X.; Wang, Y.; Zhang, R.; Wu, Y.; Wager, Y.Z. Antibiotics and Antibiotic-Resista Genes in Municipal Solid Waste Landfills: Current Situation and Perspective. *Curr. Opin. Environ. Sci. Health* **2023**, *31*, 10042 [CrossRef]
24. Ma, Y.; Wilson, C.A.; Novak, J.T.; Riffat, R.; Aynur, S.; Murthy, S.; Pruden, A. Effect of Various Sludge Digestion Conditions o Sulfonamide, Macrolide, and Tetracycline Resistance Genes and Class i Integrons. *Environ. Sci. Technol.* **2011**, *45*, 7855–786 [CrossRef] [PubMed]
25. Wang, J.; Chu, L.; Wojnárovits, L.; Takács, E. Occurrence and Fate of Antibiotics, Antibiotic Resistant Genes (ARGs) and Antibiot Resistant Bacteria (ARB) in Municipal Wastewater Treatment Plant: An Overview. *Sci. Total Environ.* **2020**, *744*, 140997. [CrossRe [PubMed]
26. Burch, T.R.; Firnstahl, A.D.; Spencer, S.K.; Larson, R.A.; Borchardt, M.A. Fate and Seasonality of Antimicrobial Resistance Gene during Full-Scale Anaerobic Digestion of Cattle Manure across Seven Livestock Production Facilities. *J. Environ. Qual.* **2022**, 5 352–363. [CrossRef] [PubMed]
27. Sun, W.; Qian, X.; Gu, J.; Wang, X.J.; Duan, M.L. Mechanism and Effect of Temperature on Variations in Antibiotic Resistanc Genes during Anaerobic Digestion of Dairy Manure. *Sci. Rep.* **2016**, *6*, 30237. [CrossRef]
28. Mortezaei, Y.; Williams, M.R.; Demirer, G.N. Effect of Temperature and Solids Retention Time on the Removal of Antibioti Resistance Genes during Anaerobic Digestion of Sludge. *Bioresour. Technol. Rep.* **2023**, *21*, 101377. [CrossRef]
29. Flores-Orozco, D.; Levin, D.; Kumar, A.; Sparling, R.; Derakhshani, H.; Cicek, N. Metagenomic Comparison of Effects c Mesophilic and Thermophilic Manure Anaerobic Digestion on Antimicrobial Resistance Genes and Mobile Genetic Element *Environ. Adv.* **2024**, *15*, 100472. [CrossRef]
30. Lorine, D.; Céline, D.; Caroline, L.M.; Frédéric, B.; Lorette, H.; Julie, B.; Laure, M.; Christine, Z.; Typhaine, P.; Sandra, R.; et a Influence of Operating Conditions on the Persistence of E. Coli, Enterococci, Clostridium Perfringens and Clostridioides Difficil in Semi-Continuous Mesophilic Anaerobic Reactors. *Waste Manag.* **2021**, *134*, 32–41. [CrossRef] [PubMed]
31. Bai, W.; Tang, R.; Wu, G.; Wang, W.; Yuan, S.; Xiao, L.; Zhan, X.; Hu, Z.H. Role of Suspended Solids on the Co-Precipitation c Pathogenic Indicators and Antibiotic Resistance Genes with Struvite from Digested Swine Wastewater. *J. Hazard. Mater.* **202** *459*, 132235. [CrossRef] [PubMed]

Cerrada-Romero, C.; Berastegui-Cabrera, J.; Camacho-Martínez, P.; Goikoetxea-Aguirre, J.; Pérez-Palacios, P.; Santibáñez, S.; José Blanco-Vidal, M.; Valiente, A.; Alba, J.; Rodríguez-Álvarez, R.; et al. Excretion and Viability of SARS-CoV-2 in Feces and Its Association with the Clinical Outcome of COVID-19. *Sci. Rep.* **2022**, *12*, 7397. [CrossRef] [PubMed]

Official Journal of the European Union Regulation (EU) No 2019/1009 of the European Parliament and of the Council Laying down Rules on the Making Available on the Market of EU Fertilising Products and Amendig; The Publications Office of the European Union: Luxembourg, 2019.

Haynes, E.; Ramwell, C.; Griffiths, T.; Walker, D.; Smith, J. *Review of Antibiotic Use in Crops, Associated Risk of Antimicrobial Resistance and Research Gaps*; Department for Environment, Food and Rural Affairs (Defra) & The Food Standards Agency (FSA): London, UK, 2020.

European Medicines Agency. *European Surveillance of Veterinary Antimicrobial Consumption Sales of Veterinary Antimicrobial Agents in 31 European Countries in 2022*; European Medicines Agency: Luxemburg, 2022.

Wang, F.; Fu, Y.H.; Sheng, H.J.; Topp, E.; Jiang, X.; Zhu, Y.G.; Tiedje, J.M. Antibiotic Resistance in the Soil Ecosystem: A One Health Perspective. *Curr. Opin. Environ. Sci. Health* **2021**, *20*, 100230. [CrossRef]

Yang, J.; Xiang, J.; Goh, S.G.; Xie, Y.; Nam, O.C.; Gin, K.Y.H.; He, Y. Food Waste Compost and Digestate as Novel Fertilizers: Impacts on Antibiotic Resistome and Potential Risks in a Soil-Vegetable System. *Sci. Total Environ.* **2024**, *923*, 171346. [CrossRef]

Zhang, R.; Gong, C.; Li, J.; Zhuang, H.; Lan, L.; Zhou, L.; Shan, S.; Wang, Y. Tracing the Transfer Characteristics of Antibiotic Resistance Genes from Swine Manure to Biogas Residue and Then to Soil. *Sci. Total Environ.* **2024**, *912*, 169181. [CrossRef]

Gurmessa, B.; Milanovic, V.; Foppa Pedretti, E.; Corti, G.; Ashworth, A.J.; Aquilanti, L.; Ferrocino, I.; Rita Corvaglia, M.; Cocco, S. Post-Digestate Composting Shifts Microbial Composition and Degrades Antimicrobial Resistance Genes. *Bioresour. Technol.* **2021**, *340*, 125662. [CrossRef]

Wang, M.; Shen, W.; Yan, L.; Wang, X.H.; Xu, H. Stepwise Impact of Urban Wastewater Treatment on the Bacterial Community Structure, Antibiotic Contents, and Prevalence of Antimicrobial Resistance. *Environ. Pollut.* **2017**, *231*, 1578–1585. [CrossRef] [PubMed]

European Council. *Council Recommendation on Stepping up EU Actions to Combat Antimicrobial Resistance in a One Health Approach*; European Council: Brussels, Belgium, 2023.

Ministero della Salute Piano Nazionale di Contrasto dell'antibiotico-Resistenza; Ministero della Salute: Rome, Italy, 2017; pp. 1–80.

Regional Commitee for Europe Roadmap on Antimicrobial Resistance for the WHO European Region 2023–2030; Elsevier Ltd.: Astana, Kazakhstan, 2023; Volume 7.

Fernández-Rodríguez, M.J.; Mancilla-Leytón, J.M.; Jiménez-Rodríguez, A.; Borja, R.; Rincón, B. Reuse of the Digestate Obtained from the Biomethanization of Olive Mill Solid Waste (OMSW) as Soil Amendment or Fertilizer for the Cultivation of Forage Grass (Lolium Rigidum Var. Wimmera). *Sci. Total Environ.* **2021**, *792*, 148465. [CrossRef]

sustainability

MDPI

Article

A Proposed Circular Economy Model for Hospital Bio-Waste Management in Municipal Settings

Anastasios Sepetis [1,*], Konstantinos Georgantas [2] and Ioannis Nikolaou [3]

[1] Department of Business Administration, University of West Attica, 12243 Egaleo, Greece
[2] Health and Social Care Management, University of West Attica, 12243 Egaleo, Greece; kgeorgantas@uniwa.gr
[3] Business and Environmental Technology Economics Laboratory, Department of Environmental Engineering, Democritus University of Thrace, 67100 Xanthi, Greece; inikol@env.duth.gr
* Correspondence: tsepet@uniwa.gr; Tel.: +30-6944423044

Abstract: This paper presents a proposed circular economy (CE) model for hospital bio-waste management, using Athens as a case study to demonstrate its applicability in urban environments. The model incorporates waste segregation at the source, resource recovery methods such as composting and anaerobic digestion, and data-driven tools to enhance the efficiency and sustainability of healthcare waste management. The study investigates the transition from linear to CE practices, focusing on structured collection strategies, collaborative efforts between hospitals and municipal authorities, and continuous tracking of waste flows. A comprehensive analysis of bio-waste volumes from participating hospitals over a three-year period is conducted, utilizing multi-criteria decision-making (MCDM) tools such as TOPSIS to evaluate the system's effectiveness. The results indicate a significant increase in hospital participation, improved waste separation, and optimized resource recovery, offering a scalable framework for other municipalities seeking to implement CE-based waste management practices in healthcare settings.

Keywords: climate change; sustainable healthcare; sustainable cities; circular economy; urban sustainability; sustainable development goals; bio-waste management; decision-making methodology; TOPSIS method

1. Introduction

Transitioning to a CE model in the context of hospital bio-waste management is far from straightforward, as it demands not only a shift in waste processing but also a fundamental change in how hospitals approach sustainability. By addressing the challenges of resource recovery and waste minimization, this study demonstrates how CE principles could be more realistically applied in the hospital setting. Hospital bio-waste, including infectious, hazardous, and organic materials, poses significant environmental and public health risks if not managed effectively [1]. Traditional linear models, characterized by waste disposal without resource recovery, are inadequate for addressing the rising waste volumes generated by healthcare facilities [2]. This research delves into how circular economy (CE) principles can revolutionize hospital bio-waste management within cities, responding to the limitations of conventional models by prioritizing resource recovery and sustainable practices.

While the traditional "take–make–dispose" model has long dominated waste management, its environmental and economic shortcomings are becoming increasingly apparent, especially in sectors like healthcare where waste streams are often hazardous. This model, particularly in the healthcare sector, is becoming increasingly unsustainable due to the growing volume of bio-waste generated by hospitals. Linear approaches often lead to the incineration or landfilling of bio-waste, causing significant environmental impacts, including greenhouse gas emissions and the depletion of valuable resources [3]. The adoption of CE principles within hospital waste management contributes to urban sustainability by reducing environmental impacts, enhancing resource efficiency, and supporting circular resource flows [4]. By adopting CE principles, the model transforms hospital bio-waste into resources like compost or energy, contributing to the broader objectives of urban sustainability and circular resource use [5].

The selection of the TOPSIS method in this study is driven by its capability to address multi-criteria decision-making challenges, particularly in complex systems such as hospital bio-waste management. TOPSIS offers a systematic framework to evaluate and rank alternatives based on their closeness to an ideal solution, making it particularly suitable for comparing the performance of waste management strategies across multiple sustainability criteria. This approach ensures robust and objective decision-making, aligning with the primary objectives of circular economy principles. The choice of TOPSIS is further justified by its ability to manage complex decision-making processes involving multiple, often conflicting criteria. In the context of hospital bio-waste management, where factors such as environmental impact, cost efficiency, and resource recovery hold varying levels of priority, TOPSIS ensures a balanced and transparent evaluation. Its computational efficiency and proven applicability in sustainability-related studies make it particularly suitable for urban waste management scenarios. This method allows stakeholders to assess alternatives systematically, thereby aligning decisions with circular economy principles while addressing both environmental and operational goals.

In the context of sustainable cities, it is essential for healthcare facilities to integrate with urban bio-waste management systems to enhance overall efficiency and effectiveness. Partnerships between healthcare providers and municipal waste management services are crucial for aligning waste management practices with broader sustainability goals [6]. Furthermore, sustainable procurement practices in hospitals, such as choosing suppliers that prioritize eco-friendly products and sustainable practices, can significantly contribute to reducing the overall environmental impact of healthcare facilities [7]. This holistic approach promotes a culture of sustainability within the healthcare sector, ensuring a sustainable future for cities and communities.

Municipalities adopting CE models in hospital waste management can see reductions in greenhouse gas emissions and improvements in public health. This approach not only reduces the environmental footprint of waste management but also aligns with broader sustainability goals, such as reducing greenhouse gas emissions and promoting resource efficiency [8]. The healthcare sector, with its large volume of hazardous and organic waste, presents a significant opportunity for municipalities to apply CE models effectively [9].

The transition to a circular model in hospital bio-waste management also offers potential economic benefits for municipalities. By investing in composting facilities and advanced waste segregation systems municipalities can reduce the cost of waste disposal while generating new revenue streams from recovered resources [10]. Moreover, the integration of CE models fosters stronger collaboration between hospitals and municipal authorities, encouraging innovation in waste handling practices and improving overall public health outcomes [11].

This research uniquely contributes to the existing literature by providing a comprehensive framework that integrates circular economy principles with advanced decision-making methodologies, specifically TOPSIS, to address the critical challenges of hospital bio-waste management in urban contexts. The study raises an important question: Can the principles of circular economy truly transform the way hospital bio-waste is managed, especially within densely populated urban settings? Our objective is to assess if and how these principles can contribute not only to resource recovery but also to creating healthier and more resilient cities. The study utilizes a three-year dataset of bio-waste volumes from hospitals in Athens to assess the CE model's impacts on waste management efficiency and sustainability. By analyzing these trends through the TOPSIS multi-criteria decision-making tool, we aim to quantify the model's effectiveness in fostering sustainable practices in urban healthcare waste systems

Following this introduction, Section 2 delves into the existing literature, focusing on the limitations of linear models and the opportunities that circular systems present for hospital waste management. Section 3 elaborates on the circular economy (CE) model applied in Athens, describing the bio-waste collection system, the collaboration between hospitals and municipal authorities, and the integration of advanced tracking technologies. Section 4 presents the methodology, outlining data collection procedures and the application of the TOPSIS decision-making tool to evaluate the impact of the CE model. Section 5 offers a discussion of the results, situating the findings within the broader context of related research, and identifying the primary successes and ongoing challenges of the model's implementation. Finally, Section 6 provides conclusions and explores the implications for expanding CE-based waste management practices in urban healthcare systems.

Through this analysis, we aim to demonstrate the feasibility and benefits of integrating circular economy principles into healthcare waste management and to provide a framework for other cities and healthcare systems to adopt similar models. The study underlines how a holistic CE approach can facilitate the transformation of waste management practices, ensuring both environmental and economic gains.

2. Literature Review

Municipalities today face mounting pressures to adopt sustainable waste management systems, making the transition from a linear economy to a circular economy (CE) model especially urgent when it comes to managing hospital bio-waste. This shift responds directly to the pressing need for environmental and economic resilience within urban settings. Unlike linear models, CE emphasizes resource recovery, reducing landfill dependency and environmental impacts, which are particularly critical in urban hospital settings. This transition not only promotes recycling and waste reduction but also aligns with broader urban sustainability goals, helping cities address the environmental challenges posed by increased waste volume [12].

Amid intensifying regulatory mandates and public calls for greener practices, municipalities are facing increasing pressure to rethink traditional waste approaches—especially in high-risk areas like healthcare waste. Linear disposal methods fail to fully mitigate the hazards posed by bio-waste, often compromising public health and ecological integrity. Circular economy models offer a solution by promoting resource recovery and reducing landfill dependency, thus minimizing the environmental footprint and enhancing urban resilience [13,14]. Moreover, multi-criteria decision-making (MCDM) tools such as TOPSIS have been widely utilized to evaluate and prioritize waste management strategies based on sustainability criteria, providing a systematic approach to addressing complex urban challenges [15,16].The high costs associated with hazardous waste disposal and the need for safer, more efficient waste management further drive municipalities towards CE principles, which enable economic savings and environmental protection through sustainable bio-waste processing.

Furthermore, studies and efforts underscore the environmental and economic advantages of shifting from linear to circular waste management, especially for sectors like healthcare where waste generation is substantial and complex [17,18]. In adopting CE models, municipalities can implement advanced segregation and treatment technologies that not only ensure safe disposal but also facilitate the recovery of resources, such as energy from biogas or organic fertilizers from compost [19]. TOPSIS has also been employed to evaluate trade-offs among conflicting criteria, such as cost efficiency, environmental impact, and resource recovery, providing actionable insights for implementing circular economy principles [19]. This shift aligns with global sustainability targets and helps municipalities address the dual challenges of increasing waste volumes and stringent environmental standards [20,21].

Healthcare bio-waste, though not classified as hazardous waste, raises significant concerns due to its potential environmental and public health impacts [22]. European researchers emphasize that improper disposal practices can adversely affect urban ecosystems and human health, particularly in densely populated cities, where the risk of contamination and pollution is heightened [23]. In response, a shift toward circular economy (CE) models is essential, emphasizing waste treatment systems that not only ensure safe disposal but also facilitate resource recovery. CE models that have been implemented and several future scenarios encourage municipalities to adopt structured, sustainable waste management practices, including segregation, treatment, and the repurposing of bio-waste into valuable resources like compost or energy [24,25].

Assessment strategies play a pivotal role in optimizing waste collection schemes in urban areas, enabling municipalities to refine their approaches and improve efficiency [26]. The literature highlights several key strategies that municipalities should adopt when transitioning from linear to CE models in hospital bio-waste management. These include collection systems, waste segregation at source, and frequent collection schedules to minimize contamination and ensure efficient processing [27]. For instance, Seruga (2016) outlines the benefits of door-to-door waste collection in the case of rural areas with various standard bins being used, where bio-waste from multiple spots is processed at a single facility equipped with advanced technologies such as anaerobic digestion and composting [28].

Policy frameworks are pivotal in establishing a foundation for CE in healthcare waste management, as documented in numerous case studies [19]. In addition to robust policy frameworks, MCDM tools like TOPSIS play a crucial role in aligning municipal waste management practices with sustainability goals, offering a structured framework to assess and optimize diverse waste management [29]. Such frameworks not only drive compliance but also encourage proactive innovation, pushing hospitals and municipalities toward sustainable practices through incentives and regulatory requirements. Governments must establish regulations that encourage sustainable waste management practices, such as providing financial incentives for hospitals that adopt eco-friendly technologies The successful implementation of CE models in hospital bio-waste management is also closely tied to innovation and technological advancements. According to the World Economic Forum (2020), the use of advanced waste treatment technologies, such as autoclaving and anaerobic digestion, has been crucial in reducing the environmental impact of bio-waste [30]. These technologies not only ensure the safe disposal of hazardous waste but also enable the recovery of valuable resources, such as biogas [31].

Athens, for instance, illustrates the tangible benefits of integrating hospital bio-waste with city-wide waste systems, achieving notable improvements in sustainability and significant cost savings. The Athens model also highlights the efficiencies gained through municipal oversight, helping ensure regulatory compliance while enhancing operational synergy [32]. In conclusion, the CE model for managing hospital bio-waste not only presents a pathway to sustainable urban development but also sets a compelling precedent for cities worldwide. By adopting CE principles, cities can reduce waste, promote recycling, and minimize the environmental risks associated with hospital bio-waste. The literature strongly supports the assertion that municipalities need to prioritize CE models, particularly in the healthcare sector, to ensure long-term sustainability [33].

Globally, several cities and municipalities have adopted circular economy (CE) principles to manage hospital bio-waste, showcasing a wide range of strategies and technologies. Municipal authorities have integrated composting facilities to handle bio-waste from healthcare institutions, significantly reducing the volume of waste sent to landfills [34]. In Germany, the integration of bin collection systems with local treatment facilities has significantly enhanced waste diversion rates in urban hospitals. This approach has led to improved segregation and recycling outcomes, aligning with the country's stringent waste management regulations [35]. The "Closed Substance Cycle Waste Management Act" mandates the prevention, recycling, and disposal of waste, emphasizing the importance of proper waste segregation at the source. Hospitals are required to designate responsible personnel to ensure adherence to these regulations, incorporating waste management into their quality management systems. By implementing these measures, German hospitals have achieved more efficient waste processing and increased recycling rates, contributing to environmental sustainability [36].

In Spain, government-led initiatives have fostered public–private partnerships to promote circular economy (CE) models in hospital waste management. The "España Circular 2030" strategy outlines objectives to reduce resource consumption and waste generation, encouraging collaboration between public and private sectors [37]. Public–private partnerships have proven instrumental in advancing CE practices, especially in urban settings where resource constraints and logistical challenges require collaborative solutions. These partnerships enable cities to leverage private sector expertise and resources, enhancing the scalability and efficiency of waste management systems. As seen in other CE implementations, such collaborations facilitate smoother integration of circular practices, fostering a culture of sustainability across sectors.

While technological advancements in bio-waste treatment remain crucial to the success of the CE, their efficacy depends heavily on adaptation to local contexts and ongoing stakeholder engagement. Anaerobic digestion (AD), a process that converts organic waste into biogas, has been widely adopted across Europe. In Finland, companies have integrated waste management systems that utilize sorted bio-waste to produce electricity and heat, thereby creating a closed-loop system. Several biogas plants operate in Finland and Sweden, processing biodegradable waste into renewable energy and recycled fertilizers; this approach not only reduces landfill waste but also contributes to energy production, exemplifying the principles of a circular economy [38].

In recent years, the integration of digital tools into waste management systems has become crucial for improving efficiency. Smart waste tracking technologies enable municipalities to monitor waste flows in real time, optimizing collection schedules and ensuring that waste is treated at appropriate facilities. Several studies highlight the role of data-driven decision-making in reducing operational costs and improving waste management outcomes. For instance, IoT-based intelligent waste management systems utilize sensors to gather data on waste bin levels, facilitating timely collection and reducing overflow incidents. Additionally, data analytics frameworks support decision-makers in implementing, monitoring, and optimizing smart waste management systems by analyzing trends and patterns in waste generation. These data-driven approaches have been shown to reduce operational costs and improve waste management outcomes, contributing to more sustainable urban environments [39–41].

In the Netherlands, hospitals have implemented advanced waste management systems that monitor bio-waste from its point of origin to its final treatment, ensuring adherence to circular economy (CE) principles and reducing waste volumes. For example, the University Medical Center Utrecht (UMC Utrecht) has adopted the Sterilwave solution, which decontaminates medical waste on-site, transforming it into a safe, inert material suitable for recycling or disposal. This system enhances waste traceability and minimizes the need for external waste processing, aligning with CE objectives [42]. Additionally, initiatives like the Pharmafilter system have been installed in several Dutch hospitals, including Erasmus University Medical Center. This system processes hospital waste and wastewater on-site, converting organic waste into biogas and reducing the overall waste footprint [43]. These examples demonstrate healthcare institutions' commitment to utilizing smart waste management technologies, thereby promoting sustainability and resource efficiency.

The economic benefits of transitioning to circular economy (CE) models have been well documented in various studies. By adopting resource recovery methods, such as composting and biogas generation, municipalities can offset the costs associated with waste disposal while generating revenue from the sale of these by-products. For instance, the European Biogas Association highlights that several municipalities across the EU separately collect and digest organic municipal waste in biogas plants, resulting in multiple greenhouse gas emission savings and the production of renewable energy and organic fertilizers [44]. Additionally, the integration of composting innovations has been shown to transform waste into valuable products, contributing to a circular economy. These practices not only reduce landfill usage but also create economic opportunities through the commercialization of compost and biogas, thereby enhancing municipal revenue streams [45].

The transition to circular economy (CE) models significantly diminishes dependence on landfills and incineration, thereby reducing both the environmental and the economic burdens associated with waste management. For instance, the NHS Clinical Waste Strategy aims to transform waste management by eliminating unnecessary waste and finding innovative ways to reuse materials, ensuring waste is processed in the most cost-effective, efficient, and sustainable manner. Implementing such bio-waste treatment systems based on CE principles has led to substantial cost savings and environmental benefits [46].

In light of the growing environmental and economic challenges associated with hospital bio-waste, the transition to circular economy (CE) models has emerged as a necessary step for municipalities and healthcare institutions. The existing literature demonstrates the effectiveness of CE principles in improving resource recovery, reducing waste, and minimizing environmental risks. However, despite the advancements in technologies and policies, there remains a gap in understanding with regard to how these models can be systematically evaluated and optimized, particularly in urban settings.

This study addresses this gap by focusing on the application of a CE model for hospital bio-waste management in Athens. The TOPSIS (Technique for Order Preference by Similarity to Ideal Solution) methodology will be employed to evaluate the performance of this model, comparing various factors. The use of TOPSIS in this study provides a validated approach for prioritizing hospital bio-waste management strategies, ensuring that the selected solutions align with the principles of the circular economy. Through this approach, we aim to provide a comprehensive framework that can be adapted by other municipalities seeking to adopt CE models for sustainable urban development.

3. Material and Methods

This section outlines the strategic methodology applied to evaluate and optimize the bio-waste collection network in Athens's hospitals, designed within the framework of the circular economy (CE) model. Emphasizing waste minimization, resource recovery, and sustainability, our analysis employs the TOPSIS multi-criteria decision-making methodology to prioritize management strategies based on critical factors, including environmental impact, cost efficiency, resource recovery, scalability, and stakeholder acceptance.

3.1. Method Research Structure—Scientific Questions and Research Objectives

The focus of this research is to evaluate the application of a circular economy (CE) model for hospital bio-waste management within an urban context, with the city of Athens serving as the primary case study. The primary objective is to measure the environmental and economic benefits that can be achieved through CE principles in hospital bio-waste management, specifically examining stakeholder collaboration and resource recovery. The methodology's structured multi-criteria decision-making process aids in comparing the relative effectiveness of composting and landfill as bio-waste strategies. The TOPSIS (Technique for Order Preference by Similarity to Ideal Solution) methodology was selected to prioritize bio-waste management strategies based on multiple criteria, including environmental impact, cost efficiency, resource recovery, scalability, and stakeholder acceptance [47].

The research methodology follows a structured multi-criterion decision-making approach. This involves defining the relevant criteria and alternatives, assigning weights to each criterion, applying the TOPSIS method, and ranking the alternatives based on their relative closeness to the ideal solution. The choice to evaluate only two alternatives, composting and sanitary landfill, stems from their dominance in the existing Greek waste management systems. Composting represents an advanced, CE-compliant method that prioritizes resource recovery, while sanitary landfill reflects the basic option that, despite its drawbacks, remains a prevalent practice in urban settings. This binary focus ensures a clear and focused comparison between the innovative CE-aligned approach and the widely used traditional method.

First, the decision matrix was constructed, capturing the performance of each alternative against the defined criteria. The decision-making matrix was constructed based on expert consultations and publicly available data from similar waste management initiatives in European urban contexts. Experts from the Municipality of Athens, waste management authorities, and healthcare professionals provided insights into the performance of composting and sanitary landfill across the five criteria. The decision matrix was normalized to ensure comparability across different criteria, followed by the application of a weighting system. Each criterion was assigned a weight based on its importance to overall sustainability. The weights for the criteria were determined using the Analytical Hierarchy Process (AHP), a structured decision-making framework designed to quantify subjective judgments in multi-criteria scenarios. AHP was particularly suitable for this study due to its ability to integrate expert opinions systematically while maintaining transparency and consistency in the weighting process. Experts from diverse fields—including environmental science, healthcare management, and municipal waste operations—were consulted, ensuring a balanced reflection of environmental, economic, and operational priorities. This approach aligns with recent applications of AHP in sustainable planning, where criteria such as environmental impact and stakeholder acceptance are frequently prioritized to achieve long-term sustainability goals [48]. Next, the ideal solution (best possible outcomes) and negative-ideal solution (worst outcomes) were calculated. Finally, the Euclidean distance between each alternative and the ideal/negative-ideal solutions was determined, and the alternatives were ranked based on their relative closeness to the ideal solution.

The diagram below (Figure 1) provides a visual representation of the research methodology employed in this study, highlighting the key components and processes involved in our approach.

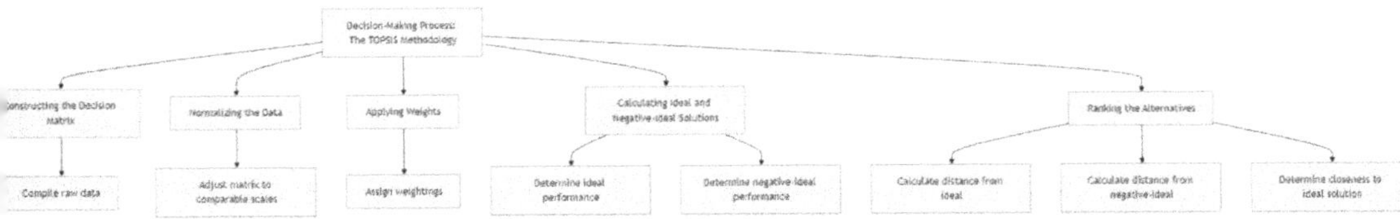

Figure 1. A schematic flowchart of the TOPSIS.

The TOPSIS (Technique for Order Preference by Similarity to Ideal Solution) methodology was employed in the city of Athens to evaluate and rank bio-waste management strategies in hospitals, framed within the circular economy (CE) model. The two alternatives under consideration—composting and sanitary landfill—were assessed using five carefully selected criteria: environmental impact, cost efficiency, resource recovery, scalability, and stakeholder acceptance. Each criterion within the TOPSIS framework was carefully selected and weighted based on its relevance to sustainability goals. The weightings were determined through stakeholder and expert consultations, reflecting the priority of minimizing environmental impact, maximizing resource recovery, and ensuring cost-efficiency. Specifically, environmental impact and resource recovery were assigned higher weights, underscoring the importance of these factors in sustainable bio-waste management within urban hospital settings.

The initial decision matrix contains the raw performance data for each alternative under each criterion. The values represent the performance of composting and sanitary landfill on a scale relative to each criterion (Table 1).

Table 1. Alternatives' performance against each criterion.

Criterion	Composting	Sanitary Landfill
Environmental Impact	85	45
Cost Efficiency	65	80
Resource Recovery	90	25
Scalability	55	95
Stakeholder Acceptance	70	50

The decision matrix was normalized to adjust for different scales of measurement, allowing for a fair comparison across all criteria (Table 2).

Table 2. Normalized decision values.

Criterion	Composting	Sanitary Landfill
Environmental Impact	0.875	0.462
Cost Efficiency	0.622	0.766
Resource Recovery	0.955	0.265
Scalability	0.500	0.866
Stakeholder Acceptance	0.777	0.555

The weighted normalized matrix, obtained by multiplying each normalized value by its corresponding weight, is shown below (Table 3):

Table 3. Weighted normalized values.

Criterion	Composting	Sanitary Landfill
Environmental Impact	0.262	0.138
Cost Efficiency	0.093	0.115
Resource Recovery	0.191	0.053
Scalability	0.050	0.087
Stakeholder Acceptance	0.194	0.138

The weighting of criteria such as environmental impact, cost efficiency, and resource recovery was instrumental in guiding the decision-making process towards sustainable options. By assigning higher importance to environmental sustainability and stakeholder engagement, this methodology aligns with CE principles, prioritizing options that contribute to long-term urban sustainability.

The first step is shown below. The following five criteria were chosen to evaluate the alternatives based on their relevance to hospital bio-waste management, sustainability goals, and practical applicability. The goal is to rank alternatives for hospital bio-waste management in Athens according to key sustainability criteria that align with the principles of the circular economy (CE).

Environmental impact: Environmental sustainability is central to bio-waste management and the CE model. This criterion evaluates the potential of each alternative to reduce environmental harm, focusing on greenhouse gas emissions, pollution, and reducing landfill use. Composting, for example, is known to reduce methane emissions and return nutrients to the soil, whereas landfills are major contributors to environmental degradation.

Cost efficiency: Financial feasibility is crucial for municipalities and hospitals wishing to adopt any bio-waste management strategy. This criterion considers the overall costs of implementation, including capital investment, operational expenses, and long-term savings. Alternatives that offer cost-effective solutions while maintaining high sustainability standards are rated more favorably.

Resource recovery: A key aspect of the CE model is the reuse and recovery of resources from waste. This criterion evaluates the potential of each alternative to recover valuable materials or produce energy. Composting, which converts organic waste into valuable compost, ranks highly here, while landfills offer minimal resource recovery opportunities.

Scalability: The scalability criterion assesses the ability of each alternative to adapt to different scales of waste generation. Hospitals generate varying amounts of bio-waste depending on their size, so scalability is an important factor. A scalable solution ensures that smaller and larger hospitals alike can participate in the program with ease.

Stakeholder acceptance: The success of any waste management strategy depends on the participation and acceptance of stakeholders, including hospital staff, waste management authorities, and the public. This criterion measures the willingness and ability of key stakeholders to adopt and support the bio-waste management alternative. Composting, for instance, enjoys strong support in communities that prioritize sustainability, while landfill use faces more resistance due to its environmental drawbacks.

The second step of the TOPSIS method is to assign weights to the identified criteria. The weights reflect the relative importance of each criterion. The following two alternatives were implemented and analyzed based on their effectiveness in the local context.

Composting: Composting is a process that breaks down organic materials, such as food waste and biodegradable hospital waste, into nutrient-rich compost. This alternative was chosen due to its alignment with the CE principles of reducing landfill use, recovering valuable resources, and minimizing environmental harm. Composting has been successfully implemented in various urban settings, and its potential to integrate with existing waste management systems in Greece makes it a viable solution for hospitals.

Sanitary landfill: Despite its environmental drawbacks, sanitary landfill remains a common practice in many countries, including Greece. This alternative was included to provide a comparison with composting, focusing on its cost-effectiveness and scalability. Sanitary landfill is less resource-intensive initially, but it contributes significantly to greenhouse gas emissions and offers no resource recovery. As a baseline alternative, it provides insight into the performance of composting when evaluated against conventional waste disposal methods.

In the TOPSIS method, the positive-ideal solution (PIS) represents the best possible outcome for each criterion, while the negative-ideal solution (NIS) represents the worst possible outcome. These solutions are calculated by selecting the maximum and minimum normalized values for each criterion:

- PIS: This alternative maximizes benefits (offering higher scores for positive criteria like resource recovery) and minimizes costs (lower scores for negative criteria like environmental impact).
- NIS: This alternative minimizes benefits and maximizes costs, representing the least favorable outcome.
- For each criterion, we state the following:
- PIS = {max value for benefit criteria, min value for cost criteria}.
- NIS = {min value for benefit criteria, max value for cost criteria}.

In continuation, the separation values (PIS and NIS) are presented in Table 4. Based on the separation values from the PIS and NIS, we are able to calculate the relative closeness of each alternative to the ideal solution. The formula for calculating the relative closeness is as follows:

$$R_i = \frac{d_i^-}{d_i^- + d_i^+}$$

Table 4. Separation values (PIS and NIS).

Alternative	Distance to PIS	Distance to NIS
Composting	0.345	0.781
Sanitary Landfill	0.652	0.529

To validate the robustness of the TOPSIS results, a sensitivity analysis was conducted by varying the weights of the criteria within a ±10% range [49]. This variation tested the impact of altering the relative importance of environmental impact, cost efficiency, resource recovery, scalability, and stakeholder acceptance on the final ranking. The analysis consistently identified composting as the superior alternative, with its ranking remaining unchanged across all scenarios. This robustness underscores the reliability of the model and its alignment with CE principles, offering a systematic approach to decision-making in urban bio-waste management systems.

3.2. *Applying the Decision-Making Methodology to Hospital Bio-Waste Management in the City of Athens*

The CE-driven decision-making approach enables an objective evaluation of each bio-waste management strategy, focusing on sustainability impacts and resource efficiency. By employing TOPSIS, a clear, data-driven comparison emerges between composting and landfill. Composting aligns with CE principles by promoting resource recovery, while landfill remains tied to the traditional linear model. To evaluate these strategies, the following weights were assigned based on sustainability priorities: environmental impact

(30%), cost efficiency (15%), resource recovery (20%), scalability (10%), and stakeholder acceptance (25%).

After establishing the ideal and negative-ideal solutions, the Euclidean distance for each alternative was calculated, providing a clear ranking of bio-waste management strategies based on their closeness to the ideal solution.

Composting: Relative closeness = 0.781.
Sanitary landfill: Relative closeness = 0.529.

The analysis identified composting as the more sustainable option for hospital bio-waste, with a relative closeness of 0.781 to the ideal solution, compared to 0.529 for landfill. Composting's performance aligns closely with CE objectives, particularly in terms of environmental impact and resource recovery. The practice of converting hospital bio-waste into valuable organic matter supports Athens's broader sustainability goals, moving away from linear disposal models towards a regenerative, CE-aligned system.

While sanitary landfill scored higher in terms of cost efficiency and scalability, these benefits are overshadowed by its significant environmental drawbacks. The landfill method, though initially more affordable, contributes to greenhouse gas emissions and long-term land use issues, which run counter to the CE objectives of reducing the ecological footprint of urban waste management. For hospitals in Athens, particularly those participating in the bio-waste program, composting presents a far more sustainable solution that not only reduces environmental harm but also supports the city's ambition to integrate waste management into a circular, resource-recovering system.

The close collaboration between hospitals and the Municipality of Athens has been critical in implementing composting as a viable alternative. The high stakeholder acceptance of composting reflects the strong collaboration between healthcare providers and municipal authorities, reinforcing the city's commitment to achieving sustainable waste management under a circular economy framework Composting emerges as the superior choice within a CE framework due to its resource recovery potential and lower environmental impact, while sanitary landfill, despite its cost efficiency, does not align with sustainability targets due to its high emissions and lack of resource utilization. The results suggest that, as the city of Athens continues to scale its circular economy initiatives, composting will play a central role in transforming hospital bio-waste into valuable resources, further integrating hospitals into the city's sustainability framework.

3.3. Evaluation Methodology Description

The evaluation methodology aligns with CE goals by focusing on two core parameters: bio-waste production and stakeholder engagement. These parameters form a comprehensive framework to gauge the network's scalability and sustainability, ensuring that bio-waste is consistently diverted from traditional disposal methods while fostering growing stakeholder participation.

The first evaluation parameter involves tracking the total bio-waste volume produced by participating hospitals. By systematically collecting data over time, this metric reflects the success of waste separation practices and the program's overall impact on reducing landfill dependency. Data gathered at regular intervals provide insight into bio-waste trends across participating hospitals, highlighting whether improved separation practices have led to increased bio-waste diversion from traditional disposal methods.

The second key parameter evaluates the rate of hospital participation in the bio-waste network. Tracking the number of hospitals joining the network over the years offers a measure of the program's scalability, capturing its ability to integrate new facilities within a CE model. Systematic recording of hospital entries, supplemented by surveys and interviews, sheds light on the motivations behind engagement and hospitals' commitment to CE principles.

Evaluating hospital participation offers insight into the program's scalability and reflects the network's success in integrating healthcare facilities into the municipal CE model, enhancing the overall sustainability of urban bio-waste management. In terms of stakeholder engagement, the methodology involves tracking participation rates, identifying hospitals that have joined or exited the program, and determining the impact of these changes on overall bio-waste production. The participation trend will be evaluated using both quantitative measures (number of hospitals involved, bio-waste contributions) and qualitative feedback from hospital stakeholders.

This methodology provides a robust framework for assessing the success of the hospital bio-waste collection network, focusing on key performance indicators such as the volume of bio-waste produced and the expansion of hospital participation. To further validate the model, a comparative analysis was performed against historical data from European CE initiatives in urban bio-waste management. The outcomes were consistent with those reported in similar studies, reinforcing the validity of the methodology [13]. Future work will aim to incorporate real-time data tracking to enhance accuracy and adaptability.

Applying this methodology to real-world data will help refine the network's operational efficiency, providing a feedback loop that supports continuous improvement and strengthens the city's commitment to sustainable, CE-aligned waste management.

4. Strategic Implementation and Evaluation of the Bio-Waste Collection System for Hospitals in Athens: A Case Study

This section examines the strategic phases and operational challenges involved in implementing the hospital bio-waste collection system in Athens, with a focus on integrating circular economy principles into municipal waste management frameworks. Effective bio-waste management in healthcare facilities is crucial for ensuring environmental sustainability and mitigating the risks associated with improper waste disposal. This section discusses the implementation of bio-waste collection models in hospitals within the city of Athens, focusing on practical strategies and their integration with municipal waste management systems.

Optimizing Processes: The Methodological Framework for a Hospital Bio-Waste Collection System

To ensure effective bio-waste management in hospitals, Athens designed its collection program around phased implementation and source separation strategies. These processes aim to reduce contamination and support sustainability, which are key CE objectives. The Athens bio-waste program was rolled out in phases, initially focusing on organic waste collection from major producers, including hospitals. Phase one prioritized plant-based organic waste, which consistently creates a large waste stream within these facilities.

A primary strategy of the program emphasizes source separation, particularly of food waste, which constitutes a significant portion of hospital bio-waste. Source separation is critical for minimizing contamination and maximizing the quality of recovered materials, aligning directly with CE principles. Hospitals implemented an interior-yard collection system to ensure organic waste was isolated from other waste streams at the source. To support this, the Municipality of Athens provided specialized tools, such as compostable bags and designated bins for food waste, helping to maintain clean collection points. This setup not only minimized contamination but also enhanced hygiene, increasing staff participation in bio-waste management efforts. Recognizing food waste as a major component, the program emphasized segregation at the kitchen level. Participating hospitals received small

(140–240 L) bins and compostable bags to facilitate collection within kitchens. Strategically placed bins encouraged convenient and hygienic collection practices. The compostable bags further reduced contamination risk, minimizing cleaning frequency—an important factor in food waste handling, as spoilage and pests are common issues. These measures led to marked improvements in both waste separation quality and overall kitchen hygiene.

The mandatory nature of this system, combined with ongoing training and public awareness campaigns, significantly increased compliance among hospitals and healthcare providers.

Hospitals collaborated closely with the Municipality of Athens, which played a central role in providing infrastructure, including 660 L bins and specialized collection vehicles. The Municipality's support in providing infrastructure has been pivotal in scaling the program, ensuring that hospitals have consistent access to necessary collection tools and transportation resources. These vehicles were designed to handle organic waste safely and efficiently, ensuring timely collection from each hospital. The program's success is largely attributable to the effective integration of hospitals within the broader municipal waste collection framework.

The frequency of bio-waste collection varies depending on climatic conditions and the volume of organic waste generated. In Mediterranean climates, such as that of Athens, frequent collection is necessary to prevent the accumulation of bio-waste that can lead to leachate production and pest infestations. Collection frequencies in Athens ranged from three to six times per week, depending on the hospital's output and the time of year.

Collection vehicles were also adapted to meet the specific needs of bio-waste transport. For example, food waste, which has a low bulk density and high moisture content, required the use of non-compacting vehicles to preserve the integrity of the waste during transportation. The municipality acquired specialized milled-type waste collection vehicles equipped with onboard washing systems to clean the bins immediately after emptying, ensuring high hygiene standards.

Despite the successes, several challenges have emerged during the implementation of bio-waste collection models in Athens' hospitals. One key issue is maintaining consistent participation from all hospitals and ensuring that bio-waste is properly segregated at the source. Challenges such as the high cost of compostable bags and the need for regular maintenance of collection vehicles underscore the ongoing financial and operational investment required to sustain the program.

Continuous education and training sessions were conducted for hospital staff, emphasizing the importance of proper waste segregation and the environmental benefits of bio-waste recycling. These initiatives significantly improved compliance rates and enhanced the overall effectiveness of the program. Further investments in infrastructure, including the expansion of the vehicle fleet and improved bag distribution systems, are also being considered to enhance the overall efficiency of the program.

The implementation of bio-waste collection models in Athens's hospitals, in collaboration with the Municipality of Athens, has provided a valuable case study in sustainable waste management practices. By focusing on source separation, specialized tools, and frequent collection, the program has succeeded in reducing contamination and improving the overall handling of hospital-generated bio-waste. Moving forward, continuous investment in infrastructure and education will be essential to maintaining the program's success and expanding its scope.

The following diagram (Figure 2) visually depicts the implementation framework for the hospital bio-waste collection system in Athens, detailing the strategic phases and core components of the program, from initial design to future planning and evaluation.

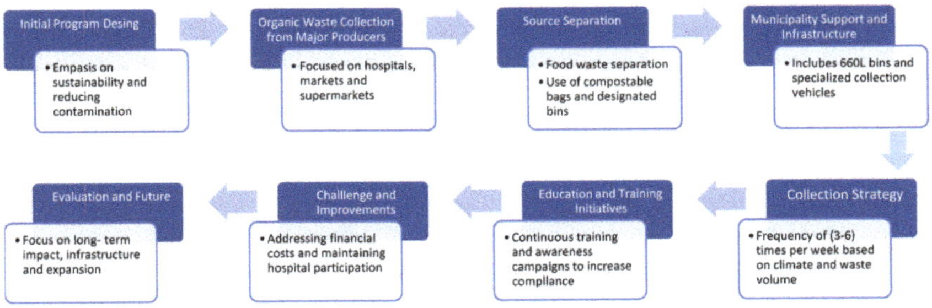

Figure 2. A flowchart of the phased implementation and optimization processes for hospital bio-waste collection in Athens.

Future efforts should focus on enhancing the program's financial sustainability and operational efficiency to ensure its long-term impact on hospital bio-waste management in Athens.

5. Findings. How Is CE Implemented in Hospital Bio-Waste Management?

This section presents the findings from the Athens hospital bio-waste management network, emphasizing trends in participant engagement and bio-waste production. The findings highlight the network's alignment with circular economy (CE) goals, focusing on how stakeholder participation and internal waste practices have evolved from 2021 to 2023.

The following Table 5 provides an overview of the annual bio-waste generation per hospital bed across Athens's hospitals that are participating in the management program.

Table 5. Bio-waste generation per hospital bed per year (2021–2023).

Year	Hospitals	Total Beds	Bio-Waste (Kg)	Bio-Waste Per Bed (Kg/Bed)
2021	18	7.224	599.456	83
2022	20	7.857	667.224	85
2023	20	7.843	808.044	103

A detailed analysis of bio-waste production per bed highlights further insights into the network's efficiency. The average annual bio-waste generated per hospital bed was calculated to be approximately 83 kg in 2021, 85 kg in 2022, and 103 kg in 2023 [50]. This steady increase in bio-waste generation per bed, despite the relatively stable number of beds in participating hospitals, underscores the effectiveness of enhanced segregation and internal waste management practices. These findings provide additional evidence of the network's scalability and its ability to optimize resource recovery within existing hospital infrastructure.

The phased initiation began as a pilot program in 2020, with 18 hospitals actively participating by 2021. This gradual inclusion enabled adaptive logistics and infrastructure upgrades, even amid the COVID-19 pandemic's heightened waste handling requirements.

In 2022, with municipal support, the network grew to 20 hospitals. Collection efficiency improved through optimized routes and additional infrastructure, reflecting ongoing refinements in waste management logistics and bin allocation. In 2023, participation was steady, and the network saw a substantial increase in bio-waste volume due to refined segregation practices within hospitals, indicating enhanced internal compliance and resource recovery. These year-on-year trends underline the network's capacity for sustainable bio-waste diversion, demonstrating that a focus on training, infrastructure, and compliance can significantly enhance resource recovery without expanding the network's scope.

Across these years, distinct trends emerge. The phased integration in 2021 led to a gradual increase in bio-waste volumes as more hospitals were brought into the network. By 2022, the focus shifted to refining operations, with collection efficiency improving through optimized routes and frequent collections. In 2023, with participation steady, bio-waste volume continued to rise due to effective segregation practices, underlining the network's success in sustainably managing waste without needing to expand its participant base. This consistent increase despite stable participation reflects the scalability and efficiency of the network, signaling that enhanced training, infrastructure, and compliance measures are key to maximizing bio-waste recovery.

The Municipality of Athens' support was instrumental in these outcomes, through investments in specialized collection vehicles, strategically placed bins, and streamlined routes. The continuous backing from the Municipality provided the structural backbone needed to manage increasing volumes of bio-waste, demonstrating the critical role of municipal support in sustainable CE-aligned waste management. Hospitals found it easier to comply with the collection model, which was strengthened by ongoing staff training and awareness initiatives. This continuous municipal backing allowed the program to absorb increased waste volumes and integrate new hospitals effectively, thereby reinforcing the CE objectives. Stakeholder acceptance, highlighted by high compliance rates among healthcare providers, further contributed to the program's overall success.

The graphical representation (Chart 1) highlights these trends visually, depicting the year-on-year growth in bio-waste volumes and underscoring the network's efficiency gains over time.

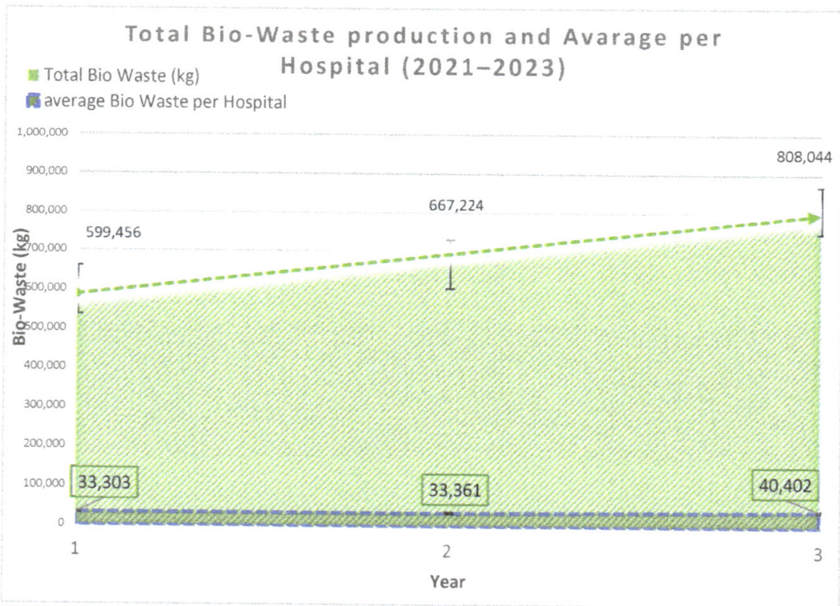

Chart 1. Bio-waste production 2021–2023.

The steady increase in bio-waste volumes, combined with consistent participant numbers, illustrates the model's scalability and adaptability, offering a robust foundation for other cities looking to implement CE-based hospital waste management. Athens' experience showcases a well-integrated, adaptable model that continuously improves through training, infrastructure investment, and alignment with sustainability principles. In summary, the strategic implementation of the bio-waste collection system in Athens's hospitals demonstrates the practical applicability of CE principles in urban waste management. The collaborative approach, continuous stakeholder engagement, and adaptive strategies were key factors contributing to the program's success. These findings set the stage for the subsequent discussion on the broader implications and potential for replication in other urban contexts. Additionally, the bio-waste per bed metric offers a replicable indicator for assessing the efficiency of hospital waste management programs. By incorporating this metric, the Athens model provides a benchmark for comparative analyses with similar CE-based programs globally, allowing municipalities to assess their progress and refine their waste management strategies.

6. Discussion

The analysis of bio-waste collection trends in hospitals across Athens over the three-year period underscores the success of the city's strategic approach to sustainable waste management under the principles of the circular economy (CE). This section discusses the effectiveness of the Athens model, comparing it to similar CE implementations in other urban contexts, and highlights the key factors contributing to its success. The consistent increase in bio-waste volumes, despite the stable number of participating hospitals between 2022 and 2023, highlights the effectiveness of the program's strategies, including enhanced segregation practices, staff training, and integration with municipal waste management systems. This adaptability, even during the COVID-19 pandemic, reflects the model's resilience and scalability in urban waste management contexts [12].

This alignment resonates with broader findings from European municipal waste systems, where CE models have demonstrated the potential for significant improvements in waste diversion and resource recovery [51]. The success of the Athens bio-waste program aligns with general trends observed in other European contexts where CE models have been implemented for urban waste management. Studies emphasize the importance of continuous stakeholder engagement, training, and infrastructure investment as fundamental factors that drive successful CE adoption in waste management systems [30].

The phased integration adopted in Athens, began with a smaller network of 18 hospitals in 2021 and expanded during 2023, reflects the effective expansion strategies observed in other CE implementations. Gradual integration allows for logistical adaptation and incremental optimization of processes, minimizing resistance among stakeholders and facilitating smoother adoption of CE practices [27].

The transition from a linear to a CE model for bio-waste management has resulted in multiple environmental and economic benefits. By promoting segregation at the source, the Athens program has reduced contamination, improved the quality of collected waste, and facilitated more efficient recycling and composting processes. Such practices align with broader CE objectives and have shown to increase waste diversion rates and reduce greenhouse gas emissions in other studies [33].

A notable trend that has been observed is the steady rise in bio-waste volumes between 2022 and 2023, even without the addition of new hospitals. This trend suggests that participating hospitals have optimized their internal waste management processes, contributing to more effective segregation and collection. The ongoing training and awareness campaigns have reinforced these practices, supporting Athens' sustainability goals and aligning with similar efficiencies observed in other urban CE initiatives [37].

The Athens hospital bio-waste program offers a compelling case for how CE principles can transform urban waste management systems. Its adaptability, scalability, and focus on continuous improvement establish a benchmark for similar initiatives. Future developments, including the potential integration of real-time monitoring, are expected to enhance operational efficiency further, supporting Athens' broader sustainability goals and offering a replicable model for other municipalities aiming to adopt CE-based waste management practices [41,44].

The Athens hospital bio-waste program serves as a compelling example of how CE principles can be effectively implemented in urban waste management. Its emphasis on adaptability, scalability, and ongoing improvement underscores the potential for long-term impact in meeting sustainability goals. As other municipalities look to adopt similar CE-based practices, Athens' experience offers valuable insights and a replicable framework that highlights the importance of stakeholder engagement, infrastructure investment, and the integration of advanced monitoring technologies.

7. Conclusions

The hospital bio-waste collection network implemented in Athens serves as a significant case study illustrating the effective integration of circular economy (CE) principles into urban waste management systems. Over the course of three years, the program has achieved steady growth, now involving 20 hospitals, and has consistently increased the volume of bio-waste diverted from conventional disposal methods. The stable hospital participation from 2022 to 2023, coupled with the notable rise in bio-waste volumes, underscores the success of enhanced segregation practices and optimized collection processes.

The strategic expansion that began in 2021—during the challenging COVID-19 pandemic—demonstrated the program's adaptability and resilience. Initially involving 18 hospitals, the phased integration of additional participants allowed the program to develop a robust network that could accommodate scaling and operational adjustments. This approach, marked by careful planning, continuous training, and strong municipal support, serves as a replicable model for other cities transitioning from linear to circular waste management systems.

Transitioning to a CE model in Athens has resulted in multiple environmental and economic benefits, including improved resource recovery, reduced reliance on landfills, and enhanced environmental stewardship. By promoting the separation and composting of bio-waste, the program supports sustainability goals and contributes to a cleaner and healthier urban environment. The observed increase in bio-waste volumes, particularly between 2022 and 2023, suggests that hospitals are actively engaging in Athens' broader initiative to foster resource efficiency and waste minimization.

Despite its successes, the study has certain limitations. First, the evaluation considered only two alternatives—composting and sanitary landfill—which, while reflective of current Greek waste management systems, may not capture the full spectrum of potential strategies. Expanding the analysis to include additional alternatives, such as waste-to-energy, could provide a more comprehensive understanding of CE applications. Second, the dataset, while spanning three years, is limited in its temporal scope. A longer-term dataset might offer greater insights into the program's scalability and resilience under varying conditions. Finally, the reliance on expert opinions for determining criteria weights, while methodologically sound, introduces a degree of subjectivity that could impact the generalizability of the findings.

Building on the findings of this study, future research should explore several avenues to enhance the understanding and application of CE principles in urban waste management. First, incorporating additional waste management alternatives, such as advanced waste-to-energy technologies or decentralized treatment systems, could broaden the analysis and could allow us identify innovative solutions. Second, conducting longitudinal studies that extend beyond three years could better capture the long-term impacts of CE models on urban sustainability. Third, integrating digital tools like IoT-based monitoring systems and predictive analytics could improve the accuracy and efficiency of waste management processes. Lastly, expanding the scope of research to include comparative analyses across multiple cities or regions would help validate the findings and establish universally applicable best practices for CE implementation in healthcare waste systems.

The experiences from Athens' hospital bio-waste collection program underscore the broader potential for CE models in urban settings. By adhering to CE principles, cities can build sustainable waste management systems that address environmental concerns while yielding economic advantages. This program provides valuable insights for municipalities looking to strengthen their waste management practices, highlighting the importance of stakeholder engagement, continuous improvement, and scalability. As more cities adopt similar models, the approach developed in Athens can serve as a blueprint for effective, sustainable bio-waste management strategies in healthcare systems.

In conclusion, the shift from linear to CE-based sustainable bio-waste management models in healthcare systems has proven successful in Athens, demonstrating the feasibility and benefits of integrating CE principles into urban waste management. The consistent growth in bio-waste volumes, effective stakeholder participation, and scalable implementation model underscore the program's achievements and potential for future expansion. As cities worldwide explore CE practices, the Athens CE bio waste model for healthcare system offers an exemplary framework for sustainable urban development that can be adapted to diverse urban contexts.

Author Contributions: Resources, A.S. and K.G.; writing—original draft preparation, A.S. and K.G.; writing—review and editing, A.S and K.G. supervision, A.S. and I.N. All authors have read and agreed to the published version of the manuscript.

Funding: This work has been supported by the CARING NATURE project that has received funding from the European Union's Horizon Europe research and innovation programme under the Grant Agreement No. 101101322.

Institutional Review Board Statement: Not applicable.

Informed Consent Statement: Not applicable.

Data Availability Statement: Data are contained within the article.

Conflicts of Interest: The authors declare no conflicts of interest.

eferences

Sepetis, A.; Zaza, P.N.; Rizos, F.; Bagos, P.G. Identifying and Predicting Healthcare Waste Management Costs for an Optimal Sustainable Management System: Evidence from the Greek Public Sector. *Int. J. Environ. Res. Public Health* **2022**, *19*, 9821. [CrossRef] [PubMed]

Ferronato, N.; Torretta, V. Waste Mismanagement in Developing Countries: A Review of Global Issues. *Int. J. Environ. Res. Public Health* **2019**, *16*, 1060. [CrossRef] [PubMed]

MacNeill, A.J.; Hopf, H.; Khanuja, A.; Alizamir, S.; Bilec, M.; Eckelman, M.J.; Hernandez, L.; McGain, F.; Simonsen, K.; Thiel, C.; et al. Transforming the medical device industry: Road map to a circular economy: Study examines a medical device industry transformation. *Health Aff.* **2020**, *12*, 2088–2097. [CrossRef] [PubMed]

D'Alessandro, C.; Szopik-Depczyńska, K.; Tarczyńska-Łuniewska, M.; Silvestri, C.; Ioppolo, G. Exploring Circular Economy Practices in the Healthcare Sector: A Systematic Review and Bibliometric Analysis. *Sustainability* **2024**, *16*, 401. [CrossRef]

Klemeš, J.J.; Fan, Y.V.; Tan, R.R.; Jiang, P. Minimising the present and future plastic waste, energy and environmental footprints related to COVID-19. *Renew. Sustain. Energy Rev.* **2020**, *127*, 109883. [CrossRef]

Suárez, M.Á.; Gambuzzi, E.; Soriano Disla, J.M.; Castejón, G.; Poggiaroni, G.; Ling, J. ROOTS—Circular policies for changing the biowaste system. *Open Res. Eur.* **2023**, *3*, 78. [CrossRef] [PubMed]

Bolton, A. From Health Sector Waste Minimisation Towards a Circular Economy. Available online: https://www.esr.cri.nz/media/jejnzskx/esr-environmental-health-report-dhb-waste-report.pdf (accessed on 10 November 2024).

Mahjoob, A.; Alfadhli, Y.; Omachonu, V. Healthcare waste and sustainability: Implications for a circular economy. *Sustainability* **2023**, *15*, 7788. [CrossRef]

9. MacNeill, A.J.; Lillywhite, R.; Brown, C.J. The impact of surgery on global climate: A carbon footprinting study of operati theatres in three health systems. *Lancet Planet. Health* **2017**, *1*, e381–e388. [CrossRef]
10. Rangga, J.U.; Josfirin Syed Ismail, S.; Rasdi, I.; Karuppiah, K. Waste Management Costs Reduction and the Recycling Pro Estimation from the Segregation Programme in Malaysia. *Pertanika J. Sci. Technol.* **2022**, *30*, 1457–1478. [CrossRef]
11. World Economic Forum. Circular Economy in Cities: Evolving the Model for a Sustainable Urban Future. White Pap Available online: https://www.weforum.org/publications/circular-economy-in-cities-evolving-the-model-for-a-sustainab urban-future/ (accessed on 10 November 2024).
12. Rajeswari, B. Procedural Bio-Medical Waste Management-A Comparison with International Standards. *Bonfring Int. J. Ind. E Manag. Sci.* **2012**, *2*, 174–180. [CrossRef]
13. Sharma, H.B.; Vanapalli, K.R.; Samal, B.; Cheela, V.R.S.; Dubey, B.K.; Bhattacharya, J. Circular economy approach in solid was management system to achieve UN-SDGs: Solutions for post-COVID recovery. *Sci. Total Environ.* **2021**, *800*, 149605. [CrossR [PubMed]
14. Kenny, C.; Priyadarshini, A. Review of Current Healthcare Waste Management Methods and Their Effect on Global Heal *Healthcare* **2021**, *9*, 284. [CrossRef] [PubMed]
15. Vego, G.; Kučar-Dragičević, S.; Koprivanac, N. Application of multi-criteria decision-making on strategic municipal solid was management in Dalmatia, Croatia. *Waste Manag.* **2008**, *28*, 2192–2201. [CrossRef]
16. Kharat, M.G.; Murthy, S.; Kamble, S.J.; Raut, R.D.; Kamble, S.S.; Kharat, M.G. Fuzzy multi-criteria decision analysis f environmentally conscious solid waste treatment and disposal technology selection. *Technol. Soc.* **2019**, *57*, 20–29. [CrossRef]
17. Circular Economy Strategies for Healthcare Sustainability: Some Insights from Italy. In *Grand Challenges: Companies and Universiti Working for a Better Society—Referred Electronic Conference Proceedings, Proceedings of the Sinergie-SIMA 2020 Conference, Pisa, Italy, 7- September 2020*; FONDAZIONE CUEIM: Verona, Italy, 2020; Part I; p. 91. Available online: https://iris.unica.it/retrieve/handl 11584/303938/431246/FP-CONFERENCE-PROCEEDINGS-2020-PARTE-I.pdf#page=105 (accessed on 10 November 2024).
18. Sasmoko; Zaman, K.; Malik, M.; Awan, U.; Handayani, W.; Jabor, M.K.; Asif, M. Environmental Effects of Bio-Waste Recycling c Industrial Circular Economy and Eco-Sustainability. *Recycling* **2022**, *7*, 60. [CrossRef]
19. Marasca, S.; Montanini, L.; D'Andrea, A.; Cerioni, E. The how and why of integrated reporting in a public health care organizatio The stakeholders' perspective. *Bus. Strategy Environ.* **2020**, *29*, 1714–1722. [CrossRef]
20. Kaur, M. Biowaste: Introduction, Origin, and Management. In *Sustainable Food Waste Management*; Aslam, R., Mobin, M., Aslar J., Eds.; Materials Horizons: From Nature to Nanomaterials; Springer: Singapore, 2024. [CrossRef]
21. Domingo, J.L.; Nadal, M. Domestic waste composting facilities: A review of human health risks. *Environ. Int.* **2009**, *35*, 382–38 [CrossRef] [PubMed]
22. Han, S.; Jeong, Y.; Lee, K.; In, J. Environmental sustainability in health care: An empirical investigation of US hospitals. *Bu Strategy Environ.* **2024**, *33*, 6045–6065. [CrossRef]
23. Janik-Karpinska, E.; Brancaleoni, R.; Niemcewicz, M.; Wojtas, W.; Foco, M.; Podogrocki, M.; Bijak, M. Healthcare Waste— Serious Problem for Global Health. *Healthcare* **2023**, *11*, 242. [CrossRef] [PubMed]
24. Rolewicz-Kalińska, A.; Lelicińska-Serafin, K.; Manczarski, P. The Circular Economy and Organic Fraction of Municipal Soli Waste Recycling Strategies. *Energies* **2020**, *13*, 4366. [CrossRef]
25. D'Adamo, I.; Marcello Falcone, P.; Huisingh, D.; Piergiuseppe, M. A circular economy model based on biomethane: What are th opportunities for the municipality of Rome and beyond? *Renew. Energy* **2021**, *163*, 1660–1672. [CrossRef]
26. Ferreira, F.; Avelino, C.; Bentes, I.; Matos, C.; Teixeira, C.A. Assessment strategies for municipal selective waste collection scheme *Waste Manag.* **2017**, *59*, 3–13. [CrossRef]
27. Health Care Without Harm (Europe). Measuring and Reducing Plastics in the Healthcare Sector. Available online: https:/ europe.noharm.org/sites/default/files/documents-files/6608/2020-11_HCWH-Europe-position-paper-waste.pdf (accessed o 10 November 2024).
28. Seruga, P. The Municipal Solid Waste Management System with Anaerobic Digestion. *Energies* **2021**, *14*, 2067. [CrossRef]
29. Mir, M.A.; Ghazvinei, P.T.; Sulaiman, N.M.N.; Basri, N.E.A.; Saheri, S.; Mahmood, N.Z.; Jahan, A.; Begum, R.A.; Aghamohammad N. Application of TOPSIS and VIKOR improved versions in a multi criteria decision analysis to develop an optimized municip solid waste management model. *J. Environ. Manag.* **2016**, *166*, 109–115. [CrossRef]
30. World Economic Forum (2020). Circular Economy in Urban Waste Management. Available online: https://www.weforum.org stories/circular-economy/ (accessed on 10 November 2024).
31. Moya, D.; Aldás, C.; López, G.; Kaparaju, P. Municipal solid waste as a valuable renewable energy resource: A worldwid opportunity of energy recovery by using Waste-To-Energy Technologies. *Energy Procedia* **2017**, *134*, 286–295. [CrossRef]
32. Sepetis, A.; Georgantas, K.; Nikolaou, I. Innovative Strategies for Bio-Waste Collection in Major Cities during the COVID-1 Pandemic: A Comprehensive Model for Sustainable Cities—The City of Athens Experience. *Urban Sci.* **2024**, *8*, 80. [CrossRef]
33. Singh, N.; Ogunseitan, O.A.; Tang, Y. Medical waste: Current challenges and future opportunities for sustainable managemen *Crit. Rev. Environ. Sci. Technol.* **2021**, *52*, 2000–2022. [CrossRef]

Separate Collection of Municipal Waste. Available online: https://edsna.gr/ergostasio-mixanikis-anakiklosis/ (accessed on 10 November 2024).

Schwarz-Herion, O.; Omran, A.; Rapp, H.P. A case study on successful municipal solid waste management in industrialized countries by the example of Karlsruhe city, Germany. *J. Eng. Ann. Fac. Eng. Hunedoara* **2008**, *5*, 266–273.

Hansen, D.; Mikloweit, U.; Ross, B.; Popp, W. Healthcare waste management in Germany. *Int. J. Infect. Control* **2014**, *10*, 1–5. [CrossRef]

España Circular 2030: The New Circular Economy Strategy for a #FuturoSostenible in Spain. Available online: https://circulareconomy.europa.eu/platform/en/strategies/espana-circular-2030-new-circular-economy-strategy-futurosostenible-spain (accessed on 10 November 2024).

Gasum. Biogas Plants—Supply Chain. Available online: https://www.gasum.com/en/gasum/supply-chain/biogas-plants/ (accessed on 10 November 2024).

Ncube, M.M.; Ngulube, P. Enhancing environmental decision-making: A systematic review of data analytics applications in monitoring and management. *Discov. Sustain.* **2024**, *5*, 290. [CrossRef]

Anagnostopoulos, T.; Zaslavsky, A.; Kolomvatsos, K.; Medvedev, A.; Amirian, P.; Morley, J.; Hadjieftymiades, S. Challenges and Opportunities of Waste Management in IoT-Enabled Smart Cities: A Survey. *IEEE Trans. Sustain. Comput.* **2017**, *2*, 275–289. [CrossRef]

Nižetić, S.; Djilali, N.; Papadopoulos, A.; Rodrigues, J.J.P.C. Smart technologies for promotion of energy efficiency, utilization of sustainable resources and waste management. *J. Clean. Prod.* **2019**, *231*, 565–591. [CrossRef]

HPRC. A Medical Waste Recycling Best Practice of The Netherlands. Available online: https://www.hprc.org/a-medical-waste-recycling-best-practice-of-the-netherlands/ (accessed on 10 November 2024).

CORDIS. Simplifying Hospital Waste with Bio-Based Disposables. Available online: https://cordis.europa.eu/article/id/123222-simplifying-hospital-waste-with-biobased-disposables (accessed on 10 November 2024).

EBA. Circular Economy and Waste Management. Available online: https://www.europeanbiogas.eu/circular-economy-and-waste-legislation/ (accessed on 10 November 2024).

Sharma, A.; Soni, R.; Soni, S.K. From waste to wealth: Exploring modern composting innovations and compost valorization. *J. Mater. Cycles Waste Manag.* **2024**, *26*, 20–48. [CrossRef]

NHS Health Technical Memorandum 07-01: Safe and Sustainable Management of Healthcare Waste. Available online: https://www.england.nhs.uk/wp-content/uploads/2021/05/B2159iii-health-technical-memorandum-07-01.pdf (accessed on 10 November 2024).

Uzun Ozsahin, D.; Gökçekuş, H.; Uzun, B.; LaMoreaux, J. (Eds.) *Application of Multi-Criteria Decision Analysis in Environmental and Civil Engineering*, 1st ed.; Professional Practice in Earth Sciences; Springer: Berlin/Heidelberg, Germany, 2021.

Dabija, D.; Nastase, C.-E.; Chetrariu, A.; Dabija, A. Multi-Criteria Analysis in Circular Economy Principles: Using AHP Model for Risk Assessment in Sustainable Whisky Production. *Computation* **2024**, *12*, 206. [CrossRef]

Hassan, I.; Alhamrouni, I.; Azhan, N.H. A CRITIC–TOPSIS Multi-Criteria Decision-Making Approach for Optimum Site Selection for Solar PV Farm. *Energies* **2023**, *16*, 4245. [CrossRef]

Greek Ministry of Health. Available online: https://www.moh.gov.gr/articles/bihealth/stoixeia-noshleytikhs-kinhshs/11344-klines-noshleythentes-hmeres-noshleias-2022 (accessed on 10 November 2024).

Bala, A.; Raugei, M.; Teixeira, C.A.; Fernández, A.; Pan-Montojo, F.; Fullana-i-Palmer, P. Assessing the Environmental Performance of Municipal Solid Waste Collection: A New Predictive LCA Model. *Sustainability* **2021**, *13*, 5810. [CrossRef] sustainability MDPI

Review

Role of the Circular Economy Framework for Sustainable Wast Management and Climate Change Mitigation

Stephen James Ijimdiya [1], Muthukrishna Vellaisamy Kumarasamy [1,2,*], Joy Tuoyo Adu [1] and Dinagarapandi Pandi

[1] Civil Engineering Programme, School of Engineering, University of KwaZulu-Natal, Durban 4041, South Africa; stephenj@ukzn.ac.za (S.J.I.); aduj@ukzn.ac.za (J.T.A.); pandid@ukzn.ac.za (D.P.)
[2] Saveetha School of Engineering, Saveetha Institute of Medical and Technical Sciences, Chennai 600072, India
* Correspondence: kumarasamy@ukzn.ac.za; Tel.: +27-78-218-4383

Abstract

The rapid increase in global waste generation and its associated greenhouse gas (GHG) emissions pose increasing challenges to sustainable development and climate mitigation. Current estimations indicate that municipal solid waste generation will increase from approximately 2.3 billion tonnes in 2023 to nearly 3.8 billion tonnes by 2050, with waste-generated emissions, mainly methane, accounting for an important share of bi-temporal climate impact. This review examines the circular economy (CE) as a systemic framework for transforming waste management practices to reduce emissions while improving resource efficiency. Based on a structured narrative synthesis of more than 170 peer-reviewed studies and authoritative international reports, the paper evaluates CE-integrated strategies for waste prevention, recycling, recovery, and valorization. Reported estimates suggest that integrating CE principles into waste management systems could reduce proected GHG emissions growth by approximately 30–39% by 2050 through material efficiency gains, energy savings, landfill diversion, and organic waste valorization. Sector-specific considerations, including applications in high-emission industries such as the energy and chemical sectors, highlight both the mitigation potential and technical constraints of CE implementation. Overall, the review demonstrates that CE-based waste management can deliver quantifiable climate benefits when supported by appropriate policies, technologies, and context-specific system design.

Keywords: circular economy; sustainable waste management; greenhouse gas emissions; resource efficiency; material circularity; environmental sustainability

1. Introduction

1.1. Background

Global waste generation has extended to unparalleled heights, causing major environmental, economic, and social challenges. Municipal solid waste (MSW) generation is estimated to rise from roughly 2.3 billion tonnes in 2023 to nearly 3.8 billion tonnes by 2050, and this can be attributed to population growth, increased industrialization, and consumption patterns that are not in line with sustainable goals [1]. This huge waste generation will result in serious environmental and economic problems, as the cost of controlling it is set to increase globally from USD 252 billion in 2020 to over USD 640 billion annually by 2050 if this trend continues [1]. This rise in financial burden emphasizes the weakness of existing waste systems, particularly in low- and middle-income areas where collection and treatment volume remain insufficient [2]. Beyond its scale, waste generation

https://doi.org/10.3390/su1804194

directly impacts climate change. For instance, landfilling of organic waste is a main cause of methane emissions, a greenhouse gas that has a global warming potential of about 28–36 times more carbon dioxide (CO_2) gases released over 100 years [3]. Additionally, open burning of waste, mostly in developing countries, releases black carbon, a short-term pollutant, which contributes significantly to immediate climate change, affects health conditions, and is considered more harmful than CO_2 [4,5]. Therefore, waste management systems are increasingly acknowledged as a serious involvement socket for greenhouse gas mitigation in national and global climate approaches.

Although the availability of various waste management methods designed to reduce environmental impacts is prevalent, linear economic models based on production, consumption, disposal, and removal have proven to be basically unsustainable. These models encourage the exhaustion of resources, unnecessary waste generation, and contribute to environmental degradation [6,7]. In particular, the linear model supports dependence on unprocessed inputs and high-carbon means of production, contributing to both scarcities of resources and overall emissions [8]. On the flip side, substitute systemic methods underline resource efficiency, waste prevention, and material recovery as ways toward sustainability. In this case, the circular economy (CE) has materialized as a strategic retort to the limits of linear waste management systems. CE supports the preservation of material value via reuse, recycling, and reformative design, permitting materials and products to continue in circulation for a long time [9,10]. Research shows that recycling and material recovery, mostly for materials like paper, metals, and plastics, can result in significant energy savings and reduced greenhouse gas emissions as compared to major production from raw materials [11,12]. In addition, the integration of CE into urban waste management has the potential to reduce greenhouse gases by up to 39% by 2050 via waste prevention, organic waste valorization, and reduced dependence on landfilling [13].

Although the CE frameworks are progressively discussed in several sustainability and waste management studies, it is being treated in a general way without the integration of CE concepts with waste management, climate mitigation, and sector-specific technical facts, most especially in industries with high emissions like chemical and energy [14]. This review discloses that gap by producing evidence from CE theory, waste systems engineering, and climate policy, giving an integrated outlook on how CE- based waste management strategies can produce measurable greenhouse gas decreases. By placing CE as a technically practical and climate-related framework instead of merely a resource-efficiency idea, this review supports the scientific basis for CE-driven waste management involvement.

1.2. Perspective of CE

The circular economy is a philosophy that considers growth as reducing dependence on already scarce resources. Contrary to the current linear economy model, CE emphasizes the reuse, recycle, and manufacture of materials to form a closed system, such that less waste is generated and environmental degradation is reduced [15]. CE is now gaining relevance as a better sustainable option to the present economic models [15].

1.2.1. Defining the CE

There is no universally accepted definition of circular economy, but the major thrust lies in circularity and the sustainability of materials. The European Commission [16] defines it as 'an economy where the cost of products, materials, and resources is fixed in the economy for as long as imagined, and the production of waste is reduced'. According to the Ellen MacArthur Foundation [17], "it is the system of the industry that restores and redevelops by design". Kirchherr et al. [18] delimit it to "a system of the economy that

https://doi.org/10.3390/su18041946

substitutes the end of a product's life with reducing, reusing, recycling, and improving materials in the production or distribution, and consumption processes". This implies that the circular economy covers environmental, economic, and social dimensions, entailing sustainability and efficiency as its foundation and main focus.

1.2.2. Core Principles of CE

The Circular Economy model works based on the following principles:

- Waste/product and pollution system design should be carried out to prevent the production of waste and pollution [19].
- Products/materials should be kept in continuous use to prolong their life through reuse, repair, and recycling [20].
- Practices that increase natural resources, like composting and ecosystem renewal, should be encouraged [20].
- The circular economy reconsiders value chains and business models through logical thinking that requires innovation [21].
- The circular economy gives priority to the use of renewable energy and sustainable use of resources [22].

CE principles emphasize reducing material inputs, extending product lifecycles, recovering value from waste, and regenerating natural systems. When embedded within waste management systems, these principles operate through interconnected material and energy flows supported by feedback mechanisms across multiple sectors. Figure 1 presents a conceptual framework explaining how CE principles—reduce, reuse, recycle, and regenerate—are rooted within waste management systems through material and energy response loops connecting industrial, energy, and chemical sectors.

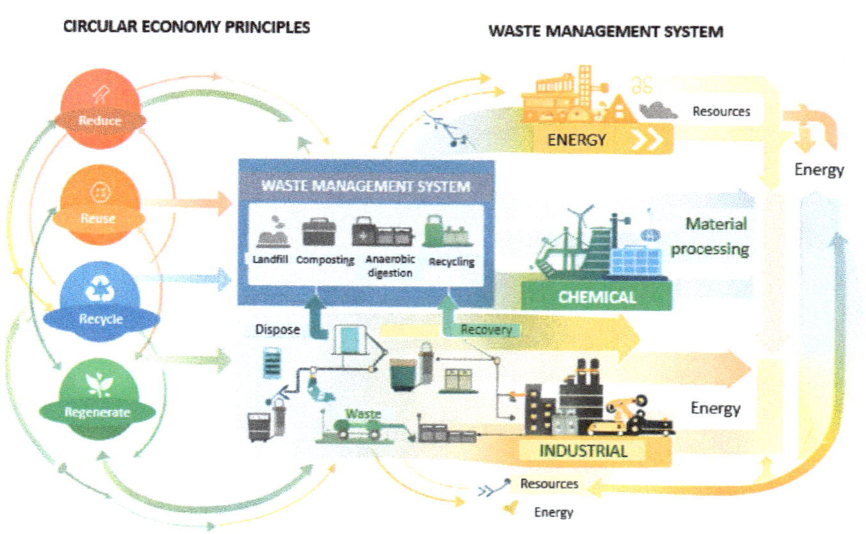

Figure 1. Conceptual framework of circular economy principles within waste management systems [1].

1.2.3. The CE as a Response to Environmental and Social Challenges

The global economy is currently confronted by serious environmental challenges, such as climate change, biodiversity loss, scarcity of resources, and pollution, to mention a few [23]. CE tackles environmental and resource challenges mainly by curtailing resource extraction, reducing pollution, and closing material loops. Each action or principle

https://doi.org/10.3390/su1804194

contributes directly to the sustainability of the environment. The CE model provides a general and logical response to these challenges by reforming economic events to support ecological rims as follows:

- Dealing with the scarcity of resources and the security of materials: CE makes best use of present materials, thereby increasing the security of materials [24].
- Cutting down on greenhouse gas emissions: CE encourages energy productivity and lessens greenhouse gas emissions [12].
- Curtailing waste and pollution: CE converts waste into resources and cuts down on pollution [8].
- Improving economic effectiveness and Innovation: CE promotes innovation, cost productivity, and market benefit [25].
- Supporting social presence and creating jobs: CE generates employment opportunities through recycling, repair, and service-based models [26].

1.3. Scope and Limitations

This review centers around the assumption that realizing sustainable waste management and significant climate mitigation entails a systemic transition from linear models of production and disposal to circular, sustainable, and waste-minimizing systems. The study also assumes CE strategies impact several interrelated areas—technological, environmental, economic, policy, and sector-specific—and that comprehending these connections is critical for measuring the practical possibility and climate benefits of CE implementation. These assumptions steered the choice of thematic areas examined in the manuscript. Even though this review highlights the role of CE strategies for sustainable waste management and climate mitigation, it also identifies the fact that their implementation is subject to technical, economic, and institutional limitations, which are examined in other sections. This review adopts a system-level perspective on CE-driven waste management and climate mitigation across different sectors. Although it does not give a detailed sector-by-sector techno–economic analysis, a directed discussion of high-emission and waste-intensive industries, mainly the energy and chemical sectors, is conducted where sector-specific waste streams, hazardous materials, and continuous production limitations analytically impact the application of CE strategies.

1.4. Objectives of the Review

The primary objective of this review is to evaluate how CE principles, strategies, and technologies can increase sustainable waste management while contributing to quantifiable decreases in greenhouse gas emissions. To achieve this objective, the review synthesizes insights from CE theory, waste systems engineering, climate mitigation frameworks, and industry-specific technical pathways. Consideration is also given to high-emission sectors like the energy and chemical industries, where CE implementation has both substantial prospects and operational issues.

The review scope covers the following:

- The conceptual and theoretical foundations of the CE;
- The integration of CE in contemporary waste management hierarchies;
- Material flow systems and technological involvements that enable emission reductions;
- The contribution of CE approaches to climate mitigation, as well as emissions reduction possibilities;
- Aiding policy, economic, and digital drivers;
- Current barriers and future outlooks.

In general, it bridges conceptual CE discussions with technical evidence and applied considerations, presenting an inclusive and integrated outlook on how CE strategies

https://doi.org/10.3390/su18041946

can be used as a practical pathway for attaining climate-resilient and sustainable waste management systems.

1.5. Literature Search and Selection Criteria

This review embraces an organized narrative approach to study the role of CE frameworks in sustainable waste management and climate change mitigation. A methodical search was carried out using major scientific databases, including Scopus, and ScienceDirect, to ensure a detailed and qualitative analysis. The search approach used combinations of keywords connected to circular economy, waste management, and climate change, including "circular economy", "waste management", "resource recovery", "greenhouse gas emissions", "climate mitigation", "industrial waste", and "life cycle assessment". The search approach was led by the assumption that the implementation of CE is fundamentally interdisciplinary and collaborative and hence entails combining perspectives from environmental engineering, industrial ecology, and sustainability science. Peer-reviewed journal articles, authoritative reports from international organizations like IPCC, UNEP, World Bank, and high-impact review papers published between 2000 and 2025 were deliberated. Studies were included if they (i) clearly addressed CE principles, (ii) studied waste management systems or material recovery paths, and (iii) delivered qualitative or quantitative understandings of environmental or climate impacts. Conference abstracts, non-peer-reviewed articles, and studies that are not applicable to CE or waste systems were not included.

To improve methodological accuracy and repetition, recovered accounts were screened on the basis of titles and then by full-text search. Identical records were removed, and regressive reference tracing was carried out to recognize additional applicable studies. This structured collection process confirms transparency, reproducibility, and broad coverage of the literature and at the same time maintains the integrity of a technical review [27,28].

Therefore, this review follows a structured narrative review method, enabling interdisciplinary synthesis, at the same time preserving methodological transparency and thoroughness.

2. Theoretical Foundation of the Circular Economy

2.1. Historic Development

The concept of the CE originated several decades ago, drawing on diverse ideas and environmental perspectives. Its origin is traceable to the 1960s and 1970s, as a result of the increasing consciousness of ecological risks and the necessity for more sustainable management of resources [29]. The work of Kenneth Boulding on *The Economics of the Coming Spaceship Earth* (1966) hypothesized the notion of a closed-loop economy where material inputs and waste outputs are reduced, forming the bedrock of CE.

In the years 1980s and 1990s, theories like industrial ecology and cradle-to-grave design received attention. Industrial ecology presented the correlation between industrial systems and natural ecosystems, where waste from one practice benefits the input for another [30]. The cradle-to-grave method, promoted by McDonough and Braungart [31], underscores the design of materials with their whole life cycle in mind, encouraging materials that can be continuously cycled in either biological or technical loops.

The word 'Circular Economy' became more conventional in the early 2000s, most especially in China's policy framework. The Chinese government added the principles of the circular economy into its national development policy in 2002, intending to increase resource productivity and reduce pollution [32]. In Europe, the circular economy received importance with the European Commission's Circular Economy Action Plan [16], which underscored sustainable material design, prevention of waste, and resource productivity

https://doi.org/10.3390/su1804194

throughout the economy of the European Union. Generally, the idea of CE is grounded in many disciplines: ecological economics, systems thinking, and the theory of sustainable development [33].

2.2. Pillars and Strategies of CE

The real-world application of CE is directed by an order of approaches usually shortened to the 3Rs: Reduce, Reuse, and Recycle. These central ideologies are designed to reduce resource input and waste output [34], though the contemporary understanding of circular economy goes past the 3Rs, integrating wider approaches such as Reuse, Rethink, Repair, Remanufacture, Repurpose, and Recuperate [35]. These extended approaches are usually shown as a fragment of the '10Rs' framework and reveal a deeper move to product longevity, material optimization, and complete improvement. Two innovative business models are fundamental to CE:

i. Product-as-a-Service (PaaS): this model changes the attention from product possession to admittance, where clients pay for the service that a product offers rather than the product itself. For instance, enterprises like Philips offer lighting as a service instead of selling light bulbs. This motivates producers to produce strong, repairable, and recyclable products [36].
ii. Industrial synergy: this approach involves the teamwork between industries where the waste or by-products of one practice function as the raw materials for another. Kalundborg, Denmark, is a main instance of a positive industrial synergy system that has considerably reduced emissions and resource depletion [37].

Instead of aiding merely as conceptual frameworks, CE models such as the 3Rs/10Rs grading, Product-as-a-Service (PaaS), and cradle-to-cradle design basically modify material and energy streams within industrial systems, in that way impacting waste generation and greenhouse gas (GHG) emissions. Technologically, multi-level strategies (e.g., refuse, reduce, and rethink) lead to the utmost emissions declines by averting material mining and major processing, which are usually the most energy-demanding phases of product life cycles [38].

PaaS models encourage producers to use resilient materials, integrated designs, and prognostic maintenance technologies (e.g., sensors and digital twins), prolonging product lifespans and decreasing material output. Empirical studies show that extending product lifespans by 20–30% can reduce lifecycle emissions by up to 25%, particularly for energy-intensive goods such as electronics and appliances [39]. Similarly, repair and remanufacturing technologies, enabled by standardized components and advanced disassembly processes, significantly reduce waste flows while retaining up to 80% of the embodied energy of original products [40].

Cradle-to-cradle and engineering synergy frameworks further redesign waste streams by permitting circular material recovery and multi-sector resource exchange. Industrial enablers such as material flow analysis (MFA), actual waste categorization, and process combination permit industrial secondary products (e.g., fly ash, waste heat, digestate) to replace raw materials, decreasing landfill disposal and related methane emissions [41]. In such systems, waste is converted into a subordinate resource, directly reducing overall emissions at the system level.

On the other hand, basic strategies such as recycling, even though important, regularly convey lesser emission decreases, owing to energy necessities for collection, sorting, and reprocessing. Innovative recycling technologies (e.g., chemical recycling, solvent-based separation, pyrolysis) can increase material recovery rates, but their climate benefits are strongly dependent on energy sources and process efficiency [42]. As a result, CE frameworks underline waste prevention and lifespan extension even more over phase-

https://doi.org/10.3390/su18041946

out recycling. As displayed in Figure 2, CE principles are explained in practice through combined strategies and business models that simplify circular material and energy flows.

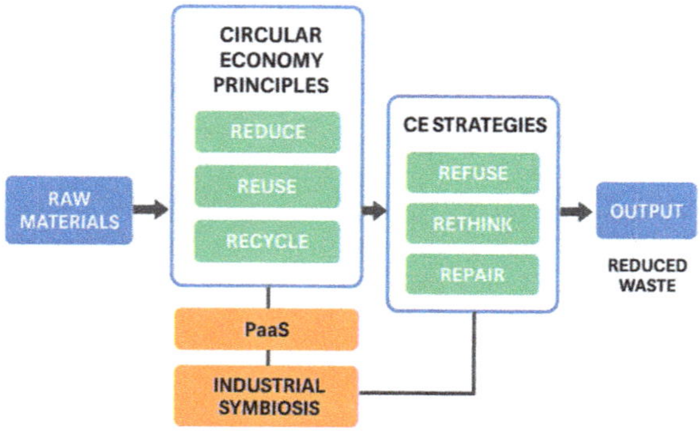

Figure 2. Conceptual framework showing CE principles, strategies, and enabling business models.

2.3. *Performance Measure and Pointers of CE*

Evaluating the usefulness and evolution of CE approaches needs vigorous parameters and pointers. Some parameters have been established to calculate diverse facets of circularity at product, organizational, and macroeconomic levels, as follows:

- The Material Circularity Indicator (MCI) established by the Ellen MacArthur Foundation and Granta Design. It measures the degree to which material flows of a product are recuperative or reformative by design, on the basis of factors such as virgin against recycled input, product lifetime, and the value of output [43]. The MCI takes into consideration four major components:

  i. Fraction of virgin feedstock (V).
  ii. Fraction of content of the material recycled (R).
  iii. Fraction of output of the material reused/recycled (E).
  iv. Product lifecycle utility (U) in relation to industry coverage.

  The MCI can be calculated thus:

$$\text{Linear Flow Index(LFI)} = \text{V} + \text{W} - \text{R} \quad (1)$$

where W = unrecovered waste portion

At that point:

$$Utility\ Factor\ (X) = \frac{u_P}{u_{av}} \quad (2)$$

Lastly, the MCI is given as

$$\text{MCI} = 1 - \frac{LFI}{2X} \quad (3)$$

where LFI = fraction of virgin material + unrecovered waste − recycled content, and X = ratio of product lifespan to industry average.

Higher MCI indicates reduced dependency on virgin materials, lower waste generation, and, indirectly, lower emissions due to avoided primary production.

A correction factor is useful to make sure MCI is within 0–1, with 1 considered as completely circular and 0 completely linear.

- Life Cycle Assessment (LCA) is another thoughtful way used to assess the environmental effects of materials and methods during the course of their life cycle, from raw material removal to disposal. Though conventionally used in linear systems, it can be adjusted for circular systems by adding metrics for resource productivity, waste

https://doi.org/10.3390/su1804194

decline, and unimportant material usage [44]. Although traditional LCA does not have a specific formula for determining the 'circularity' of a material, revised circular LCA comprises indicators such as Resource Efficiency (RE) and Recycled Content (RC). These LCA-interrelated metrics permit contrast between circular and linear product systems.

(a) Resource Efficiency (RE)

$$RE = \frac{Functional\ output}{Total\ material\ input} \tag{4}$$

In this case, a higher value of RE means a more efficient use of material

(b) Recycled Content (RC)

$$RC = \frac{Mass\ of\ recycled\ feedstock}{Total\ product\ mass} \tag{5}$$

(c) End-of-Lifecycle Recovery Rate (RR)

$$RR = \frac{Mass\ recorvered\ \ at\ end-of-lifecycle}{Total\ mass\ of\ product} \tag{6}$$

These metrics quantify how efficiently materials are utilized and recovered, enabling assessment of GHG emission reductions due to reduced primary resource extraction [45].

Other key parameters include the following:

- Circular Economy Performance Indicators (CEPIs): CEPIs evaluate resource productivity, recycling rates, and waste intensity at organizational or sectoral levels [46]. These consist of parts such as resource efficiency, rates of recycling, and the amount of waste generation. These systems of measurements are used to quantify efficiency, not to calculate circularity; nonetheless, they are commonly used in a policy framework [46]. Key metrics for CEPI formulas include the following:

(a) Resource Productivity (RP)

$$RP = \frac{GDP}{Domestic\ Material\ Consumption\ \ (DMC)} \tag{7}$$

(b) Recycling Rate (RR)

$$RR = \frac{Recycled\ waste}{Total\ waste\ generated} \tag{8}$$

(c) Waste Generation Intensity (WGI)

$$WGI = \frac{Total\ waste}{Population\ or\ GDP} \tag{9}$$

CEPIs provide macro-level insights into material efficiency, waste minimization, and economic circularity, linking directly to emission mitigation strategies [47].

Material Flow Analysis (MFA): MFA indicators assist in measuring and envisioning the physical flows of materials in a system, finding wastefulness and chances for circular intrusions. These indicators give insight into the circularity of natural or sector-level circularity [48]. Major formulas for indicators are the following:

(a) Circularity Rate (CR)

$$CR = \frac{Secondary\ recycled\ material\ utilized\ in\ a\ system}{Total\ material\ inputed\ in\ a\ system} \tag{10}$$

https://doi.org/10.3390/su18041946

(b) Stock Accumulation (SA)

$$\text{SA} = Inputs - Outputs \tag{11}$$

(c) Dependency Ratio (DR)

$$\text{DR} = \frac{Imports}{Domestic\ Extraction} \tag{12}$$

MFA helps visualize material loops and quantify potential emission reductions by identifying where waste can be reintegrated into production [49].

Circularity Gap Indicator (CGI): this metric assesses the amount of the global economy that is circular, giving a macro-level perspective, and is usually applied in cities, nations, and globally [50]. CGI, as of 2023, was estimated to be approximately 72%, which shows that most of the world is considered highly linear. The CGI is calculated thus:

$$\text{CGI} = \frac{Total\ materials\ recycled\ and\ reused}{Total\ material\ inputed\ into\ the\ economy} \tag{13}$$

CGI provides a macro-level perspective on circularity, and highlights how linear material flows contribute to emissions and resource depletion [51].

Table 1 summarizes the various CE indicators. Developing consistent, sector-precise, and accessible metrics remains a task. Nevertheless, precise measurement tools are needed for making policy and business plans, and for following progress to sustainability goals.

Table 1. CE Indicators.

Indicator	Formula	Scale	Environmental Relevance
MCI	1 − LFI/(2X)	Product	Circularity, condensed virgin material usage
LCA (RE, RC, RR)	Proportions	Product/System	Resource efficiency, condensed lifespan emissions
CEPIs (RP, RR, WGI)	Proportions	Sector/National	Waste decrease, material output
MFA (CR, SA, DR)	Proportions	Sector/National	Material loops, emission discount prospective
CGI	Recycled and reused/over-all input	Global/National	Total circularity, system-level emission mitigation

3. Waste Management in CE Framework

3.1. CE in Waste Hierarchy

The waste hierarchy (Figure 3) functions as a guiding principle for sustainable waste management. CE redefines this hierarchy by prioritizing systemic resource efficiency over downstream solutions. In a circular economy structure, this pyramid is rearranged to emphasize waste prevention and reuse, and as a result, make the most of resource productivity and reduce environmental impacts [52]. Prevention has to do with the generation of waste at the design and production levels using environmentally friendly design, product life postponement, and consumption models that are sustainable [53].

Reuse has to do with product repair, renovation, and remanufacturing, extending product lifecycles, and lessening the demand for raw material. Mechanical and chemical recycling are vital to improving materials for rehabilitation into construction cycles. Recovery, like energy recovery using anaerobic digestion or incineration with energy detention, serves as a last-option valorization stage. Disposal into landfill is the least preferred option, and is not accepted in the circular economy associated guidelines. Circular economy motivated guidelines gradually include Extended Producer Responsibility (EPR) and product

https://doi.org/10.3390/su1804194

stewardship structures to move liability to producers, encouraging upstream design for recyclability and reparability [54].

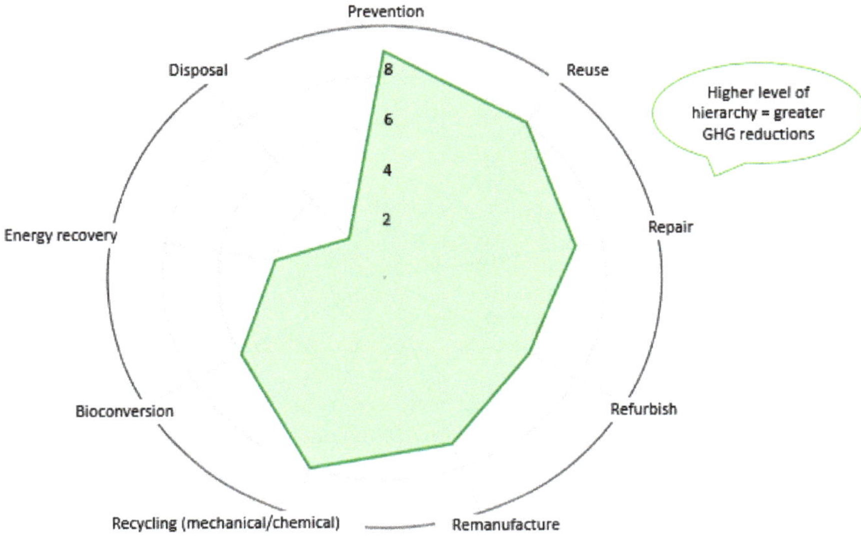

Figure 3. CE waste hierarchy (The scale represents relative effectiveness).

3.2. Material Flow Analysis in Major Sectors

Material flow analysis (MFA) is a major means in the implementation of circular economy. Table 2 shows the material flow analysis in the main sectors. It gives the picture and quantity of the stock of materials that flows across different stages of the product life cycle. It pinpoints areas for intervention and discloses the possibility of closed material circles [55].

Table 2. Material Flow in Major Sectors.

Sector	Generated Waste	Main Materials	Current Challenges	CE Prospects
Packaging	Plastic	Polyethylene Terephthalate (PET), High-Density Polyethylene (HDPE), Low-Density Polyethylene (LDPE)	Contamination, down cycling	Chemical recycling, design for recyclability
Electronics	E-waste	Rare earths, valuable metals, plastics	Informal processing, toxicity	Urban mining, modular design, producer responsibility
Agriculture and Food	Food waste	Organic matter, water, nutrients	Methane emissions, nutrient loss	Anaerobic digestion, composting, bioplastics
Construction	Construction and demolition waste	Concrete, metals, timber	Mixed streams, low recycling rate	Careful demolition, reuse of aggregates
Textiles	Textile waste	Cotton, polyester, blends	Fiber separation, microfibers	Fiber-to-fiber recycling, bio-based textiles

https://doi.org/10.3390/su18041946

Table 2. *Cont.*

Sector	Generated Waste	Main Materials	Current Challenges	CE Prospects
Shipbuilding and Marine vessels (e.g., yachts, large boats)	Ship parts, scraps, hulls, superstructures	Glass-fiber armored plastics (GFRP), carbon fiber compounds, resins, metals	Scarce recycling alternative for composites; harmful resin additives; lack of controlling frameworks	High-temperature pyrolysis for fiber retrieval; mechanical milling for filler applications; design for dismantling; composite replacement strategies
Renewable Energy Structures (wind turbines, photovoltaics)	Spoilt turbine blades, nacelles, PV panels	Glass fiber composites, carbon fibers, epoxy resins, rare earth magnets, silicon cells, aluminum frames	Compound blades are difficult to recycle; landfill bans are evolving; PV modules hold poisonous materials (Pb, Cd)	Solvolysis and pyrolysis for composite retrieval; circular PV design; closed-loop glass recycling; remanufacturing of turbine materials; digital-twin asset supervision

- Packaging: packaging waste is one of the most persistent CE challenges. The global plastic value chain shows low circularity, with about ~9% of plastic waste being recycled, and the rest either being landfilled, incinerated, or outflowing into the environment [56]. This low recovery amount is mostly a result of the prevalence of transitory plastic packaging materials such as PET, HDPE, and LDPE, which are vastly susceptible to contamination, downcycling, and degradation. Even though mechanical recycling is conventional, it is frequently mired by contamination, breakdown of polymer properties, and the downcycling of materials into inferior products. Chemical recycling methods such as pyrolysis, depolymerization, and solvolysis are gradually accepted as key in overcoming contamination and renewing monomers or chemical input materials, hence allowing accurate closed-loop cycles [57]. Moreover, design innovations like mono-material packaging, digital watermarks, and eco-labeling can considerably advance collection and recyclability [58]. MFA studies also identify packaging as one of the main areas, besides the construction and automotive sectors, where interventions can produce the highest fundamental impact [56].
- Electronics: e-waste is one of the fastest emerging and direst waste streams, globally. It embodies a valuable material flow, comprising metals like gold, silver, copper, and important rare earth elements needed for renewable energy technologies, batteries, and electronics production. Despite its importance, only 17.4% of global e-waste is properly collected and recycled in environmentally rigorous conditions, and the rest either landfilled, casually processed, or lost in the overall waste stream [59,60]. MFA enables tracking of electronic components across their life cycle—from production, consumption, and collection to end-of-life disposal. Such insights are critical in designing collection systems, refurbishment schemes, and urban mining strategies that optimize recovery rates and minimize environmental externalities. CE involvement in electronics management must synchronize policy, design, and technology. Extended producer responsibility (EPR) systems embolden producers to collect used devices, generating incentives for integrated design, repairability, and reuse of components. Environmentally designed approaches, such as removing unsafe additives, integrated architectures, and standardized borders, will enable easier repair, progress, and recovery. Additionally, advanced recycling approaches like hydrometallurgy, bioleaching, and pyrometallurgy allow the recovery of valuable and rare earth metals, reducing dependence on original extraction [60]. By introducing circularity into electronics supply chains, it is possible to concurrently mitigate environmental burdens and lessen reliance on virgin materials.

- Food and Agriculture: food waste and agricultural waste remains consist mainly of organic matter and nutrients. About one-third of all food generated globally is wasted. This amounts to 1.3 billion tonnes yearly [61]. MFA in food systems shows wastefulness across the supply chain, from manufacture to consumption. When not properly managed, these wastes contribute to methane emissions, a powerful GHG, and result in the loss of surrounding resources like water and energy [62]. CE-motivated valorization approaches, comprising anaerobic digestion, bioconversion, and composting, can produce biogas, fertile soil amendments, and even substitute proteins [63]. Moreover, source interventions such as enhanced storage, digital boards for redistribution, and changes in consumer behavior are needed to reduce the generation of waste in the first instance.
- Construction: construction and demolition waste (CDW) comprise one of the principal waste streams globally, predominantly encompassing concrete, metals, and timber. A key obstacle is the varied and miscellaneous nature of CDW, which obscures sorting and recovery. Nevertheless, CE prospects lie in design for disassembly, reuse of structural components, and careful demolition practices that preserve the quality of material. Additionally, recycled collections and ancillary raw materials can considerably balance virgin extraction, contributing to climate mitigation (European Environment Agency [64]).
- Textiles: textile waste shows challenges related to mixed fibers, synthetic microfibers, and excessive fast-fashion making. Fiber separation equipment is unfledged, and hence, the recycling of mixed fabrics into superior-quality fibers is inadequate. CE involvements center on fiber-to-fiber recycling—mechanical and chemical, bio-based textiles, and circular business models like rental, resale, and repair services [65]. Tackling microfiber pollution also involves innovation in washing technologies and textile finishing.

The analysis of these combined sectors reveals that even though CE offers prospects for recovery of waste and resource use, each sector faces distinctive challenges—such as contamination in packaging, informal recycling in e-waste, nutrient loss in food waste, material diversity in construction and demolition waste, and fiber separation issues in textiles. However, there are strong opportunities for cross-sectional collaboration. For example, remains from food-waste bioconversion could be used in the production of bio-based textile fibers, whereas recycled plastics from packaging could be used as inputs in construction composites. These connections reveal the systemic nature of the CE, where waste from one sector becomes a resource for another, resulting in more resilient, low-carbon, and resource-efficient systems.

3.3. Waste Management Innovations in the Circular Economy

Innovation plays a critical role in transitioning to a CE, particularly in the conversion of waste into valuable resources. Table 3 presents the technological innovations supporting CE-based waste management. However, the success of these technologies is impacted by technical limitations such as process efficiencyand emission exchanges, and energy requirements change with scale.

https://doi.org/10.3390/su18041946

Table 3. Waste Management Technologies Based on CE.

Technology	Description	Area of Application	Contribution of CE	Demerits	Reference
Waste sorting (AI/optical)	Automatic sorting using sophisticated sensors, optical, and AI-based systems	MSW, plastics	Improves the efficiency of recycling	High cost; involves maintenance and a skilled process	[66]
Mechanical recycling	Physical reprocessing of materials	Plastics, textiles	Encourages closed-loop cycles and reuse	Loss of quality due to contamination; inadequate recyclability after several cycles	[67]
Chemical recycling	Depolymerization and re-synthesis	Diverse plastic streams, composites	Disables contamination problems, supports materials difficult to recycle	High energy contribution; possible environmental impacts from residues	[58]
Bioconversion	Use of microbes of micro-organisms/insects	Food waste, agricultural remains	Transforms organics into compost, bioenergy, and proteins	Influenced by the type of feedstock, requires optimization of process parameters	[68]
Pyrolysis	Thermal decomposition of organic waste into bio-oil, syngas, and char	Plastics, diverse organic waste	Offers importance in waste-to-energy and the recovery of materials	Costly set-up; involves pre-sorting control of emissions	[69]

The use of sophisticated sensors for sorting, mechanical and chemical recycling, and conversion of wastes using biological means, is important in making circular economy principles operational in managing waste. The sorting of waste using optical and AI-based technologies improves the secondary materials [66]. The use of mechanical means for the recycling of materials physically offers prospects, but could face restrictions as a result of contamination [67]. Chemical recycling, like pyrolysis and gasification, is useful in recovering materials that are difficult to recycle [57]. The conversion of waste using biological techniques like anaerobic digestion and black soldier fly processing helps in recovering energy and nutrients from organic waste [68]. Pyrolysis changes organic waste into bio-oil and syngas, giving value in combined waste-to-energy systems [69].

Mechanical recycling mostly has the lowest energy intensity and GHG emissions as compared to other recycling alternatives, as long as purity of material is high and conveyance expanses are restricted [70]. Life-cycle assessments constantly demonstrate that mechanical recycling of plastics can lessen GHG emissions by 30–80%, as compared to raw polymer production, though these benefits drop abruptly with emergent contamination and downward cycling [71].

In comparison, chemical recycling procedures are intrinsically energy-demanding owing to the high temperatures and endothermic reactions prerequisite to break polymer chains. Pyrolysis and gasification usually function at temperatures exceeding 400–800 °C, resulting in higher energy depletion and related emissions, except when joined with renewable energy sources or operational heat-addition systems [72]. Thermodynamic controls, together with entropy generation and conversion inadequacies, limit general material and energy recovery yields, underscoring the essentials for process optimization and addition within industrial energy systems [73]. However, when applied to non-

https://doi.org/10.3390/su1804194

recyclable waste streams, chemical recycling can provide overall climate advantages by balancing fossil-based feedstocks in petrochemical manufacture [74].

Biological treatment paths such as anaerobic digestion usually display lesser process energy requirements, though their emission performance is dependent intensely on methane detention efficiency and management of digestate [75]. Inefficient systems can result in elusive methane emissions that considerably offset climate benefits [76]. Innovative process control and gas capture technologies are, as a result, important for exploiting both circularity and climate mitigation results.

3.4. Case Studies of CE Waste Management Models

Some regions and countries have achieved success in including the principles of the circular economy in their waste management frameworks. Table 4 shows the summary of the comparative analysis of CE waste management models across countries.

Table 4. Comparative Analysis of CE Waste Management Models.

Region	Main Plans/Policies	Successes	CE Model Element	Challenges
European Union (EU)	Circular Economy Action Plan (2020); Waste Framework Order; Plastics Policy; Extended Producer Responsibility (EPR)	Municipal recycling rates > 50% in top states; 11% decline in landfill reliance (2010–2023)	Governing organization; environmentally friendly design standards; material passports; inverse logistics	Irregular enactment among member states; defiance cracks in weaker economies
China	Circular Economy Promotion Law; National Sword policy (2018); CE model cities and environmentally friendly industrial parks	Official collection of ~65% e-waste; extensive industrial synergy in parks	Integrated governance; top-down arrangement investment; industrial synergy	Casual recycling continues; regional differences in implementation and defiance
Netherlands	National CE Plan (100% CE by 2050); Circular procurement strategies; City-level CE blueprint	A 32% decline in raw material use since 2010; effective zero-waste areas	Multi-investor partnership; indigenous testing; community commitment	Upgrading local attainments nationally; managing echo effects; reliance on global supply chains

The European Union (EU) is one of the most established states in the implementation of CE plans. Through its Circular Economy Action Plan, the Waste Framework Directive, and the Plastics Strategy, the EU has provided full regulatory and policy frameworks for the circulation of material [77]. Successes comprise municipal recycling rates above 50% in top states such as Germany and Austria, together with a stated 11% reduction in dependence on landfill between 2010 and 2023 [78]. Main structures of the EU model comprise controlling harmonization, the implementation of mandatory eco-design standards under the Eco-design for Sustainable Products Regulation (ESPR), and the emergent use of Digital Product Passports (DPPs) to advance traceability of resources across supply chains [79,80]. Nonetheless, challenges continue, mainly in the irregular advancement among Member States, with numerous weaker economies at risk of missing the 2025 municipal and packaging-waste recycling targets, mainly as a result of infrastructural cracks and weak implementation of extended producer responsibility [81,82]. These differences highlight the necessity for upgrading measures and investment maintenance to ensure reasonable implementation of CE objectives across the Union [82].

https://doi.org/10.3390/su18041946

In China, CE policies have been established through the Circular Economy Promotion Law, eco-industrial park plans, and borders such as the National Sword Policy that barred imports of inferior recyclables [83]. The official recycling of waste electrical and electronic equipment (WEEE) has stretched rapidly, increasing from ~10.1 million units in 2012 to ~87.85 million units in 2021 for five main classes (TVs, ACs, refrigerators, washing machines, and PCs), with official recycling amounts exceeding 40% [84]. Industrial synergy initiatives in regions such as Tianjin demonstrate the potential of centralized CE planning. However, an insistent dependence on the informal recycling sector and regional enforcement differences limit uniformity across provinces [85,86].

The Netherlands is a member of the members' state of the European Union, but is considered a separate case study because its national-level CE initiatives, governance structures, and performance metrics offer deeper perceptions that match the bigger EU-wide outlook, owing to her detailed national policy differences. Although the EU provides an all-embracing regulatory framework like the Circular Economy Action Plan (CEAP), the Waste Framework Directive (WFD), and the Ecodesign for Sustainable Product Regulation (ESPR), the Netherlands has cutting-edge country-specific CE strategies that go significantly past EU-wide obligation. The Netherlands has positioned itself as a CE leader, targeting 100% circularity by 2050. Provisional areas include halving primary abiotic raw material use by 2030 [87]. In spite of early progress like the approval of circular procurement policies and the implementation of zero-waste pilots, current valuations propose that the country is not yet on the path to meet its 2030 target [88]. Its circular material-use level (~24.5%) is somewhat high as compared to the EU average, but inadequate for its determined goals [89]. The Dutch model stands out for accentuating multi-stakeholder partnership and city-level CE roadmaps, but its dependency on international supply chains poses challenges for scaling a completely circular system.

Mutually, these case studies show that although the EU prioritizes regulatory organization, China centers on industrial synergy and integrated design, and the Netherlands stresses innovation and partnership, shared challenges continue. These include irregular implementation (EU, China), informal waste handling (China), and dependence on international trade (Netherlands). Meanwhile, each model suggests lessons that can be adapted: the EU's implementation mechanisms, China's scaling capability, and the Netherlands' combined innovations show that mixed, context-thoughtful approaches are important to attaining strong, low-carbon, and resource-efficient CE systems.

The success and borders of CE-based waste management models in the countries mentioned can be mapped out as combining governance design, policy implementation ability, market enticements, and socio-technical position. For instance, In the European Union, achievement is mainly as a result of robust harmonization of regulation, recycling goals that are bounded by law, and extended producer obligation structures that adopt waste management costs into the design of products. On the other hand, the same layered governance that permits harmonization also contributes to irregular results, since implementation capability and application differ considerably amongst member states, mostly in emerging economies.

China's model makes evident that regional governance and significant state investment can speedily hasten CE infrastructure implementation, particularly from end-to-end of eco-industrial parks and industrial synergy. The usefulness of this line lies in its capability to line up CE goals with national industrial policy and economic development. However, determined dependence on informal recycling segments and erratic regional implementation sideline environmental performance and traceability of material, specifying that centralized approaches alone are inadequate without harmonizing market mechanisms and social inclusion policies.

https://doi.org/10.3390/su1804194

In the Netherlands, success is principally due to robust institutional harmonization, stakeholder engagement, and investigation at the city and regional stages, buttressed by circular procurement and invention-responsive policy frameworks. However, challenges abound in scaling resident successes on a national scale and decreasing necessity on inclusive material supply chains, which limits total decreases in use of virgin material. These outcomes point to the fact that CE usefulness capitalizes on the fact that regulatory precision, economic motivations, implementation ability, and shareholder involvement are at the same time present, and that the absence of any one of these essentials can considerably impact boundary performance. Summarily, CE results are not dependent on policy ambition alone, but on the level of placement between governance structures, market inclination, technological capacity, and social engagement. This explains why no solitary model can be generally replicated, and underlines the need for tailored CE strategies rather than a generic policy approach.

3.5. CE in the Energy and Chemical Industry

The energy and chemical sectors are major contributors to total greenhouse gas emissions and the generation of harmful waste materials, accounting for almost 30% of global industrial CO_2 emissions, which emphasizes their essential role in climate mitigation efforts [90]. CE application in these sectors demands more than just a general recycling approach, but rather technical solutions. Attaining circularity can be through ways like the chemical recycling of plastics to naphtha or the addition of oils generated during the process of pyrolysis into steam crackers [91,92], the reuse of solvents recovered from closed loops [93], and hydrometallurgical retrieval of dangerous metals from used hydrotreating catalysts [94]. Other major ways for increasing circularity and decreasing the demand for energy are process-strengthening methods like reactive distillation, membrane reactors, and heat-integration networks [95]. Moreover, evolving CE implementation, such as closed-loop process-water systems, bio-based feedstock replacement, and the reuse of CO_2 as a chemical feedstock (for example, methanol synthesis or polymer precursors), further validates the potential for circularity within fundamental reaction pathways.

Notwithstanding this, the implementation of CE in the chemical industry is limited by continuous extensive production models that involve constant purity of feedstock, stringent control of contamination, and consistent process flow [96]. These limitations hinder the direct addition of secondary raw materials. Harmful waste streams like sludges from refineries, tar remains/residues, expended catalysts, and wastewater which is high in salt content need specific CE-compatible ways like vitrification, zero-liquid-discharge systems, advanced oxidation, gasification, and metal recovery [97]. Additionally, the long lifetime of process equipment, constricted process integration, and high thermochemical constancy requirements requires CE strategies to be well-matched with continuous and high-cost operations, frequently depending on digital monitoring, predictive maintenance, modular retrofits to avoid process interruptions, including AI-based process control, IoT sensors, and digital twins, to conserve process constancy while improving circularity. Addressing these limitations and prospects is critical for translating CE values into industrially actionable approaches.

4. CE Strategies in Mitigating Climate Change

4.1. Emissions from Waste Treatment Processes

Waste management contributes significantly to global GHG emissions, particularly methane (CH_4), carbon dioxide (CO_2), and nitrous oxide (N_2O), which are mainly emitted from landfills, the treatment of wastewater, and the burning of waste. Approximately 3% of total human-influenced GHG emissions come from waste, with methane from landfills

https://doi.org/10.3390/su18041946

representing a considerable portion due to the anaerobic decay of organic matter [98]. While this percentage seems reasonable in comparison to other sectors like energy or transport, the high warming potential of methane and the likely progression of MSW in urban areas present the sector as a serious part of climate mitigation strategies [99].

Methane emissions from landfills are the main problem, considering the methane has a global warming potential (GWP) 28–36 times more than CO_2 over 100 years and more than 80 times greater over 20 years [100,101]. These emissions add explicitly to proximate climate warming, worsening reaction mechanisms like the increase in heatwaves, changes in precipitation forms, and the quickening of polar ice melt [102]. Moreover, landfill leachates are discharged and volatile organic compounds can interact with atmospheric chemistry, contributing ultimately to the development of the tropospheric ozone, a strong climate propeller and air pollutant [103].

Incineration reduces the volume of waste, but mostly releases CO_2 and may produce pollutants like dioxins and furans if not well controlled [104]. The CO_2 produced from incineration, particularly when fossil-derived plastics form a large part of the waste stream, adds to atmospheric carbon burdens. Equally, open burning of waste majorly releases CO_2, although it may also release dioxins and furans, contingent on the type of waste and the effectiveness of the burning method [105]. In addition, open burning also produces black carbon, a momentary climate pollutant with a GWP hundreds of times more than CO_2 in shorter periods [106]. Black carbon deposit on ice and snow speeds up albedo loss, and promotes increasing global warming and regional climate disorder [107].

Conversely, composting and anaerobic digestion have lower emissions profiles, though both can produce high GWP gases like CH_4 and N_2O in inadequate conditions. However, controlled aerobic composting and circular anaerobic digestion process with accurate gas-capture equipment can lessen these emissions efficiently [108,109]. Also, recycling, though not a direct treatment method for organic waste, can ultimately lessen emissions by avoiding the need for virgin material mining and handling, which usually require high energy, thereby dropping source carbon footprints [110]. Generally, emissions from waste systems not only contribute to the global carbon budget, but also increase climate-associated threats like a rise in sea level, loss of biodiversity, and food and water security. Controlling emissions from waste sectors is hence an important strategy for global climate-change mitigation.

Figure 4 compares the greenhouse gas emission profiles of major waste treatment pathways, demonstrating the comparative climate impacts of landfilling, incineration, recycling, composting, and anaerobic digestion. The contrast highlights the mitigation potential of CE-aligned paths, mostly those related to industrial and high-emission waste streams.

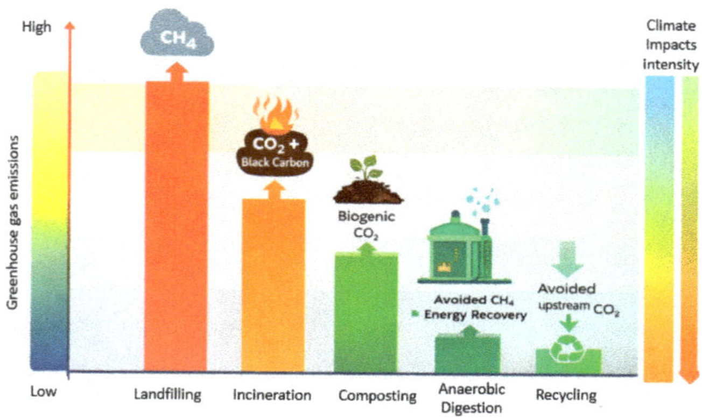

Figure 4. Greenhouse gas emission profiles of major waste treatment paths [100,101].

https://doi.org/10.3390/su1804194

4.2. Reduction in Carbon via Circular Economy

Circular economy principles are employed to close material loops and reduce input of resources and environmental side effects, as well as greenhouse gas emissions. Different circular economy strategies are usually used in the management of waste to reduce emissions, as well as material exchange, recovery of energy, composting, and production of biochar. Material exchange via recycling and reuse considerably lessens greenhouse gas emissions. For example, recycling aluminum saves up to 95% of the energy that is needed to produce it from raw bauxite, whereas recycling plastics can save energy of up to 60–70% as compared to pure plastic manufacture [111,112]. Replacing traditional construction materials with recycled aggregate materials or industrial wastes also leads to declines in tangible carbon emissions [113].

The saving of energy is one principal benefit of the idea of the circular economy. Take, for example, waste-to-energy burning to recover energy, which can transfer fossil fuel utilization in the generation of electricity, though the amount of energy that will be recovered is vastly reliant on the composition of the feedstock material and efficiency of the energy recovery system [114]. Also, anaerobic digestion of organic waste not only hinders organic waste from going into landfills, but could produce biogas, a renewable source of energy that can substitute natural gas [115].

4.3. Accounting for Carbon in the Circular Economy

The addition of circular economy approaches into nationwide carbon accounting structures and climate strategies is not only a promising, but a fast-emerging area of concern. Outmoded greenhouse gas accounts time and again play down the mitigation possibility of circular economy approaches because they center primarily on direct emissions within sectoral borders, and do not consider life-cycle emissions and avoided emissions from material substitution and recycling [116,117].

In order to close this gap, new structures are developing to include circular economy actions into carbon accounting. To cite an instance, the European Union's Circular Economy Action Plan inspires member states to measure greenhouse gas savings from the prevention of waste, reuse, and recycling under extended producer responsibility (EPR) and eco-design strategies [118]. Also, the addition of circular economy policies into Nationally Determined Contributions (NDCs) under the Paris Agreement is being encouraged by organizations like the Ellen MacArthur Foundation [119], which underscores the fact that circularity in material flows can decrease global greenhouse gas emissions by 20–30%. Accounting for carbon in the circular economy entails a change from process-based accounts to consumption-based or systems-based accounting methods, which reflect upstream and downstream emissions through product life cycles [120]. Such methods are mainly vital for circular economy approaches that depend on worldwide supply chains and transboundary material flows.

Conventional tools and procedures, such as Material Flow Analysis (MFA), Life Cycle Assessment (LCA), and Hybrid Input–Output (IO) models, are progressively being used to assess the carbon impacts of circular economy involvement on urban, local, and nationwide scales [121,122]. Even though these approaches offer vigorous awareness into the environmental impacts of waste systems, they have certain procedural limitations like direct assumptions, absence of vigorous temporal resolution, and feeble incorporation of behavioral factors [123,124]. These limitations highlight the fact that no single approach is enough for capturing the complete range of CE's role in carbon mitigation; consequently, there is a need for a combination of frameworks [125].

Recent advances in carbon accounting approaches go beyond traditional tools. Dynamic Material Flow Analysis (DMFA) permits investigators to find resource stocks and

https://doi.org/10.3390/su18041946

flows in time, thus detecting "carbon lock-ins" connected with long-lasting structure and predicting impending waste-associated emissions [126]. Likewise, Environmentally Extended Multi-Regional Input–Output (EE-MRIO) models detail the cosmopolitan nature of material trade and consumption-based emissions, offering insights into the carbon embodied in imports and exports [127]. These methodologies are progressively applicable for countries with high dependence on material imports, where consumption-based emissions can deviate considerably from production-based records.

Furthermore, evolving behavioral and systems-based modelling tools offer capable leeway. Agent-Based Modelling (ABM) can mimic household-, industrial-, and policy-performer connections to evaluate the adoption and dispersal of CE approaches, allowing examination of diverse behavioral reactions to interferences like recycling orders or carbon estimating [128]. Similarly, System Dynamics Modelling (SDM) is current in capturing response loops, delays, and anomalies in waste–energy–emission systems, offering a vibrant perception on the lasting impacts of CE procedures [129].

Globally, Integrated Assessment Models (IAMs), which have been extensively utilized to evaluate pathways to attain 1.5 °C and 2 °C goals, are now being expanded to include CE involvement, comprising resource efficiency, material replacement, and waste valorization [130]. By implementing CE approaches into IAMs, investigators and policymakers can assess their total contributions to global climate stabilization efforts and lessen dependence on harmful emission technologies. Lastly, Product Environmental Footprint (PEF) tools developed by the European Commission offer uniform, product-level carbon accounting frameworks that can improve the parity and adaptability of CE approaches at both industry and policy levels [131]. Jointly, these tools change carbon accounting from a stationary, reviewing application into a dynamic, systemic, and progressive framework. Their adoption into general GHG lists and climate policies can advance the recognition of CE approaches as genuine mitigation ways under the Paris Agreement. Nevertheless, procedural coordination and the availability of data are a serious challenge for safeguarding uniformity across countries and sectors.

5. Implementation of the Circular Economy

In order to transition to a circular economy, there is a need for a multidimensional and combined method that is driven by robust supporting factors. These factors comprise both policy and regulatory frameworks, economic instruments, stakeholder engagement, and scientific novelties and innovations. Each of these factors is important in creating the general changes that are necessary for a successful transition from linear to circular models of production and consumption.

5.1. Policy and Regulatory Frameworks

Policy and regulation are the foundation upon which the circular economy is implemented by building the supporting environment, setting the needed goals, and promoting innovation. National and regional circular-economy frameworks provide a plan for lasting visions, objectives, and priority sectors. For example, the European Union's Circular Economy Action Plan under the European Green Deal has set a pattern by forming governmental and non-governmental actions that advance the principles of circular economy across the lifecycle of products [132].

Extended Producer Responsibility (EPR) is one of the foundational controlling mechanisms for the circular economy. It makes it mandatory for producers to take up responsibility for the management of their product's final life, and, by so doing, encourage eco-design and reduce the generation of waste. Nations like Japan, Germany, and South Korea have employed robust EPR systems that have considerably improved rates of recycling and re-

https://doi.org/10.3390/su1804194

source recovery [133]. The taxes and bans placed on landfilling are other monitoring levers that discourage the disposal of recyclable and biodegradable waste, thus encouraging other waste management strategies like reuse, recycling, and composting. For instance, Sweden's landfill tax has led to a significant reduction in landfilling, with over 99% of domestic waste now being recycled or used for energy recovery [134].

5.2. Economic Levers

Economic motivations and deterrents are crucial for adopting environmental externalities and promoting savings in circular economy practices. Green Public Procurement (GPP) is a mandatory policy tool that enables governments to order the acquisition of environmentally desirable goods and services. GPP can encourage important market demand for circular products and services, as seen in countries like the Netherlands and Finland, where green earning is included in national circular economy plans [135].

Carbon valuing mechanisms, plus carbon taxes and emission exchange systems, incentivize emission drops and energy efficiency advances across supply chains. By allocating a cost to carbon emissions, these tools encourage firms to embrace resource-effective and low-carbon technologies important to circular economy models [136]. Additionally, directed grants and economic incentives for circular economy innovations, like tax credits for material recycling services or low-interest loans for sustainable product design, can lessen economic risks connected with circular economy transitions, mostly for small and medium-sized enterprises. Investment in CE launches and expansion is also increasing, regularly sustained by green bonds and impact investment assets. These economic tools station capital into projects that add to CE results like closed-loop recycling, bio-based materials, and the circular product as a service model [137].

5.3. Engagement with Stakeholders

The general nature of CE requires the dynamic engagement of different participants, like governments, industry, academia, and indigenous communities. Governments play a directing part by developing policy consistency, promoting cross-sectoral teamwork, and funding research and development in circular advances. Additionally, local governments are gradually using household-level motivations to boost circular practices. Instances include pay-as-you-throw (PAYT) curricula, reductions for households operating specialized home composters, and discounts for purchasing refillable items or products that are repairable. Such economic tools allow citizens to directly reduce the cost of municipal waste management.

Public–private partnerships (PPPs) have been shown to be effective in building CE infrastructure, such as industrial cooperation parks and waste-to-energy services.

Industries play a pivotal role in the implementation of CE practices at the operating level. Prominent companies are rewriting business models to incorporate circular principles such as remanufacturing, hiring, and reverse logistics. For example, Philips and HP have embraced product-as-a-service models, moving from title to service delivery, thus decreasing the material amount [138,139].

Academic institutes contribute via research, education, and innovation. Different disciplines, circular-economy research centers, and university–industry alliances are aiding the development of innovative materials, life-cycle assessment tools, and circular economy metrics. Furthermore, capacity-building creativities like curricula focused on the circular economy and training programs are promoting a workforce prepared for the circular transition [140]. Communities and civil society organizations (CSOs) also play a significant part by promoting behavioral change and pinpointing CE initiatives. Community-driven repair workshops, zero-waste actions, and indigenous exchange organizations illustrate

https://doi.org/10.3390/su18041946

common circular economy actions, regularly supported by community policies and non-governmental organizations [141].

Several positive CE-oriented systems globally depend on inclusive citizen cooperation reinforced by regulatory and economic tools. Obligatory waste-sorting systems, employed in countries like Germany, South Korea, and Slovenia, have been revealed to immensely improve rates of material recovery and reduction of contaminants in recyclables [142]. These systems depend on a strong guiding principle, color-coded bins, and municipal administration to guarantee compliance. Deposit-return schemes (DRS) for beverage containers, extensively approved in the EU, form strong economic motivations for households to return PET bottles, cans, and glass containers, by contributing refunds per unit. DRS has attained return rates beyond 90% in countries like Norway and Lithuania [143].

5.4. Innovation and Digitalization

Technological innovation and digitalization are pivotal facilitators of the circular economy, improving the efficiency of resources, traceability of products, and optimization of systems. Digital technologies like Artificial Intelligence (AI), the Internet of Things (IoT), and blockchain are principal to this change. AI is being engaged in predictive maintenance, intelligent sorting of waste, and enhanced logistics during recycling processes. For instance, machine-learning algorithms can pinpoint components that are valuable at the end of the life of electronics, allowing efficient recovery of material [144]. IoT devices enable actual checking of the use of the product and the generation of waste, aiding circular interventions that are data-driven, like automatic record management and tracking of waste [145].

Blockchain technology safeguards transparency and tracking through supply chains by tightly recording dealings and material flows. This is mainly helpful for confirming recycled content, guaranteeing compliance with extended producer-responsibility regulations, and allowing product permits that notify consumers about product source, structure, and recyclability [146]. Digital platforms and markets are also evolving to enable the exchange of resources, reuse of products, and collaborative consumption. Examples include B2B platforms for industrial cooperation, digital twins for circular product design, and networked membership applications that extend the lifecycles of products [147].

6. Obstacles and Challenges

Despite the potential of CE strategies, their implementation is restricted by numerous factors that hinder practicability and scalability. Some of these factors include the high energy amount required for certain recycling and recovery processes, thermodynamic productivity restrictions that constrain material and energy recovery vintages, feedstock purity necessities for inferior materials, and challenges connected with mixing circular processes into extensive uninterrupted industrial systems. Additionally, the practical development of unindustrialized CE technologies, infrastructure accessibility, and process constancy necessities also impact their environmental performance and climate mitigation potential. Identifying these limitations is necessary for avoiding lack of nuance and for aligning CE strategies with accurate engineering and functioning situations. This section discusses specific obstacles and challenges that limit the implementation of CE, in detail.

6.1. Technical and Structural Limitations

Technical and infrastructural shortages result in a major barrier to the application of circular approaches, especially in developing nations. Many places lack suitable recycling amenities, waste segregation systems, and innovative material-recovery skills, which are essential for allowing high-quality circular flows [110,148]. For example, poor sorting and collection structures regularly result in the pollution of recyclable materials, and,

https://doi.org/10.3390/su1804194

subsequently, in lower-quality secondary raw materials, low market value, and increased rates of landfills and burning.

The lack of uniform processes and inadequate digital infrastructure hinders the growth of reverse logistics, product tracking, and lifecycle data structure, which are essential for circular supply chains [149]. Innovative technologies like AI, IoT, and blockchain could potentially enable the actual monitoring of waste and tracking of materials, but their implementation is limited due to cost, skills gaps, and compatibility issues. Besides, several industrial processes were designed based on a linear "take–make–dispose" approach, making upgrading or changing current systems with circular options both expensive and technically difficult. The change necessitates not only upgrades to infrastructure, but also innovation in product design, materials engineering, and rebuilding technologies [150].

Another serious technical and infrastructural issue in several countries is the lack of standardized and clear labelling of packaging waste. Products are often marked with vague or confusing signs, like the "green dot", which consumers wrongly understand as signifying recyclability [151]. This discrepancy across producers and counties causes contamination in recycling streams and raises the costs of processing. Furthermore, evolving materials, such as multilayer packaging, biopolymers, and composite materials, are frequently promoted as circular; however, most of them currently lack practical, scalable recycling or composting pathways. Therefore, these materials are commonly landfilled or incinerated because current waste-management systems cannot process their mixed structures, additives, or unsuitable polymer combinations [152].

6.2. Economic and Market Limitations

Economic constraints play an important role in preventing the change to circularity. One of the main issues is the lack of financial motivations and the domination of linear economic models that highlight temporary profits over lasting sustainability. In many cases, non-recycled materials cost less than those recycled, due to market subsidizations, economies of scale, or the lack of environmental cost internalization [153]. Small and medium-sized enterprises, which form the support of many economies, regularly lack the money and risk-appetite to invest in circular innovations or green technologies [154]. Moreover, the lack of mature subordinate markets for recycled materials restricts demand, depressing investment in recycling set-up and circular business models.

Price instability in ancillary raw-material markets is also a challenge, as it makes it difficult for businesses to strategize and scale processes. For example, inconsistent oil prices directly influence the affordability of recycled plastics compared to new ones [155]. Also, the lack of consistent certifications and product quality assurance for recycled materials further weakens market assurance and trust of consumers. Public procurement systems and savings structures also remain subjective toward linear infrastructure, stressing the need for circular principles in green financing mechanisms and public–private corporations [156].

6.3. Institutional and Policy Fissures

Institutional breakup and policy disorganization are among the most prevalent problems that hinder the implementation of the circular economy. Many governments do not have inclusive circular economy plans, or the current strategies do not align well across sectors like waste management, energy, agriculture, and industrial growth [157]. Furthermore, strategies frequently focus intently on end-of-life recycling instead of encouraging upstream prevention, product design, and complete innovation. In some nations, governing frameworks are obsolete or unclear, and are unable to give clear descriptions, tasks, and implementation mechanisms for the practices of the circular economy. For example,

https://doi.org/10.3390/su18041946

the grouping of waste resources as "hazardous" or "non-waste" frequently lacks precision, hindering their reuse or cross-boundary trade [158].

Weak institutional ability and exploitation can further grind down the efficiency of CE programs, particularly in LMICs. Inadequate monitoring, poor implementation of environmental standards, and a lack of inter-agency organization can lead to policy operation gaps [159]. There is also a need for policies that encourage innovation and entrepreneurship in the circular economy, like tax inducements for eco-design, subsidies for circular startups, and governing sandboxes for model projects. Also, without incorporating CE into nationwide development and climate plans, the efforts remain disjointed and less impactful.

6.4. Behavioral and Social Sprints

Social and cultural factors considerably impact the success of CE changes, frequently acting as indirect but deeply rooted obstacles. Consumer behaviors such as overconsumption, low input in recycling programs, and opposition to product-reuse models (for example, overhauled electronics or shared services), are the main problems to demand-side circularity [160].

Absence of awareness of, and education about, CE principles among consumers, industries, and legislators is a frequent challenge. Studies have reported that many investors are not familiar with circular policies like product-as-a-service, cradle-to-cradle design, or industrial synergy [161]. This knowledge gap hinders the implementation of CE practices, even when organizational and policy support are present. Traditional defiance of waste, cleanliness, and possession also plays a part. In some cultures, there is dishonor associated with second-hand things or products that are produced from recycled materials, which restricts market infiltration [162]. Also, the "offhand culture" encouraged by reckless fashion, single-use packaging, and planned desuetude opposes the central philosophy of CE. Possibly the most underrated barrier to the implementation of CE is the lack of consumer knowledge, awareness, and commitment to circular practices. Consumers often fail to properly segregate waste at source, even when there is the availability of infrastructure, which results in high rates of contamination that weaken the efficiency of recycling [163].

In many cultures, waste management is still perceived as the responsibility of the government instead of a collective common responsibility. This attitude pointedly limits the participation of households in composting, source-segregation, and reuse structures [164]. Public education campaigns are inadequate, inconsistent, or poorly implemented, leaving big knowledge gaps across groups of the population. Overcoming these social obstacles needs complete changes in education, marketing, and public engagement. Communal innovation, pushing techniques, and alertness promotions can help change values and practices to more sustainable consumption and production designs.

7. Future View and Research Directions

7.1. Total Integration of CE and Climate Policy

One of the foremost needs is the complete integration of the principles of the circular economy into domestic and global climate policies. At present, climate frameworks, like Nationally Determined Contributions (NDCs) under the Paris Agreement, regularly manage the mitigation possibility of circular policies, concentrating mainly on renewable energy and efficiency measures [165]. Nonetheless, studies have established that CE practices like material reuse, recycling, and product-life extension can considerably lessen GHG emissions across numerous sectors, as well as construction, transportation, and food systems [166]. Incorporating CE into climate policy needs the bringing together of waste management, industrial strategy, and environmental guidelines. For example, life-cycle-

https://doi.org/10.3390/su1804194

centered strategy tools and carbon rating mechanisms can be extended to include resource productivity metrics [167]. Moreover, international bodies must encourage policy unity by supporting CE strategies with the Sustainable Development Goals (SDGs) and climate adaptation efforts.

7.2. CE in Developing Economies and Informal Sectors

The CE changeover must be all-encompassing and circumstance-specific, particularly for developing economies where informal sectors dominate handling of waste and resource recovery. In many developing countries, informal pickers of waste add considerably to recycling efforts, but remain disregarded in policy and structure development [168]. A fair changeover requires formal acknowledgement, communal protections, and capacity building for these players.

Developing economies face distinctive challenges, as well as inadequate institutional capacity, financial limitations, and a lack of technical structure [169], though these areas also present prospects for advancing into circular practices, avoiding the linear industrial paths of the Overall North. Trial initiatives in Africa and Asia show that dispersed CE models like community-centered composting or moderate plastics recycling can produce great environmental and socio-economic benefits when sustained by supporting policies and funding [170]. We therefore call for research on CE scalable models that consider indigenous socio-economic dynamics and integrate the voices of informal sector investors. Relative situation studies and South–South information exchange will be important in familiarizing CE strategies with various overall frameworks.

7.3. Circular Design and Sustainable Consumption

The upstream addition of circularity through design novelty and change in consumer behavior is another serious area that requires innovation. Circular design involves the development of products that are strong, flexible, repairable, and recyclable, ensuring the efficiency of resources and prolonging the life cycles of products [150]. This method must be rooted in the starting point of products and sustained by digital tools like product passports and material-tracing systems.

Similarly essential is the advancement of sustainable consumption designs such as demand-side measures like product-as-a-service models, mutual-economy platforms, and consumer education. This can help decouple economic development from resource use, [171], though behavioral apathy, cultural inclinations, and affordability are still major barriers to extensive acceptance. To spread circular design and consumption, research across many disciplines is needed to bridge design thinking, social science, and lifecycle engineering. Regulatory tools like Extended Producer Responsibility (EPR), eco-labeling, and green procurement policies must also be cultivated to encourage source circularity.

7.4. Call for Interdisciplinary Research and All-Inclusive Assessment

The intricacy of CE enactment and its link with climate and sustainability goals requires complete, collaborative approaches. Current CE valuations frequently lack uniformity and overlook complete responses, balances, and reflection effects [172]. For example, recycling creativity may reduce waste but increase energy consumption and emissions in the supply chain. We therefore call on future research to highlight the growth of joined assessment models that synchronize material flow analysis (MFA), life cycle assessment (LCA), and system dynamics. These tools should be able to assess not only environmental effects, but also social and economic consequences, through numerous measures and time distances. Furthermore, CE research should be collaboratively produced with stakeholders, comprising legislators, industries, the public, and informal players, to ensure importance and applicability. Research across disciplines and laboratories can serve as trial platforms for testing CE innovations and governance mechanisms in practical settings.

8. Conclusions

This review provides a comprehensive synthesis of existing studies on integrating the circular economy (CE) framework into sustainable waste management as a strategic pathway for climate change mitigation. By examining theoretical basics, operational practices, and enabling mechanisms, the review reveals that CE can transform production and consumption structures while lessening greenhouse gas (GHG) emissions. The circular economy offers a strong platform for the reduction of environmental impacts and the enablement of climate resilience. Including the principles of the circular economy into waste management will result in significant reductions in greenhouse gas emissions and also promote economic and material efficiency. Notwithstanding this, a supportive policy and governing environment, financial motivations, cross-sector collaboration, and the employment of digital technologies are essential to its effective implementation. In realizing its full benefits and implementation, challenges like technical restrictions, economic and policy reservations, and social resistance, mainly in developing and informal regions, must be overcome.

Beyond synthesizing existing studies, the review contributes scientifically by framing the circular economy as an integrated climate mitigation approach instead of just a waste management concept. Through linking the principles of CE together with emissions mitigation strategies, sector-focused engineering solutions, and climate change policy tools, the review gives an inclusive framework for understanding how CE can support computable progress in the direction of climate goals. The review also discusses CE in high-emission sectors, such as the chemical and energy industries, which are scarcely considered in the literature.

Future work should emphasize evolving quantifiable CE–GHG indicators, modelling CE possibility in continuous industrial processes, and integrating CE approaches into national climate strategies. Prioritizing digitalization (together with digital twins, AI-driven tracking, and decentralized material tracking) and interdisciplinary collaboration will be important for scaling CE in energy and emission-intensive industries. Consolidating these parts will guarantee CE transitions from a conceptual ideal into a real-world, sector-specific climate resilience and waste-system revolution model.

In summary, the objectives of this review were fully realized. The analysis combined CE theory, waste management practices, a roadmap to decarbonization, and sector-focused technical approaches, indicating how CE strategies can promote CE and reduce emissions. By leveraging technological innovations, policy support, practical challenges, and the needs of future research, the review provides an inclusive assessment consistent with the stated objectives.

Author Contributions: Conceptualization, S.J.I. and M.V.K.; validation, S.J.I., M.V.K., J.T.A. and D.P.; investigation, S.J.I. and M.V.K.; writing—original draft preparation, S.J.I.; writing—review and editing, S.J.I., M.V.K., J.T.A. and D.P.; visualization, M.V.K., J.T.A. and D.P.; supervision, M.V.K. and J.T.A.; project administration, M.V.K. and J.T.A.; funding acquisition, M.V.K. All authors have read and agreed to the published version of the manuscript.

Funding: This research was funded by National Research Foundation, South Africa, through the NRF SARChI Chair (Waste and Climate Change).

Data Availability Statement: No new data were created.

Acknowledgments: Authors acknowledge all the support from the University of KwaZulu-Natal and the National Research Foundation, South Africa, for this research.

https://doi.org/10.3390/su180419

Conflicts of Interest: The authors declare no conflicts of interest. The funders had no role in the design of the study; in the collection, analyses, or interpretation of data; in the writing of the manuscript; or in the decision to publish the results.

Abbreviations

The following abbreviations are used in this manuscript:

CE	Circular economy
GHG	Greenhouse gas
MSW	Municipal solid waste
LCA	Life Cycle Assessment
CEPIs	Circular Economy Performance Indicators
MFA	Material Flow Analysis
PET	Polyethylene Terephthalate
HDPE	High-Density Polyethylene
LDPE	Low-Density Polyethylene
EPR	Extended producer responsibility
CDW	Construction and demolition waste
ESPR	Ecodesign for Sustainable Products Regulation
IO	Input–Output
EE-MRIO	Environmentally Extended Multi-Regional Input–Output
ABM	Agent-Based Modelling
SDM	System Dynamics Modelling
IAMs	Integrated Assessment Models
PEF	Product Environmental Footprint
GPP	Green public procurement
PPPs	Public–private partnerships (PPPs)
CSOs	Communities and civil society organizations
AI	Artificial Intelligence
IoT	Internet of Things
NDCs	Nationally Determined Contributions
SDGs	Sustainable Development Goals
MCI	Material Circularity Indicator
RC	Recycled Content
RE	Resource Efficiency
RR	Recovery Rate
RP	Resources Productivity
WGI	Waste Generation Intensity
SA	Stock Accumulation
DR	Dependency Ratio
CIG	Circularity Indicator Gap

eferences

United Nations Environment Programme (UNEP). *Global Waste Management Outlook 2024*; UNEP: Nairobi, Kenya, 2024. Available online: https://www.unep.org/resources/global-waste-management-outlook-2024 (accessed on 20 November 2025).
Sakaya, R.; Niwagaba, C.B.; Katukiza, A.Y.; Lüthi, C. Influence of National Government and Development Partners on Uganda's Water, Sanitation, and Solid Waste Services Delivery: A Review. *World Water Policy* **2025**, *11*, 939–952. [CrossRef]
Kiehbadroudinezhad, M.; Merabet, A.; Hosseinzadeh-Bandbafha, H. Landfill source of greenhouse gas emission. In *Advances and Technology Development in Greenhouse Gases: Emission, Capture and Conversion*; Elsevier: Amsterdam, The Netherlands, 2024; pp. 123–145.
Reyna-Bensusan, N.; Wilson, D.C.; Davy, P.M.; Fuller, G.W.; Fowler, G.D.; Smith, S.R. Experimental measurements of black carbon emission factors to estimate the global impact of uncontrolled burning of waste. *Atmos. Environ.* **2019**, *213*, 629–639. [CrossRef]
Hadei, M.; Yousefian, F. Global Air Quality and Climate Change. In *Air Pollution, Air Quality, and Climate Change*; Elsevier: Amsterdam, The Netherlands, 2025; pp. 175–197.

https://doi.org/10.3390/su18041946

6. Hidalgo, D.; Martín-Marroquín, J.M.; Corona, F. A multi-waste management concept as a basis towards a circular economy mod *Renew. Sust. Energy Rev.* **2019**, *111*, 481–489. [CrossRef]
7. Zhang, Z.; Malik, M.Z.; Khan, A.; Ali, N.; Malik, S.; Bilal, M. Environmental impacts of hazardous waste, and managemer strategies to reconcile circular economy and eco-sustainability. *Sci. Total Environ.* **2022**, *807*, 150856. [CrossRef] [PubMed]
8. Ji, M.; Li, J.; Zhang, M. What Drives the Agricultural Carbon Emissions for Low-Carbon Transition? Evidence from Chin *Environ. Impact Assess. Rev.* **2024**, *105*, 107440. [CrossRef]
9. Reddy, V.V.; Kumar, P.; Rao, A.L.N.; Kumar, R.; Singh, S.; Asha, V.; Kareem, S.H. Waste to wealth generation: Innovativ methodologies in resource utilization and minimization in circular economy. In *E3S Web of Conferences*; EDP Sciences: Les Ul France, 2023; Volume 453, p. 1035. [CrossRef]
10. Korhonen, J.; Honkasalo, A.; Seppälä, J. Circular economy: The concept and its limitations. *Ecol. Econ.* **2018**, *143*, 37–46. [CrossRe
11. Yang, M.; Chen, L.; Wang, J.; Msigwa, G.; Osman, A.I.; Fawzy, S.; Rooney, D.W.; Yap, P.S. Circular economy strategies fo combating climate change and other environmental issues. *Environ. Chem. Lett.* **2023**, *21*, 55–80. [CrossRef]
12. Van Ewijk, S.; Stegemann, J.A.; Ekins, P. Global life cycle paper flows, recycling metrics, and material efficiency. *J. Ind. Ecol.* **201** *22*, 686–693. [CrossRef]
13. Yamada, K.; Ii, R.; Yamamoto, M.; Ueda, H.; Sakai, S. Japan's greenhouse gas reduction scenarios toward net zero by 2050 in th material cycles and waste management sector. *J. Mater. Cycles Waste Manag.* **2023**, *25*, 1807–1823. [CrossRef]
14. Bouramdane, A.A. How Can Multi-Criteria Decision-Making Optimize Adaptation Strategies for Climate Change-Induce Migration? *Afr. Cities J.* **2025**, *5*, 119–158. [CrossRef]
15. Kirchherr, J.; Yang, N.H.N.; Schulze-Spüntrup, F.; Heerink, M.J.; Hartley, K. Conceptualizing the circular economy (revisited): A analysis of 221 definitions. *Resour. Conserv. Recycl.* **2023**, *194*, 107001. [CrossRef]
16. European Commission. *Closing the Loop—An EU Action Plan for the Circular Economy; COM (2015) 614 Final*; European Commissio Brussels, Belgium, 2015.
17. Ellen MacArthur Foundation. *Towards the Circular Economy: Economic and Business Rationale for an Accelerated Transition*; Elle MacArthur Foundation: Cowes, UK, 2013.
18. Kirchherr, J.; Reike, D.; Hekkert, M. Conceptualizing the circular economy: An analysis of 114 definitions. *Resour. Conserv. Recy* **2017**, *127*, 221–232. [CrossRef]
19. Tambovceva, T.; Titko, J. *Introduction to Circular Economy*; Ekonomikas un Kultūras Augstskola: Riga, Latvia, 2020.
20. Kandpal, V.; Jaswal, A.; Santibanez Gonzalez, E.D.; Agarwal, N. Circular economy principles: Shifting towards sustainab prosperity. In *Sustainable Energy Transition: Circular Economy and Sustainable Financing for Environmental, Social and Governan (ESG) Practices*; Springer: Cham, Switzerland, 2024; pp. 125–165.
21. Ishaq, S.; Tanveer, U.; Hoang, T.G.; Wamba Samuel, F. From linear to circular: Capabilities, defensive reasoning and supply cha collaborations in manufacturers' circular economy transition. *Br. J. Manag*, 2025, *Online*. [CrossRef]
22. Wang, Y.; He, Y.; Gao, X. Synergizing Renewable Energy and Circular Economy Strategies: Pioneering Pathways to Environment Sustainability. *Sustainability* **2025**, *17*, 1801. [CrossRef]
23. Calisto Friant, M.; Vermeulen, W.J.; Salomone, R. Transition to a sustainable circular society: More than just resource efficienc *Circ. Econ. Sustain.* **2024**, *4*, 23–42. [CrossRef]
24. Petelin, E. Security Priorities in Circular Economy: A Conceptual Review. *Sustain. Prod. Consum.* **2024**, *47*, 655–669. [CrossRef]
25. Jensen, H.H., 3rd. R–Revamping Business Through Repair. In *Circular Economy Opportunities and Pathways for Manufacturer Manufacturing Renewed*; Springer Nature Switzerland: Cham, Switzerland, 2025; pp. 137–160.
26. Jensen, H.H. Circular Economy as a Catalyst for Innovation and Growth. In *Circular Economy Opportunities and Pathways f Manufacturers: Manufacturing Renewed*; Springer: Cham, Switzerland, 2025; pp. 429–442.
27. Grant, M.J.; Booth, A. A typology of reviews: An analysis of 14 review types and associated methodologies. *Health Inf. Libr.* **2009**, *26*, 91–108. [CrossRef]
28. Snyder, H. Literature review as a research methodology: An overview and guidelines. *J. Bus. Res.* **2019**, *104*, 333–339. [CrossRe
29. Geissdoerfer, M.; Savaget, P.; Bocken, N.M.P.; Hultink, E.J. The circular economy—A new sustainability paradigm? *J. Clean. Pro* **2017**, *143*, 757–768. [CrossRef]
30. Lowe, E.A.; Evans, L.K. Industrial ecology and industrial ecosystems. *J. Clean. Prod.* **1995**, *3*, 47–53. [CrossRef]
31. McDonough, W.; Braungart, M. *Cradle to Cradle: Remaking the Way We Make Things*; North Point Press: New York, NY, USA, 200
32. Yuan, Z.; Bi, J.; Moriguichi, Y. The Circular Economy: A New Development Strategy in China. *J. Ind. Ecol.* **2006**, *10*, 1–2. [CrossRe
33. De Angelis, R.; Ianulardo, G. Circular economy principles as a basis for a sustainability management theory: A systems thinkir and moral imagination approach. *Bus. Strategy Environ.* **2024**, *33*, 4861–4870. [CrossRef]
34. Mirzaei, S.; Shokouhyar, S. Applying a thematic analysis in identifying the role of circular economy in sustainable supply cha practices. *Environ. Dev. Sustain.* **2023**, *25*, 4691–4722. [CrossRef]
35. Khaw-ngern, K.; Peuchthonglang, P.; Klomkul, L.; Khaw-ngern, C. The 9Rs strategies for the circular economy 3.0. *Psychol. Edu J.* **2021**, *58*, 1440–1446. [CrossRef]

https://doi.org/10.3390/su180419

Moura, P.B.D.S. Circular Economy's Role in Waste Reduction and New Business Generation. Master's Thesis, ISCTE—Lisbon University Institute, Lisbon, Portugal, 2023.

Capucha, F.; Henriques, J.; Ferrão, P.; Iten, M.; Margarido, F. Analysing industrial symbiosis implementation in European cement industry: An applied life cycle assessment perspective. *Int. J. Life Cycle Assess.* **2023**, *28*, 516–535. [CrossRef]

Creutzig, F.; Simoes, S.G.; Leipold, S.; Berrill, P.; Azevedo, I.; Edelenbosch, O.; Wilson, C. Demand-Side Strategies Key for Mitigating Material Impacts of Energy Transitions. *Nat. Clim. Change* **2024**, *14*, 561–572. [CrossRef]

Kutscher, C.F.; Milford, J.B.; Kreith, F. *Principles of Sustainable Energy Systems*; CRC Press: Boca Raton, FL, USA, 2025.

Khan, W.A.; Esat, V.; Hammad, M.; Ali, H.; Zafar, M.Q.; Ali, R. Remanufacturing. In *Computer Aided Engineering Design and Manufacturing: A Fourth Industrial Revolution Perspective*; Springer: Cham, Switzerland, 2025; pp. 187–211.

Soni, A.; Gupta, S.K.; Rajamohan, N.; Yusuf, M. Waste-to-Energy Technologies: A Sustainable Pathway for Resource Recovery and Materials Management. *Mater. Adv.* **2025**, *6*, 4598–4622. [CrossRef]

Xu, Z.; Sanchez-Rivera, K.; Granger, C.; Zhou, P.; Munguia-Lopez, A.D.C.; Ikegwu, U.M.; Huber, G.W. Solvent-Based Plastic Recycling Technologies. *Nat. Chem. Eng.* **2025**, *2*, 407–423. [CrossRef]

Jiang, L. Measuring Product-Level Circularity Performance Based on the Material Circularity Indicator: An Economic Value-Based Metric with the Indicator of Residual Value. Master's Thesis, University of Twente, Enschede, The Netherlands, 2020.

Brändström, J.; Saidani, M. Comparison between circularity metrics and LCA: A case study on circular economy strategies. *J. Clean. Prod.* **2022**, *371*, 133537. [CrossRef]

Sonderegger, T.; Berger, M.; Alvarenga, R.; Bach, V.; Cimprich, A.; Dewulf, J.; Young, S.B. Mineral Resources in Life Cycle Impact Assessment—Part I: A Critical Review of Existing Methods. *Int. J. Life Cycle Assess.* **2020**, *25*, 784–797. [CrossRef]

Sánchez-Ortiz, J.; Rodríguez-Cornejo, V.; Del Río-Sánchez, R.; García-Valderrama, T. Indicators to Measure Efficiency in Circular Economies. *Sustainability* **2020**, *12*, 4483. [CrossRef]

Munonye, W.C. Towards Circular Economy Metrics: A Systematic Review. *Circ. Econ. Sustain.* **2025**, *5*, 4093–4135. [CrossRef]

Nakamura, S. Material Flow Analysis. In *A Practical Guide to Industrial Ecology by Input-Output Analysis: Matrix-Based Calculus of Sustainability*; Springer International Publishing: Cham, Switzerland, 2023; pp. 285–334.

Velenturf, A.P.; Archer, S.A.; Gomes, H.I.; Christgen, B.; Lag-Brotons, A.J.; Purnell, P. Circular Economy and the Matter of Integrated Resources. *Sci. Total Environ.* **2019**, *689*, 963–969. [CrossRef] [PubMed]

Corona, B.; Shen, L.; Reike, D.; Carreón, J.R.; Worrell, E. Towards sustainable development through the circular economy—A review and critical assessment on current circularity metrics. *Resour. Conserv. Recycl.* **2019**, *151*, 104498. [CrossRef]

de Souza, V.M.; Fröhling, M.; Pigosso, D.C.A. A Multi-Level Resource Circularity Index Based on the European Union's Circular Economy Monitoring Framework. *Waste Biomass Valorization* **2024**, *15*, 615–636. [CrossRef]

Pires, A.; Martinho, G. Waste hierarchy index for circular economy in waste management. *Waste Manag.* **2019**, *95*, 298–305. [CrossRef] [PubMed]

Westkämper, E. Life cycle management and assessment: Approaches and visions towards sustainable manufacturing (keynote paper). *CIRP Ann.* **2000**, *49*, 501–526. [CrossRef]

Frisk Jonsson, E.; Kylin, J. Towards a Circular Economy, Enabling the Development of Circular Products by Investigating Additional Parameters in a Profitability Mode. Master's Thesis, Chalmers University of Technology, Gothenburg, Sweden, 2024.

De Angelis, R.; Howard, M.; Miemczyk, J. Supply Chain Management and the Circular Economy: Towards the Circular Supply Chain. *Prod. Plan. Control* **2018**, *29*, 425–437. [CrossRef]

Goga, T. A Life Cycle-Based Investigation into the Potential of a Circular and Low-Carbon Plastics Economy in South Africa. Ph.D. Thesis, University of Cape Town, Cape Town, South Africa, 2024. Available online: http://hdl.handle.net/11427/40950 (accessed on 18 February 2025).

Jeswani, H.; Krüger, C.; Russ, M.; Horlacher, M.; Antony, F.; Hann, S.; Azapagic, A. Life cycle environmental impacts of chemical recycling via pyrolysis of mixed plastic waste in comparison with mechanical recycling and energy recovery. *Sci. Total Environ.* **2021**, *769*, 144483. [CrossRef]

de Jong, E.; Goumans, I.; Visser, R.; Puente, Á.; Gruter, G.J. The opportunities and challenges of biobased packaging solutions. *Polymers* **2025**, *17*, 2217. [CrossRef]

Vuppaladadiyam, S.S.V.; Thomas, B.S.; Kundu, C.; Vuppaladadiyam, A.K.; Duan, H.; Bhattacharya, S. Can e-waste recycling provide a solution to the scarcity of rare earth metals? An overview of e-waste recycling methods. *Sci. Total Environ.* **2024**, *924*, 171453. [CrossRef]

Baldé, C.P.; Forti, V.; Gray, V.; Kuehr, R.; Stegmann, P. *The Global E-Waste Monitor 2017*; United Nations University, International Telecommunication Union, International Solid Waste Association: Bonn, Germany; Geneva, Switzerland; Vienna, Austria, 2017. Available online: https://ewastemonitor.info (accessed on 18 February 2025).

Abbade, E.B. Estimating the nutritional loss and the feeding potential derived from food losses worldwide. *World Dev.* **2020**, *134*, 105038. [CrossRef]

https://doi.org/10.3390/su18041946

62. Urugo, M.M.; Teka, T.A.; Gemede, H.F.; Mersha, S.; Tessema, A.; Woldemariam, H.W.; Admassu, H. A comprehensive review current approaches on food waste reduction strategies. *Compr. Rev. Food Sci. Food Saf.* **2024**, *23*, e70011. [CrossRef]
63. Vasileiadou, A. From Organic Wastes to Bioenergy, Biofuels, and Value-Added Products for Urban Sustainability and Circul Economy: A review. *Urban Sci.* **2024**, *8*, 121. [CrossRef]
64. Kopsidas, O.N.; Giakoumatos, S.D.V. Economics of Recycling and Recovery. *Nat. Resour.* **2021**, *12*, 73–90. [CrossRef]
65. Sandin, G.; Peters, G.M. Environmental impact of textile reuse and recycling–A review. *J. Clean. Prod.* **2018**, *184*, 353–36 [CrossRef]
66. Lakhouit, A. Revolutionizing Urban Solid Waste Management with AI and IoT: A Review of Smart Solutions for Waste Collectio Sorting, and Recycling. *Results Eng.* **2025**, *25*, 104018. [CrossRef]
67. Dumée, L.F. Circular materials and circular design—Review on challenges towards sustainable manufacturing and recyclir *Circ. Econ. Sustain.* **2022**, *2*, 9–23. [CrossRef]
68. Liu, T.; Klammsteiner, T.; Dregulo, A.M.; Kumar, V.; Zhou, Y.; Zhang, Z.; Awasthi, M.K. Black soldier fly larvae for organ manure recycling and its potential for a circular bioeconomy: A review. *Sci. Total Environ.* **2022**, *833*, 155122. [CrossRef]
69. Gad, M.S.; Panchal, H.; Ağbulut, Ü. Waste to energy: An experimental comparison of burning the waste-derived bio-oils produce by transesterification and pyrolysis methods. *Energy* **2022**, *242*, 122945. [CrossRef]
70. Roosen, M.; Tonini, D.; Albizzati, P.F.; Caro, D.; Cristóbal, J.; Lase, I.S.; De Meester, S. Operational Framework to Quanti "Quality of Recycling" across Different Material Types. *Environ. Sci. Technol.* **2023**, *57*, 13669–13680. [CrossRef]
71. Gabisa, E.W.; Ratanatamskul, C.; Gheewala, S.H. Recycling of Plastics as a Strategy to Reduce Life Cycle GHG Emissio Microplastics and Resource Depletion. *Sustainability* **2023**, *15*, 11529. [CrossRef]
72. Ungureanu, N.; Vlăduț, N.V.; Biriș, S.Ș.; Gheorghiță, N.E.; Ionescu, M. Biomass Pyrolysis Pathways for Renewable Energ and Sustainable Resource Recovery: A Critical Review of Processes, Parameters, and Product Valorization. *Sustainability* **202** *17*, 7806. [CrossRef]
73. Yasuri, A.K. Harnessing Entropy: Innovations in Energy Efficiency and Sustainability. *Sustain. Futures* **2025**, *10*, 100952. [CrossRe
74. Chari, S.; Paulillo, A.; Materazzi, M. Exploring the Potential of Chemical Recycling Using a Distributed Model in the UK—A Li Cycle Assessment Perspective. *Waste Manag.* **2025**, *199*, 13–24.
75. Timonen, K.; Sinkko, T.; Luostarinen, S.; Tampio, E.; Joensuu, K. Life Cycle Assessment of Anaerobic Digestion: Emissi Allocation for Energy and Digestate. *J. Clean. Prod.* **2019**, *235*, 1567–1579. [CrossRef]
76. Oyewunmi, T. Natural Gas in a Carbon-Constrained World: Examining the Role of Institutions in Curbing Methane and Oth Fugitive Emissions. *La. State Univ. J. Energy Law Resour.* **2021**, *9*, 87.
77. Agovino, M.; Cerciello, M.; Musella, G.; Garofalo, A. European waste management regulations and the transition towards circul economy: A shift-and-share analysis. *J. Environ. Manag.* **2024**, *354*, 120423. [CrossRef]
78. Solomatina, I. Waste management as an element of reverse logistics in the circular economy. *Management/Vadyba* **2025**, *41*, 71.
79. European Commission. Ecodesign for Sustainable Products Regulation (ESPR). 2024. Available online: http //commission.europa.eu/energy-climate-change-environment/standards-tools-and-labels/products-labelling-rules-and-requirements/ecodesign-sustainable-products-regulation_en (accessed on 20 November 2025).
80. Council of the EU. Products Fit for the Green Transition: Council and Parliament Conclude a Provisional Agreement on the Ecod sign Regulation. 2023. Available online: https://www.consilium.europa.eu/en/press/press-releases/2023/12/05/products-for-the-green-transition-council-and-parliament-conclude-a-provisional-agreement-on-the-ecodesign-regulation (accessed on November 2025).
81. Kamali Saraji, M.; Torabi, M. Progress Towards a Circular Economy. A Comparative Analysis of EU Member States. *Sustainabili* **2025**, *17*, 8448. [CrossRef]
82. European Commission. Circular Economy: Commission Recommends Actions to Boost Recycling in 18 Member States at Risk Missing Waste Targets. 2023. Available online: https://cyprus.representation.ec.europa.eu/news/circular-economy-commissio recommends-actions-boost-recycling-18-member-states-risk-missing-waste-2023-06-08_en (accessed on 20 November 2025).
83. Schulz, Y.; Lora-Wainwright, A. In the Name of Circularity: Environmental Improvement and Business Slowdown in a Chine Recycling Hub. *Worldw. Waste* **2019**, *2*, 1–13. [CrossRef]
84. Deng, Y.; Wu, W.; Zhang, X.; Li, S.; Song, X.; Wang, J. Overview of China's waste electrical and electronic equipment recycling the last two decades. *Sustainability* **2024**, *16*, 10683. [CrossRef]
85. Izuchukwu Precious, O.; Zino Izu, O.; Frank Chudi, A.; Theresa Ojevwe, A.; Chinwe Sheila, N. Evaluating the Success of Circul Economy Models in Reducing Waste: Evidence from Europe and China. *Am. J. Biol. Nat. Sci.* **2025**, *2*, 72–96.
86. Guo, J. Explaining Industrial Cluster Evolution: A Case Study of E-Waste Clusters in China. Doctoral Dissertation, University Southampton, Southampton, UK, 2025.
87. Longsdorf, S.; Duin, L. *The Circular Economy and Its Impact on Developing and Emerging Countries*; Ecologic Institute: Berlin Germany, 2022. Available online: https://www.ecologic.eu/sites/default/files/publication/2022/50068-Circular-Econom and-Developing-Countries-final.pdf (accessed on 27 October 2025).

https://doi.org/10.3390/su180419

- PBL Netherlands Environmental Assessment Agency. *Integral Circular Economy Report 2025: Assessment for the Netherlands 2025*; PBL: The Hague, The Netherlands, 2025. Available online: https://www.pbl.nl/en/publications/integral-circular-economy-report-2025-summary (accessed on 27 October 2025).
- European Environment Agency (EEA); PBL Netherlands Environmental Assessment Agency. *Circular Economy Country Profile 2024—Netherlands*; European Environment Agency: Copenhagen, Denmark, 2024. Available online: https://www.eea.europa.eu/en/topics/in-depth/circular-economy/country-profiles-on-circular-economy/circular-economy-country-profiles-2024/netherlands_2024-ce-country-profile_final.pdf (accessed on 27 October 2025).
- Bian, R.; Zhang, T.; Zhao, F.; Chen, J.; Liang, C.; Li, W.; Yuan, L. Greenhouse Gas Emissions from Waste Sectors in China during 2006–2019: Implications for Carbon Mitigation. *Process Saf. Environ. Prot.* **2022**, *161*, 488–497. [CrossRef]
- Caballero, M.; Sroisamut, T.; Kiss, A.A.; Somoza-Tornos, A. Integrating Chemical Recycling into Brownfield Processes: Waste Polyethylene Pyrolysis and Naphtha Steam Cracking. *Syst. Control Trans.* **2025**, *4*, 668–673. [CrossRef]
- Surendhran, R. Catalyst Development and Reaction Engineering for Chemical Upgrading of Emerging Feedstocks. Ph.D. Thesis, University of Delaware, Newark, DE, USA, 2024.
- Dehiya, B.S. Recycling within the Chemical Industry: The Circular Economy Approach—Industrial Perspectives. In *Energy Materials*; CRC Press: Boca Raton, FL, USA, 2024; pp. 329–346.
- Le, M.N.; Lee, M.S. A Review on Hydrometallurgical Processes for the Recovery of Valuable Metals from Spent Catalysts and Life Cycle Analysis Perspective. *Miner. Process. Extr. Metall. Rev.* **2021**, *42*, 335–354. [CrossRef]
- Reddy, C.C. Applications and Potential of Process Intensification in Chemical Process Industries. In *Control and Safety Analysis of Intensified Chemical Processes*; Wiley-VCH: Weinheim, Germany, 2024; pp. 15–46.
- Walmsley, T.G.; Ong, B.H.; Klemeš, J.J.; Tan, R.R.; Varbanov, P.S. Circular Integration of Processes, Industries, and Economies. *Renew. Sustain. Energy Rev.* **2019**, *107*, 507–515. [CrossRef]
- Rahman, H.U.; Khan, M.; Ditta, A. Recent Advances in Sustainable Waste Management Practices. In *Bioremediation and Nanotechnology for Climate Change Mitigation*; Springer: Berlin/Heidelberg, Germany, 2025; pp. 103–136. [CrossRef]
- Singh, A.; Tirpude, A.; Mathew, N.; Arfin, T. Gaseous Emissions from Solid Waste Disposal. In *Waste Derived Carbon Nanomaterials*; American Chemical Society: Washington, DC, USA, 2025; Volume 1, pp. 95–114.
- Kaza, S.; Yao, L.; Bhada-Tata, P.; Van Woerden, F. *What a Waste 2.0: A Global Snapshot of Solid Waste Management to 2050*; World Bank Publications: Washington, DC, USA, 2018.
- Mor, S.; Ravindra, K. Municipal solid waste landfills in lower- and middle-income countries: Environmental impacts, challenges and sustainable management practices. *Process Saf. Environ. Prot.* **2023**, *174*, 510–530. [CrossRef]
- Manheim, D.C.; Yeşiller, N.; Hanson, J.L.; Blake, D.R. Climate impacts of landfill gas emissions: Analysis for 20-year and 100-year time horizons. *Waste Manag.* **2024**, *186*, 318–330. [CrossRef] [PubMed]
- Shukla, P.R.; Skeg, J.; Buendia, E.C.; Masson-Delmotte, V.; Pörtner, H.O.; Roberts, D.C.; Zhai, P.; Slade, R.; Connors, S.; van Diemen, R.; et al. *Climate Change and Land: An IPCC Special Report on Climate Change, Desertification, Land Degradation, Sustainable Land Management, Food Security, and Greenhouse Gas Fluxes in Terrestrial Ecosystems*; IPCC: Geneva, Switzerland, 2019.
- Salami, L.; Popoola, L.T. A comprehensive review of atmospheric air pollutants assessment around landfill sites. *Air Soil Water Resour.* **2023**, *16*, 11786221221145379. [CrossRef]
- Singh, D.; Wohlleben, W.; Roche, R.D.L.T.; White, J.C.; Demokritou, P. Thermal decomposition/incineration of nano-enabled coatings and effects of nanofiller/matrix properties and operational conditions on byproduct release dynamics: Potential environmental health implications. *NanoImpact* **2019**, *13*, 44–55. [CrossRef]
- Bano, A. Direct environmental pollution from solid waste. In *Waste-to-Energy*; Elsevier: Amsterdam, The Netherlands, 2025; pp. 249–269. [CrossRef]
- Nartey, B.A. A Study of Short-Lived Climate Pollutants Associated with Solid Waste Management Activities in Accra. Doctoral Dissertation, University of Ghana, Accra, Ghana, 2019.
- Zhang, Z.; Zhou, L.; Zhang, M. A progress review of black carbon deposition on Arctic snow and ice and its impact on climate change. *Adv. Polar Sci.* **2024**, *35*, 178–191. [CrossRef]
- Dietrich, M.; Fongen, M.; Foereid, B. Anaerobic digestion affecting nitrous oxide and methane emissions from the composting process. *Bioresoure Technol. Rep.* **2021**, *15*, 100752. [CrossRef]
- Elsayed, A.; Kakar, F.L.; Abdelrahman, A.M.; Ahmed, N.; AlSayed, A.; Zagloul, M.S.; Elbeshbishy, E. Enhancing anaerobic digestion efficiency: A comprehensive review on innovative intensification technologies. *Energy Convers. Manag.* **2024**, *320*, 118979. [CrossRef]
- Srivastava, A.; Pandey, S.; Shahwal, R.; Sur, A. Recycling of Waste into Useful Materials and Their Energy Applications. In *Microbial Niche Nexus Sustaining Environmental Biological Wastewater and Water-Energy-Environment Nexus*; Springer: Cham, Switzerland, 2025; pp. 251–296.
- Trowell, K.A.; Goroshin, S.; Frost, D.L.; Bergthorson, J.M. Aluminum and its role as a recyclable, sustainable carrier of renewable energy. *Appl. Energy* **2020**, *275*, 115112. [CrossRef]

https://doi.org/10.3390/su18041946

112. Shen, L.; Worrell, E. Plastic recycling. In *Handbook of Recycling*; Elsevier: Amsterdam, The Netherlands, 2024; pp. 497–510.
113. Almusaed, A.; Yitmen, I.; Myhren, J.A.; Almssad, A. Assessing the impact of recycled building materials on environment sustainability and energy efficiency: A comprehensive framework for reducing greenhouse gas emissions. *Buildings* **2024**, *14*, 156 [CrossRef]
114. Lisbona, P.; Pascual, S.; Pérez, V. Waste to energy: Trends and perspectives. *Chem. Eng. J. Adv.* **2023**, *14*, 100494. [CrossRef]
115. Atelge, M.R.; Krisa, D.; Kumar, G.; Eskicioglu, C.; Nguyen, D.D.; Chang, S.W.; Unalan, S. Biogas production from organic wast Recent progress and perspectives. *Waste Biomass Valorization* **2020**, *11*, 1019–1040. [CrossRef]
116. Munir, M.T.; Naqvi, M.; Li, B. A converging path: A decade's reflection on net zero emissions and the circular economy. *Fron Energy Res.* **2024**, *12*, 1332174. [CrossRef]
117. García-Navarro, J.; Poltronieri, O.C.C. Circular economy and GHG emissions, current situation. In *Circular Economy on Energy an Natural Resources Industries: New Processes and Applications to Reduce, Reuse and Recycle Materials and Decrease Greenhouse Gas Emissions*; Springer International Publishing: Cham, Switzerland, 2024; pp. 1–18. [CrossRef]
118. Pouikli, K. Concretising the role of extended producer responsibility in European Union waste law and policy through the lens the circular economy. In *ERA Forum*; Springer: Berlin/Heidelberg, Germany, 2020; Volume 20, pp. 491–508. [CrossRef]
119. Ellen MacArthur Foundation. *Completing the Picture: How the Circular Economy Tackles Climate Change*; Ellen MacArthu Foundation: Cowes, UK, 2021. Available online: https://ellenmacarthurfoundation.org/completing-the-picture (accesse on 20 November 2025).
120. Papageorgiou, A.; Björklund, A.; Sinha, R.; de Almeida, M.L.R.; Steubing, B. Coupling material and energy flow analysis with li cycle assessment to support circular strategies at the urban level. *Int. J. Life Cycle Assess.* **2024**, *29*, 1209–1228. [CrossRef]
121. Talwar, S. Circular Economy Initiatives in Indian Manufacturing: An Ecosystem View. Doctoral Dissertation, Macquar University, Sydney, Australia, 2021.
122. Boral, S.; Black, L.; Velis, C. Conceptualizing Systems Thinking and Complexity Modelling for Circular Economy Quantificatio A Systematic Review and Critical Analysis. *Res. Sq.* **2025**, *preprint*.
123. Guinée, J.B.; Heijungs, R.; Huppes, G.; Zamagni, A.; Masoni, P.; Buonamici, R.; Rydberg, T. Life cycle assessment: Past, preser and future. *Environ. Sci. Technol.* **2011**, *45*, 90–96. [CrossRef]
124. Brunner, P.H.; Rechberger, H. *Handbook of Material Flow Analysis: For Environmental, Resource, and Waste Engineers*; CRC Pres Boca Raton, FL, USA, 2016.
125. Gautam, P.; Salunke, D.; Lad, D.; Gautam, A. Convergent synergy of carbon capture within the circular economy paradigm: nexus for realizing multifaceted sustainable development goals. *Carbon Res.* **2025**, *4*, 3. [CrossRef]
126. Liu, L.; Qu, J.; Li, X.; Liao, Q.; Niu, Y. Review of material flow analysis and its application under carbon neutralization target: bibliometric perspective. *Carbon Footpr.* **2024**, *3*, 3. [CrossRef]
127. Yuan, Y.; Zhao, Y.; Yang, H.; Chong, C.H.; Ma, L.; Chang, S.; Li, Z. Mapping the transmission of carbon emission responsibili among multiple regions from the perspective of the energy supply chain: EA-MRIO method and a case study of Chin *Sustainability* **2025**, *17*, 8166. [CrossRef]
128. Walzberg, J.; Frayret, J.M.; Eberle, A.L.; Carpenter, A.; Heath, G. Agent-based modeling and simulation for the circular econom Lessons learned and path forward. *J. Ind. Ecol.* **2023**, *27*, 1227–1238. [CrossRef]
129. Costa, D.G.D. A Systems View on Circular Economy Transitions: Examining the Deceleration of Resource Flows. Doctor Dissertation, Universidade de São Paulo, São Paulo, Brazil, 2020.
130. Van Vuuren, D.P.; Stehfest, E.; Gernaat, D.E.; Van Den Berg, M.; Bijl, D.L.; De Boer, H.S.; Daioglou, V.; Doelman, J.C.; Edelenbosc O.Y.; Harmsen, M.; et al. Alternative pathways to the 1.5 °C target reduce the need for negative emission technologies. *Nat. Cli Chang.* **2018**, *8*, 391–397. [CrossRef]
131. Mäkinen, M. Impact Assessment of Product Development in the Circular Economy. Master's Thesis, Lappeenranta–Lah University of Technology LUT, Lappeenranta, Finland, 2023. Available online: https://lutpub.lut.fi/handle/10024/16658 (accessed on 20 November 2025).
132. European Commission. *A New Circular Economy Action Plan for a Cleaner and More Competitive Europe; COM (2020) 98 Fin* European Commission: Brussels, Belgium, 2020.
133. Van Yken, J.; Boxall, N.J.; Cheng, K.Y.; Nikoloski, A.N.; Moheimani, N.R.; Kaksonen, A.H. E-Waste Recycling and Resourc Recovery: A Review on Technologies, Barriers and Enablers with a Focus on Oceania. *Metals* **2021**, *11*, 1313. [CrossRef]
134. Bergquist, A.K.; Lindmark, M.; Petrusenko, N. Creating value out of waste: The transformation of the Swedish waste an recycling sector, 1970s–2010s. *Bus. Hist. Rev.* **2023**, *97*, 3–31. [CrossRef]
135. Alhola, K.; Ryding, S.O.; Salmenperä, H.; Busch, N.J. Exploiting the Potential of Public Procurement: Opportunities for Circul Economy. *J. Ind. Ecol.* **2019**, *23*, 96–109. [CrossRef]
136. World Bank. *State and Trends of Carbon Pricing 2023*; World Bank Group: Washington, DC, USA, 2023. Available onlin https://openknowledge.worldbank.org/handle/10986/39796 (accessed on 21 November 2025).

https://doi.org/10.3390/su180419

7. Ellen MacArthur Foundation. *Financing the Circular Economy: Capturing the Opportunity*; Ellen MacArthur Foundation: Cowes, UK, 2022. Available online: https://ellenmacarthurfoundation.org (accessed on 20 November 2025).
8. Verleye, K.; De Keyser, A.; Raassens, N.; Alblas, A.A.; Lit, F.C.; Huijben, J.C. Pushing forward the transition to a circular economy by adopting an actor engagement lens. *J. Serv. Res.* **2024**, *27*, 69–88. [CrossRef]
9. Woldeyes, T.D. Emerging Circular Economy Business Models in the Waste Electrical and Electronic Equipment (WEEE) Sector. Ph.D. Thesis, University of Padova, Padova, Italy, 15 April 2024. Available online: https://hdl.handle.net/11577/3514284 (accessed on 20 November 2025).
0. Trevisan, A.H.; Boscarato, A.; Acerbi, F.; Terzi, S.; Sassanelli, C. Enhancing Circular Economy education and training for the manufacturing sector: A holistic skills framework. *J. Environ. Manag.* **2025**, *380*, 124982. [CrossRef]
1. Atabakifardtehrani, Y. Exploring Barriers in Promoting Circular Economy Practices: Insights on Consumer Durable Goods in Remote and Indigenous Communities in Newfoundland and Labrador. Ph.D. Thesis, Memorial University of Newfoundland, St. John's, NL, Canada, 2025.
2. Ameer, M.J. A Comparative Study of Waste Management Systems in Developing Countries and Western World: Observations from Pakistan, Finland, Sweden, and Germany. Bachelor's Thesis, Centria University of Applied Sciences, Kokkola, Finland, 2025. Available online: https://urn.fi/URN:NBN:fi:amk-2025052918136 (accessed on 20 November 2025).
3. Reshetnikova, I.; Lieskovska, V.; Polous, O.; Murin, V. The Importance of the Deposit-Return System of Packaging in the Circular Economy System: Slovak Innovation Experience for Ukraine. *Mark. Manag. Innov.* **2025**, *16*, 1–14. [CrossRef]
4. Simaei, E.; Rahimifard, S. AI-based decision support system for enhancing end-of-life value recovery from e-wastes. *Int. J. Sustain. Eng.* **2024**, *17*, 80–96. [CrossRef]
5. Turskis, Z.; Šniokienė, V. IoT-Driven Transformation of Circular Economy Efficiency: An Overview. *Math. Comput. Appl.* **2024**, *29*, 49. [CrossRef]
6. Saberi, S.; Kouhizadeh, M.; Sarkis, J.; Shen, L. Blockchain Technology and Its Relationships to Sustainable Supply Chain Management. *Int. J. Prod. Res.* **2019**, *57*, 2117–2135. [CrossRef]
7. Lieder, M.; Rashid, A. Towards Circular Economy Implementation: A Comprehensive Review in Context of Manufacturing Industry. *J. Clean. Prod.* **2016**, *115*, 36–51. [CrossRef]
8. Kumar, R.; Kamboj, N.; Mitra, D.; Ray, A. Strengthening Local Communities Through Advanced Recycling Technology for Sustainable Growth. In *AI Technologies for Enhancing Recycling Processes*; IGI Global Scientific Publishing: Hershey, PA, USA, 2025; pp. 393–420.
9. Jean, G. Effect of E-Supply Chain Coordination on Reverse Logistics and Circular Supply Chains; Preprint or Working Paper. October 2024. Available online: https://www.researchgate.net/publication/384883079_Effect_of_E-supply_Chain_Coordination_on_Reverse_Logistics_and_Circular_Supply_Chains (accessed on 20 November 2025).
0. Bocken, N.M.P.; de Pauw, I.; Bakker, C.; van der Grinten, B. Product design and business model strategies for a circular economy. *J. Ind. Prod. Eng.* **2016**, *33*, 308–320. [CrossRef]
1. Raźniewska, M. Compostable Packaging Waste Management—Main Barriers, Reasons, and the Potential Directions for Development. *Sustainability* **2022**, *14*, 3748. [CrossRef]
2. Ciawi, Y.; Tonyes, S.G.; Dwipayanti, N.M.U. Multilayer Packaging Recycling: Challenges, Current Practices, and Future Prospects. *Acad. Environ. Sci. Sustain.* **2025**, *2*, 1–12. [CrossRef]
3. Van Beukering, P.J. *Recycling, International Trade and the Environment*; Springer Science & Business Media: Dordrecht, The Netherlands, 2001.
4. Dey, P.K.; Malesios, C.; De, D.; Budhwar, P.; Chowdhury, S.; Cheffi, W. Circular Economy to Enhance Sustainability of Small and Medium Sized Enterprises. In *Supply Chain Sustainability in Small and Medium Sized Enterprises*; Routledge: London, UK, 2022; pp. 10–45.
5. Tejaswini, M.S.S.R.; Pathak, P.; Ramkrishna, S.; Ganesh, P.S. A comprehensive review on an integrative approach for sustainable management of plastic waste and its associated externalities. *Sci. Total Environ.* **2022**, *825*, 153973. [CrossRef]
6. Sönnichsen, S.D.; Clement, J. Review of green and sustainable public procurement: Towards circular public procurement. *J. Clean. Prod.* **2020**, *245*, 118901. [CrossRef]
7. Schröder, P. *Promoting a Just Transition to an Inclusive Circular Economy*; Royal Institute of International Affairs: London, UK, 2020; pp. 1–33.
8. Thapa, K. Closing the Circular Economy Loophole: On Fair Governance for European Waste Exports. Ph.D. Thesis, Utrecht University, Utrecht, The Netherlands, 2023.
9. Nasriani, I. Systemic Barriers and Strategic Solutions in Waste Governance. *Sinergi Int. J. Account. Tax.* **2024**, *2*, 239–251. [CrossRef]
0. Chapman, D.A.; Eyckmans, J.; Van Acker, K. The impact of demand-side strategies to enable a more circular economy in private car mobility. *Sustain. Prod. Consum.* **2024**, *49*, 263–275. [CrossRef]
1. Zahran, S. Investigating the nexus between green supply chain practices and sustainable waste management in advancing the circular economy. *Sustainability* **2024**, *16*, 3566. [CrossRef]

https://doi.org/10.3390/su18041946

162. Panju, M. What Goes Around Comes Around? The Sustainability Paradox of Second-Hand Clothing Marketplaces in a Cros Cultural Context. *East Afr. J. Interdiscip. Studies.* **2024**, *7*, 1–27. [CrossRef]
163. Getanda, K.O. *Managing Plastic Waste Crisis in Kenyan Urban Centers with Circular Economy Approach*; Tampere University Applied Sciences: Tampere, Finland, 2023. Available online: https://urn.fi/URN:NBN:fi:amk-2023121135847 (accessed c 20 November 2025).
164. Hirschhorn, J.; Jackson, T.; Baas, L. Towards Prevention—The Emerging Environmental Management Paradigm. In *Cle Production Strategies: Developing Preventive Environmental Management in the Industrial Economy*; CRC Press: Boca Raton, FL, US 2024; pp. 125–142.
165. Dewasiri, N.J.; Gnanapala, W.K.A.C.; Rathnasiri, M.S.H.; Yadav, M.; Tewari, L.M.; Mishra, U. The way forward for net ze Aligning nationally determined contributions (NDCs) and holistic sustainable development in South Asia. In *Transition Towards Sustainable Future: Net Zero Policies and Environmental Sustainability*; Springer: Singapore, 2024; pp. 95–116. [CrossRef]
166. Mora, P.; Fernandez, F.G.A. (Eds.) *Circular Economy on Energy and Natural Resources Industries: New Processes and Applications Reduce, Reuse and Recycle Materials and Decrease Greenhouse Gases Emissions*; Springer Nature: Cham, Switzerland, 2024.
167. Picardo, A.; Soltero, V.M.; Peralta, E. Life Cycle Assessment of Sustainable Road Networks: Current State and Future Directio *Buildings* **2023**, *13*, 2648. [CrossRef]
168. Almansour, M.; Akrami, M. Towards Zero Waste: An In-Depth Analysis of National Policies, Strategies, and Case Studies Waste Minimisation. *Sustainability* **2024**, *16*, 10105. [CrossRef]
169. Debrah, J.K.; Teye, G.K.; Dinis, M.A.P. Barriers and Challenges to Waste Management Hindering the Circular Economy i Sub-Saharan Africa. *Urban Sci.* **2022**, *6*, 57. [CrossRef]
170. Oyinlola, M.; Kolade, O.; Schröder, P.; Odumuyiwa, V.; Rawn, B.; Wakunuma, K.; Abolfathi, S. The environment for a digital enabled circular plastics economy in Africa: Lessons from cross-sectional stakeholder engagements. *J. Strategy Manag.* **202** [CrossRef]
171. Mont, O.; Heiskanen, E. Breaking the stalemate of sustainable consumption with industrial ecology and a functional perspecti *Ecol. Econ.* **2015**, *116*, 12–20. [CrossRef]
172. Oliveira, M.; Miguel, M.; van Langen, S.K.; Ncube, A.; Zucaro, A.; Fiorentino, G.; Genovese, A. Circular Economy and th Transition to a Sustainable Society: Integrated Assessment Methods for a New Paradigm. *Circ. Econ. Sustain.* **2021**, *1*, 99–1 [CrossRef]

https://doi.org/10.3390/su180419

MDPI AG
Grosspeteranlage 5
4052 Basel
Switzerland
Tel.: +41 61 683 77 34

Sustainability Editorial Office
E-mail: sustainability@mdpi.com
www.mdpi.com/journal/sustainability

www.ingramcontent.com/pod-product-compliance
Lightning Source LLC
Chambersburg PA
CBHW061811210926
51850CB00006B/668
* 9 7 8 3 7 2 5 8 7 5 4 9 8 *